T0326597

Physics of Cancer

Physics of Cancer

Claudia Tanja Mierke

Biological Physics Division, University of Leipzig, Germany

IOP Publishing, Bristol, UK

ISBN   978-0-7503-1134-2 (ebook)
ISBN   978-0-7503-1135-9 (print)
ISBN   978-0-7503-1136-6 (mobi)

DOI   10.1088/978-0-7503-1134-2

Version: 20150901

IOP Expanding Physics
ISSN 2053-2563 (online)
ISSN 2054-7315 (print)

British Library Cataloguing-in-Publication Data: A catalogue record for this book is available from the British Library.

Published by IOP Publishing, wholly owned by The Institute of Physics, London

IOP Publishing, Temple Circus, Temple Way, Bristol, BS1 6HG, UK

US Office: IOP Publishing, Inc., 190 North Independence Mall West, Suite 601, Philadelphia, PA 19106, USA

This book was written with the support of Thomas M L Mierke.

Contents

Preface

This book is about the novel and promising field of physics of cancer, which has become the focus of many biophysical research groups all over the world. The book aims to present several biophysical approaches used in cancer research. The major findings that contribute significantly to the field are presented from a biophysical point of view. The intended readership includes everyone interested in cancer research, especially those readers studying biological physics, physics or tumor-biology. The book is suitable for upper undergraduate, graduate, doctoral and post-doctoral students, senior scientists, lecturers, principal investigators and professors. I wrote this book because I was kindly invited to by IOP Publishing and because there is currently no such book available in any format. Many parts of it are supported by figures that I have drawn myself or designed. I hope that the book will promote understanding of why physics of cancer is so important for cancer research. All types of readers, including students and researchers, should be guided through the physics of cancer field by it. Every chapter is a complete part, with references and further reading suggestions. The book also has a glossary that will help the reader to follow the book easily. I hope that the book will help to establish physics of cancer as an essential part of cancer research.

Claudia Tanja Mierke

About the author

Claudia Tanja Mierke

 Claudia Tanja Mierke studied biology at TU Braunschweig in Germany and received her doctoral degree from the Medical School of Hannover in March 2001; her thesis concerned human endothelial and mast cell cell–cell interactions. Her postdoctoral research was performed in different institutes of the University of Erlangen-Nuremberg, and covered a number of fields: cancer research; inflammation of endothelial cells and molecular cancer research; and biophysical cancer research. In 2012 she habilitated in biophysics. Her scientific results were presented at many international conferences and published in leading international peer-reviewed journals in this field. Since 2010, she has worked as a professor at the University of Leipzig and she has published many review articles and book chapters on the subject of physical-driven cancer research. As a professor of soft matter physics and biological physics, her research focus is on biophysical cancer research, including cell motility and transmigration through endothelial vessel linings.

With a background in biology and molecular oncology, Professor Mierke is now head of the Biological Physics Division at the University of Leipzig, where she regularly teaches molecular and cell biology, biophysics and soft matter physics to physicists, and is concerned with various research areas that contribute to our understanding of the physical aspects of cancer.

Introduction

The following aspects of the physics of cancer are considered in this book: a biophysical view of tumor biology, the mechanical properties of cancer cells, microenvironmental effects and the interaction of cancer cells with other cells during transmigration.

The first part of this book will help readers to get into the field by introducing them to tumor biology. Moreover, it will explain important aspects of classical tumor biology from a biophysical point of view and will provide the basis for understanding the following three parts of the book. Whenever possible, the restrictively tumor-biological viewpoint available in many review articles on classical cancer research is criticized and the missing biomechanical aspects are included. State of the art physical cancer research is included and integrated into the classical field of tumor biology.

The impact of mechanical properties on cellular behavior, such as cell motility, during the progression of cancer is discussed in part 2. Several biophysical methods are presented and their impact on physics-driven cancer research is highlighted. Mechanical properties of cells such as force transmission and the generation of cells play an important and basic role in numerous biological processes, including cell motion, replication and survival. Single-molecule force spectroscopy methods, including the most common methods such as magnetic and optical tweezers, as well as atomic force microscopy (AFM), have developed into powerful tools for investigating forces and their impact on cell motility.

The role of the tumor microenvironment during cellular invasion is explored in part 3. Cancer cell invasion is not only regulated by the cancer cell's capacity to migrate through the connective tissue. The microenvironment also impacts on the motility of cancer cells through the extracellular matrix, with respect to structural confinement and mechanical properties. In turn, cancer cells also affect the matrix's mechanical properties by degrading the extracellular matrix or secreting extracellular matrix proteins and growth factors. Thus, cancer cells are able to alter the structure, composition and, consequently, the mechanical properties of the extracellular matrix of connective tissues. In addition, cancer-associated fibroblasts embedded in the extracellular matrix of primary tumors can restructure the local microenvironment of tumors by contracting the matrix, thus altering its mechanical properties, such as stiffness. Similarly to cancer cells, these cancer-associated fibroblasts (CAFs) degrade the extracellular matrix and secrete cytokines, chemokines, growth factors and extracellular matrix proteins.

In part 4 the impact of the mechanical and biochemical interaction of cancer cells with other cells in transendothelial migration is considered. Most cancer-related deaths are not evoked by the primary tumor, they are caused by the process of metastasis. One of the main steps of cancer metastasis is the transendothelial migration of cancer cells, a complex event in which cell adhesion and the transmigration of cancer cells need to be precisely regulated. In particular, the transmigration involves biochemical and biomechanical interactions of metastatic cancer cells with the endothelial cell lining of blood or lymph vessels. Under normal conditions, the endothelium acts as a barrier against the invasion of cancer cells in order to decrease cancer cell migration and consequently cancer metastasis. However, certain cancer cells can overcome the endothelial cell monolayer by activating alterations within endothelial cells, including reduction of endothelial cell stiffness, regulation of adhesion molecules and the remodeling of the endothelial cytoskeleton. In turn, cancer cells need to dynamically alter their cytoskeleton and their cell shape and they may apply forces toward the endothelium in order to facilitate their transmigration. Thus, the biomechanical properties of cancer and endothelial cells seem to play an important role in trans-endothelial migration. Moreover, it has been suggested that mechanical alterations are necessary in both cancer and endothelial cells in order to regulate cancer cell invasiveness. However, the detailed regulatory mechanisms are not well understood and specifically the role of force exerted by aggressive and invasive cancer cells on the endothelium and its specific force application mechanisms remains elusive. Finally, the question of how such a mechanism may facilitate cancer cell invasion, and in the case of certain cancer cells, how it may even increase invasiveness after transendothelial migration is considered.

IOP Publishing

Physics of Cancer

Claudia Tanja Mierke

Chapter 1

Initiation of a neoplasm or tumor

Summary

Chapter 1 describes the classical, tumor-biological viewpoint on the initiation of a tumor, its further growth, the process of neoangiogenesis and its importance for tumor growth and malignancy. Then the process of malignant cancer progression is presented and the main steps are described in detail. The focus is on the motility of cancer cells, especially their ability to transmigrate through barriers such as basement membranes and endothelial cells. The hallmarks of cancer are presented from a biophysical point of view and the missing mechanical aspect is described and included as a novel hallmark. Finally, the impact of mechanical properties on cancer cell invasion is explained, providing the basis for understanding the later chapters in this book.

1.1 Initiation of a neoplasm, tumor growth and neoangiogenesis

What promotes the initiation of a neoplasm? What evokes tumor growth and thus malignant progression of tumors? Why can a non-vascularized tumor only grow to a restricted tumor size and not be able to grow further? Why is vascularization of a tumor so important for its survival and malignancy? Why is tumor angiogenesis called neoangiogenesis? All these questions will be answered in the following sections of chapter 1.

1.1.1 Initiation of a neoplasm and tumor growth

The onset of a neoplasia starts as a complex scenario in a complicated process consisting of multiple steps that basically involve alterations in proto-oncogenes and tumor suppressor genes. In particular, proto-oncogenes are activated, while tumor suppressor genes are inactivated (Knudson 1971). Proto-oncogenes consist of a group of genes transforming normal cells to cancerous cells when they are mutated (Adamson 1987, Weinstein and Joe 2006). Typically, mutations in proto-oncogenes stay dominant, however, the mutated proto-oncogene is named an oncogene. Proto-oncogenes encode proteins related to cell division stimulation, inhibition of cell differentiation and reduction of cell death (Weinstein and Joe 2006). These processes promote normal human development and help to maintain tissues and organs. Nonetheless, oncogenes regulate the elevated production of these proteins, causing increased cell division, suppression of cell differentiation and omission of cell death. All these signs build up the phenotypes of cancer cells. For this reason, oncogenes are regarded as a potential molecular target for anti-cancer drugs.

Five to six independent mutational events usually contribute to the formation of human solid tumors, whereas three to four mutational events involving different genes are sufficient to cause leukemias in humans (Thomas *et al* 2007). In animals, carcinogenesis can be induced by chemical mutagens such as 7, 12-dimethylbenzanthracene (DMBA) (Balmain and Brown 1988) and a simultaneously administered chemical promotor to stimulate the growth of mutated cells such as 12-O-tetradecanoylphorbol-13-acetate (TBA), which belongs to the group of phorbol esters (Slaga 1983). First, precancerous papillomas are formed over a period of months and then they progress to skin carcinomas. DMBA may lead to a mutation in codon 61 of the H-ras oncogene, but still the growth needs to be stimulated by TPA (Balmain and Brown 1988, Slaga 1983). Chemical carcinogenesis in animals such as mice is used to model skin carcinogenesis, which is divided into three phases: initiation, promotion and progression (Moolgavkar and Knudson 1981). In addition to chemical mutagenesis, a virus-induced mutagenesis is also able to cause carcinogenesis. In particular, a single oncogene is able to facilitate tumor formation by infection with some rapidly transforming retroviruses, such as RSV (Temin 1988). However, the virus may also carry two different oncogenes that work together in order to cause a neoplastic phenotype (Temin 1988). An example is an avian erythroblastosis virus carrying the erbA and erbB oncogenes (Damm *et al* 1987). Transformation studies using non-immmortalized cell lines showed cooperation between the myc protein in the nucleus and the ras-protein associated with the cytoplasmic-membrane site (Wang *et al* 2011) in transforming rat embryo fibroblasts. The cooperation between SV40 large T product and the mutated H-ras gene is necessary to transform 'normal' human epithelial cells and fibroblast cells, if these cells constitutively express the catalytic subunit of the telomerase enzyme (Wang *et al* 2011). Thus, these cells display a complex pattern leading to the neoplastic transformation of human cells. Taken together, the interplay between two different types of oncogenes (nuclear and cytoplasmic) has been demonstrated several times, but is not strictly necessary for the malignant transformation of cells. For example, single myc oncogene expression leads to multiple genetic alterations that facilitate tumor formation. This results in increased incidence of clonal neoplasias and tumors (Wang *et al* 2011). Even other events are necessary for neoplasia and subsequently tumor formation.

The onset and progression of human neoplasia is associated with the activation of oncogenes and the inactivation or complete loss of tumor suppressor genes. Due to the high variability of the tumor types, the mechanisms of oncogene activation are highly variable. The common feature is that the activation of oncogenes leads to genetic alterations of cellular proto-oncogenes. This is generally associated with an advantage in cellular growth. Three genetic mechanisms for the activation of proto-oncogenes in human neoplasms are possible: mutations, gene amplifications and chromosome rearrangements. All of these lead to alterations of the proto-oncogene structure or to an increase in the expression of the protooncogene. Given the multistep nature of the neoplasia process, we expect that more than one mechanism accounts for the formation of human tumors through alteration of the numbers of cancer-associated genes. For the whole expression of the neoplastic phenotype and the total capacity to metastasize, a combination of the activation of proto-oncogenes and the inactivation or loss of tumor suppressor genes is necessary (Adamson 1987, Weinstein and Joe 2006). In particular, firstly, the mutations are in critical regulatory domains or regions of the gene and mediate structural alterations. Examples are retroviral oncogenes that have deletions contributing to their activation (Temin 1988). There can also be substitutions (called point mutations), which means only a single amino acid is altered within the protein (Temin 1988). Secondly, the gene amplification is increased by expansion of the copy numbers of a single gene providing resistance to growth-inhibiting drugs (Temin 1988). Thirdly, chromosomal rearrangements can occur, as frequently observed in hematologic malignancies, soft-tissue sarcomas and certain solid tumors (Temin 1988). In the Burkitt lymphoma, chromosomal translocations are often observed and less pronounced chromosomal inversions have been detected (Dalla-Favera et al 1982). In the latter case, chromosomal breakpoints between two genes support these inversions. This may then lead on the one hand to transcriptional activation of certain proto-oncogenes and on the other hand to the fusion of genes, for example in chronic myelogenous leukemia (CML) (Lozzio and Lozzio 1975).

Although there is much variability in the pathways for the initiation and progression of tumors in humans, numerous studies of different types of malignancy have revealed the multistep character of human cancer. All the above-mentioned mechanisms for initiating neoplasm and promoting tumor progression may affect the mechanical properties of cancer cells and subsequently alter their physiological function in a certain microenvironment, for example they may become more motile compared to non-transformed 'normal' cells.

1.1.2 Neoangiogenesis

Neoangiogenesis occurs in a tumor of a certain size and enhances its malignancy state. This process of neoangiogenesis in a tumor is called tumor angiogenesis. A tumor without a vasculary system or nearby blood vessels will stop growing. This indicates that the process of angiogenesis is necessary to overcome the restriction of a tumor to a diameter of 1–2 mm (Folkmann et al 1963). This means that a tumor without angiogenesis is not malignant and will probably cause no damage to surrounding or hosting organs. In particular, tumor angiogenesis is the proliferation of endothelial cells lining blood vessels that break through the tumor and grow to novel vessels in order to supply the cancer cells of the inner tumor mass with nutrients and oxygen (Gimbrone et al 1972, 1974). The vessels can also serve to remove waste products, such as toxic substances that reduce tumor proliferation (Folkmann et al 1963).

The onset of tumor angiogenesis begins with a cancerous tumor consisting of cells that secrete molecules to their surrounding microenvironment of 'normal' host tissue, such as the extracellular matrix of connective tissue. The signaling molecules activate genes in the cells of the microenvironment to produce proteins promoting or inducing the growth of new blood vessels (Gimbrone *et al* 1974). The gradients of these signaling molecules are most commonly directed around the primary tumor and thus the new vessel expands in the direction of the tumor in order to grow and migrate into it (Gimbrone *et al* 1974). This behavior indicates that tumor growth needs angiogenesis.

A pioneering experiment was performed to test the importance of angiogenesis, in which a cancerous tumor from an animal was removed and a 'normal' healthy organ from an animal of the same strain not carrying a tumor was isolated. Some cancer cells from the isolated tumor were injected into the healthy organ and cultured in a glass chamber containing a nutrient solution that was pumped into the organ to keep it healthy. After one or two weeks only small tumors had formed, not larger than 1–2 mm in diameter and without any connection to the organ's vascular system (Gimbrone *et al* 1974). This result indicated that without angiogenesis the tumor stops growing at an early stage and at a diameter of 1–2 mm.

The process of angiogenesis is regulated by the amount of activating and inhibitory proteins. The number of inhibitors is normally significantly higher than the number of activators, leading to the inhibition of vessel growth (Otrock *et al* 2007). In the case of blood vessel injury or organ growth, the quantity of angiogenesis activators increases, whereas the number of inhibitory proteins decreases, leading to neoangiogenesis by the division of the vascular endothelial cell lining of 'older' blood vessels. The outgrowth of these endothelial cells is the onset of new blood vessel formation (Otrock *et al* 2007). The walls of blood vessels consist of vascular endothelial cells that normally do not divide, but if they do, they perform this on average every three years upon stimulation through angiogenesis (Otrock *et al* 2007). Angiogenesis does not just occur in tumors, it may occur normally during developmental stages and growth; if it occurs within a developing embryo, it needs to build up a primitive network of capillaries, veins and arteries, and this is called vasculogenesis (Patan 2004). At later stages angiogenesis can remodel this network by creating new blood vessels and capillaries that build up the circulatory system.

A basic experiment was conducted in order to discover which molecules are involved in inducing angiogenesis and what their origin is. Are these molecules provided by the cancer cells of the neoplasm or the primary tumors themselves, or are they rather produced from the surrounding tissue microenvironment? The experiment was performed by implanting cancer cells in a chamber with a membrane that served as a permeable border for molecules, but not for cells, which could not pass through (Gimbrone *et al* 1973). The result was that angiogenesis was induced in the regions surrounding the implant, indicating that small activating molecules exerted from the cancer cells passed through the membrane and induced angiogenesis in the local microenvironment (Gimbrone *et al* 1973). Key players in the angiogenesis process are vascular endothelial growth factor (VEGF) and basic fibroblast growth factor (bFGF), which are produced and secreted from cancer cells within the primary tumor in the local microenvironment. Endothelial cells possess receptors on their cell surface for these two molecules that bind and induce a signaling cascade in order to transmit a signal in the endothelial cell's nucleus (Mierke *et al* 2011, Bergers *et al* 2003). In the nucleus several genes are now transcribed, which are necessary to facilitate new endothelial cell growth and subsequently new vessel formation. In particular, the activation of endothelial cells through VEGF and bFGF induces several consecutive steps for building new blood vessels (Gimbrone *et al* 1973, Patan 2004). The first event is the production of matrix metalloproteinases (MMPs), which can degrade the surrounding extracellular matrix when released into the micro-environment (Patan 2004). The degradation of the extracellular matrix leads to the existence of a space where endothelial cells can migrate to and divide in order to form hollow tubes and finally mature blood vessels (Patan 2004).

Other inhibitors of angiogenesis exist, such as angiostatin, endostatin, interferons and TIMP-1, -2 and -3 (Ribatti 2009). These substances are very promising for inhibiting tumor growth. Unfortunately, only endostatin reveals such therapeutic effects on cancer growth as the primary tumor disappearing after several rounds of treatment (Eder *et al* 2002). In addition, no resistance effect to the endostatin treatment was observed in mice after repeated treatments. Moreover, it has been suggested that these inhibitors are able to reduce the speed of cancer metastasis. Whether this hypothesis holds true has been analyzed by injecting different types of mouse cancer cells under the skin of mice. The cells were grown for up to two weeks and then the primary tumor was removed by surgery. The mice were monitored for weeks to assess whether they developed secondary tumors (metastases). Normally, the mice form up to fifty visible tumors that spread in the lungs even before the primary tumor resection, whereas mice treated with angiostatin displayed on average only 2–3 tumors, indicating an approximately twenty-fold reduction in the spreading rate (metastasis) of the cancer cells from the primary tumor (Kirsch *et al* 1998, Bergers *et al* 1999). Why do certain metastases remain dormant for years? One possible answer is that no angiogenesis has occurred and thus no further tumor growth as blood vessels are missing (Gimbrone *et al* 1974). An explanation for this result may be that certain primary tumors secrete angiostatin into the blood fluid, inhibiting blood vessel growth throughout the whole body in other tissues. Then, the preliminary tiny tumors are no longer visible and cannot grow into secondary tumors, unless the primary tumor is removed and the angiostatin is no longer released into the blood fluid.

Besides the above-mentioned role of TIMP-1 as an inhibitor of pro-tumorigenic matrix metalloproteinases, TIMP-1 has recently been reported as a pro-metastatic factor that is strongly associated with poor prognosis in many cancer types (Kuvaja *et al* 2007, McCarthy *et al* 1999, Cui *et al* 2014). There seems to be a disparity between this finding and the inhibitory function of TIMP-1, but this new function of TIMP-1 is independent of and additional to its inhibitory function. TIMP-1 can signal as a molecule regulating cancer progression. In particular, it has been found that in lung adenocarcinoma cells an increase of exogenous and endogenous TIMP-1 up-regulates microRNA-210 (miR-210) by using an CD63/PI3K/AKT/HIF-1-dependent signal transduction pathway (Cui *et al* 2014, Wang *et al* 2014). This miR-210 belongs to the short RNAs that regulate the expression levels of other genes and is strongly linked to the hypoxia pathway; to be more specific, TIMP-1 induced P110/P85 PI3K-signalling and the phosphorylation of AKT. This then induces an increase of hypoxia-inducible factor-1α (HIF-1α) protein levels, together with an increase of HIF-1-regulated mRNA expression, and subsequently the up-regulation of the microRNA miR-210 is facilitated. If TIMP-1 is overexpressed in cancer cells, miR-210 accumulates in exosomes *in vitro* and *in vivo* (Cui *et al* 2014). In turn, these exosomes induce tube formation activity in human endothelial cells of the umbilical vein (HUVECs), as indicated by the enhanced angiogenesis activity in A549L-derived tumor xenografts (Cui *et al* 2014). In summary, TIMP-1 has a new pro-tumorigenic signaling function that may explain why elevated TIMP-1 levels are found in lung cancer patients and why these are associated with poor prognosis for the patients.

The network of proteases, their inhibitors, and effector molecules is balanced and is a determinant of tissue homeostasis. For example, imbalances of this network and the tissue homeostasis caused by elevated levels of the host tissue inhibitor TIMP-1 increase the susceptibility of target organs to support metastasis by activation of the hepatocyte growth factor (HGF) pathway (Schelter *et al* 2011). In addition, up-regulated expression of HIF-1α is associated with cancer progression and has been found to induce HGF-signaling through the up-regulation of the HGF-receptor Met via canonical means of stress induction, for example lack of oxygen (Hellmann *et al* 2002). However, it has long been supposed that there is a connection between TIMP-1, HIF-1α and HGF-signaling in the promotion of metastasis. Indeed, it has been reported that HIF-1α and HIF-1-signaling were enhanced during the liver metastasis of L-CI.5s T-lymphoma cells in syngeneic (genetically identical, or sufficiently identical and immunologically compatible to permit transplantation) DBA/2 mice that overexpress TIMP-1 (Schelter *et al* 2011). Moreover, the addition of recombinant TIMP-1 to L-CI.5s cells *in vitro* induced HIF-1α and HIF-1-signaling. In line with this, the knock-down of HIF-1α within L-CI.5s cells did not induce HIF-1α and HIF-1-signaling and thus the cells were not invasive. *In vivo* experiments showed that HIF-1α knock-down pronouncedly impaired Met receptor expression and Met receptor phosphorylation, reducing liver metastasis (Lee *et al* 2008). Moreover, the HGF-dependent TIMP-1 induced phosphorylation of Met and thus the increased invasiveness *in vitro* were facilitated by HIF-1α (Comito *et al* 2011). Finally, increased levels of TIMP-1 in the local microenvironment of cancer cells caused metastasis by inducing HIF-1α-dependent HGF-signaling. The finding that there is a connection between the protease inhibitor TIMP-1 and the stress-related factor HIF-1α is novel and impacts on the tissue homeostasis regulating cancer metastasis (Schelter *et al* 2011).

The hypothesis that interfering with the process of angiogenesis restricts tumor growth was further supported by genetic studies of mice lacking the two genes Id1 and Id3. The absence of these two genes inhibits angiogenesis. Angiogenesis-deficient mutant mice were injected with mouse breast cancer cells and observed for tumor growth. There was indeed only a short period of tumor growth and even then the whole tumor vanished completely after several weeks, resulting in the mice becoming healthy again (Li *et al* 2004). However, if cancer cells of the lung are injected into these angiogenesis-deficient mutant mice, the results are slightly different. In particular, the lung cancer cells develop slow growing tumors in these mice. Moreover, these tumors do not to spread to other organs and thus do not form metastases, resulting in a prolonged lifetime for these tumor-bearing mice compared to normal mice carrying tumors (Li *et al* 2004).

As these experiments were very promising, the following question was raised. Can the inhibition of angiogenesis slow down or prevent the growth and spread of a tumor even in humans? To answer this question many angiogenesis inhibitors belonging to different categories have been tested to cure cancer patients. Among these inhibitors are those inhibiting endothelial cells in a direct manner, whereas other inhibitors block the angiogenesis signaling cascade or abolish the endothelial cell's ability to degrade the surrounding extracellular matrix (El-Kenawi and El-Remessy 2013, Lee *et al* 2011). In more detail, one class of inhibitors, for example endostatin for the angiogenesis, contains molecules that directly inhibit the endothelial cell's growth. Another inhibitory drug is combretastatin A4, which induces apoptosis (programmed cell death) specifically in endothelial cells, whereas other drugs can interact with cell surface receptors such as integrins and subsequently destroy selectively proliferating endothelial cells (Ding *et al* 2011, Wu *et al* 2014). A second group of angiogenesis inhibitors is composed of molecules that interfere with steps in the angiogenesis signaling cascade of humans, for example anti-VEGF antibodies inhibiting the binding of growth factors to VEGF receptors. One anti-VEGF monoclonal antibody, bevacizumab (Avastin), has been demonstrated to impair tumor growth and thus extend the survival of cancer patients. A third agent is interferon-alpha (IFNα, which can counteract the production of bFGF and VEGF and thus impair the initiation of the growth-factor-driven signal transduction cascade (Frey *et al* 2011).

The fourth group of angiogenesis inhibitors contains substances that are directed against the endothelial produced MMPs (enzymes that initiate the breakdown of the local microenvironment). As the breakdown of the surrounding extracellular matrix is required for the migration of endothelial cells into surrounding tissues and the endothelial cell proliferation for the outgrowth of new blood vessels, inhibitory drugs targeting endothelial MMPs impaired angiogenesis and hence tumor growth and malignant tumor progression.

A fifth group of drugs is being investigated intensively for inhibition of angiogenesis and subsequent tumor growth; these drugs are either non-specific or not clearly understood, for example carboxyamidotriazole (CAI) works by inhibiting the calcium ion influx into all kinds of cells, including endothelial cells. As this restriction of calcium uptake specifically suppresses the growth of endothelial cells, it is expected that such a general mechanism can also affect other cell types and many other cellular processes. What is still not under discussion? The mechanical impact of endothelial cells on the mechanical properties of cancer cells and their function. This is described in more detail under transmigration, a process in which cancer cells can migrate through an endothelial cell monolayer in order to migrate into or out of blood or lymph vessels.

What role do the biomechanical properties of the primary tumor play in tumor angiogenesis?
In addition to the growth factors and cytokines regulating neoangiogenesis already discussed, the mechanical properties of a primary tumor may also facilitate endothelial

vessel growth within the tumor to increase the primary tumor size and effect the malignant progression of cancer. The tumor stiffness and the high interstitial pressure within the primary tumor blocks the diffusion of metabolites (Jain 1987). In turn, the tumor can regulate cellular angiogenesis in order to induce neoangiogenesis within the primary tumor. In particular, the local vascular system in the tumor microenvironment is induced by the outgrowth of new capillaries from preexisting vessels to grow in the primary tumor in order to supply the proliferating cancer cells with metabolites (Ruddell et al 2014). Even within the tumor, the maturation and remodeling of new microvessels needs the perfect coordination of many diverse processes in the microvasculature (Klagsbrun and Moses 1999). For the induction of new blood vessel sprouts, the pericytes located around endothelial vessel linings must be moved from the branching vessel. Then, the endothelial cell basement membrane and extracellular matrix surrounding the blood vessels must be degraded and restructured by MMPs (Kräling et al 1999). In the next step, the new extracellular matrix is synthesized and secreted by neighboring stromal cells (Lu et al 2012). This newly designed extracellular matrix, together with several soluble growth factors, evokes the migration and proliferation of neighboring endothelial cells. In the last step, endothelial cells build up a monolayer that results in a tube-like structure. Next, mural cells such as pericytes wrapping microvessels and smooth muscle cells wrapping large vessels are recruited to the non-luminal surface of the novel endothelial cell lining. The remaining uncovered vessels regress, showing that the process of angiogenesis is highly ordered and strongly regulated, as quiescent mature endothelial cells within the endothelial cell lining of vessels need to divide and branch out of an existing vessel by omitting excessive endothelial growth. Thus, cell–cell and cell–matrix adhesions of endothelial cells are crucial for normal survival within vessels and for tumor neoangiogenesis.

In tumors, the 'tumor' endothelium is dysregulated regarding hypoxia and chronic growth factor stimulation (such as vascular endothelial growth factor (VEGF)): tumor blood vessels possess irregular vessel diameters, are fragile, leaky and the blood flow is abnormal, suggesting that the tumor endothelium regulates tumor growth and metastasis. In particular, chaotic networks of the endothelium lacking the normal hierarchical arrangement of artery–arteriole–capillary have been found (Warren et al 1978, Konerding et al 1999). This leads to a poor stability of tumor endothelial vessels, together with lower numbers of pericytes, which are even less tightly attached to tumor endothelial cells than they are to normal endothelial cells (Baluk et al 2005). The vessel stability has a major impact on the blood flow and its directionality, which can be measured even at single-capillary resolution in primary tumors (Kamoun et al 2010). The blood vessel density increases during the early tumor formation, but decreases in larger tumors. The tumor vessel system itself can serve as a marker for malignant tumor progression as the poor quality of tumor vessels with irregular pericytes on the vessel wall and multiple basement membranes are associated with cancer metastasis (Nagy et al 2009, McDonald and Choyke 2003).

Tumor endothelial cells are identified and isolated using cell-surface markers such as PECAM-1 (CD31) (Hida et al 2004), ICAM-2 (Dudley et al 2008) and CD146 (St Croix et al 2000), which are all still markers of endothelial cells from all vascular beds (capillary, venous, arterial and lymphatic). Among the isolated tumor endothelial cells are the normal endothelial cells of the host organ. Defective, discontinuous endothelial monolayers have been detected in tumors where the gaps within the endothelial layer are filled with cancer cells mimicking endothelial cells as they express VE-cadherin (Maniotis *et al* 1999). Tumor endothelial cells have a high turnover rate and are more motile than normal endothelial cells located in healthy tissues. All this may contribute to the numerous gap formations in tumor endothelia. Moreover, even the intercellular cell–cell junctions are loosely formed, as well as the basement membranes, which results in large holes in the monolayers. These holes are supposed to be the entry sites for cancer cells to migrate through the human body and subsequently to metastasize at targeted sites. How these abnormalities of the tumor endothelial blood vessels are caused is not yet understood. Tumors are in fact dysfunctional organs: metabolic pathways are corrupted, cancer cells withdraw their microenvironment of nutrients by secreting toxic waste products such as lactate into their local tumor microenvironment, and tumors render local microenvironments areas of non-perfusion, leading to hypoxia. In this 'abnormal' microenvironment it is likely that abnormal endothelial vessels grow within the tumor (Merlo *et al* 2006). In line with this, most primary tumors produce high concentrations of vascular endothelial growth factor-A (VEGF-A), which serves as a potent vasodilator that is able to cause fluid leakage and high interstitial pressures, abnormal branching morphogenesis of the endothelium as well as small gaps in the endothelial monolayer (Nagy *et al* 2009). VEGF stimulates the endothelium, leading to a breakdown of the entire endothelial barrier function. Within a tumor, the vessels are squeezed and compressed by surrounding cancer cells, which causes external mechanical tension, strain and may finally alter the blood flow, leading to abnormal endothelial walls (Padera *et al* 2004, De Val and Black 2009). What is the function of these abnormal endothelial cells within the tumor? Do abnormal endothelial cells promote tumor growth and cancer progression to support metastasis? The inter-action between tumor endothelial cells and other cell types such as leukocytes may indeed be altered by the endothelial cell abnormalities within the tumor vasculature. In more detail, special adhesion molecules may be decreased on the cell surface of the tumor endothelial cells, helping primary tumors to escape immune surveillance due to impaired crosstalk between T-lymphocytes and the endothelial vessel linings (Griffioen *et al* 1996, Dirkx *et al* 2006). This hypothesis can be further supported by a study showing that the penetration and efficacy of primed T-lymphocytes used for tumor immunotherapy was increased by the addition of proinflammatory cytokines, up-regulating ICAM-1 and VCAM-1 on the tumor endothelial cells (Garbi *et al* 2004). More support is provided by the detection of leukocytes at the periphery of tumor vessels that secret factors required for tumor angiogenesis (Dudley *et al* 2010). In addition, it has been demonstrated that proinflammatory cells such as macro-phages facilitate metastasis (Qian and Pollard 2010).

1.2 Malignant progression of cancer (metastasis)

A worse outcome for a cancer patient is that the cancer has metastasized. The process of metastasis involves many consecutive steps as well as some optional side steps or side pathways. Basically, each neoplasm has in principle a possibility of metastasizing, but not all of them do this. This is related to the special micro-environment, the aggressiveness of a special subtype of cancer cells within the primary tumor and the start of medication, for example resection of the primary tumor before metastasis has occurred and subsequently some chemo- or radio-therapy. Nonetheless, there are still cases in which the metastasis has started and it finally leads to the death of the cancer patient, although the patient has undergone chemo- or radiotherapy. Indeed, metastasis is the main cause of cancer deaths and still no major progress has been made to reduce the number of cancer-related deaths.

Thus, many cancer research projects are needed on this particular subject that start from an alternative prospective and from another viewpoint than that of the classical one of tumor biologists. Here, the novel viewpoint of a biophysicist investigating cancer research should be helpful to overcome classical barriers and to reach novel ground. Thus, a new field called simply 'physics of cancer' has become increasingly important in cancer research and many researchers are interested in this novel viewpoint. In the following, the metastasis cascade is described briefly from a biophysical viewpoint.

What kinds of mechanisms are existent in the malignant progression of cancer?
Based on the biological and molecular mechanisms that cause primary tumors, a clear picture can be drawn, but for the mechanisms causing the subsequent invasion and metastasis no clear picture arises and many aspects remain elusive. Research to generate conceptual outlines for cancer malignancies and their causes has begun. However, the new physics of cancer field has arrived to try to fill the gap and bring another dimension to cancer research in order to understand why certain cancers become malignant and metastasize and others do not. There is a classical under-standing of how metastatic dissemination of cancer cells occurs that is called the invasion-metastasis cascade (Fidler 2003). This cascade involves the local invasion of primary cancer cells into the vascular system, such as the blood or lymph vessels (intravasation), their transport through the whole body via the vessel flow, their accumulation in microvessels of distant tissues, their possible transmigration through the vessel lining (extravasation), their invasion of the parenchyma of targeted tissue and their building of micrometastatic cell clusters that may eventually grow to macroscopic metastases. This process is called colonization and is the starting point for the worst outcome of cancer (figure 1.1).

What determines the 'early determination' of a primary tumor?
A major point is the timing of the individual cancer cell's acquisition of the capacity to invade and finally metastasize. Can every cancer cell within a primary tumor reach the ability to become malignant? Or does only a small subpopulation of cancer cells in this tumor gain the ability to invade the surrounding tissue and metastasize? What about the majority of cancer cells within a primary tumor, do they have an identical ability to invade and metastasize, or different abilities? If the ability to invade and metastasize is identical, it is suggested that the acquisition of this ability occurs early in the tumor development. By contrast, if only a small subset of cancer

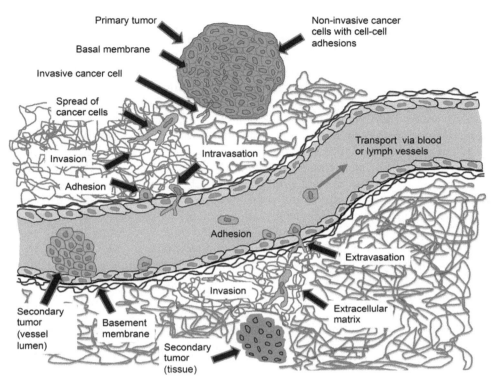

Figure 1.1. Cancer metastasis cascade.

cells acquire this ability to invade and metastasize, they must have obtained this in a later phase of the primary tumor formation and thus cancer progression.

How can this issue be addressed in an experiment? One possible approach is to analyze whether there is a subpopulation that can be found by clonal expansion favoring successive mutation and after several selection cycles we will obtain subclones. An alternative way of obtaining subclones is selection for a special marker, such as a cell–matrix or cell–cell surface receptor through cell sorting. Another alternative is sorting for special mechanical properties, such as the deformability of cancer cells that may support tumor pathogenesis through the selection of an aggressive cancer cell phenotype. A big problem of the model is: why should an invasive or metastatic phenotype be an advantage for a cell confined within the primary tumor?

Indeed, it seems to have no advantage for the primary tumor at first glance. This suggests that the occurrence of metastasis is no phenotype of the primary tumor selected by the formation of the primary tumor (Bernards and Weinberg 2002). Instead, there seems to be another phenotype that may arise as an inadvertent consequence of the acquisition of alleles providing an advantage for the growth of the primary tumor mass. This finding means that the alleles behave pleiotropically by encoding a selected phenotype of the primary tumor (growth and survival of the initial tumor) and an unselected phenotype (malignant tumor progression, including invasiveness and metastatic behavior). However, there may be another possibility, that the selection of highly invasive and metastatic cancer cells arises from the microenvironment that is the tumor itself or the surrounding extracellular matrix. In particular, the selection process may be driven by cellular mechanical properties that are altered within the cells of the primary tumor (Huang and Ingber 2005) and also of the surrounding microenvironment (Lokody 2014). This view is supported by the complexity of the cellular invasiveness and the metastatic cascade that completes the early steps of the primary tumor formation. In line with this, a question may be raised as to whether the metastatic traits require several accumulated mutations that compete for the variety of genetic lesions causing the primary tumor formation. However, there is some evidence that the metastatic dissemination of cancer cells does not rely on the acquisition of additional genetic lesions beyond those of the primary tumor. There exist mutations that are drivers of biological phenotypes and mutations that are passengers, solely reflecting the increased mutability of the cancer cell's genome and thus representing random genetic background noise of the primary tumor (Wood *et al* 2007).

In the following, the proposition that metastasis-specific mutations acquired during malignant cancer progression have a role is argued against. The first argument is that the prognosis of primary tumors, including their ability to metastasize, can be determined by analyzing the gene expression profile of the tumors. In particular, the gene expression profiles are altered by acquired somatic mutations and promotor methylation events that a normal cell has not acquired during its differentiation. In order to analyze the transcription patterns of primary breast carcinomas using gene expression arrays, it was possible to predict which primary tumors would undergo a malignant progression and which would stay non-malignant (van de Vijver *et al* 2002, Fan *et al* 2006). However, this prediction is not always possible and includes some errors. If this method for predicting the malignant tumor progression works, it implies that the mechanisms regulating cancer metastasis require a majority of neoplastic cells in a primary tumor to display an altered expression profile. This is in contrast to the finding that only a small number of neoplastic cells in the primary tumor have the ability to metastasize, suggesting that additional parameters may drive malignant cancer progression, and finally metastasis, such as mechanical alterations to these metastatic cells. Moreover, the metastatic spread of cancers seems to be determined in the early phase of cancer progression in order to be expressed in the majority of cells in the primary tumor (Bernards *et al* 2002).

A second argument for the early determination of the cells' ability to show a metastatic spread in the primary tumor is that only the early passage of human mammary epithelial cells cultured in a defined culture medium and transfected with a defined set of genes, such as the SV40 virus early region (specifying the large T and small t oncoproteins), the hTERT gene (encoding the catalytic subunit of the telomerase holoenzyme) and a ras oncogene developed metastasis, whereas the same cells cultured in a different medium did not. In particular, these two cell culture populations exhibit different gene expression patterns and upon transformation with one of the three genes (see above) developed histopathologically distinct tumors: the first is a squamous cell carcinoma and the second is an invasive ductal adenocarcinoma of the breast (Ince *et al* 2007).

These results support the idea that the differentiation program of the normal cell of origin is strongly determinant for the behavior of cancer cells upon transformation. The expression patterns of the two transformed cell types are closely related to those of their respective normal precursors and display relatively minimal similarity to one another. This may be a reason why one type of tumor, the invasive ductal carcinoma, metastasized to the lungs, whereas the other type, the squamous cell carcinoma, did not metastasize (Ince *et al* 2007). Both types of tumor had undergone the same set of experimentally introduced genes representing somatic mutations, but they still had different metastasizing abilities. This finding show that the differentiation state of the normal cell of origin plays a major role in determining the metastatic spread of a tumor. In summary, the properties of the normal cell of origin that is the progenitor of all the neoplastic cells of a primary tumor determine whether its offspring will have the ability to metastasize.

Another example is normal human melanocytes that have been transformed with the same oncogenes. The transformed melanocytes (the model of spontaneous melanomas) built up primary tumors with a large number of metastases in different targeted organs (Gupta *et al* 2005). Thus, the differentiation program of the normal cell of origin has a strong impact on the probability of metastasis occurring.

A third argument for the lack of metastasis-specific genes is that the same set of genetic lesions (mutations) in the genomes of primary cancer cells is generally found in the genomes of their derived metastases (Jones *et al* 2008).

Thus, the driving force for metastasis is not based on specific genetic lesions that are evolved during the multistep formation of a primary tumor. Instead, the dissemination of cancer cells from a primary tumor occurs as a side-effect of the primary tumor formation and does not seem to be a property achieved during the multistep process of primary tumor formation.

How is the invasiveness determined and how does metastases arise?
It has been established that a complex invasion-metastasis cascade is necessary. What drives metastasis, if there are no additional mutations required beyond those necessary for the formation of the primary tumor? A classical tumor-biological answer would be that cancer cells utilize complex biological programs used by normal, healthy cells and organismic physiology. In many steps of the morphogenesis of normal cells the epithelial–mesenchymal transition (EMT) plays an important role (Thiery 2002) (figure 1.2). In more detail, several distinct morphogenetic steps involve the local migration of epithelial cells as well as their migration to distant sites during embryonic development. However, normal epithelial cells are incapable of these translocations, evoked by active movement, as they only move laterally in the epithelial plane while maintaining adhesion to the underlying basement membrane or basal lamina. Under certain conditions, there is a special case when these epithelial cells acquire active movement and invade the extracellular matrix: the shedding by epithelial cells and the switch to mesenchymal properties. In summary, EMTs play an important role during embryogenesis during gastrulation and emigration of cells to distinct targeted sites within the embryo (Thiery 2003). EMTs are induced by a number of transcription factors (TFs) that are transiently

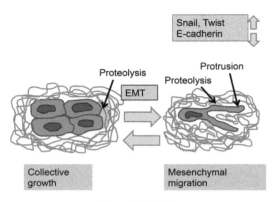

Figure 1.2. EMT.

active during special stages of embryogenesis and at specific sites within the embryo (Thiery and Sleeman 2006, Batlle *et al* 2000, Yang *et al* 2004, Gumbiner 2005, Peinado *et al* 2007, Hartwell *et al* 2006). The down-regulation of the TF expressions leads to the reversion of EMT, the so-called MET, so that the cells are now back in the epithelial state (the ground state), whereas the mesenchymal state seems to be the activated state. As EMTs occur at specific sites of the embryo, they need to be induced via the local microenvironment, for example neighboring cells and extracellular matrix protein signals, suggesting that EMT-inducing TFs may be regulated by the neoplastic cell's local extracellular-induced signaling. However, these descriptions do not show the relevance of EMT during cancer progression, but two aspects of cancer cells support the relevance of EMT induced by TFs. First, many phenotypes of embryogenic cells are imitated by aggressive and invasive cancer cells. Second, numerous embryonic TFs, such as Slug, Snail, Twist, Goosecoid, SIP-1, FOXC2 and ZEB1, which regulate EMTs during embryogenesis are also detected in human cancer cells and their expression correlates with the aggressiveness of these cancer cells.

Moreover, most of the TFs were identified during the investigation of the development of model organisms, such as Xenopus and Drosophila. As these results are transferable to humans, these TFs must be strongly conserved in the genomes of distantly related animals, indicating their critical roles in the embryogenesis of diverse organisms. All these observations lead to the suggestion that TFs enable carcinoma cells to obtain highly malignant properties, such as cell invasiveness, resistance to apoptosis and secretion of proteases that degrade extracellular matrix confinement. Furthermore, the EMT and hence the expression of these TFs is not restricted to initial embryonic development, it also plays a prominent role in wound healing processes in adults and re-epithelialization processes (Savagner *et al* 2005).

What role does the induction of EMT-inducing TF expression play?
As described above, various EMT-inducing TFs act during embryogenesis upon certain signals from nearby cells. Thus, it seems to be the case that the same type of signal (for example, Wnts and Hedgehogs, members of the transforming growth

factor beta family, as well as the ligands of tyrosine kinase receptors) affects various carcinoma cells during malignant cancer progression. In particular, none of these ligands has the capacity to trigger EMT alone, but the combination is able to induce EMT of cancer cells. The exact rules are still under intensive investigation. It has been suggested that during cancer progression these signals for EMT are provided by the tumor-associated stroma containing mesenchymal cells. These stromal cells originate from the stroma of the tissue in which the tumor grows or are directed from the bone marrow, which generates many distinct types of mesenchymal progenitor cells and releases them into the circulation for them to be available for local recruitment by carcinoma cells (Direkze and Alison 2006). Indeed, these cells are then located in the tumor-associated stroma and subsequently differentiate into a variety of mesenchymal cell types, such as myofibroblasts and endothelial cells. Additionally, there might be EMT-inducing signals that are not released by the tumor stroma of early stage tumors. During tumor progression the tumor stroma grows and is activated (reactive) in a similar way to tissues exposed to active wound healing processes or chronic inflammatory processes. What role do the mechanical properties of the tumor stroma play? In most biological and medical studies this question is simply omitted and in the physics of cancer field this topic is still under investigation.

Besides EMT-induction, the tumor stroma may have another impact on the behavior of cancer cells outside of the primary tumor, further enhancing cancer cell motility and guiding cancer cells toward their targeted tissue (celled secondary sites). On their way to the targeted tissue cancer cells migrate through other vastly different non-tumor-associated tissues. At these secondary sites, the stroma is not altered compared to the primary tumor associated stroma through the permanent stimulation by cancer cells and may hence not have these activated states that induce further EMT. Thus, EMT will be reversed and the cancer cells reassemble to a neoplasm that has similar properties to the primary tumor. However, it has been suggested that the absence of mesenchymal phenotypes in cancer metastasis disproves the idea that cancer cells must undergo an EMT to disseminate from primary tumors (Rhim *et al* 2012). This is in contrast to the argument that the reversibility of the EMT may have finished, so that the cancer cells returned to their ground state (of zero motility) in these secondary tumors. As heterotypic signals can induce EMT in carcinoma cells, these cells do not need to undergo additional mutations in order to become highly aggressive and invasive. Hence, primary carcinoma cells are able to perform EMT if the extracellular signals for EMT are recognized. What variable factors define whether cancer cells within a primary tumor will undergo an EMT? Is it really the appropriate mix of heterotypic signals that must be detected by cancer cells and that are essential for their EMT? It has been suggested that the differentiation program of the normal cell of origin represents one critical determinant of this responsiveness to the different signals regulating EMT as this program is supposed to set the stage for the malignant progression of cancers. Accumulated somatic mutations and promotor methylations during primary tumor progression also favor the responsiveness to EMT-inducing signals. However, the exact mechanism of the EMT induction and reversion is still not yet well understood, although many factors have been identified as regulating EMT. What role do mechanical properties play in inducing EMT?

There is currently no answer, as this is still under intensive investigation. As these malignant features are manifested due to signals from activated stroma they are not thought to be the objects of selection leading to the primary tumor formation. In particular, the somatically generated alleles provide the responsiveness to these signals, whereas the alleles selected during primary tumor formation directly specify malignant features. Thus, the expression of highly malignant features occurs as an accidental consequence of the initial actions of alleles that are 'unrelated' to the phenotypes of cell invasion and cancer metastasis.

When carcinoma cells undergo an EMT, they adopt mesenchymal phenotypes, invade the 'activated' surrounding tumor stroma and then move into adjacent normal tissues outside of the primary tumor borders during metastasis; they are nearly indistinguishable from normal mesenchymal cells. This raises the question of whether the EMT is solely an artifact. There are at least two arguments against this. The first argument is that most carcinoma cells undergo EMT incompletely, for example E-cadherin and cytokeratins are down-regulated and mesenchymal markers such as N-cadherin, vimentin and fibronectin are up-regulated (Schramm 2014). The coexistence of epithelial and mesenchymal markers may provide evidence for this hypothesis. The second argument is that carcinoma cells, which have undergone EMT, express certain markers not expressed by 'true' mesenchymal cells and may be detectable within the normal tissue as aggregates (Van Aarsen *et al* 2008).

The impact of the EMT and the invasion during the progression
of the metastasis cascade
How do these pleiotropically acting EMT-inducing TFs enable cancer cells to succeed with invasion and the metastasis cascade? In addition, these TFs provide increased resistance to apoptotic cell death, cell movement, secretion of matrix degrading enzymes and tissue invasiveness (Jiang *et al* 2014). Can a single TF enable the metastatic cascade? Can this disseminated cancer cell survive in the new microenvironment, where it does not fit in and to which it is not adapted? The success rate for building up a secondary tumor is relatively low. A reason may be the poor adaptation of the cancer cell to its target site. Thus, the growth from micrometastasis to a macroscopic metastasis and hence colonization is a rare event as only one out of thousands succeeds. It has emerged that the colonization is not a problem, it is rather the increased resistance to apoptosis that is critical for the metastasis cascade. Are the steps of the metastatic cascade governed by one of the TFs? If this holds true, all the multiple steps of the metastatic cascade would appear to be quite simple when regulated by one TF or a small group of them. However, the regulation of these TFs would then be highly critical in order to keep them under controlled tissue homeostasis.

Is there a special type of permanent EMT?
Cancer cells have been found locked in the mesenchymal state, not displaying any plasticity (Gregory *et al* 2011). This result indicates that the reversion to an epithelial phenotype is no longer possible. How may this irreversible EMT be caused? Possible explanations for this are: genetic or other biochemical alterations, or altered mechanical properties of cells due to their microenvironment or stimulation. In recent decades

it has turned out that cell surface protein E-cadherin has a prominent role in mediating cell–cell adherence junctions between neighboring epithelial cells (Tania *et al* 2014). This role has determined E-cadherin as a typical, canonical epithelial marker. In line with this, the promoter of the E-cadherin encoding gene exhibits binding sites for several EMT-inducing TFs that are able to repress E-cadherin transcription (Bolos *et al* 2003, Cano *et al* 2000, Comijn *et al* 2001). E-cadherin repression seems to be one of the main functions of these TFs. However, via an unknown mechanism cytokeratin expression is down-regulated, whereas various mesenchymal genes are up-regulated or induced (Tania *et al* 2014). Taken together, E-cadherin expression is a main target of the regulation by EMT-inducing TFs.

By contrast, certain tumors show alterations in E-cadherin expression, containing point-mutations or deletions in the E-cadherin gene that lead to a production of truncated or unstable proteins (Kanai *et al* 1994). Another associated change of functional E-cadherin is that the cytoplasmic proteins that usually serve to physically link E-cadherin to the acto-myosin cytoskeleton are translocated (Kam and Quaranta 2009) and hence rapidly degraded (Gerlach *et al* 2014). An exception is β-catenin, this molecule can survive by escaping phosphorylation by the non-phosphorylated glycogen synthase kinase-3β (GSK-3β) and hence proteosomic degradation (Gerlach *et al* 2014). The GSK-3β is hyperphosphorylated by Akt kinases and thus inactive. The free β-catenin can localize to the nucleus when it is associated with a T-cell factor group of TFs and together with other signals triggers a large number of downstream target genes that are mostly involved in the regulation of the EMT switch (Onder *et al* 2008). In particular, gene expression analysis has been shown that the loss of E-cadherin leads to the induction of multiple transcription factors, such as Twist. Twist in turn is necessary for the loss of E-cadherin and finally induces metastasis, indicating that the loss of E-cadherin in primary tumors facilitates metastatic dissemination through transcriptional and functional alterations (Onder *et al* 2008). However, in principle there is a permanent EMT as cancer cell lines exist that are able to stay in their mesenchymal phenotype.

1.2.1 Spreading of cancer cells

The onset of the malignant progression starts with the spreading of cancer cells from the primary tumor into the surrounding microenvironment. The principles behind this phenomenon are not yet clear. Several questions remain unanswered. How do these malignant and highly aggressive cancer cells manage to migrate out of the primary tumor where intercellular junctions usually exist between adjacent cancer cells? What does the differential adhesion hypothesis contribute to the understanding of how certain cancer cells migrate out of the primary tumor? Do these special cancer cells walk out by a mechanism called jamming? Do these 'highly aggressive' cancer cells possess different mechanical properties, such as cellular deformability?

Differential adhesion hypothesis
The differential adhesion hypothesis (DAH) was initially postulated by Foty and Steinberg (Foty and Steinberg 2004, Steinberg 1970, 1978) and provides a physical explanation for the spontaneous liquid-like tissue segregation, mutual envelopment

and sorting-out behaviors of embryonic tissues and embryonic cells. This DAH was reported earlier to depend upon tissue affinities (Holtfreter 1939, Townes and Holtfreter 1955). In particular, the DAH explains the segregation, envelopment and sorting-out behaviors and the rounding-up of irregular embryonic tissue fragments as rearrangements of cells that seek to decrease the cell population's adhesive-free energy, whereas the overall cell–cell bonding events increase. This thermodynamic hypothesis has been formulated based on experiments comparing the behavior of cell populations during the cell sorting process and mutual tissue spreading, with expectations based upon each of the hypotheses, but it cannot explain these processes (Steinberg 1962a, 1962b, 1962c, 1963, 1964, 1970). Up to now, only the DAH makes correct predictions. Analysis of embryonic tissues capable of the morphogenetic behavior revealed that these tissues could be characterized macroscopically as elasticoviscous liquids whose elemental components are motile, mutually adhesive cells. Ultrastructural and mechanical studies of rearranging cell aggregates were able to confirm these findings (Forgacs *et al* 1998, Gordon *et al* 1972, Phillips and Steinberg 1978, Phillips *et al* 1977, Steinberg and Poole 1982).

The relative surface tensions of two immiscible liquids determine which liquid will envelop the other. The DAH concludes that the mutual spreading ability of tissues (Davis 1984, Davis *et al* 1997, Foty *et al* 1994, 1996, Phillips and Davis 1978) is specified by their relative surface tensions (Steinberg 1970), which was proved by newly developed tissue surface tensiometers. Thus, using these surface tensiometers it has been shown that in every mutually adhesive tissue pair tested, the tissue with lower surface tension always envelops that with higher surface tension. Moreover, it has been reported that this behavior is independent of the identities of the adhesion molecules used by the interacting cells (Duguay *et al* 2003, Foty *et al* 1994, 1996). What has to be confirmed by experiment is the postulation that the tissue surface tensions underlying mutual tissue segregation, spreading and cell sorting are obtained solely from the intensities of the cell–cell adhesions building up these tissues. Indeed, it has been suggested that cells sort out because of a different cadherin expression level, for example N-cadherin (Foty and Steinberg 2005).

Cell jamming

Cell jamming has been observed during studies of inert soft condensed matter. In particular, spontaneous intermittent fluctuations, dynamic heterogeneity, coopera-tivity, force chains and kinetic arrest are the hallmarks of the glass transition supposed to be associated with jamming (Garrahan 2011, Bi *et al* 2011, Vitelli and van Hecke 2011, Trappe *et al* 2001, Liu and Nagel 1998, Liu *et al* 1995). Although jamming remains debatable and not very well understood, the concept has become a focus of much research as it tries to unite understanding of a broad range of soft matter forms, such as foams, pastes, colloids, slurries and suspensions, which can flow in some situations but jam in others.

The same hallmarks can be found in the dynamics of cell monolayers. These dynamics conform quantitatively to the Avramov–Milchev equation, which describes the rate of structural rearrangements (Angelini *et al* 2011) and shows the growing length and time scales that are quantified using the more rigorous four-point susceptibility (Tambe *et al* 2011; Berthier *et al* 2005). As inert and living condensed systems dynamics are constrained by many of the same physical factors, the assertion of cell jamming may be suitable. In more detail, looking at the basic unit such as a living cell, a foam bubble or a colloidal particle, they all include volume exclusion (two particles cannot occupy the same space at the same time), volume (size) (Zhou *et al* 2009), deformability (Mattsson *et al* 2009), mutual crowding, mutual caging (Schall *et al* 2007, Segre *et al* 2001), mutual adhesion or repulsion (Trappe *et al* 2001) and evoked mechanical deformation, such as stretch or shear (Trepat *et al* 2007, Krishnan *et al* 2009, Oliver *et al* 2010, Wyss *et al* 2007). In a cell monolayer, the cells move more freely as their size, crowding, stiffness or mutual adhesion decreases, or as their motile forces or imposed stretch become greater. However, as adhesion or crowding increases, or as motile forces decrease, cellular rearrangements might slow, cooperativity will increase, and the monolayer will be topologically frozen and all cells will be caged by their neighboring cells (Ladoux 2009, Tambe *et al* 2011, Angelini *et al* 2011, Angelini *et al* 2010). It seems that the jamming hypothesis can unify the effects of diverse biological factors that have been considered to be separate and independent of each other.

Does a jamming phase diagram exist?
As jamming phase diagrams are known from inert soft matter (Trappe *et al* 2001; Liu *et al* 1995), these effects may also play a role in the living systems within a supposed jamming phase diagram. In particular, we have the first axis for cellular crowding (expressing the inverse as the reciprocal of cellular density with infinite density in the origin). On the second axis we have the cell–cell adhesion (expressed as a reciprocal with infinitely sticky cells mapped to the origin). On the third axis we display the effects of cell motile forces. Additional axes may be possible, for example for imposed stretch or shear loading (Trepat *et al* 2007; Krishnan *et al* 2009), cellular volume (Zhou *et al* 2009), cellular stiffness (Mattsson *et al* 2009) and substrate stiffness (Krishnan *et al* 2011, Angelini *et al* 2010), but often they are not available. In a phase diagram with multi-dimensional space, the origin, and regions near the origin, are jammed and rearrangements are not possible as each cell is fully caged (confined) by its neighbors, or even glued to its neighbors, or it cannot generate the necessary driving motile force. However, away from the origin, especially along certain trajectories, structural rearrangements are possible with increasing probability. Unjamming can take place if cells are stretched, undergo apoptosis or are extruded from the monolayer; these all lead to a decrease in cell density (Eisenhoffer *et al* 2012). In line with this, if adhesive interactions are weak enough, or if the stretch becomes large enough in order to rupture cell–cell adhesions, the unjamming mode of the cellular system is preferred. Unjamming is present if motile forces are strong enough to pull individual cells away and dissociate from neighboring cells to reach a loose and disaggregated system.

An example is MCF10A human breast cancer cells that overexpress ErbB2, which promotes proliferation and cell crowding, pushing the system toward a glassy and jammed state, whereas the overexpression of the potential tumor associated and L1-CAM interacting protein 14-3-3ζ degrades cell–cell junctions, moving the system out of the jamming state toward a fluidized and unjammed state. Another example is hepatocyte growth factor (HGF) facilitated scattering of MDCK cells leading to the disruption of cadherin-dependent cell–cell adherence junctions depending upon integrin adhesion and the phosphorylation of the myosin regulatory light chain (MRLC) (de Rooij *et al* 2005). All these findings fit perfectly into the jamming phase diagram.

Uniting biology and physics

The binary alternatives of jammed and unjammed states are not the only possibilities, in inert systems the jamming phase diagram includes fragile intermediate states with possibly no less relevance to the biology of the monolayer (Bi *et al* 2011, Vitelli and van Hecke 2011). The jamming hypothesis states that specific events at the molecular scale necessarily modulate and respond to cooperative heterogeneities caused by jamming at a much larger scale of organization, but specific events at the molecular scale can never by themselves explain these cooperative large-scale events. In this physical view the specific molecular events are ignored or they belong to an integrative framework that was unanticipated and overwritten.

This leads to novel questions that have not been asked before. Does an epithelial monolayer build a solid-like aggregated sheet displaying excellent barrier function with little possibility of cell invasion or escape due to the phenomenon that the constituent cells are jammed (Eisenhoffer *et al* 2012)? Do certain specific cell populations become fluid-like and hence permissive of paracellular leakage, transformation, cell escape or cell invasion due to the phenomenon that they become unjammed? Do the formation of pattern and wound-healing processes require that the cells are in an unjammed state (Serra-Picamal *et al* 2012)? If the answer is yes, what represents the critical physical threshold? What are the signaling events at the level of gene expression and signaling and the resulting physical changes that favor or hinder the jamming of cells? One possibility may be force-dependent thresholds and novel pathways that regulate cellular polarization (Prager-Khoutorsky *et al* 2011), but can these thresholds and pathways work in collective processes? However, trials targeting adhesion molecules in order to decrease tumor progression have been reported to be ineffective, indicating that migration events are somehow reprogrammed by not yet defined mechanisms to maintain invasiveness through morphological and functional dedifferentiation (Friedl *et al* 2004, Friedl and Wolf 2003, Rice 2012). Does jamming allow certain cancer cell subpopulations to unjam, awake from dormancy and thus evolve so as to maintain invasiveness by selection for adhesive interaction, compressive stress and cyclic deformation? Each of these complex questions cannot be answered solely by physics or biology. One possibility seems to be to combine these approaches to answer these important questions and reach new horizons in cancer research by incorporating the physics of cancer.

1.2.2 Migration of cancer cells into the microenvironment

The motility of cells in three-dimensional (3D) extracellular matrices is a prerequisite for tissue assembly and regeneration, immune cell trafficking and diseases such as cancer. In more detail, the process of metastasis depends on the migration of single cancer cells that migrate out of the primary tumor. The migration of cancer cells through the extracellular matrix of connective tissue is a cyclic process comprising multiple steps, including: (i) the actin polymerization-dependent protrusion of pseudopods at the leading edge; (ii) the integrin-mediated adhesion to the extracellular matrix; (iii) the contact-dependent degradation of the extracellular matrix evoked by its cleavage through cell surface proteases; (iv) the actomyosin-facilitated contraction of the cell's body, increasing longitudinal tension; and (v) the retraction of the cell's rear part followed by the translocation of the whole cell body (Doyle *et al* 2013). All these steps describe only one specific mode of migration that can be chosen by metastatic cancer cells: the protrusive mode (Maruthamuthu and Gardel 2014). However, there are still other modes of cancer cell migration, such as the blebbing mode (Laser-Azogui *et al* 2014). Which mode of migration is favored by a certain type of cancer cell and how the choice of mode is altered by the specific microenvironmental conditions is still under discussion. In addition, there is an intermediate lobopodial migration mode, so far only detected for fibroblasts, and it has been suggested that cancer cells also choose this under certain circumstances. Hence, further investigation is required to determine whether cancer cells are able to use this fibroblastoid lobopodial mode.

The ability of cancers to metastasize depends on the cell's ability to migrate to and invade connective tissue, adhere, and possibly transmigrate through a barrier such as basal membrane and the endothelium. However, what determines a special mode of invasion and how the appearance or the switch between the different modes is regulated is not yet understood. In more detail, the invasion mode is supposed to play an important role for the regulation of the basement membrane or endothelial barrier-crossing transmigration potential of cancer cells and has a major impact on their invasion speed.

How powerfully a migrating and invading cell overcomes the different obstacles found in dense 3D matrices depends strongly on its mechanical properties and how it is able to generate and transmit its protrusive forces. Hence forces and material properties determine which mode of invasive cell migration is favored for a special cancer cell type or cancer cell subpopulation. Indeed, it has been established that cancer cells with certain mechanical properties such as contractile force transmission and generation are able to invade 3D extracellular matrices more efficiently than less contractile cancer cells (Mierke *et al* 2008a, Mierke *et al* 2011). Nonetheless, what kind of migration is chosen when cells try to squeeze through narrower spaces, such as vascular endothelial cell walls? In this special case of transmigration blebbing motion may be more favorable. Additionally, in preliminary experiments it has been found that the stiffness of the plasma membrane drastically softens in primary human mamma and human cervix carcinoma cells, favoring the blebbing process of cell migration (figure 1.3). Thus, one may hypothesize that the mechanical properties and the type of force generation of cancer cells determine their invasion mode and may also regulate the switch between the different migration modes. The following questions remain unanswered. What are the major mechanisms that regulate the invasion mode of cancer cells? What role do microenvironmental properties such as the mechanics and structure of the extracellular matrix play regarding the migration mode of cancer cells? In order to investigate this, one needs to dissect the crosstalk between the invasion modes and the environmental confinements such as mesh or

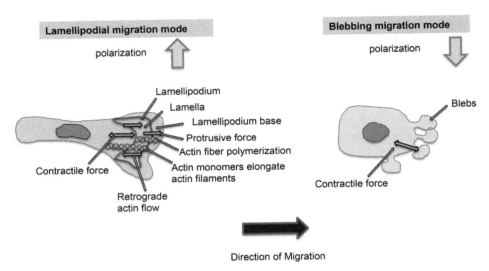

Figure 1.3. Modes of migration in 3D.

pore size, stiffness of the entire cell, the plasma membrane and the extracellular matrix, and the proteomics of the cellular adhesion machinery as well as the extracellular protein composition.

To understand the interaction between invading cancer cells and their microenvironmental confinement and why certain cancer cells use a particular invasion mode such as blebbing or protrusive modes, the following major questions should be answered. What roles do the influence of cytoskeletal stiffness and cell contractility play regarding the invasive cell motility? To what extent do these factors favor either the protrusive or blebbing-based migration modes of cancer cells? How does the stiffness of the plasma membrane impact on the cellular invasive behavior, such as protrusive or blebbing-based motion? How strongly do adhesive cancer cells differ from weakly adhesive cells with respect to their preferred invasive motility modes? Is the blebbing mode of invasion preferred by small mesh sizes of the microenvironmental confinement, such as the connective tissue matrix scaffold, whereas the protrusive mode is supported by large mesh sizes? Can the blebbing or protrusive modes of cancer cells support transmigration through barriers such as vascular endothelial cell linings and basal membranes?

Knowing the answers will contribute significantly to the understanding of how cancer cells utilize a certain invasion mode in order to migrate through the 3D microenvironment and what role the material properties of cancer cells and their microenvironments play. Moreover, these data will help to reveal the respective contributions of the mechanical properties of cancer cells and their microenvironments in supporting the invasive behavior of epithelial-derived carcinomas, in particular metastasis.

Malignant cancer progression involves the process of metastasis, which makes it a systemic disease that leads to death. The complex process of metastasis is composed of many steps, which follow a linear propagation. The metastatic cascade can be delayed by stopping at special steps in order to start again after some relapse time in which the aggressive cancer cells are dormant. How this phenomenon is regulated or induced is not yet well understood and thus needs further investigation. The metastatic cascade begins with the spreading of cancer cells from the primary tumor (dissemination), which migrate into the surrounding tumor microenvironment that is locally 'transformed' by the primary tumor (Bizzarri and Cucina 2014). Then these cancer cells transmigrate into blood or lymph vessels (intravasation) through the basal membrane and endothelial barriers, are transported through the vessel flow, adhere to the endothelial cell lining, grow and build up a secondary tumor (Al-Mehdi et al 2000). Thereby the new tumor is initiated either in the blood or lymph vessel, or cancer cells possibly transmigrate through the endothelial vessel lining (extravasation) into the extracellular matrix of the connective tissue. In the latter case, these cancer cells then migrate further into the targeted tissue, divide and form a secondary tumor, which means that the primary tumor has metastasized.

During the last two decades, the typical migration modes, such as mesenchymal and amoeboid motion, have been investigated (Wolf et al 2003a, Taddei et al 2014, Friedl and Wolf 2003). To date, these migration modes have not been clearly defined. Although these migration modes are reported to be a mechanistically well-described concept, they are solely a morphological description of the migration mode rather than an exactly defined and hence distinguishable invasion mode (Lämmermann and Sixt 2009). This needs to be clarified, in order to compare the migration modes and to define their occurrence and regulation. In particular, we will analyze whether a cancer cell prefers a certain mode of invasion and which mechanical phenotype defines this mode exactly and makes it distinguishable from the other invasion modes. The term amoeboid is not precisely defined, as migrating cells with roundish shape are categorized as amoeboid without analyzing the mechanistic aspects of their cytoskeletal remodeling dynamics and their mechanical forces. In line with this, a definition of the amoeboid migration has been based on the cell morphology, adhesiveness and proteolytic remodeling of the microenvironment for the interstitial migration of leukocytes (Wolf et al 2003b, Sabeh et al 2009). Proteolytic degradation is linked to the amoeboid migration mode depending on cellular deformability (Sabeh et al 2004, Wolf et al 2007, Rowe and Weiss 2009). In addition, proteolytic degradation is also associated with interstitial invasion of mesenchymal cells into collagen-rich and hence dense 3D extracellular matrices, and these mesenchymal cells migrate by using proteolytic and non-proteolytic degradation through narrow constrictions. Important questions remain unanswered. How do cellular forces such as contractile or protrusive forces facilitate the switch between the different invasion modes? How does the definition of protrusive or blebbing-based invasion modes fit into this scenario?

In previous studies the mechanical analysis was centered solely on actomyosin mechanics, whereas regulatory aspects of the actomyosin networks or the mechanical impact of other cytoskeletal elements such as actin-crosslinkers, or intermediate filaments such as vimentin are still elusive (Brown et al 2001, Laevsky and Knecht 2003, Tooley et al 2008, Wei et al 2008). In particular, a connection between the actomyosin cytoskeleton and the intermediate cytoskeleton has been proposed, but the proteins mediating this linkage have not yet been discovered (Seltmann *et al* 2013). However, understanding the regulatory function of these cytoskeletal factors in well-defined migration models will be necessary and will encourage understanding of the diverse mechanics involved in the different migration modes. There are still many open questions that should be answered. Which types of physical invasion strategies can cancer cells utilize and are they indeed derived from leukemias or fibroblasts? How do cancer cells facilitate the switch between different invasion modes, such as the protrusive and blebbing-based modes? How fast is this switch and is it permanent? How are these two 'novel' invasion modes related to the classical epithelial mesenchymal transition that supports the invasive behavior of cancer cells?

Despite these open questions regarding the cellular migration of cancer cells, a general conceptual model has been developed for how a lamellipodium can support protrusive cell motility. In more detail, actin filaments polymerizing below the leading edge of the cell membrane generate a pushing force (like a thermal ratchet) toward the cell membrane required for the formation of cellular protrusions. The surface tension of the cell membrane is able to oppose the free anterograde expansion (outward) of the actin network and thus actin filaments are then pushed back into the cytoplasm of the cell, which is detectable as a retrograde (inward) flow of actin (Mierke 2014, Ponti *et al* 2004, Gardel *et al* 2008). In more detail, cell–matrix adhesion receptors such as integrins connect the internal cytoskeleton to the external extracellular matrix through focal adhesions containing focal adhesion proteins such as vinculin (Mierke *et al* 2008b; Mierke *et al* 2010), focal adhesion kinase (FAK) (Mierke 2013), paxillin (Schaller 2001) and talin (Ziegler *et al* 2008). These adhesive contacts ensure that the retrograde-directed forces, which are enforced by actomyosin contraction, are transformed into outward locomotion of the cell's body in the migration direction (clutch hypothesis) (Vicente-Manzanares *et al* 2009). Taken together, the basic concept of the lamellipodial motion relies on cellular mechanics. In particular, the actin polymerization facilitates the formation of membrane protrusions. This particular migration mode of cancer cells is defined as the protrusive mode. Besides these lamellipodium-supported outward forces, filopodia-like extensions are required for cell invasion, which are suggested to be more force sensing rather than force generating, as well as invadosomes, which are needed for 3D invasion of tissue barriers (Ridley 2011). Lamellipodial-driven motion has predominately been investigated on two-dimensional (2D) substrates. Recent studies have found that many aspects differ significantly in three dimensions, supposing that the results obtained in 2D assays should be confirmed in 3D assays (Meyer *et al* 2012, Mierke *et al* 2010, Mierke 2013).

One alternative way for cancer cells to migrate without protrusive forces is migration through membrane blebbing. Blebs are cellular spherical extensions that are instrumental for cell migration in developmental processes as well as diseases like cancer. These cellular blebs are anterior cellular extensions and devoid of actin filaments. For a long time they have only been considered to be exclusively a hallmark of apoptosis (Mills *et al* 1998). In the last decade, they have become a hallmark for a special migration mode (Charras *et al* 2006, Charras and Paluch 2008, Paluch and Raz 2013). The phenomenon of blebbing has been seen in 2D

(Yoshida and Soldati 2006) and 3D migration assays (Charras and Paluch 2008). In particular, the intracellular hydrostatic pressure generated by actomyosin contraction causes (i) the rupture of the actin cortex (Tinevez *et al* 2009) and/or (ii) the rupture of the focal adhesion protein mediated linkage between actin cytoskeleton and the cell membrane (Charras *et al* 2006). These two proposed mechanisms depend on the cellular mechanics. In particular, if the membrane loses its mechanical anchor to the extracellular matrix, the intracellular pressure facilitates the formation of a membrane bleb (Charras and Paluch 2008). These mechanisms are hard to determine, as they may act in combination. The blebbing lifecycle is divided into three phases: initiation, growth and retraction. The membrane bleb grows until a new actin cortex is reassembled, which then may also contract, repeating the whole blebbing cycle (Charras and Paluch 2008). The site of the bleb initiation during migration has not yet been discovered. The blebbing strategy of cells is physiologically relevant motility, in which the cells migrate by a directed and persistent motion (Blaser *et al* 2006). Moreover, membrane blebbing may alter the surface tension of the cell membrane and subsequently the mechanical properties of the entire cell.

However, the precise mechanisms when cancer cells use either the protrusive or the blebbing-based invasion mode and how cancer cells can switch between these two are not clear. Furthermore, what impact does the microenvironmental confinement have on the determination of the invasion mode? What we know is that in tissue and cell cultures, several cells are able to switch between blebbing and protrusive motility due to microenvironmental confinements, in response to genetic alterations or pharmacological drugs (Lämmermann and Sixt 2009, Diz-Munoz *et al* 2010, Poincloux *et al* 2011). A clear definition of the individual conditions for the switch between these two invasion modes has not yet been postulated.

However, the key points for analysis of the blebbing migration compared to the protrusive mode are the effect of the cell–matrix adhesion, matrix geometry, matrix mechanics and the regulation of the signaling processes between the front and the rear of a motile cell. For both motility modes, the blebbing and the protrusive mode of migration depend on stabilized cell–cell or cell–matrix adhesions (Renkawitz and Sixt 2010). The cell adhesion provides stability and has a broad physiological importance, but seems to be significantly diminished in fast-migrating amoeboid cells such as immune cells, which are able to migrate through the tissue in the blebbing and in the protrusive migration mode (called synonymously lamellipodial mode) (Lämmermann and Sixt 2009). For example, the slow mesenchymal movement of cells through the extracellular matrix, which relies completely on focalized cell–matrix adhesions, is a solely protrusive migration mode.

Another important feature of cellular motility is the dimensionality of the surrounding microenvironment in which the cells will migrate; this is supposed to have a broad impact on the motility of cells (Rubashkin *et al* 2014, Mierke *et al* 2010), and in particular, the blebbing and protrusive migration modes may occur on 2D substrates and in 3D microenvironments. A main difference that has been reported between 2D and 3D motility assays is that 3D, but not 2D, microenvironments support cellular motility and invasion under minimal adhesion forces and thus decreased adhesion strength (Friedl and Wolf 2010). Moreover, how cellular polarity supports cellular motility in 2D and 3D migration systems and whether cellular polarity is exclusively associated with the protrusive migration mode is not yet known.

In addition, it has been found that cells migrate by forming cylindrical-shaped lobopodia with protrusions containing multiple tiny blebs at their tips, which might be an intermediate invasion mode between the blebbing and protrusive modes (Petrie *et al* 2012). Until now, this lobopodial mode has only been described for fibroblasts, whether cancer cells are also able to use this intermediate invasion mode has not been determined in a more precise way. The classical method to analyze cell migration was to use 2D migration systems. On these planar substrates the cells form flat lamellipodia, whereas when seeded into 3D collagen matrices of non-cross-linked bovine collagen type I these cells exert some kind of 3D lamellipodia, the so-called invadopodia. In several reports invadopodia have been associated with the proteolytic degradation of the extracellular matrix (Linder *et al* 2011, Destaing *et al* 2011). However, if the collagen fibers are crosslinked, the cells start to migrate further in a lobopodial mode, in which they sense the mechanical properties of the surrounding microenvironment, such as the elastic properties of the matrix's scaffold. In particular, if the 3D matrix stiffness is low then the cells utilize a lobopodial mode of invasion. Whereas the actomyosin contraction causes stress hardening of the extracellular matrix that may automatically activate the blebbing mode, where cells need to squeeze through the pores of the extracellular matrix. Finally, it can be summarized that the stiffness of the local microenvironment is not the major driving factor in lobopodial-based migration. It seems that it is rather the shape of the stress–strain curve that is the driving factor, as lobopodia are only formed in linearly elastic microenvironments such as the skin and the cell-derived matrix, and not in non-cross-linked 3D collagen matrices showing strain stiffening, where the formation of lamellipodia is promoted (protrusive invasion mode) (Petrie *et al* 2012). These findings are in contrast to the hypothesis that matrix stiffness increases lamellipodia/invadopodia-driven cancer cell invasion (Leventalx *et al* 2009, Alcaraz *et al* 2011, Pathak and Kumar 2013, Mouw *et al* 2014). There are still open questions, for instance regarding the stabilized functional polarity of cells that perform persistent directional migration, and how cells manage to sense linear elasticity and distinguish it from pure strain stiffening? How is the extracellular proteolysis of the matrix confinement connected or associated with lobopodial migration? Do cells migrating in a lobopodial mode secrete extracellular matrix proteins such as fibronectin to remodel their microenvironment? Can cells other than fibroblasts, for example cancer cells, use this lobopodial migration?

However, matrix geometry has been suggested as a means to determine the invasion strategy of cancer cells (Tozluoglu *et al* 2013). In addition, matrix geometry, surface tension and cell–cell coupling through adherence junctions may play a major role in selecting the optimal and efficient migration mode under the specific constraints (Campinho *et al* 2013, Maître *et al* 2012).

Another mode of migration is the degradation of the extracellular matrix
Another important mode of migration is the one in which the cancer cells degrade the surrounding extracellular matrix in order to migrate deeper into it. What we have

discussed up to now is that cancer invasion is a multi-step process, which regulates the interplay between cancer cells and the underlying mechanotransduction signaling processes toward the surrounding extracellular matrix. This results in morphologically well-defined cancer invasion strategies. As refined by the microenvironment, in mesenchymal tumors such as melanoma and fibrosarcoma either single-cell or collective invasion modes can be detected. What regulates such plasticity of mesenchymal invasion programs on the mechanical and molecular levels is still not well studied. Do tissue properties regulate the switch of invasion modes? To investigate this issue, spheroids of MV3 melanoma and HT1080 fibrosarcoma cells are embedded and cultured into 3D collagen fiber matrices of varying density and stiffness and analyzed for their migration type. In addition, the migration efficacy is analyzed with the MMP-dependent collagen degradation enabled or pharmacologically inhibited. It has been suggested that with increasing density of the collagen matrices, and dependent on the ability to proteolytically break down collagen fibers and provide track clearance, but independent of matrix stiffness, cells are able to switch from a single-cell to a collective mode of invasion (Haeger *et al* 2014). The conversion from single cell to collective invasion included the gain of cell-to-cell adherence junctions, supracellular polarization and possibly joint guidance along the pre-defined cellular migration tracks. In general, the network density of the extracellular matrix regulates the type of invasion mode used by mesenchymal cancer cells. It has been observed that fibrillar, high porosity extracellular matrices enable the dissemination of single cells, whereas dense extracellular matrices induce cell–cell interactions, a leader to follower cell behavior leading to collective migration with obligatory protease-degradable migration. Taken together, these results show that the plasticity of cancer invasion modes in response to the extracellular matrix properties such as matrix porosity and confinement plays an important role in providing the efficient invasion strategy, thereby recapitulating invasion patterns of mesenchymal tumors *in vivo*. In particular, the conversion from single to collective cell invasion due to increasing extracellular matrix confinement favors the concept of cell jamming as a guiding principle for certain cancer cell types such as the melanoma and fibrosarcoma into dense connective tissue.

Even when migrating as a collective cluster of motile cells, each individual cell undergoing the cellular collective senses signals or gradients in order to mobilize physical forces. These forces themselves regulate local cellular motions from which the collective cellular migrations subsequently arise. However, these forces cannot explain spontaneous noisy fluctuations (even large ones). In particular, the implicit assumption is linear causality in which systematic local motions, on average, are caused by local forces, and these local forces are in turn evoked by local signals. A novel hypothesis states that strong and dominant mechanical events are not solely local. This indicates that the mechanical causality cascade is not so linear, and the fluctuations may not be seen as noise, they are rather an essential feature of this mechanism. Fluctuations and non-local cooperative events defining the cellular collective motion can be described by the process of cell jamming. Cellular jamming can even unite several factors, including cellular crowding, force transmission, cadherin-dependent cell–cell adhesion, integrin-dependent cell–substrate adhesion, myosin-dependent motile forces and contractility, actin-dependent deformability, proliferation, compression and cell stretch, which are reported to act together but still mostly act separately and independently.

1.2.3 Transendothelial migration of cancer cells

What impact do transmigrating and invasive cancer cells have on the biomechanical properties of the endothelial cell lining?
The role of endothelial cells in the regulation of cancer cell invasiveness into 3D extracellular matrices is still not yet fully understood. Moreover, the coordination and regulation of cancer cell transendothelial migration may be a complex scenario, which is not yet fully characterized. In numerous or almost all previous studies, the endothelium acts as a barrier against the migration and invasion of cancer cells (Al-Mehdi *et al* 2000, Zijlstra *et al* 2008). In more detail, the endothelium decreases pronouncedly the invasion of cancer cells and subsequently metastasis (Van Sluis *et al* 2009). Contrary to these studies, several recent reports propose a novel paradigm in which endothelial cells regulate the invasiveness of special cancer cells by increasing their dissemination through vessels (Kedrin *et al* 2008) or even by increasing the invasiveness of cancer cells (Mierke *et al* 2008a). Although special cell adhesion molecules have been identified to facilitate tumor–endothelial cell inter-actions and subsequently support metastasis formation, the role of endothelial cells' mechanical properties during the transmigration and invasion of cancer cells is still not well understood. However, it has been supposed that the altered mechanical properties of endothelial cells can support one of the two main functions of the endothelium in cancer metastasis, regulation of tumor growth and metastasis: firstly, endothelial cells act as a barrier inhibiting transmigration and secondly, they serve as an enhancer for cancer cell invasion. How this behavior of endothelial cells is actually regulated is not yet known. Moreover, the conditions that determine whether the endothelium acts as an enhancer or a repressor of cancer cell trans-migration are still elusive. A main biochemical pathway of the tumor–endothelial interaction has so far been reported to involve cell adhesion receptors and integrins such as platelet endothelial cell adhesion molecule-1 (PECAM-1) and $\alpha v \beta 3$ integrins (Voura *et al* 2000). As integrins facilitate a linkage between the extracellular matrix and the cell's actomyosin cytoskeleton (Neff *et al* 1982, Damsky *et al* 1985, Riveline *et al* 2001), the outside–inside connection is facilitated through the mechano-coupling focal adhesion and cytoskeletal adaptor protein vinculin (Mierke *et al* 2008b). In particular, this connection determines the amount of cellular counter-forces that maintain the cell shape, cell morphology and cell stiffness (Rape *et al* 2011). Finally, an overall biomechanical approach investigating the endothelial barrier breakdown in the presence of co-cultured invasive cancer cells is still elusive. Microrheologic measurements, such as the magnetic tweezer microrheology, have been adequate for determining the endothelial cell's mechanical properties, such as cellular stiffness during co-culture with invasive or non-invasive cancer cells compared to mono-cultured endothelial cells (Mierke 2011). Indeed, the endothelial cell's stiffness is influenced during the co-culture with cancer cells, depending on their aggressiveness. In particular, highly invasive human breast cancer cells

influence the mechanical properties of co-cultured microvascular endothelial cells by lowering the stiffness of endothelial cells, whereas non-invasive cancer cells have no effect on endothelial cell stiffness (Mierke 2011). Whether a direct cell–cell contact between endothelial cells and cancer cells is necessary is still under intensive investigation. In addition, the nanoscale particle tracking method is used for diffusion measurements of actomyosin cytoskeletal-bound beads serving as intra-cellular markers for structural changes of the intercellular cytoskeletal scaffold. As expected, this method has turned out to be suitable to analyze the actomyosin driven cytoskeletal remodeling dynamics (Mierke 2011). Thus, the cytoskeletal remodeling dynamics of endothelial cells are reported to increase in co-culture with highly invasive cancer cells, when the co-culture is performed by direct cell–cell contact, whereas they are unchanged by co-culture with non-invasive cancer cells (Mierke 2011). Finally, these results demonstrate that highly invasive breast cancer cells, but not less invasive breast cancer cells, actively alter the biomechanical properties of co-cultured endothelial cells. Thus, these findings may provide an explanation for the dramatic breakdown of the endothelial barrier function of vessel walls.

What impact do the biomechanical properties of cancer cells have in intravasation? During the entry into and the exit from the blood or lymph vessels, cancer cells experience pronounced morphological shape changes, which are regulated by contractile forces transmission and the generation or cytoskeletal remodeling dynamics, which empower them to migrate through endothelial cell–cell adhesions. The cytoplasm of a cell is regarded as a highly complex system that behaves like an elastic material such as a spring at high deformation rates, whereas at low deformation rates it acts like a viscous material such as honey that exhibits a yield stress (Wirtz 2009). In particular, the cellular elasticity mirrors the ability of the cytoplasm to restructure and remodel upon the application of external stresses such as forces. In turn, the cellular viscosity describes the capability of the cytoplasm to flow upon external shear stresses. Thus, cancer cells possess viscoelastic properties prepared to react to all kinds of mechanical stimulation from their local micro-environment. The MMP-facilitated degradation of the extracellular matrix is only to a certain degree a rate-limiting step for the invasion speed and migration efficiency of cancer cells, instead the deformability of the interphase cell nucleus is decisive, as it is the largest organelle within the cell (Friedl *et al* 2010). Moreover, the nucleus is at least ten-fold stiffer than the soft cytoplasm of cancer cells (Dahl *et al* 2004, Tseng *et al* 2004). In particular, the elasticity of the cellular nucleus depends on the mechanical properties of the nuclear lamina underneath the nuclear envelope consist-ing of intermediate filaments (Dahl *et al* 2004), the chromatin organization (Gerlitz and Bustin 2010) and special linker proteins connecting the nucleus and the cytoskeleton via LINC complexes (Crisp *et al* 2006, Stewart-Hutchinson *et al* 2008, Hale *et al* 2008, Lee *et al* 2007). The precise nucleus position within the cell body is important for many cellular processes such as cell motility and invasion through narrow pores of the extracellular matrix. LINC complexes consist of special proteins that span over the nuclear envelope and mediate mechanical connections between the nuclear lamina and the cytoskeleton (Crisp *et al* 2006). In more detail, these tight connections between the nucleus and the cytoplasm are facilitated by interactions between SUN domain-containing proteins such as SUN1 and SUN2 and Klarsicht homology (KASH) domain-containing proteins at the outer nuclear membrane such as nesprin 2 and nesprin 3, which are able to bind actin directly as well as indirectly (Starr and Han 2003, Starr *et al* 2001, Technau and Roth 2008, Lei *et al* 2009). Subsequently, the loss of LINC complex components leads to nuclear shape defects, a softening of the nucleus and the cytoplasm (Lammerding *et al* 2004) and has a major impact on the cellular motility on 2D substrates (Hale *et al* 2008, Lee *et al* 2007). Although mutations of nesprins and lamins A/C have been found in breast cancer (Wood *et al* 2007), their role in the 3D motility of cancer cells is still not yet well understood.

In line with this, the mechanical properties have been shown to be softer in cancer cells compared to normal, healthy cells and thus are associated with an increased metastatic potential (Rösel *et al* 2008, Bloom *et al* 2008, Guck *et al* 2005). The softer cytoplasm of aggressive cancer cells can be explained as a loss of cytoskeletal organization. The softening of the cytoskeleton should be further investigated in *in vivo* models for cancer metastasis and in a 3D extracellular matrix invasion assay. The development and major improvement of novel biophysical methods such as nanoscale particle-tracking, magnetic tweezer microrheology, traction force measurements in 3D, optical cell stretching and twisting microrheology will help to develop stiffness measurements that can be carried out directly in animal models or 3D environments (Wirtz 2009, Bausch *et al* 1999, Guck *et al* 2005, Legant *et al* 2010, Raupach *et al* 2007). The knowledge of the values of these mechanical properties may be used as a biophysical diagnostic marker of cancer and the determination of the metastatic potential (Cross *et al* 2007). Currently, there is no clue as to why aggressive cancer cells should be softer than non-transformed healthy cells. One hypothesis is that the invasion of cancer cells through 3D extracellular matrices and transendothelial migration needs precisely regulated mechanical properties. If the cancer cells are too stiff to deform themselves or too soft to generate and transmit forces to their microenvironment, they may no longer be able to squeeze through the pores of highly crosslinked collagen fibers of the connective tissue. In has been suggested that the mechanical properties of invasive cancer cells may depend on the current cell–cycle phase, the impact of neighboring cells, the regulatory role of the connective tissue's physical properties, cytokine or chemokines concentration around the cell, growth factor stimulation, receptor ligands and receptor shedding enzymes, membrane surface tension and membrane domains such as lipid rafts and their mobility. Indeed, single cell measurements have consistently revealed that individual cancer cells display a wide range of mechanical properties. These findings indicate that only a certain sub-type of the cancer cells is able to migrate and to intravasate into the endothelial cell lining of blood or lymph vessels. These special invasive cancer cells possess mechanical properties that facilitate cancer cell motility in connective tissue and intravasation into blood or lymph vessels. These invasive properties of cancer cells as well as their mechanical properties are stably associated even over a huge number of passages in cell culture, after injection into animals and re-isolation from animal tumor models. How can physical properties of cancer cells such as stiffness be passed on from generation to generation? When these physical properties are based on genomic mutations, they can be altered by the addition of pharmacological drugs in order to regulate the proteins affecting cellular mechanics, invasion and trans-endothelial migration. In more detail, cancer cells are supposed to vary their mechanical properties according to their progression level in the metastatic process.

Moreover, it is suggested that the optimal mechanical properties for the invasion of cancer cells into the surrounding tumor stroma near the primary tumor are different from those for the majority of stationary cancer cells within the solid primary tumor. In particular, these special mechanical properties seem to be essential for the cancer cells to overcome the steric restrictions of 3D extracellular matrices such as the extracellular matrix or endothelial vessel walls. Hence, the mechanical properties of cancer cells need to be adopted during the metastatic process in order to react to the altered environmental properties. Cancer cells must overcome microenvironmental restrictions to get to the next step of the metastatic process. These alterations in the mechanical properties of cancer cells seem to be regulated by biochemical gradients (Tseng *et al* 2004), interstitial flow rates (Swartz and Fleury 2007) and neighboring cells such as the endothelial cells lining the vessel wall (Mierke 2011).

The impact of shear stress on endothelial lining and transmigrating cells

During their travel through the blood or lymphoid system, cancer cells experience external dynamic forces, immunological stimulation and collisions with other cell types, such as red blood cells, and interact with the endothelial lining of the vessels. All of which may have an impact on the survival of motile cancer cells and the cancer cells' capacity to metastasize in targeted organs. Those circulating cancer cells that have been able to withstand fluid shear stress and immunosurveillance can adhere to the endothelial cell monolayer of the vessel, transmigrate and invade distant organs and then enter targeted tissues. However, only a small and specialized subtype fraction of the cancer cells survives and generates metastases, whereas the majority of the cancer cells die or remain dormant within the vessels or tissues (Fidler *et al* 2002).

Upon entry into the circulatory system (intravasation), the migration path and cell fate of a cancer cell is influenced by a number of physical and mechanical parameters: the flow rate and the diameter of the blood vessels as well as the interplay between shear flow and intercellular adhesion leads to the abolition of cancer cell movement in larger vessels. In more detail, the shear stress arises between adjacent layers of the blood fluid (of certain viscosity) moving at different velocities: the velocity of a liquid in a cylinder (the blood vessel) is maximal in the center, but zero at the border of the cylinder. The total shear stress is the product of fluid's viscosity and shear rate. The blood fluid behaves as a Newtonian fluid at shear rates higher than $100\,s^{-1}$, with a shear stress that increases linearly with the shear rate. In contrast to cylinders, the maximum shear stress is experienced at the vessel wall. The shear flow influences the translational and rotational motion of cancer cells and regulates the orientation of cancer cells as well as the speed of receptor–ligand interactions during cell adhesion. What we do not yet know is whether the shear flow induces the deformation of cancer cells or even affects the magnitude of adhesion. Moreover, the impact of the shear flow on the viability and proliferation rates of cancer cells is not yet well understood.

The extravasation of cancer cells from blood or lymph vessels
If a cancer cell within the vessel lumen wants to exit, it must first adhere to the endothelial blood vessel lining. There are two proposed mechanisms of cancer cell trapping: firstly, the physical arrest (named occlusion) and secondly, the cell adhesion. Which of these two mechanisms is preferred depends on the actual diameter of the blood vessel.

The physical arrest means that a cancer cell intravasates into a vessel whose diameter is less than that of the cancer cell's diameter. Thus, the cancer cell arrests within the vessel by mechanical trapping (physical occlusion). Circulating cancer cells of epithelial origin are normally larger than 10 μm in size, hence physical occlusion only plays a role in small vessels, such as capillaries smaller than 10 μm. In line with this, cancer cells have been found at branches in blood vessels of the brain. In particular, cancer cells' extravasation through the endothelium and finally their metastasis has been analyzed *in vivo* by intravital microscopy using a mouse model (Kienast *et al* 2010).

In the cell adhesion mechanism, the extravasation of a cancer cell from a large blood vessel possessing a vessel diameter that is larger than that of the cancer cell initially requires the adhesion of the cancer cell to the endothelial vessel wall through intercellular cell–cell adhesion. The probability of getting caught in a large vessel is the collision frequency between membrane-bound receptors on the surface of the cancer cells and endothelial ligands exposed on the surface as well as the residence time, which depends on the shear force and adhesive forces between the circulating cancer cell and the endothelial vessel lining (Zhu *et al* 2008).

During their journey within the vessels, cancer cells are exposed to translational and tangential velocity. In particular, the translational velocity of a cell travelling through the vessel is always larger than the surface tangential velocity, which leads to a rolling (slipping) motion relative to the endothelial vessel wall. The slipping motion increases the binding rate (depending on the rotation rate of the cancer cell) between a single receptor on a cancer cells in the blood stream and ligands on the surface of endothelial cells lining the vessel wall (Chang and Hammer 1999). The total adhesion strength is defined by the tensile strength of the particular receptor–ligand connection and the number of receptor–ligand interactions involved, whereas possible cooperative effects are excluded (Duguay *et al* 2003, Niessen and Gumbiner 2002). What can help to reveal the cooperative effects? The further development and invention of biophysical methods for measuring the mechanical properties of ligand–receptor bonds may help to distinguish between single-molecule affinity and multi-molecular avidity exactly (Huang *et al* 2010, Marshall *et al* 2003). In particular, the kinetic and mechanical properties, such as the tensile strength of a single receptor–ligand connection, regulate whether a bond will form at a certain shear stress and subsequently lead to cell adhesion. The initiation of a receptor–ligand-dependent adhesion of cancer cells to the endothelium is facilitated by selectins under shear stress and requires a sufficient fast binding rate, sufficient tensile strength in order to resist the dispersive hydrodynamic force and a slow rupture rate of receptor–ligand bonds. Thus, this all helps to promote the formation of additional receptor–ligand interactions. As integrins have slower binding rates than selectins, they are only able to bind after selectin–ligand mediated cancer cell binding. Finally, these integrin receptors have cooperative effects: for example, integrin clustering leads to the formation of multiple bonds and hence to strong adhesion of cancer cells (Hynes 2002).

Cancer cells are supposed to escape immune surveillance through association with platelets when migrating through the blood vessels (McCarty et al 2000). This association may additionally provide guidance for the extravasation step of cancer cells (McCarty et al 2000). Nonetheless, the role of these platelets in promoting cancer metastasis is still not clearly understood and thus needs further investigation through using an established mouse model (Gasic et al 1968, Camerer et al 2004). There are still open questions. Are cancer cells masked by platelets in the blood stream in order to be protected from immune response reactions such as the clearance by phagocytosis? Can platelets support the cancer cell's adhesion to the endothelial vessel lining? Does the secretion of VEGF by platelets alter the vascular hyperpermeability and subsequently the extravasation of cancer cells? What is the reason for having leukocytes in close neighborhood to metastatic and highly aggressive cancer cells? Why do transmigrating cancer cells use similar mechanisms as transmigrating leukocytes, such as neutrophils? What regulates the specificity of cancer cells to metastasize exclusively in targeted organs and not in other sites (Steeg 2006, Weiss 2000, Jacob et al 2007, Fidler 2003)? There are currently at least three hypotheses. The first is the one called the seed and soil hypothesis, which states that a cancer cell will only metastasize to a tissue site where the microenvironment is suitable and favors cancer cell survival and secondary tumor growth (Paget 1989). The second is called the mechanical hypothesis and is based on structural restrictions; this states that the process of cancer metastasis occurs at sites related to the directional pattern of blood flow, as the blood fluid flows from most organs to the heart, then to the lungs by the venous system and back via the arterial system (Weiss 2000). The third is called the stiffness hypothesis, which states that cancer metastasis occurs only at sites where the mechanical properties of the targeted tissue are similar or even equal to the mechanical conditions of the primary tumor.

Does the endothelial cell's actomyosin cytoskeleton serve as a migration grid for transmigrating cancer cells?

In the classical view, the endothelial cell monolayer of vessels acts as a barrier and as such it is a key rate-limiting step for the transmigration, invasion and metastasis of invasive cancer cells (Zijlstra et al 2008). The endothelial vessels wall is commonly presented as a strong tissue barrier toward the dissemination of cancer cells (Wittchen et al 2005). However, recent reports demonstrate a novel role for the endothelial vessel wall, in which the endothelial cell lining increases the invasiveness of special cancer cells. In the first report, breast cancer cells display enhanced dispersion through hematogeneous dissemination in close neighborhood to blood vessels (Kedrin et al 2008). In the second report, the invasiveness of certain cancer cell lines is endothelial-cell dependent and is even enhanced for special cancer cell types (Mierke et al 2008a). Nevertheless, the process of cancer cell invasion has been investigated in numerous research studies, but the molecular and mechanical mechanisms of cancer cell transendothelial migration are not yet well understood.

The intravasation of cancer cells in the vessels includes as a physical process the interaction of at least three cell types: the special invasive cancer cell, the nearby macrophage and the opposing endothelial cell. This interaction will engage the mechano- and biochemical-transduction cytoskeletal properties of these three neighboring cells. To reveal whether signals from cancer cells stimulate endothelial cells, a 3D co-culture assay with a collagen fiber matrix can be used in which the real-time intra-endothelial signaling events evoked by invasive cancer cells or macrophages are analyzed and compared to monocultured endothelial cells (Khuon *et al* 2010). In more detail, this assay involves the assembly of a vasculature network in a 3D collagen matrix using endothelial cells expressing a fluorescent resonant energy-transfer-based biosensor that reports the activity of myosin light chain kinase (MLCK) inside the endothelial cell in real time (Chew *et al* 2002). As expected, endothelial cells sensed their environment such as the 3D collagen fiber matrix mechanically. Indeed, the 3D microenvironment induces lumen formation in endothelial cells, which then display basal–apical polarity in the proper orientation supported by α4 laminin deposition. As suggested, invasive cancer cells are able to alter the MLCK-mediated actomyosin function in the surrounding underlying endothelium. Subsequently, cancer cells are able to transmigrate through the endothelial cell monolayer in at least two ways: by a transmigration mode using a transcellular path (through neighboring endothelial cells) or by a transmigration mode using a paracellular (though endothelial cell–cell junctions) path (Khuon *et al* 2010). When cancer cells use a transcellular mode to invade, they trigger MLCK activation inside the endothelial cell, which correlates with enhanced local phosphorylation of the myosin-II regulatory light chain (RLC) and endothelial myosin contraction. This has been functionally investigated using endothelial cells expressing a RLC mutant that cannot be phosphorylated; thus, the transcellular modes of cancer cell intravasation are reduced. Finally, the transcellular transmigration mode is indeed used and can be characterized: (i) invasive cancer cells undergo transcellular migration; (ii) cancer cells undergoing the transcellular transmigration mode induce transient and local MLCK activation as well as myosin contraction of adjacent endothelial cells at the transmigration site and subsequently tissue invasion; and (iii) the transcellular transmigration path depends on the phosphorylation of myosin-II RLC. In summary, these findings demonstrate that the endothelium plays an active rather than a passive role in supporting cancer cells' intravasation.

1.2.4 Secondary tumor in targeted tissues

If a secondary tumor occurs in the progression of cancer disease, the primary tumor has metastasized in a targeted tissue. How the tissue specificity is determined is still elusive and under debate. It is still common knowledge that the hypothesis formulated by Stephen Paget in 1989 holds true, in which the process of metastasis depends on a crosstalk between special cancer cells (the seeds) and the specific microenvironments of organs (the soil) (Paget 1989). In established reports it is stated that the metastatic potential of a cancer cell depends on its interactions with the homeostatic factors that facilitate primary tumor-cell growth, survival, angiogenesis, invasion and metastasis. What major findings are most likely to answer the question of why certain cancer cells metastasize in a special target organ? It is suggested that the mechanical properties of the primary tumor tissue are similar to those of the targeted tissue (Kumar and Weaver 2009). In order to see how these mechanical properties fit into the process of metastatic cancer progression, we need to discuss the classical and established hallmarks of cancer.

1.3 Hallmarks of cancer

In the last 25 years, many medical and biological aspects of the field of classical tumor biology have been investigated and much effort has been put into this field of cancer research. Thus, it is not surprising that in 2000 six 'classical' hallmarks of cancer were proposed in order to describe the major steps of the malignant cancer progression cascade process in a more detailed and precise mode (Hanahan and Weinberg 2000): (i) sustaining proliferative signaling, (ii) evading growth suppressors, (iii) activating invasion and metastasis, (iv) enabling replicative immortality, (v) inducing angiogenesis and (vi) resisting cell death (figure 1.4). After defining these six hallmarks, the following question may arise: how many of these regulatory circuits must a cancer-candidate cell deregulate to be finally cancerous? However, this question cannot be fully answered yet, as the six proposed hallmarks describe only the first set of regulatory circuits and have been revealed to be insufficient to describe all possible types of cancer. Thus, this first set of hallmarks has to be refined. Indeed, 11 years later the same two

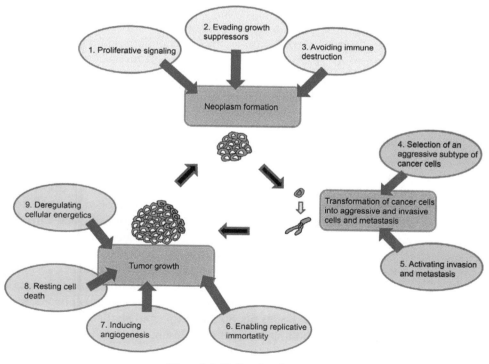

Figure 1.4. Hallmarks of cancer.

researchers who proposed the first set of six hallmarks of cancer defined the next two novel, additional hallmarks to improve the principles for classification of malignant cancer progression (Hanahan and Weinberg 2011): (i) the utilization of abnormal metabolic pathways and (ii) the repression of the immune system (figure 1.4). The addition of the seventh and eighth hallmarks of cancer to their list showed that the immune system was viewed as being more important in malignant cancer progression.

Many molecules relevant for the motility and invasion of cancer cells, such as $\alpha6\beta4$, $\alpha v\beta3$, $\alpha v\beta5$, $\alpha5\beta1$, E-cadherin, N-cadherin, Notch1-4 receptors, CXCR2 and CXCR4, have been reported to play a regulatory role in cancer disease (Gong *et al* 1997, Bauer *et al* 2007, Mierke *et al* 2008a, 2008b, 2008c, Sawada *et al* 2008, Gilcrease *et al* 2009, Ricono *et al* 2009, Teicher and Fricker 2010). Despite all of these major findings based on novel molecular or biochemical approaches such as genomics and proteomics, these novel approaches were not able to fundamentally change patients' survival rates in the field of cancer research. After the initial enthusiastic reception, it has been seen that these molecular biological technology-based approaches have not yet reached their goals, although they have contributed to the discovery of significant insights into cancer biology, cancer diagnosis and prognosis. The detailed classification of tumors due to numerous novel marker proteins for special cancer types and the detailed mapping of specific human cancer types has begun. Nonetheless, a major criticism of these novel approaches is the high variation among the gene and protein expression levels during the progression of cancer, possibly depending on the cancer disease stage, but also varying between cancer patients, possibly due to altered environmental conditions. With this knowledge, it is not yet well understood how these novel biological approaches contribute in a regulatory way to the progression of cancer and to what extent. In these genomic- and proteomic-based analysis methods the spatial localization of the proteins in special compartments such as lipid rafts (Runz *et al* 2008), their state of activation such as phosphorylation and assembly, and finally their lifetime, turnover rate, modification rate (for glycosylation and recycling, for example) are still not included (Veiga *et al* 1997, Garcia *et al* 1998, Caswell *et al* 2008, Gu *et al* 2011, Liu *et al* 2011).

Current classical biological and biochemical approaches to reveal insights into cancer biology have not yet covered the whole complexity of the cancer disease and have so far failed, in particular, to give more appreciation to the malignant progression of cancer. Cancer research has adapted classical physical approaches. Moreover, novel biophysical analysis methods have been developed in order to be suitable for use in the field of cancer research. To date, these promising new directions for physical-based cancer research have significantly challenged the field and have demolished the classical biological and biochemical view on cancer disease. In addition, cancer disease is closely associated with inflammatory processes and hence the physical view has changed and needs to be adapted to inflammatory diseases. In addition to solid epithelial originated tumors, 'soft' leukemias will also be addressed in order to compare them with the results obtained for epithelial cancer cells.

As there is still at least one hallmark missing, including the mechanical properties of cancer cells, this book will focus on the impact of the physical properties of cancer cells, their physical microenvironment such as the extracellular matrix, and on neighboring cells such as endothelial cells, cancer-associated macrophages and fibroblasts. These processes or regulatory circuits are presented here from a biophysical point of view and thus break down the classical and well-known eight-hallmarks-based view of cancer.

Do physical aspects affect the classical hallmarks of cancer?

All the eight postulated hallmarks of cancer miss the aspect of the mechanical properties of cancer cells and their local microenvironment, indicating that the common view on cancer is not from a physical viewpoint. Here, the physical view on cancer disease is addressed and is the focus of the discussion.

There is agreement that physical aspects in cancer research can no longer be excluded and hence must be adapted into the current understanding of cancer research. In particular, this means that a ninth hallmark, which incorporates physics into classical cancer research, is needed to fully reveal the complex regulatory scenario of cancer metastasis. Indeed, this ninth hallmark states that the primary tumor and the tumor microenvironment can alter the survival conditions and cellular properties of a special type of cancer cells, which then subsequently supports the selection of an aggressive (highly invasive) subtype of cancer cells (figure 1.4). Moreover, this aggressive subtype of cancer cells is supposed to decrease cell–cell adhesions to neighboring cells, affecting the mechanical properties of neighboring and interacting cells, migrate through the tumor boundary of the primary tumor, including the tumor-surrounding basement membrane, and invade the local tumor stroma. In summary, each of these steps affects either the mechanical properties of the migrating cancer cells, their microenvironment, the neighboring cells representing a stable barrier and the interacting endothelial cells lining blood or lymph vessels. Indeed, this novel ninth hallmark can be included after the hallmark of avoiding immune destruction and before the hallmark of activating invasion and metastasis (Mierke 2013). In more detail, the novel ninth hallmark, the selection of an aggressive subtype of cancer cells, may have the ability to reduce cell–cell adhesions, possibly even increase cell–matrix adhesions and adapt the mechanical properties of cancer cells, facilitating cancer cell transmigration through the basement membrane and migration into the connective tissue. Moreover, these nine hallmarks can then be put into three groups: the neoplasm formation group (including hallmarks 1–3), the transformation of cancer cells into aggressive and invasive cells group (including hallmarks 4–5) and the tumor growth group (including hallmarks 6–9) (Mierke 2013) (figure 1.4). Finally, when re-investigating the eight classical hallmarks, the physical aspect should be included in each of them to refine this particular hallmark.

1.4 The impact of the mechanical properties of cancer cells on their migration

During the last decade, many studies have suggested that the mechanical properties of cancer cells are a key player in facilitating cancer cell migration. Novel biophysical methods have been developed and further improved in order to measure the mechanical properties of cancer cells precisely.

The biomechanical properties of cancer cells determine their own aggressiveness
What determines the cellular motility in 2D and 3D migratory environments is still elusive. Nonetheless, the capacity of cancer cells to migrate and invade 3D connective tissues depends on certain mechanical parameters determining the overall migration speed through a dense 3D extracellular matrix (with a typical pore-size of 1–2 μm). Among these parameters are (i) cell adhesion and de-adhesion dynamics, such as turnover of focal adhesions and adhesion strength; (ii) cyto-skeletal remodeling dynamics, cellular fluidity and cellular stiffness; (iii) extracellular matrix remodeling and enzymatic degradation of the extracellular matrix; and (iv) protrusive (contractile) force generation (Webb *et al* 2004, Friedl and Brocker 2000, Mierke *et al* 2008c). The exact values do not restrict the migratory potential of cancer cells, it is rather the balance between the parameters that is crucial for the efficiency of cancer cell invasion, invasion speed and invasion depths into 3D extracellular matrices (Zaman *et al* 2006). These parameters can be altered and shifted toward a single parameter, whereas they are still all-important for regulating the invasiveness and invasion efficiency of the cancer cells. An imbalance between these parameters may switch the mode of cell invasion from epithelial to mesen-chymal, or even to the amoeboid mode with or without the presence of traction forces. Taken together, the invasion strategy is determined by the utilization of these parameters. How cancer cells keep these parameters precisely balanced, and how the microenvironment, such as growth factors, cytokines, chemokines, matrix-protein composition, structure and concentration, and matrix mechanical stiffness impacts cell invasiveness is still elusive or the subject of controversial discussion (Steeg 2006). There are at least two mechanisms. The first is that invasive cancer cells digest the dense 3D extracellular matrix through the secretion of enzymes such as MMPs (Wolf *et al* 2003a, 2007, Friedl and Wolf 2009). The second mechanism is that sheddases exposed on the cell surface such as the secretases ADAM-10 and ADAM-17 cut cell–cell adhesion molecules such as NOTCH receptors, ephrins or E-cadherins of cancer cells or facilitate signaling, leading to nuclear translocation and the induction of gene expression, and finally induce cancer cell invasiveness (Li *et al* 2008, Itoh *et al* 2008, Riedle *et al* 2009, Singh *et al* 2009, Bozkulak and Weinmaster 2009, Brou *et al* 2000, van Tetering *et al* 2009). Additionally, ADAM-17 cuts pro-TNF-α exposed on the cell surface and releases mature TNF-α into the microenvironment (Black *et al* 1997). Besides proteolysis of the extracellular matrix or shedding of membrane receptors, another important parameter for regulating the invasion speed and determining the invasion depth of cancer cells in dense 3D extracellular matrices is the mechanical ability to generate and transmit contractile forces (Mierke *et al* 2008a, Rösel *et al* 2008). Novel biophysical methods measuring contractile forces in 3D collagen or fibrin matrices have been reported (Bloom *et al* 2008, Koch *et al* 2012, Gjorevski and Nelson 2012, Legant *et al* 2010). Thus, the invasion of cancer cells is analyzed by taking z-stack images. In some assays, matrix embedded beads may serve as markers, whereas in others the collagen fiber structure is the marker of the cellular force displacement evoked by invasive cancer cells. The tracking of collagen fibers seems to be more complicated, but also more reliable and it reduces the effect of marker phagocytosis as only a minor number of collagen fibers is cut by enzymes and internalized compared to the embedded beads in close

neighborhood to invasive cancer cells. One problem is that bead internalization (bead phagocytosis) may affect the mechanical properties of cancer cells and consequently may alter their invasiveness into dense 3D extracellular matrices.

However, major players in the regulation of cancer cell invasion into dense 3D extracellular matrices are the cell–matrix adhesion receptors such as integrins and their associated focal adhesion molecules such as vinculin and FAK. Integrins are key players in the regulation of the invasiveness of cancer cells (Mierke 2013), because they regulate cell–matrix adhesion and de-adhesion, adhesion strength and the generation or transmission of contractile forces upon outside–in (via ligand) or inside–out (via integral signaling through growth factors, etc) signalling, stimulation of integrin-receptors and cytoskeletal remodeling dynamics (Discher *et al* 2005, Giancotti 2000). The integrins are a family of cell–matrix adhesion receptors and are composed of two non-covalently linked α (one out of 18) and β (one out of 8) subunits. In addition, they regulate transmembrane connections between the cell's actomyosin cytoskeleton and the extracellular microenvironment, such as an extracellular matrix scaffold. This connection is mainly mediated by the focal adhesion protein vinculin, which can act as a mechano-coupling and mechano-regulating protein (Mierke 2009). The focal adhesions of cancer cells fulfill two functions in cell invasiveness: firstly, they transmit contractile forces to the extracellular matrix microenvironment and secondly, they anchor the cancer cells to its substrate to withstand de-adhesion and apoptosis (programmed cell death) (Zaman *et al* 2006, Palecek *et al* 1997, Loftus and Liddington 1997). The composition of focal adhesions may vary significantly, depending on the integrin type and the integrin activation state. In addition, the integrin signal transduction through the α integrin subunit is not uniform: the activation of α4 and α9 integrin subunits leads to decreased cell spreading, but the activation of α1, α3, α5, α6, αv and αIIb integrin subunits even enhances cell spreading (Horwitz and Parsons 1999, Rolli *et al* 2003, Schmid *et al* 2004, Pawar *et al* 2007, Rout *et al* 2004). However, the function of vinculin in the focal adhesions is supposed to vary due to the integrin type and activation state. In more detail, integrins are able to modulate the function of other integrins, for example the α5β1 integrin can regulate the αvβ3 facilitated focal adhesions and migration on extracellular matrices (Ly *et al* 2003). In line with this, the expression of the β1 integrin subunits has been reported to support cancer cell migration in 3D matrigel and 3D extracellular matrices (Mierke *et al* 2011).

Further reading

Lämmermann T *et al* 2008 Rapid leukocyte migration by integrin-independent flowing and squeezing *Nature* **453** 51–5

Mierke C T 2013a The integrin alphav beta3 increases cellular stiffness and cytoskeletal remodeling dynamics to facilitate cancer cell invasion *New J. Phys.* **15** 015003

Smith M L, Gourdon D, Little W C, Kubow K E, Eguiluz R A, Luna-Morris S and Vogel V 2007 Force-induced unfolding of fibronectin in the extracellular matrix of living cells *PLoS Biol.* **5** e268

Xu J, Wang F, Van Keymeulen A, Rentel M and Bourne H R 2005 Neutrophil microtubules suppress polarity and enhance directional migration *Proc. Natl Acad. Sci. USA* **102** 6884–9

References

Adamson E D 1987 Oncogenes in development *Development* **99** 449–71 PMID: 2822372

Alcaraz J, Mori H, Ghajar C M, Brownfield D, Galgoczy R and Bissell M J 2011 Collective epithelial cell invasion overcomes mechanical barriers of collagenous extracellular matrix by a narrow tube-like geometry and MMP14-dependent local softening *Integr. Biol. (Camb.)* **3** 1153–66

Al-Mehdi A B, Tozawa K, Fisher A B, Shientag L, Lee A and Muschel R J 2000 Intravascular origin of metastasis from the proliferation of endothelium-attached tumor cells: a new model for metastasis *Nat. Med.* **6** 100–2

Angelini T, Hannezo E, Trepat X, Fredberg J and Weitz D 2010 Cell migration driven by cooperative substrate deformation patterns *Phys. Rev. Lett.* **104** 168104

Angelini T E, Hannezo E, Trepat X, Marquez M, Fredberg J J and Weitz D A 2011 Glass-like dynamics of collective cell migration *Proc. Natl Acad. Sci. USA* **108** 4714–9

Balmain A and Brown K 1988 Oncogene activation in chemical carcinogenesis *Adv. Cancer Res.* **51** 147–82

Baluk P, Hashizume H and McDonald DM 2005 Cellular abnormalities of blood vessels as targets in cancer *Curr. Opin. Genet. Dev.* **15** 102–11

Batlle E, Sancho E, Francí C, Domínguez D, Monfar M, Baulida J and García De Herreros A 2000 The transcription factor snail is a repressor of E-cadherin gene expression in epithelial tumor cells *Nat. Cell Biol.* **2** 84–9

Bauer K, Mierke C and Behrens J 2007 Expression profiling reveals genes associated with transendothelial migration of tumor cells: a functional role for alpha v beta3 integrin *Int. J. Cancer* **121** 1910–8

Bausch A R, Möller W and Sackmann E 1999 Measurement of local viscoelasticity and forces in living cells by magnetic tweezers *Biophys. J.* **76** 573–9

Bernards R and Weinberg R A 2002 Metastasis genes: a progression puzzle *Nature* **418** 823

Bergers G, Javaherian K, Lo K M, Folkman J and Hanahan D 1999 Effects of angiogenesis inhibitors on multistage carcinogenesis in mice *Science* **284** 808–12

Bergers G, Song S, Meyer-Morse N, Bergsland E and Hanahan D 2003 Benefits of targeting both pericytes and endothelial cells in the tumor vasculature with kinase inhibitors *J. Clin. Invest.* **111** 1287–95

Berthier L, Biroli G, Bouchaud J-P, Cipelletti L, El Masri D, L'Hote D, Ladieu F and Pierno M 2005 Direct experimental evidence of a growing length scale accompanying the glass transition *Science* **310** 1797–800

Bi D, Zhang J, Chakraborty B and Behringer R P 2011 Jamming by shear *Nature* **480** 355–8

Bizzarri M and Cucina A 2014 Tumor and the microenvironment: a chance to reframe the paradigm of carcinogenesis? *Biomed. Res. Int.* **2014** 934038

Black R A, Rauch C T, Kozlosky C, Peschon J J, Slack J L and Wolfson M F 1997 A metalloproteinase disintegrin that releases tumor-necrosis factor-alpha from cells *Nature* **385** 729–33

Blaser H, Reichman-Fried M, Castanon I, Dumstrei K, Marlow F L, Kawakami K, Solnica-Krezel L, Heisenberg C P and Raz E 2006 Migration of zebrafish primordial germ cells: a role for myosin contraction and cytoplasmic flow *Dev. Cell* **11** 613–27

Bloom R J, George J P, Celedon A, Su S X and Wirtz D 2008 Mapping local matrix remodeling induced by a migrating tumor cell using 3D multiple-particle tracking *Biophys. J.* **95** 4077–88

Bolos V, Peinado H, Pérez-Moreno M A, Fraga M F, Esteller M and Cano A 2003 The transcription factor Slug represses E-cadherin expression and induces epithelial to mesenchymal transitions: a comparison with Snail and E47 repressors *J. Cell Sci.* **116** 499–511

Bozkulak E C and Weinmaster G 2009 Selective use of ADAM10 and ADAM17 in activation of Notch1 signaling *Mol. Cell Biol.* **29** 5679–95

Brown M J, Hallam J A, Colucci-Guyon E and Shaw S 2001 Rigidity of circulating lymphocytes is primarily conferred by vimentin intermediate filaments *J. Immunol.* **166** 6640–6

Brou C, Logeat F, Gupta N, Bessia C, LeBail O, Doedens J R, Cumano A, Roux P, Black R A and Israel A 2000 A novel proteolytic cleavage involved in Notch signaling: the role of the disintegrin-metalloprotease TACE *Mol. Cell* **5** 207–16

Camerer E, Qazi A A, Duong D N, Cornelissen I, Advincula R and Coughlin S R 2004 Platelets, protease-activated receptors and fibrinogen in hematogenous metastasis *Blood* **104** 397–401

Campinho P, Behrndt M, Ranft J, Risler T, Minc N and Heisenberg C P 2013 Tension-oriented cell divisions limit anisotropic tissue tension in epithelial spreading during zebrafish epiboly *Nat. Cell Biol.* **15** 1405–14

Cano A, Pérez-Moreno M A, Rodrigo I, Locascio A, Blanco M J, del Barrio M G, Portillo F and Nieto M A 2000 The transcription factor snail controls epithelial-mesenchymal transitions by repressing E-cadherin expression *Nat. Cell Biol.* **2** 76–83

Caswell P T, Chan M, Lindsay A J, McCaffrey M W, Boettiger D and Norman J C 2008 Rab-coupling protein coordinates recycling of alpha5beta1 integrin and EGFR1 to promote cell migration in 3D microenvironments *J. Cell Biol.* **183** 143–55

Chang K C and Hammer D A 1999 The forward rate of binding of surface-tethered reactants: effect of relative motion between two surfaces *Biophys. J.* **76** 1280–92

Charras G T, Hu C K, Coughlin M and Mitchison T J 2006 Reassembly of contractile actin cortex in cell blebs *J. Cell Biol.* **175** 477–90

Charras G and Paluch E 2008 Blebs lead the way: how to migrate without lamellipodia *Nature Rev. Mol. Cell Biol.* **9** 730–6

Chew T L, Wolf W A, Gallagher P J, Matsumura F and Chisholm R L 2002 A fluorescent resonant energy transfer-based biosensor reveals transient and regional myosin light chain kinase activation in lamella and cleavage furrows *J. Cell Biol.* **156** 543–53

Comijn J, Berx G, Vermassen P, Verschueren K, van Grunsven L, Bruyneel E, Mareel M, Huylebroeck D and van Roy F 2001 The two-handed E-box-binding zinc finger protein SIP1 downregulates E-cadherin and induces invasion *Mol. Cell* **7** 1267–78

Comito G, Calvani M, Giannoni E, Bianchini F, Calorini L, Torre E, Migliore C, Giordano S and Chiarugi P 2011 HIF-1α stabilization by mitochondrial ROS promotes Met-dependent invasive growth and vasculogenic mimicry in melanoma cells *Free Radic. Biol. Med.* **51** 893–904

Crisp M, Liu Q, Roux K, Rattner J B, Shanahan C, Burke B, Stahl P D and Hodzic D 2006 Coupling of the nucleus and cytoplasm: role of the LINC complex *J. Cell Biol.* **172** 41–53

Cross S E, Jin Y S, Rao J and Gimzewski J K 2007 Nanomechanical analysis of cells from cancer patients *Nat. Nanotechnol.* **2** 780–3

Cui H *et al* 2014 Tissue inhibitor of metalloproteinases-1 induces a pro-tumorigenic increase of miR-210 in lung adenocarcinoma cells and their exosomes *Oncogene* doi:10.1038/onc.2014.300

Dahl K N, Kahn S M, Wilson K L and Discher D E 2004 The nuclear envelope lamina network has elasticity and a compressibility limit suggestive of a molecular shock absorber *J. Cell Sci.* **117** 4779–86

Dalla-Favera R, Bregni M, Erikson J, Patterson D, Gallo R C and Croce C M 1982 Human c-myc onc gene is located on the region of chromosome 8 that is translocated in Burkitt lymphoma cells *PNAS* **79** 7834–7

Damm K, Beug H, Graf T and Vennström B 1987 A single point mutation in erbA restores the erythroid transforming potential of a mutant avian erythroblastosis virus (AEV) defective in both erbA and erbB oncogenes *EMBO J.* **6** 375–82 PMID: 2884103

Damsky C H, Knudsen K A, Bradley D, Buck C A and Horwitz A F 1985 Distribution of the cell substratum attachment (CSAT) antigen on myogenic and fibroblastic cells in culture *J. Cell Biol.* **100** 1528–39

Davis G S 1984 Migration-directing liquid properties of embryonic amphibian tissues *Am. Zool.* **24** 649–55

Davis G S, Phillips H M and Steinberg M S 1997 Germ-layer surface tensions and 'tissue affinities' in Rana pipiens gastrulae: quantitative measurements *Dev. Biol.* **192** 630–44

de Rooij J, Kerstens A, Danuser G, Schwartz M A and Waterman-Storer C M 2005 Integrin-dependent actomyosin contraction regulates epithelial cell scattering *J. Cell Biol.* **171** 153–64

Destaing O, Block M R, Planus E and Albiges-Rizo C 2011 Invadosome regulation by adhesion signaling *Curr. Opin. Cell Biol.* **23** 597–606

De Val S and Black B L 2009 Transcriptional control of endothelial cell development *Dev. Cell.* **16** 180–95

Ding X, Zhang Z, Li S and Wang A 2011 Combretastatin A4 phosphate induces programmed cell death in vascular endothelial cells *Oncol Res.* **19** 303–9

Direkze N C and Alison M R 2006 Bone marrow and tumor stroma: an intimate relationship *Hematol. Oncol.* **24** 189–95

Dirkx A E *et al* 2006 Antiangiogenesis therapy can overcome endothelial cell anergy and promote leukocyte-endothelium interactions and infiltration in tumors *FASEB J.* **20** 621–30

Discher D E, Janmey P and Wang Y L 2005 Tissue cells feel and respond to the stiffness of their substrate *Science* **310** 1139–43

Diz-Munoz A, Krieg M, Bergert M, Ibarlucea-Benitez I, Muller D J, Paluch E and Heisenberg C P 2010 Control of directed cell migration *in vivo* by membrane-to-cortex attachment *PLoS Biol.* **8** e1000544

Doyle A D, Petrie R J, Kutys M L and Yamada K M 2013 Dimensions in cell migration *Curr. Opin. Cell Biol.* **25** 642–9

Dudley A C, Khan Z A, Shih S-C, Kang S-Y, Zwaans B M M, Bischoff J and Klagsbrun M 2008 Calcification of multipotent prostate tumor endothelium *Cancer Cell* **14** 201–11

Dudley A C, Udagawa T, Melero-Martin J M, Shih S-C, Curatolo A, Moses M A and Klagsbrun M 2010 Bone marrow is a reservoir for pro-angiogenic myelomonocytic cells but not endothelial cells in spontaneous tumors *Blood* **116** 3367–71

Duguay D, Foty R A and Steinberg M S 2003 Cadherin-mediated cell adhesion and tissue segregation: qualitative and quantitative determinants *Dev. Biol.* **253** 309–23

Eder J P Jr *et al* 2002 Phase I clinical trial of recombinant human endostatin administered as a short intravenous infusion repeated daily *J. Clin. Oncol.* **20** 3772–84

Eisenhoffer G T, Loftus P D, Yoshigi M, Otsuna H, Chien C B, Morcos P A and Rosenblatt J 2012 Crowding induces live cell extrusion to maintain homeostatic cell numbers in epithelia *Nature* **484** 546–9

El-Kenawi A E and El-Remessy A B 2013 Angiogenesis inhibitors in cancer therapy: mechanistic perspective on classification and treatment rationales *Br. J. Pharmacol.* **170** 712–29

Fan C, Oh D S, Wessels L, Weigelt B, Nuyten D S, Nobel A B, van't Veer L J and Perou C M 2006 Concordance among gene-expression-based predictors of breast cancer *N. Engl. J. Med.* **355** 560–9

Fidler I J, Yano S, Zhang R D, Fujimaki T and Bucana C D 2002 The seed and soil hypothesis: vascularization and brain metastases *Lancet Oncol.* **3** 53–7

Fidler I J 2003 The pathogenesis of cancer metastasis: the 'seed and soil' hypothesis revisited *Nat. Rev. Cancer* **3** 453–8

Folkman M J, Long D M and Becker F F 1963 Growth and metastasis of tumor in organ culture *Cancer* **16** 453–67

Forgacs G, Foty R A, Shafrir Y and Steinberg M S 1998 Viscoelastic properties of living embryonic tissues: a quantitative study *Biophys. J.* **74** 2227–34

Foty R A, Forgacs G, Pfleger C M and Steinberg M S 1994 Liquid properties of embryonic tissues: measurement of interfacial tensions *Phys. Rev. Lett.* **72** 2298–301

Foty R A, Pfleger C M, Forgacs G and Steinberg M S 1996 Surface tensions of embryonic tissues predict their mutual envelopment behavior *Development* **122** 1611–20 PMID: 8625847

Foty R A and Steinberg M S 2004 Cadherin-mediated cell–cell adhesion and tissue segregation in relation to malignancy *Int. J. Dev. Biol.* **48** 397–409

Foty R A and Steinberg M S 2005 The differential adhesion hypothesis: a direct evaluation *Developmental Biology* **278** 255–63

Frey K, Zivanovic A, Schwager K and Neri D 2011 Antibody-based targeting of interferon-alpha to the tumor neovasculature: a critical evaluation *Integr Biol (Camb)* **3** 468–78

Friedl P and Brocker E B 2000 The biology of cell locomotion within three- dimensional extracellular matrix *Cell. Mol. Life Sci.* **57** 41–64

Friedl P, Hegerfeldt Y and Tusch M 2004 Collective cell migration in morphogenesis and cancer *Int. J. Dev. Biol.* **48** 441–9

Friedl P and Wolf K 2003 Tumor-cell invasion and migration: diversity and escape mechanisms *Nat. Rev. Cancer* **3** 362–74

Friedl P and Wolf K 2009 Proteolytic interstitial cell migration: a five-step process *Cancer Metastasis Rev.* **28** 129–35

Friedl P and Wolf K 2010 Plasticity of cell migration: a multiscale tuning model *J. Cell Biol.* **188** 11–9

Friedl P, Wolf K and Lammerding J 2010 Nuclear mechanics during cell migration *Curr. Opin. Cell Biol.* **23** 1–10

Garbi N, Arnold B, Gordon S, Hämmerling G J and Ganss R 2004 CpG motifs as proinflammatory factors render autochthonous tumors permissive for infiltration and destruction *J. Immunol.* **172** 5861–9

Gardel M L, Sabass B, Ji L, Danuser G, Schwarz U S and Waterman C M 2008 Traction stress in focal adhesions correlates biphasically with actin retrograde flow speed *J. Cell Biol. Dec.* **183** 999–1005

Garcia A J, Huber F and Boettiger D 1998 Force required to break a5b1 integrin- fibronectin bonds in intact adherent cells is sensitive to integrin activation state *J. Biol. Chem.* **273** 10988–93

Garrahan J P 2011 Dynamic heterogeneity comes to life *Proc. Natl Acad. Sci. USA* **108** 4701–2

Gasic G J, Gasic T B and Stewart C C 1968 Antimetastatic effects associated with platelet reduction *Proc. Natl Acad. Sci. USA* **61** 46–52

Gerlitz G and Bustin M 2010 Efficient cell migration requires global chromatin con densation *J. Cell Sci.* **123** 2207–17

Gerlach J P, Emmink B L, Nojima H, Kranenburg O and Maurice M M 2014 Wnt signalling induces accumulation of phosphorylated β-catenin in two distinct cytosolic complexes *Open Biol.* **4** 140120

Giancotti F G 2000 Complexity and specificity of integrin signalling *Nat. Cell Biol.* **2** E13–4

Gilcrease M Z, Zhou X, Lu X, Woodward W A, Hall B E and Morrissey P J 2009 Alpha6beta4 integrin crosslinking induces EGFR clustering and promotes EGF- mediated Rho activation in breast cancer *J. Exp. Clin. Cancer Res.* **28** 67

Gimbrone M A, Leapman S B, Cotran R S and Folkman J 1972 Tumor dormancy *in vivo* by prevention of neovascularization *J. Exp. Med.* **136** 261–76

Gimbrone M A, Leapman S, Cotran R S and Folkman J 1973 Tumor angiogenesis: iris neovascularization at a distance from experimental intraocular tumors *J. Natl Cancer Inst.* **50** 219–28 PMID: 4692862

Gimbrone M A, Cotran R S and Folkman J 1974 Tumor growth and neovascularization: an experimental model using rabbit cornea *J. Natl Cancer Inst.* **52** 413–27 PMID: 4363161

Gjorevski N and Nelson C M 2012 Mapping of mechanical strains and stresses around quiescent engineered three-dimensional epithelial tissues *Biophys. J.* **103** 152–62

Gregory P A *et al* 2011 An autocrine TGF-beta/ZEB/miR-200 signaling network regulates establishment and maintenance of epithelial-mesenchymal transition *Mol. Biol. Cell* **22** 1686–98

Gong J, Wang D, Sun L, Zborowska E, Willson J K and Brattain M G 1997 Role of alpha 5 beta 1 integrin in determining malignant properties of colon carcinoma cells *Cell Growth Differ.* **8** 83–90 PMID: 8993837

Gordon R, Goel N S, Steinberg M S and Wiseman L L 1972 A rheological mechanism sufficient to explain the kinetics of cell sorting *J. Theor. Biol.* **37** 43–73

Griffioen A W, Damen C A, Martinotti S, Blijham G H and Groenewegen G 1996 Endothelial intercellular adhesion molecule-1 expression is suppressed in human malignancies: the role of angiogenic factors *Cancer Res.* **56** 1111–7

Gu Z, Noss E H, Hsu V W and Brenner M B 2011 Integrins traffic rapidly via circular dorsal ruffles and macropinocytosis during stimulated cell migration *J. Cell Biol.* **193** 61–70

Guck J *et al* 2005 Optical deformability as an inherent cell marker for testing malignant transformation and metastatic competence *Biophys. J.* **88** 3689–98

Gumbiner B M 2005 Regulation of cadherin-mediated adhesion in morphogenesis *Nat. Rev. Mol. Cell Biol.* **6** 622–34

Gupta P B, Kuperwasser C, Brunet J P, Ramaswamy S, Kuo W L, Gray J W, Naber S P and Weinberg R A 2005 The melanocyte differentiation program predisposes to metastasis after neoplastic transformation *Nat. Genet.* **37** 1047–54

Haeger A, Krause M, Wolf K and Friedl P 2014 Cell jamming: collective invasion of mesenchymal tumor cells imposed by tissue confinement *Biochim. Biophys. Acta* **1840**(8) 2386–95

Hale C M, Shrestha A L, Khatau S B, Stewart-Hutchinson P J, Hernandez L, Stewart C L, Hodzic D and Wirtz D 2008 Dysfunctional connections between the nucleus and the actin and microtubule networks in laminopathic models *Biophys. J.* **95** 5462–75

Hanahan D and Weinberg R A 2000 The hallmarks of cancer *Cell* **100** 57–70

Hanahan D and Weinberg R A 2011 Hallmarks of cancer: the next generation *Cell* **144** 646–74

Hartwell K A, Muir B, Reinhardt F, Carpenter A E, Sgroi D C and Weinberg R A 2006 The Spemann organizer gene, Goosecoid, promotes tumor metastasis *Proc. Natl Acad. Sci. USA* **103** 18969–74

Hellman A, Zlotorynski E, Scherer S W, Cheung J, Vincent J B, Smith D I, Trakhtenbrot L and Kerem B 2002 A role for common fragile site induction in amplification of human oncogenes *Cancer Cell* **1** 89–97

Hida K, Hida Y, Amin D N, Flint A F, Panigrahy D, Morton C C and Klagsbrun M 2004 Tumor-associated endothelial cells with cytogenetic abnormalities *Cancer Res.* **64** 8249–55

Holtfreter J 1939 Gewebeaffinität, ein Mittel der embryonalen Formbildung *Arch. Exp. Zellforsch.* **23** 169–209

Horwitz A R and Parsons J T 1999 Cell migration–movin' on *Science* **286** 1102–3

Huang S and Ingber D E 2005 Cell tension, matrix mechanics, and cancer development *Cancer Cell* **8** 175–6

Huang J, Zarnitsyna V I, Liu B, Edwards L J, Jiang N, Evavold B D and Zhu C 2010 The kinetics of two-dimensional TCR and pMHC interactions determine T-cell responsiveness *Nature* **464** 932–6

Hynes R O 2002 Integrins: bidirectional, allosteric signaling machines *Cell* **110** 673–87

Ince T A, Richardson A L, Bell G W, Saitoh M, Godar S, Karnoub A E, Iglehart J D and Weinberg R A 2007 Transformation of different human breast epithelial cell types leads to distinct tumor phenotypes *Cancer Cell* **12** 160–70

Itoh Y, Ito N, Nagase H and Seiki M 2008 The second dimer interface of MT1-MMP, the transmembrane domain, is essential for ProMMP-2 activation on the cell surface *J. Biol. Chem.* **283** 13053–62

Jacob K, Sollier C and Jabado N 2007 Circulating tumor cells: detection, molecular profiling and future prospects *Expert Rev. Proteomics* **4** 741–56

Jain R K 1987 Transport of molecules in the tumor interstitium: a review *Cancer Res.* **47** 3039–51 PMID: 3555767

Jiang J, Jia P, Zhao Z and Shen B 2014 Key regulators in prostate cancer identified by co-expression module analysis *BMC Genomics* **15** 1015

Jones S *et al* 2008 Comparative lesion sequencing provides insights into tumor evolution *Proc. Natl Acad. Sci. USA* **105** 4283–8

Kam Y and Quaranta V 2009 Cadherin-bound beta-catenin feeds into the Wnt pathway upon adherens junctions dissociation: evidence for an intersection between beta-catenin pools *PLoS One* **24** e4580

Kamoun W S, Chae S S, Lacorre D A, Tyrrell J A, Mitre M, Gillissen M A, Fukumura D, Jain R K and Munn L L 2010 Simultaneous measurement of RBC velocity, flux, hematocrit and shear rate in vascular networks *Nat. Methods* **7** 655–60

Kanai Y, Oda T, Tsuda H, Ochiai A and Hirohashi S 1994 Point mutation of the E-cadherin gene in invasive lobular carcinoma of the breast *Japan J. Cancer Res.* **85** 1035–9

Kedrin D, Gligorijevic B, Wyckoff J, Verkhusha V V, Condeelis J, Segall J E and van Rheenen J 2008 Intravital imaging of metastatic behavior through a mammary imaging window *Nat. Methods* **5** 1019–21

Khuon S, Liang L, Dettman R W, Sporn P H, Wysolmerski R B and Chew T L 2010 Myosin light chain kinase mediates transcellular intravasation of breast cancer cells through the underlying endothelial cells: a three-dimensional FRET study *J. Cell Sci.* **123** 431–40

Kienast Y, von Baumgarten L, Fuhrmann M, Klinkert W E, Goldbrunner R, Herms J and Winkler F 2010 Real-time imaging reveals the single steps of brain metastasis formation *Nat. Med.* **16** 116–22

Klagsbrun M and Moses M A 1999 Molecular angiogenesis *Chem. Biol.* **6** R217–24

Kirsch M, Strasser J, Allende R, Bello L, Zhang J and Black P M 1998 Angiostatin suppresses malignant glioma growth *in vivo Cancer Res.* **58** 4654–9 PMID: 9788618

Knudson A G 1971 Mutation and cancer: Statistical study of retinoblastoma *Proc. Natl Acad. Sci. USA* **68** 820–3

Koch T M, Münster S, Bonakdar N, Butler J P and Fabry B 2012 3D traction forces in cancer cell invasion *PLoS ONE* **7** e33476

Konerding M A, Malkusch W, Klapthor B, van Ackern C, Fait E, Hill S A, Parkins C, Chaplin D J, Presta M and Denekamp J 1999 Evidence for characteristic vascular patterns in solid tumors: quantitative studies using corrosion casts *Br. J. Cancer.* **80** 724–32

Kräling B M, Wiederschain D G, Boehm T, Rehn M, Mulliken J B and Moses M A 1999 The role of matrix metalloproteinase activity in the maturation of human capillary endothelial cells *in vitro J. Cell Sci.* **112** 1599–609 PMID: 10212153

Krishnan R *et al* 2011 Substrate stiffening promotes endothelial monolayer disruption through enhanced physical forces *Am. J. Physiol. Cell Physiol.* **300** C146–54

Krishnan R *et al* 2009 Reinforcement versus fluidization in cytoskeletal mechanoresponsiveness *PLoS ONE* **4** e5486

Kumar S and Weaver V M 2009 Mechanics, malignancy, and metastasis: the force journey of a tumor cell *Cancer Metastasis Rev.* **28** 113–27

Kuvaja P, Würtz SØ, Talvensaari-Mattila A, Brünner N, Pääkkö P and Turpeenniemi-Hujanen T 2007 High serum TIMP-1 correlates with poor prognosis in breast carcinoma – a validation study *Cancer Biomark* **3** 293–300 PMID: 18048967

Ladoux B 2009 Cells guided on their journey *Nature Physics* **5** 377–8

Laevsky G and Knecht D A 2003 Cross-linking of actin filaments by myosin II is a major contributor to cortical integrity and cell motility in restrictive environments *J. Cell Sci.* **116** 3761–70

Lammerding J, Schulze P C, Takahashi T, Kozlov S, Sullivan T, Kamm R D, Stewart C L and Lee R T 2004 Lamin A/C deficiency causes defective nuclear mechanics and mechanotransduction *J. Clin. Invest.* **113** 370–8

Lämmermann T and Sixt M 2009 Mechanical modes of 'amoeboid' cell migration *Curr. Opin. Cell Biol.* **21** 636–44

Laser-Azogui A, Diamant-Levi T, Israeli S, Roytman Y and Tsarfaty I 2014 Met-induced membrane blebbing leads to amoeboid cell motility and invasion *Oncogene* **33** 1788–98

Lee J S, Hale C M, Panorchan P, Khatau S B, George J P, Tseng Y, Stewart C L, Hodzic D and Wirtz D 2007 Nuclear lamin A/C deficiency induces defects in cell mechanics, polarization, and migration *Biophys. J.* **93** 2542–52

Lee K H *et al* 2008 Role of hepatocyte growth factor/c-Met signaling in regulating urokinase plasminogen activator on invasiveness in human hepatocellular carcinoma: a potential therapeutic target *Clin. Exp. Metastasis* **25** 89–96

Lee S W, Jung K H, Jeong C H, Seo J H, Yoon D K, Suh J K, Kim K W and Kim W J 2011 Inhibition of endothelial cell migration through the down-regulation of MMP-9 by A-kinase anchoring protein 12 *Mol. Med. Rep.* **4** 145–9

Legant W R, Miller J S, Blakely B L, Cohen D M, Genin G M and Chen C S 2010 Measurement of mechanical tractions exerted by cells in three-dimensional matrices *Nat. Methods* **7** 969–71

Lei K, Zhang X, Ding X, Guo X, Chen M, Zhu B, Xu T, Zhuang Y, Xu R and Han M 2009 SUN1 and SUN2 play critical but partially redundant roles in anchoring nuclei in skeletal muscle cells in mice *Proc. Natl Acad. Sci. USA* **106** 10207–12

Levental K R *et al* 2009 Matrix crosslinking forces tumor progression by enhancing integrin signaling *Cell* **139** 891–906

Li H, Gerald W L and Benezra R 2004 Utilization of bone marrow-derived endothelial cell precursors in spontaneous prostate tumors varies with tumor grade *Cancer Res.* **64** 6137–43

Li X Y, Ota I, Yana I, Sabeh F and Weiss S J 2008 Molecular dissection of the structural machinery underlying the tissue-invasive activity of membrane type-1 matrix metalloproteinase *Mol. Biol. Cell* **19** 3221–33

Linder S, Wiesner C and Himmel M 2011 Degrading devices: invadosomes in proteolytic cell invasion *Annu. Rev. Cell. Dev. Biol.* **27** 185–211

Liu A J and Nagel S R 1998 Jamming is not just cool any more *Nature* **396** 21–2

Liu C H, Nagel S R, Schecter D A, Coppersmith S N, Majumdar S, Narayan O and Witten T A 1995 Force Fluctuations in Bead Packs *Science* **269** 513–5

Liu J, He X, Qi Y, Tian X, Monkley S J, Critchley D R, Corbett S A, Lowry S F, Graham A M and Li S 2011 Talin1 regulates integrin turnover to promote embryonic epithelial morphogenesis *Mol. Cell Biol.* **31** 3366–77

Loftus J C and Liddington R C 1997 Cell adhesion in vascular biology. New insights into integrin-ligand interaction *J. Clin. Invest.* **99** 2302–6

Lokody I 2014 Tumor-promoting tissue mechanics *Nat. Rev. Cancer* **14** 296–7

Lozzio C B and Lozzio B B 1975 Human chronic myelogenous leukemia cell-line with positive Philadelphia chromosome *Blood* **45** 321–34 PMID: 163658

Lu P, Weaver V M and Werb Z 2012 The extracellular matrix: a dynamic niche in cancer progression *J. Cell Biol.* **196**(4) 395–406

Ly D P, Zazzali K M and Corbett S A 2003 *De novo* expression of the integrin alpha5beta1 regulates alphavbeta3-mediated adhesion and migration on fibrinogen *J. Biol. Chem.* **278** 21878–85

Maître J L, Berthoumieux H, Krens S F, Salbreux G, Jülicher F, Paluch E and Heisenberg C P 2012 Adhesion functions in cell sorting by mechanically coupling the cortices of adhering cells *Science* **338** 253–6

Maniotis A J, Folberg R, Hess A, Seftor E A, Gardner L M, Peer J, Trent J M, Meltzer P S and Hendrix M J 1999 Vascular channel formation by human melanoma cells *in vivo* and *in vitro*: vasculogenic mimicry *Am. J. Pathol.* **155** 739–52

Marshall B T, Long M, Piper J W, Yago T, McEver R P and Zhu C 2003 Direct observation of catch bonds involving cell-adhesion molecules *Nature* **423** 190–3

Maruthamuthu V and Gardel M L 2014 Protrusive activity guides changes in cell–cell tension during epithelial cell scattering *Biophys. J.* **107** 555–63

Mattsson J, Wyss H M, Fernandez-Nieves A, Miyazaki K, Hu Z, Reichman D R and Weitz D A 2009 Soft colloids make strong glasses *Nature* **462** 83–6

McCarthy K, Maguire T, McGreal G, McDermott E, O'Higgins N and Duffy M J 1999 High levels of tissue inhibitor of metalloproteinase-1 predict poor outcome in patients with breast cancer *Int. J. Cancer* **84** 44–8

McCarty O J, Mousa S A, Bray P F and Konstantopoulos K 2000 Immobilized platelets support human colon carcinoma cell tethering, rolling, and firm adhesion under dynamic flow conditions *Blood* **96** 1789–97 PMID: 10961878

McDonald D M and Choyke P L 2003 Imaging of angiogenesis: from microscope to clinic *Nat. Med.* **9** 713–25

Merlo L M F, Pepper J W, Reid B J and Maley C C 2006 Cancer as an evolutionary and ecological process *Nat. Rev. Cancer* **6** 923–35

Meyer A S, Hughes-Alford S K, Kay J E, Castillo A, Wells A, Gertler F B and Lauffenburger D A 2012 2D protrusion but not motility predicts growth factor-induced cancer cell migration in 3D collagen *J. Cell Biol.* **197** 721–9

Mierke C T 2009 The role of vinculin in the regulation of the mechanical properties of cells *Cell Biochem. Biophys.* **53** 115–26

Mierke C T 2011 Cancer cells regulate biomechanical properties of human microvascular endothelial cells *J. Biol. Chem.* **286** 40025–37

Mierke C T 2013 Physical break-down of the classical view on cancer cell invasion and metastasis *Eur. J. Cell Biol.* **92**(3) 89–104

Mierke C T 2014 The fundamental role of mechanical properties in the progression of cancer disease and inflammation *Rep. Prog. Phys.* **77**(7 (35pp)) 076602

Mierke C T, Zitterbart D P, Kollmannsberger P, Raupach C, Schlotzer-Schrehardt U, Goecke T W, Behrens J and Fabry B 2008a Breakdown of the endothelial barrier function in tumor cell transmigration *Biophys. J.* **94** 2832–46

Mierke C T, Kollmannsberger P, Paranhos-Zitterbart D, Smith J, Fabry B and Goldmann W H 2008b Mechano-coupling and regulation of contractility by the vinculin tail domain *Biophys. J.* **94** 661–70

Mierke C T, Rosel D, Fabry B and Brabek J 2008c Contractile forces in tumor cell migration *Eur. J. Cell Biol.* **87** 669–76

Mierke C T, Kollmannsberger P, Zitterbart D P, Diez G, Koch T M, Marg S, Ziegler W H, Goldmann W H and Fabry B 2010 Vinculin facilitates cell invasion into three-dimensional collagen matrices *J. Biol. Chem.* **285** 13121–30

Mierke C T, Frey B, Fellner M, Herrmann M and Fabry B 2011 Integrin α5β1 facilitates cancer cell invasion through enhanced contractile force *J. Cell Sci.* **124** 369–83

Mills J C, Stone N L, Erhardt J and Pittman R N 1998 Apoptotic membrane blebbing is regulated by myosin light chain phosphorylation *J. Cell Biol.* **140** 627–36

Moolgavkar S H and Knudson A G 1981 Mutation and cancer: a model for human carcinogenesis *J. Natl Cancer Inst.* **66** 1037–52 PMID: 6941039

Mouw J K *et al* 2014 Tissue mechanics modulate microRNA-dependent PTEN expression to regulate malignant progression *Nature Med.* **20** 360–7

Nagy J A, Chang S-H, Dvorak A M and Dvorak H F 2009 Why are tumor blood vessels abnormal and why is it important to know? *Br. J. Cancer.* **100** 865–9

Neff N T, Lowrey C, Decker C, Tovar A, Damsky C, Buck C and Horwitz A F 1982 A monoclonal antibody detaches embryonic skeletal muscle from extracellular matrices *J. Cell Biol.* **95** 654–66

Niessen C M and Gumbiner B M 2002 Cadherin-mediated cell sorting not determined by binding or adhesion specificity *J. Cell Biol.* **156** 389–99

Oliver M, Kovats T, Mijailovich S, Butler J, Fredberg J and Lenormand G 2010 Remodeling of integrated contractile tissues and its dependence on strain-rate amplitude *Physical Review Letters* **105** 158102

Onder T, Gupta P B, Mani S A, Yang J, Lander E S and Weinberg R A 2008 Loss of E-cadherin promotes metastasis via multiple downstream transcriptional pathways *Cancer Res.* **68** 3645–54

Otrock Z K, Mahfouz R A R, Makarem J A and Shamseddine A I 2007 Understanding the biology of angiogenesis: review of the most important molecular mechanisms *Blood Cells Mol. Dis.* **39** 212–20

Padera T P, Stoll B R, Tooredman J B, Capen D, di Tomaso E and Jain R K 2004 Pathology: cancer cells compress intra tumor vessels *Nature* **427** 695

Paget S 1989 The distribution of secondary growths in cancer of the breast *Lancet* **1** 571–3 PMID: 2673568

Palecek S P, Loftus J C, Ginsberg M H, Lauffenburger D A and Horwitz A F 1997 Integrin-ligand binding propertiesgovern cell migration speed through cell–substratum adhesiveness *Nature* **385** 537–40

Paluch E K and Raz E 2013 The role and regulation of blebs in cell migration *Curr. Opin. Cell Biol.* **25** 582–90

Patan S 2004 Vasculogenesis and angiogenesis *Cancer Treat Res.* **117** 3–32

Pathak A and Kumar S 2013 Transforming potential and matrix stiffness co-regulate confinement sensitivity of tumor cell migration *Integr. Biol. (Camb.).* **5** 1067–75

Pawar S C, Demetriou M C, Nagle R B, Bowden G T and Cress A E 2007 Integrin alpha6 cleavage: a novel modification to modulate cell migration *Exp. Cell Res.* **313** 1080–9

Peinado H, Olmeda D and Cano A 2007 Snail, Zeb and bHLH factors in tumor progression: an alliance against the epithelial phenotype? *Nat. Rev. Cancer* **7** 415–28

Petrie R J, Gavara N, Chadwick R S and Yamada K M 2012 Nonpolarized signaling reveals two distinct modes of 3D cell migration *J. Cell Biol.* **197** 439–55

Phillips H M and Steinberg M S 1978 Embryonic tissues as elasticoviscous liquids: I. Rapid and slow shape changes in centrifuged cell aggregates *J. Cell Sci.* **30** 1–20 PMID: 649680

Phillips H M, Steinberg M S and Lipton B H 1977 Embryonic tissues as elasticoviscous liquids: II. Direct evidence for cell slippage in centrifuged aggregates *Dev. Biol.* **59** 124–34

Phillips H M and Davis G S 1978 Liquid-tissue mechanics in amphibian gastrulation: germ-layer assembly in Rana pipiens *Am. Zool.* **18** 81–93

Poincloux R, Collin O, Lizarraga F, Romao M, Debray M, Piel M and Chavrier P 2011 Contractility of the cell rear drives invasion of breast tumor cells in 3D Matrigel *Proc. Natl Acad. Sci. USA* **108** 1943–8

Ponti A, Machacek M, Gupton S L, Waterman-Storer C M and Danuser G 2004 Two distinct actin networks drive the protrusion of migrating cells *Science* **305**(5691) 1782–6

Prager-Khoutorsky M, Lichtenstein A, Krishnan R, Rajendran K, Mayo A, Kam Z, Geiger B and Bershadsky A D 2011 Fibroblast polarization is a matrix-rigidity-dependent process controlled by focal adhesion mechanosensing *Nat. Cell Biol.* **13** 1457–65

Qian B-Z and Pollard J W 2010 Macrophage diversity enhances tumor progression and metastasis *Cell* **141** 39–51

Rape A D, Guo W H and Wang Y L 2011 The regulation of traction force in relation to cell shape and focal adhesions *Biomaterials* **32** 2043–51

Raupach C, Paranhos-Zitterbart D, Mierke C, Metzner C, Müller A F and Fabry B 2007 Stress fluctuations and motion of cytoskeletal-bound markers *Phys. Rev. E.* **76** 011918 PMID: 17677505

Renkawitz J and Sixt M 2010 Mechanisms of force generation and force transmission during interstitial leukocyte migration *EMBO Rep.* **11** 744–50

Ribatti D 2009 Endogenous inhibitors of angiogenesis: a historical review *Leuk. Res.* **33** 638–44

Rice J 2012 The rude awakening *Nature* **485** S55–7

Ricono J M, Huang M, Barnes L A, Lau S K, Weis S M, Schlaepfer D D, Hanks S K and Cheresh D A 2009 Specific cross-talk between epidermal growth factor receptor and integrin alphavbeta5 promotes carcinoma cell invasion and metastasis *Cancer Res.* **69** 1383–91

Ridley A J 2011 Life at the leading edge *Cell* **145** 1012–22

Riedle S, Kiefel H, Gast D, Bondong S, Wolterink S, Gutwein P and Altevogt P 2009 Nuclear translocation and signalling of L1-CAM in human carcinoma cells requires ADAM10 and presenilin/gamma-secretase activity *Biochem. J.* **420** 391–402

Riveline D, Zamir E, Balaban N Q, Schwarz U S, Ishizaki T, Narumiya S, Kam Z, Geiger B and Bershadsky A D 2001 Focal contacts as mechanosensors: externally applied local mechanical force induces growth of focal contacts by an mDia1-dependent and ROCK-independent mechanism *J. Cell Biol.* **153** 1175–86

Rhim A D *et al* 2012 EMT and dissemination precede pancreatic tumor formation *Cell.* **148** 349–61

Rolli M, Fransvea E, Pilch J, Saven A and Felding-Habermann B 2003 Activated integrin alphavbeta3 cooperates with metalloproteinase MMP-9 in regulating migration of metastatic breast cancer cells *Proc. Natl Acad. Sci. USA* **100** 9482–7

Rösel D *et al* 2008 Up-regulation of Rho/ROCK signaling in sarcoma cells drives invasion and increased generation of protrusive forces *Mol. Cancer Res.* **6** 1410–20

Rout U K, Wang J, Paria B C and Armant D R 2004 Alpha5beta1, alphaVbeta3 and the platelet-associated integrin alphaIIbbeta3 coordinately regulate adhesion and migration of differentiating mouse trophoblast cells *Dev. Biol.* **268** 135–51

Rowe R G and Weiss S J 2009 Navigating ECM barriers at the invasive front: the cancer cell–stroma interface *Annu. Rev. Cell Dev. Biol.* **25** 567–95

Rubashkin M G, Ou G and Weaver V M 2014 Deconstructing signaling in three dimensions *Biochemistry* **53** 2078–90

Ruddell A, Croft A, Kelly-Spratt K, Furuya M and Kemp C J 2014 Tumors induce coordinate growth of artery, vein, and lymphatic vessel triads *BMC Cancer* **14** 354

Runz S, Mierke C T, Joumaa S, Behrens J, Fabry B and Altevogt P 2008 CD24 induces localization of beta1 integrin to lipid raft domains *Biochem. Biophys. Res. Commun.* **365** 35–41

Sabeh F *et al* 2004 Tumor cell traffic through the extracellular matrix is controlled by the membrane-anchored collagenase MT1-MMP *J. Cell Biol.* **167** 769–81

Sabeh F, Shimizu-Hirota R and Weiss S J 2009 Protease-dependent versus independent cancer cell invasion programs: three-dimensional amoeboid movement revisited *J. Cell Biol.* **185** 11–9

Savagner P, Kusewitt D F, Carver E A, Magnino F, Choi C, Gridley T and Hudson L G 2005 Developmental transcription factor slug is required for effective reepithelialization by adult keratinocytes *J. Cell Physiol.* **202** 858–66

Sawada K *et al* 2008 Loss of E-cadherin promotes ovarian cancer metastasis via alpha 5-integrin, which is a therapeutic target *Cancer Res.* **68** 2329–39

Schall P, Weitz D A and Spaepen F 2007 Structural rearrangements that govern flow in colloidal glasses *Science* **318** 1895–99

Schaller M D 2001 Paxillin: a focal adhesion-associated adaptor protein *Oncogene* **20** 6459–72

Schelter F, Halbgewachs B, Bäumler P, Neu C, Görlach A, Schrötzlmair F and Krüger A 2011 Tissue inhibitor of metalloproteinases-1-induced scattered liver metastasis is mediated by hypoxia-inducible factor-1α *Clin. Exp. Metastasis* **28** 91–99

Schmid R S, Shelton S, Stanco A, Yokota Y, Kreidberg J A and Anton E S 2004 alpha3beta1 integrin modulates neuronal migration and placement during early stages of cerebral cortical development *Development* **131** 6023–31

Schramm H M 2014 Should EMT of cancer cells be understood as epithelial-myeloid transition? *J. Cancer* **5** 125–32

Swartz M A and Fleury M E 2007 Interstitial flow and its effects in soft tissues *Annu. Rev. Biomed. Eng.* **9** 229–56

Segre P N, Prasad V, Schofield A B and Weitz D A 2001 Glasslike kinetic arrest at the colloidal–gelation transition *Phys. Rev. Lett.* **86** 6042–5

Seltmann K, Fritsch A, Käs J A and Magin T M 2013 Keratins significantly contribute to cell stiffness and impact invasive behavior *PNAS* **110** 18507–12

Serra-Picamal X, Conte V, Vincent R, Anon E, Tambe D, Bazellieres E, Butler J, Fredberg J and Trepat X 2012 Mechanical waves during tissue expansion *Nat. Phys.* **8** 628–34

Singh B, Schneider M, Knyazev P and Ullrich A 2009 UV-induced EGFR signal transactivation is dependent on proligand shedding by activated metalloproteases in skin cancer cell lines *Int. J. Cancer* **124** 531–9

Slaga T J 1983 Overview of tumor promotion in animals *Environ. Health Perspect.* **50** 3–14

St Croix B *et al* 2000 Genes expressed in human tumor endothelium *Science* **289** 1197–202

Starr D A and Han M 2003 ANChors away: an actin based mechanism of nuclear positioning *J. Cell Sci.* **116** 211–6

Starr D A, Hermann G J, Malone C J, Fixsen W, Priess J R, Horvitz H R and Han M 2001 unc-83 encodes a novel component of the nuclear envelope and is essential for proper nuclear migration *Development* **128** 5039–50

Steeg P S 2006 Tumor metastasis: mechanistic insights and clinical challenges *Nat. Med.* **12** 895–904

Steinberg M S 1962a On the mechanism of tissue reconstruction by dissociated cells: I. Population kinetics, differential adhesiveness, and the absence of directed migration *Proc. Natl Acad. Sci. USA* **48** 1577–82

Steinberg M S 1962b Mechanism of tissue reconstruction by dissociated cells: II. Time course of events *Science* **137** 762–3

Steinberg M S 1962c On the mechanism of tissue reconstruction by dissociated cells: III. Free energy relations and the reorganization of fused, heteronomic tissue fragments *Proc. Natl Acad. Sci. USA* **48** 1769–76

Steinberg M S 1963 Reconstruction of tissues by dissociated cells *Science* **141** 401–8

Steinberg M S 1964 The problem of adhesive selectivity in *cellular interactions Cellular Membranes in Development (22nd Symposium of the Society for the Study of Development and Growth)* ed M Locke (New York: Academic) pp 321–66

Steinberg M S 1970 Does differential adhesion govern self-assembly processes in histogenesis? Equilibrium configurations and the emergence of a hierarchy among populations of embryonic cells *J. Exp. Zool.* **173** 395–434

Steinberg M S 1978 Cell–cell recognition in multicellular assembly: levels of specificity *Symp. Soc. Exp. Biol.* **32** 25–49 PMID: 382423

Steinberg M S and Poole T J 1982 Liquid behavior of embryonic tissues *Cell Behavior* ed R Bellairs, A S G Curtis and G Dunn G (Cambridge: Cambridge University Press) pp 583–607

Stewart-Hutchinson P J, Hale C M, Wirtz D and Hodzic D 2008 Structural requirements for the assembly of LINC complexes and their function in cellular mechanical stiffness *Exp. Cell Res.* **314** 1892–905

Taddei M L, Giannoni E, Morandi A, Ippolito L, Ramazzotti M, Callari M, Gandellini P and Chiarugi P 2014 Mesenchymal to amoeboid transition is associated with stem-like features of melanoma cells *Cell Commun. Signal* **12** 24

Tambe D T *et al* 2011 Collective cell guidance by cooperative intercellular forces *Nat. Mater.* **10** 469–75

Tania M, Khan M A and Fu J 2014 Epithelial to mesenchymal transition inducing transcription factors and metastatic cancer *Tumor Biol.* **35** 7335–42

Technau M and Roth S 2008 The Drosophila KASH domain proteins Msp-300 and Klarsicht and the SUN domain protein klaroid have no essential function during oogenesis *Fly (Austin).* **2** 82–91

Teicher B A and Fricker S P 2010 CXCL12 (SDF-1)/CXCR4 pathway in cancer *Clin. Cancer Res.* **16** 2927–31

Temin H M 1988 Evolution of cancer genes as a mutation-driven process *Cancer Res.* **48** 1697–701 PMID: 3280119

Thiery J P 2002 Epithelial-mesenchymal transitions in tumor progression *Nat. Rev. Cancer* **2** 442–54

Thiery J P 2003 Epithelial–mesenchymal transitions in development and pathologies *Curr. Opin. Cell Biol.* **15** 740–6

Thiery J P and Sleeman J P 2006 Complex networks orchestrate epithelial–mesenchymal transitions *Nat. Rev. Mol. Cell Biol.* **7** 131–42

Thomas R K *et al* 2007 High-throughput oncogene mutation profiling in human cancer *Nat. Genet.* **39** 347–51

Tinevez J Y, Schulze U, Salbreux G, Roensch J, Joanny J F and Paluch E 2009 Role of cortical tension in bleb growth *Proc. Natl Acad. Sci. USA* **106** 18581–6

Tooley A J, Gilden J, Jacobelli J, Beemiller P, Trimble W S, Kinoshita M and Krummel M F 2008 Amoeboid T lymphocytes require the septin cytoskeleton for cortical integrity and persistent motility *Nat. Cell Biol.* **11** 17–26

Townes P L and Holtfreter J 1955 Directed movements and selective adhesion of embryonic amphibian cells *J. Exp. Zool.* **128** 53–120

Tozluoglu M, Tournier A L, Jenkins R P, Hooper S, Bates P A and Sahai E 2013 Matrix geometry determines optimal cancer cell migration strategy and modulates response to interventions *Nat. Cell Biol.* **15** 751–62

Trappe V, Prasad V, Cipelletti L, Segre P N and Weitz D A 2001 Jamming phase diagram for attractive particles *Nature* **411** 772–5

Trepat X, Deng L, An S, Navajas D, Tschumperlin D, Gerthoffer W, Butler J and Fredberg J 2007 Universal physical responses to stretch in the living cell *Nature* **447** 592–5

Tseng Y, Lee JS, Kole TP, Jiang I and Wirtz D 2004 Micro-organization and viscoelasticity of the interphase nucleus revealed by particle nanotracking *J. Cell Sci.* **117** 2159–67

Van Aarsen L A, Leone D R, Ho S, Dolinski B M, McCoon P E and LePage D J *et al* 2008 Antibody-mediated blockade of integrin αvβ6 inhibits tumor progression *in vivo* by a transforming growth factor-β-regulated mechanism *Cancer Res.* **68** 561–70

van de Vijver M *et al* 2002 A gene-expression signature as a predictor of survival in breast cancer *N. Engl. J. Med.* **347** 1999–2009

Van Sluis G L, Niers T M, Esmon C T, Tigchelaar W, Richel D J, Buller H R, Van Noorden C J and Spek C A 2009 Endogenous activated protein C limits cancer cell extravasation through sphingosine-1-phosphate receptor 1-mediated vascular endothelial barrier enhancement *Blood* **114** 1968–73

van Tetering G, van Diest P, Verlaan I, van der Wall E, Kopan R and Vooijs M 2009 Metalloprotease ADAM10 is required for Notch1 site 2 cleavage *J. Biol. Chem.* **284** 31018–27

Veiga S S, Elias M C Q B, Gremski W, Porcionatto M A, da Silva R, Nader H B and Brentani R R 1997 Post-translational modifications of alpha5beta1 integrin by glycosaminoglycan chains. The alpha5beta1 integrin is a facultative proteoglycan *J. Biol. Chem.* **272** 12529–35

Vicente-Manzanares M, Choi C K and Horwitz A R 2009 Integrins in cell migration—the actin connection *J. Cell Sci.* **122** 199–206

Vitelli V and van Hecke M 2011 Soft materials: Marginal matters *Nature* **480** 325–6

Voura E B, Chen N and Siu C H 2000 Platelet-endothelial cell adhesion molecule 1 (CD31) redistributes from the endothelial junction and is not required for the transendothelial migration of melanoma cells *Clin. Exp. Metastasis.* **18** 527–32

Wang H, Flach H, Onizawa M, Wei L, McManus M T and Weiss A 2014 Negative regulation of Hif1a expression and TH17 differentiation by the hypoxia-regulated microRNA miR-210 *Nat. Immun.* **15** 393–401

Wang C, Lisanti M P and Liaoc D J 2011 Reviewing once more the c-myc and Ras collaboration. Converging at the cyclin D1-CDK4 complex and challenging basic concepts of cancer biology *Cell Cycle* **10** 57–67

Weinstein I B and Joe A K 2006 Mechanisms of disease: oncogene addiction—a rationale for molecular targeting in cancer therapy *Nat. Clin. Pract. Oncol.* **3** 448–57

Warren B A, Shubik P and Feldman R 1978 Metastasis via the blood stream: the method of intravasation of tumor cells in a transplantable melanoma of the hamster *Cancer Lett.* **4** 245–51

Webb D J, Donais K, Whitmore L A, Thomas S M, Turner C E, Parsons J T and Horwitz A F 2004 FAK-Src signaling through paxillin, ERK, and MLCK regulates adhesion disassembly *Nat. Cell Biol.* **6** 154–61

Wei J et al 2008 Overexpression of vimentin contributes to prostate cancer invasion and metastasis via src regulation *Anticancer Res.* **28** 327–34

Weiss L 2000 Patterns of metastasis *Cancer Metastasis Rev.* **19** 281–301

Wirtz D 2009 Particle-tracking microrheology of living cells: principles and applications *Annu. Rev. Biophys.* **38** 301–26

Wittchen E S, Worthylake R A, Kelly P, Casey P J, Quilliam L A and Burridge K 2005 Rap1 GTPase inhibits leukocyte transmigration by promoting endothelial barrier function *J. Biol. Chem.* **280** 11675–82

Wolf K, Mazo I, Leung H, Engelke K, von Andrian U H, Deryugina E I, Strongin A Y, Brocker E B and Friedl P 2003a Compensation mechanism in tumor cell migration: mesenchymal-amoeboid transition after blocking of pericellular proteolysis *J. Cell Biol.* **160** 267–77

Wolf K, Müller R, Borgmann S, Bröcker EB and Friedl P 2003b Amoeboid shape change and contact guidance: T-lymphocyte crawling through fibrillar collagen is independent of matrix remodeling by MMPs and other proteases *Blood* **102** 3262–9 PMID: 12855577

Wolf K, Wu Y I, Liu Y, Geiger J, Tam E, Overall C, Stack M S and Friedl P 2007 Multi-step pericellular proteolysis controls the transition from individual to collective cancer cell invasion *Nat. Cell Biol.* **9** 893–904

Wood L D et al 2007 The genomic landscapes of human breast and colorectal cancers *Science.* **318** 1108–13

Wu J et al 2015 Plasminogen activator inhibitor-1 inhibits angiogenic signaling by uncoupling vascular endothelial growth factor receptor-2-αvβ3 integrin cross talk *Arterioscler Thromb Vasc. Biol.* **35** 111–20

Wyss H M, Miyazaki K, Mattsson J, Hu Z, Reichman D R and Weitz D A 2007 Strain-rate frequency superposition: a rheological probe of structural relaxation in soft materials *Phys. Rev. Lett.* **98** 238303

Yang J, Mani S A, Donaher J L, Ramaswamy S, Itzykson R A, Come C, Savagner P, Gitelman I, Richardson A and Weinberg R A 2004 Twist, a master regulator of morphogenesis, plays an essential role in tumor metastasis *Cell* **117** 927–39

Yoshida K and Soldati T 2006 *J. Cell. Sci.* **119** 3833–44

Zaman M H, Trapani L M, Sieminski A L, Mackellar D, Gong H, Kamm R D, Wells A, Lauffenburger D A and Matsudaira P 2006 Migration of tumor cells in 3D matrices is governed by matrix stiffness along with cell–matrix adhesion and proteolysis *Proc. Natl. Acad. Sci. USA* **103** 10889–94

Zhou E H, Trepat X, Park C Y, Lenormand G, Oliver M N, Mijailovich S M, Hardin C, Weitz D A, Butler J P and Fredberg J J 2009 Universal behavior of the osmotically compressed cell and its analogy to the colloidal glass transition *Proc. Natl Acad. Sci. USA* **106** 10632–7

Zhu C, Yago T, Lou J Z, Zarnitsyna V I and McEver R P 2008 Mechanisms for flow- enhanced cell adhesion *Ann. Biomed. Eng.* **36** 604–21

Ziegler W H, Gingras A R, Critchley D R and Emsley J 2008 Integrin connections to the cytoskeleton through talin and vinculin *Biochem. Soc. Trans.* **36** 235–9

Zijlstra A, Lewis J, Degryse B, Stuhlmann H and Quigley J P 2008 The inhibition of tumor cell intravasation and subsequent metastasis via regulation of *in vivo* tumor cell motility by the tetraspanin CD151 *Cancer Cell* **13** 221–34

IOP Publishing

Physics of Cancer

Claudia Tanja Mierke

Chapter 2

Inflammation and cancer

Summary

The link between inflammation and cancer was suggested a long time ago and it has become a major part of current cancer research. Even analyzing the biophysical parameters or preferences of inflammation is a growing research field and it will probably become the future focus of physics of cancer. Inflammation, especially chronic inflammation, has been shown to have protumorigenic effects and it therefore increases the probability of promoting certain tumors. What impact the inflammation has on the metastatic potential of tumors is not yet well known. It has been hypothesized that the transendothelial migration process leading to cancer metastasis is affected by inflammatory cytokines. However, the impact of acute and chronic inflammatory processes is still elusive. The mechanical properties may also change during these inflammatory processes and subsequently alter the tumorigenic potential of special cancer cells. However, this is not yet known and it requires much research. In this chapter, the principal cellular and molecular pathways promoting tumor progression and inhibiting tumor progression will be discussed in the context of inflammation. In addition, the interplay between cancer and inflammation will be presented.

2.1 Inflammation: acute and chronic

Inflammation is a protective response of an organism, employing host cells, blood vessels and certain proteins to fight against tissue injury and infection. This response aims to eliminate the initial reason for the cell injury, to clear and remove necrotic cells and tissues, and to start tissue repair. Inflammation is a potentially harmful mechanism to protect against intruders, such as bacteria, fungi and viruses, and cellular as well as tissue damage. Indeed, several components of the inflammatory process are able to destroy microbes, but they may also cause further damage to uninflamed tissue surrounding the inflammation site. The inflammatory process involves white blood cells, such as leukocytes, and plasma proteins that are normally located in the blood fluid.

doi:10.1088/978-0-7503-1134-2ch2

The goal of the inflammatory reaction is to guide these immune cells and inflammatory proteins to the sites of the tissue injury or the infection. Chemical substances secreted by the host cells facilitate the onset of the inflammatory process. These substances can be cytokines, including chemokines. The inflammatory process is precisely controlled and hence self-limited. However, there are excess inflammatory reactions that lead to an inappropriate inflammatory response even when there are no foreign presences. This may then lead to autoimmune disease. Hence, inflammatory responses need to be regulated and controlled precisely by the immune system to inhibit excessive tissue damage and the involvement of surrounding normal tissue in the inflammation. What are the signs of an inflammation? The cardinal signs are heat, red staining of the tissue, swelling, pain and then loss of function. How do you know whether you have an acute or a chronic inflammation? Firstly, the stimuli for acute inflammations may be infections, for example bacterial, fungal, parasitic or viral, and the toxins of microbes. Secondly, necrosis of tissues may be evoked by ischemia, physical or chemical injury from, for example, irradiation or environmental chemicals, and trauma. Thirdly, the cause of inflammation can be foreign bodies, such as splinters, sutures or just dirt. Fourthly, immune reactions induced by hypersensitivity reactions can lead to acute inflammation. Acute inflammation leads to vascular changes, such as vasodilation, increased vascular permeability and elevated adhesion of leukocytes to the endothelial cell layer of blood vessels. In addition, acute inflammation causes cellular events, such as the recruitment and activation of neutrophils. In more detail, vasodilation is the reaction of blood vessels to acute inflammation and leads to alterations in the vascular diameter. In the end, this decreases blood pressure. It is followed by vascular leakage and edema formation, which represents the accumulation of fluid and proteins of the blood plasma in the extravascular tissues (so-called interstitium). Finally, the leukocytes emigrate to extravascular tissues by undergoing the following sequential steps such as margination and rolling, activation of the endothelium and adhesion to the endothelial vessel lining, and transmigration. Acute inflammation induces an immediate increase in vascular permeability that is mediated by proinflammatory substances, such as histamine, complement factors, serotonin, bradykinin, tachykinins and nitrogen oxide species (Di Lorenzo *et al* 2009). This release of proinflammatory factors is simultaneously facilitated with the recruitment of neutrophils to the sites of the inflammation (within the first six hours) and is followed by the next step, the recruitment of monocytes and macrophages to these inflamed tissue sites (within one day) (Walzog and Gaehtgens 2000).

The step of vasodilation changes the vessel flow in order to remove the released histamine and reduce the increased heat and redness in this tissue. Additionally, the clearance of the released histamine is important as it can also stimulate vascular smooth muscle contraction. The change in the vessel flow omits stasis, e.g. the slowing down of the blood vessel flow caused by hyperviscosity of the vessel fluid, the margination of circulating leukocytes and the activation of the endothelial vessel lining. This alteration follows the increased permeability of the vasculature including the formation of an early transudate (protein-poor filtrate of the blood plasma with a few cells) and an exudate (protein-rich filtrate of the blood plasma with intermediate and high amount of cells) within extracellular tissues (Nagy *et al* 2012). The next step is vascular leakage and edema formation evoked by the change in vessel permeability caused by histamines, bradykinins and leukotrienes, which then lead to the contraction of endothelial cells and subsequently to a widening of the intercellular gaps of venules. Taken together, the endothelium lining of the vessels becomes leaky. The outpouring of the exudate (protein-rich) into the extracellular matrix of connective tissues reduces the intravascular osmotic pressure and increases the extracellular (interstitial) osmotic pressure. The last step causes the formation of edema, consisting mainly of water and ions.

Chronic inflammation has the following features. It lasts longer than acute inflammation, and other cell types such as lymphocytes, plasma cells and macrophages (monocytes) infiltrate the interstitium. This may finally lead to tissue destruction by inflammatory cells, which is repaired via fibrosis and angiogenesis. The persistence of an injury or infection leads to diseases such as colitis ulcerosa and tuberculosis, which are known chronic inflammations. Prolonged exposure to a toxic agent may lead to pulmonary conditions characterized by a self-perpetuating immune reaction resulting in damage to and inflammation of tissue, for example rheumatoid arthritis, systemic lupus erythematosus and multiple sclerosis (Viatte *et al* 2013, Jacob and Stohl 2011, Leray *et al* 2010).

2.1.1 Receptors involved in leukocyte activation

Several receptors are involved in the inflammation-dependent activation of leukocytes. These are Toll-like receptors (TLRs), different seven-transmembrane G-protein-coupled receptors and receptors for cytokines (such as IFN-gammaR and opsonins). In addition, antibodies against specific opsonins are recognized by Fcgamma receptors (Viegas *et al* 2012).

In particular, the complement system has been described as a heat-labile component of the normal blood plasma that supports the opsonization of bacteria by antibodies in order to kill them. This complements the antibacterial activity of the antibody. The complement does only act as an effector for the antibody mediated response, it can also be activated during the early onset of an infection, even in the absence of an antibody response. In more detail, the complement system consists of numerous distinct plasma proteins reacting with one another to opsonize pathogens in order to facilitate inflammatory responses. For example, the digestive enzyme pepsin inside cells can be secreted as an inactive precursor enzyme, the pepsinogen, which is only cleaved to pepsin in the acidic microenvironment of the stomach (called autodigestion).

The components of the complement system, such as the precursor zymogens, are widely distributed throughout the fluids and tissues of the body. At the inflamed sites, complement factors are activated in order to facilitate inflammatory progression via an enzymatic cascade. In this cascade, an active complement enzyme is obtained by cleavage of its zymogen precursor. This zymogen cleaves its desired substrate, such as a targeted complement zymogen, in order to activate it. This then cleaves to activate the next zymogen, and so on. Thus, the activation of a small group of complement proteins at the beginning of the inflammatory cascade is substantially amplified by each successive step of the enzymatic reaction, leading to a pronounced large complement response. In order to inhibit excess activation of the complement pathway there are regulatory mechanisms to omit the uncontrolled complement activation.

There are three distinct pathways by which the complement can be activated on the surfaces of pathogens. Although these pathways are initiated differently, they are able to generate the same effector molecules. In all of the three ways that have been discovered to date, the complement system is involved in protecting against the infection. Firstly, the complement system generates numerous activated complement proteins that are able to bind covalently to pathogens in order to opsonize them for the phagocytosis by phagocytes displaying complement receptors on their cell surface. Secondly, small fragments of certain complement proteins can operate as chemo-attractants to accumulate phagocytes, such as macrophages, to the site of the complement activation by stimulating these phagocytes. Thirdly, the terminal complement factors try to kill certain bacteria by generating holes (pores) in the bacterial membrane.

In summary, the complement, mainly the CR1 that recognizes breakdown products of C3 by either the classical or an alternative pathway, mediates the inflammatory process. Within the blood plasma, there are early activation proteins that are the markers of an inflammation such as the C-reactive protein (CRP), serum amyloid protein (SAP), fibronectin, fibrinogen and lectins (Mannose binding lectin (MBL)) (Janeway and Medzhitov 2002). They all contribute to the progression of the inflammatory cascade in order to fight against the inflammation and protect the healthy tissue and organs from that inflammation. In addition, cytokines (15 to 30 kD in size) are produced by stimulated lymphocytes, macrophages, the endothelium, the epithelium and the surrounding connective tissue. Chemokines (8 to 10 kD in size) fulfill a role as chemoattractants for specific types of leukocytes (chemotaxis process). As a secondary step, extracellular fibrillar networks are secreted by neutrophil and eosinophil granulocytes as well as mast cells in response to certain infectious pathogens and inflammatory mediators such as chemokines and cytokines (Poon *et al* 2010). Granule proteins such as the peptides of microbes and enzymes are embedded in this extracellular matrix framework. This framework provides an extracellular matrix to catch and kill microbes and induce the contact system. There are chemical mediators of inflammation that seem to be produced locally by cells at the site of primary inflammation. These cell-derived mediators are usually seques-tered in intracellular granules and then they are rapidly secreted upon cellular activation by an inflammation or are synthesized *de novo* in response to a stimulus. The plasma proteins, such as complement proteins and kinins (measure blood pressure and cause dilation of blood vessels), circulate in an inactive form and are activated by an inflammation through proteolytic cleavage (Poon *et al* 2010).

2.1.2 Extravasation of inflammatory cells

Leukocyte extravasation is a step that belongs to the inflammatory cascade of acute and chronic inflammation. The extravasation represents a critical step during the inflammatory response as it includes the migration of leukocytes out of the bloodstream into the targeted, inflamed tissues. At these inflamed tissue sites, leukocytes fulfill their effector function. The step of leukocyte extravasation requires the combined action of cellular adhesion receptors and chemokines in order to evoke the drastic morphological alterations of leukocytes and the endothelial cell vessel lining (Carman and Springer 2004). This is an active process for both cell types and leads to the extremely fast and totally efficient 'invasion' of leukocytes into the sites of inflammation, whereas the barrier function of the endothelium is not severely affected from a molecular point of view (Muller 2003). The role that the mechanical properties of this 'activated' endothelial barrier play is not yet understood. Thus, they have to be investigated in future research studies. What we have so far is a precise description of the novel revealed steps in the adhesion cascade. Among these steps are slow rolling motion, intraluminal crawling of leukocytes and alternative means of transcellular migration. Furthermore, the functional role of novel adhesion receptors is presented, as well as the spatiotemporal organization of receptors at the cell surface of the membrane, together with the signaling pathways regulating the phases of the whole extravasation process.

In more detail, adhesion receptors are essential in maintaining the integrity of tissues through the regulation of many important processes, such as cellular activation, migration, growth, differentiation and death (Frenette and Wagner 1996a, 1996b). The regulation is facilitated by direct signal transduction and the alteration of intracellular signaling pathways triggered by several growth factors (Aplin et al 1998). It is known that cell–cell interactions are essential for the regulation of hematopoiesis (Levesque et al 1999, Verfaillie 1998) and the inflammatory processes (Butcher 1991, Butcher and Picker 1996). Thus, adhesion receptors are also implicated in a wide variety of diseases, such as cardiovascular disorders and cancer involving the process of inflammation.

As the functioning of adhesion receptors is important, their coordination between adhesion receptors, the cytoskeleton and the signaling molecules is precisely regulated during the extravasation of leukocytes (a key process in the whole immune response reaction). In particular, the correct integration of outside–in and inside–out signaling in leukocytes and in the endothelial cell monolayer lining blood vessels during the extravasation step (called multi-step paradigm) is necessary for the proper function of this phenomenon (Butcher 1991, Springer 1994). The extravasation of leukocytes is not only observed during an inflammatory response, it is also found when lymphocytes recirculate to secondary lymphoid organs, but this is not addressed within this book.

The initial interactions facilitated by selectins and their ligands are the tethering and rolling of circulating leukocytes on the endothelium

The initiation of the inflammatory response is guided by the circulating leukocytes in the blood fluid and their ability to provide a contact (tethering) to the vascular endothelial cell lining to subsequently adhere to the endothelium during the dramatic shear forces exerted by the blood flow. The first steps of the extravasation process are the tethering and rolling of the leukocytes over the activated endothelial cell lining of vessel walls. The first contact or tethering of the leukocytes is facilitated by the cell–cell interaction through selectins and their ligands, while the blood flow is still present (Alon and Ley 2008). However, it is known that selectins and their ligands interact at variable affinity to each other in order to be able to react to the high frequency of association and dissociation between this connection by facilitating labile and transient tethers between the flowing leukocytes and the static endothelium (Mehta *et al* 1998, Nicholson *et al* 1998). The step of tethering decreases the velocity the leukocytes in order to facilitate their rolling over the endothelium and provide more stable interactions through integrins and their ligands. In the end, the adherence of leukocytes toward the endothelium is enhanced and they become tightly attached to it (Evans and Calderwood 2007).

In more detail, the P-, S- and L-selectins belong to the type 1 transmembrane glycoproteins. These selectin receptors interact in a Ca^{2+}-dependent way with fucosylated and sialylated hydrocarbon presented on the cell surface of counteracting cells. L-selectin is mainly expressed by leukocytes and the E- and P-selectins are exposed on the cell surface of endothelial cells, when stimulated by proinflammatory stimuli such as tumor necrosis factor alpha (TNF-alpha). In addition, P-selectin is expressed by activated platelets (Barreiro *et al* 2004). In particular, the leukocyte selectin (L-selectin) can interact with the endothelial selectin (P- and E-selectin) ligand, the P-selectin glycoprotein ligand-1 (PSGL1) protein. However, the binding of PSGL1 to P- and E-selectin initiates the interaction of leukocytes with the endothelial lining of the blood vessel, but the binding of PSGL1 to L-selectin on leukocytes mediates leukocyte–leukocyte interactions, whereby the already adhered leukocytes capture other non-adherend circulating leukocytes at inflamed sites of the endothelium. In this step, it is not necessary that the non-adherend floating leukocytes express ligands for endothelial selectins (called secondary recruitment process) (Eriksson *et al* 2001). Selectins are additionally able to bind to other glycoproteins, such as CD44 or E-selectin ligand-1 (ESL1), which are ligands for E-selectin (Hidalgo *et al* 2007). It is suggested that each special ligand has a certain function during the capture process of neutrophils. Hence, PSGL1 is involved in the initial tethering of the leukocytes and ESL1 converts the first transient tethering behavior to a slower and more stable rolling step. In order to support secondary recruitment of leukocytes, CD44 regulates the velocity of the rolling step and interferes in the polarization step of PSGL1 and L-selectin (Hidalgo *et al* 2007). Indeed, platelets have the ability to behave as secondary recruiters for leukocytes, as they possess the capability to interact directly with the circulating leukocytes and the endothelial cell lining of blood vessels simultaneously. Moreover, they are even able to release chemokines that were formerly immobilized on the luminal endothelial cell surface, thereby strongly supporting the adhesion process of leukocytes (von Hundelshausen *et al* 2009).

After the weak contact between leukocytes and the endothelium through selectins and their ligands, the next step is the establishment of a strong adhesion via integrins and their ligands. For example, the α4β1- and α4β7-integrins bind to their ligands such as the vascular cell adhesion molecule 1 (VCAM-1) and the mucosal addressin cell adhesion molecule 1 (MAdCAM-1), respectively, which are both able to independently facilitate the initial tethering (Alon *et al* 1995, Berlin *et al* 1993, 1995). In line with this, the interaction between lymphocyte function-associated antigen 1 (LFA-1) and the intercellular cell adhesion molecule 1 (ICAM-1) supports the function of L-selectin, thereby converting the transient contact into a stable contact by reducing the velocity of the rolling step (Henderson *et al* 2001, Kadono *et al* 2002).

In general, the adhesion receptors need to be precisely localized on the surface of cells in order to fulfill their function in the correct manner during the trafficking of leukocytes (von Andrian *et al* 1995). The selectins as well as their ligands and the α4 integrins cluster at the tips of the microvilli of the leukocytes. For the proper tethering of the selectins, their connection to the actin cytoskeleton via connecting proteins such as α-actinin or ezrin/radixin/moesin (ERM) is necessary (Dwir *et al* 2001, Ivetic *et al* 2002, Pavalko *et al* 1995, Killock *et al* 2009).

It has been reported that selectins activate signal transduction pathways that are linked to processes such as actin cytoskeletal reorganization. These pathways are the MAPK, p56lck, Ras or Rac2 cascade (Barreiro *et al* 2004). On the other hand, PSGL1 activates different intracellular signal transduction pathways that cause an inductive effect on the activation of leukocytes, increasing the expression of molecules that are supposed to act in the following steps of the extravasation process and that possess an effector function. They fulfill a surprising role in the induction of tolerogenic function in dendritic cells (Urzainqui *et al* 2002, 2007, Zarbock *et al* 2007, 2008).

The prominent role of leukocyte's integrins and their ligands present on endothelial cells during activation, arrest, strong adhesion and cell crawling
In order to develop innate and acquired immunity, leukocyte trafficking through the different tissues and organs and subsequent interaction with other immune cells at the sites of the inflammation is necessary (von Andrian and Mackay 2000). Integrins have a fundamental function in cellular motility as they regulate the cell–cell and cell–extracellular matrix interactions during recirculation of leukocytes and inflammation. One feature is that integrins can alter their adherent activity independently of their expression levels and their exposure on the cell's surface (Hynes 2002). In particular, circulating leukocytes in the bloodstream display their integrins in an inactive conformation to omit non-specific contact with the uninflamed endothelial cell linings of blood vessels. When they enter the inflammatory focus, a rapid *in situ* activation of the integrins takes place (Campbell *et al* 1998). In this light, the spatial distribution of integrins and their ligands on certain membrane regions is necessary for their full function. In order to provide the spatial distribution, a sharp regulation of the cytoskeleton is necessary, facilitating the recruitment of signaling molecules and second messengers for the activation of the cell (Vicente-Manzanares and Sánchez-Madrid 2004, Barreiro *et al* 2007).

The integrin receptors consist of α (18 different ones) and β (eight different ones) subunits building 24 heterodimeric receptors (Humphries *et al* 2006). The large number of possible integrin receptors provides dynamical adhesive properties through conformational changes (affinity) or through alterations in the spatial distribution on the cell's surface (avidity) (Carmen and Springer 2003). There are at least three different conformational states of integrins hypothesized, such as bent conformation with low affinity ('rest state'), extended conformation (pre-stimulation by ligand) with intermediate affinity and extended conformation with high affinity (after mechanical stimulation by the ligand) (Beglova *et al* 2002, Nishida *et al* 2006, Schwartz 2010). Members of the β2 integrin subfamily, such as LFA-1 (CD11a/CD18 or αLβ2) and the myeloid-specific integrin Mac-1 (CD11b/CD18 or αMβ2), and integrins such as α4 VLA-4 (α4β1) and α4β7, play an essential role in the adhesion of leukocytes to the endothelial cell lining of blood vessels. A majority of their ligands are transmembrane receptors of the immunoglobulin superfamily. For example, LFA-1 can bind to five intercellular adhesion molecules, such as ICAM-1 to ICAM-5, but the most important ones are ICAM-1 and ICAM-3 (Gahmberg *et al* 1990). It is known that ICAM-1 is expressed on leukocytes, dendritic cells, epithelial cells and endothelial cells, on which surface the expression is low under normal conditions and increased after exposure to proinflammatory cytokines (Dustin *et al* 1986). However, ICAM-3 is constitutively expressed on all kinds of leukocytes (Acevedo *et al* 1993). LFA-1, can interact with another ligand such as the junctional adhesion molecule A (JAM-A), which is located on the apical region of the endothelial tight junctions and after stimulation with special proinflammatory factors it is partially redistributed to the apical surface of the endothelial cells (Ostermann *et al* 2002). Mac-1 binds to ICAM-1, JAM-C and the receptor for advanced glycation endproducts (RAGE) (Chavakis *et al* 2003, Lamagna *et al* 2005). The integrin α4β1 (VLA-4) interacts with VCAM-1 (Elices *et al* 1990), which is an adhesion receptor that is expressed *de novo* after the activation of the endothelium (Carlos and Harlan 1994) and additionally can bind to JAM-B (Cunningham *et al* 2002). In more detail, VLA-4 additionally binds to ADAM-28, thrombospondin, osteopontin, von Willebrand factor, fibronectin and the invasin bacterial protein (Mittelbrunn *et al* 2006). The αEβ7 integrin interacts with VCAM-1 and fibronectin, but is also able to bind to MAdCAM-1, which is a receptor detected in the lymphoid tissues of the mucosa (Berlin *et al* 1993).

What role do chemokines play in the regulation of the activity of integrins?
At the onset of the tethering on the vascular endothelial cell lining, the rolling velocity of the leukocytes slows, while they are activated by the endothelial cell's luminal surface expressing immobilized chemokines and integrin ligands (Ley *et al* 2007). This leads to the arrest of leukocytes and subsequently to strong adhesion to the endothelial cell lining of vessels under normal flow characteristics (Alon *et al* 2003, Rot and von Andrian 2004). During leukocyte activation the leukocyte undergoes a morphological transformation from a rounded, circulating cell to a promigratory polarized cell with a clear cell front and rear (called uropod) (del Pozo *et al* 1995). This cellular polarization is helpful for the coordination of intracellular forces to facilitate the crawling step during the extravasation cascade (Geiger and Bershadsky 2002).

The chemokines can interact with the glycosaminoglycans of the apical endothelial membrane of the endothelial cell lining of the blood vessel by signaling via the G protein coupled receptors (GPCR) located on the surface of the leukocyte's microvilli, thereby immediately inducing the outside–in signaling cascade. This cascade triggers multiple conformational alterations in the integrins (Constantin *et al* 2000, Sanchez-Madrid and del Pozo 1999, Shamri *et al* 2005). Chemokines induce complex signaling mechanisms within a short time period, which regulate the activation of integrins and support the hypothesis that there is a compartmentalized and pre-formed protein network (called signalosomes) inside the leukocytes (Laudanna and Alon 2006). This is further supported by the presence of specific chemokines in special vascular beds that mediate the selective recruitment of different leukocyte subpopulations to the side of the primary inflammation or to secondary lymphoid organs such as lymph nodes (Luster 1998). Moreover, chemokines have differing effects on specific integrins, even if these are located within the same local microenvironment (Laudanna 2005).

How is the ligand-facilitated affinity of integrins altered?
The conformation of integrins is altered reversibly after chemokine-induced activation, as the conformation converts from an inactive (bent) state with low affinity to an extended state with intermediate affinity. This process provides the necessary conformation of the integrin to bind its ligand expressed on the endothelium efficiently. In particular, integrins possess an I domain in their α subunits that changes its conformation after binding to its ligand, leading to the complete activation of the integrin and subsequently to the arrest of the leukocyte (Cabanas and Hogg 1993, Jun *et al* 2001, Salas *et al* 2004). Taken together, the high-affinity state of the integrin's conformation for immediate arrest of the leukocyte on the surface of the endothelial cell lining of blood vessels requires the presence of bound chemokines and the ligand of the integrin (Shamri *et al* 2005, Grabovsky *et al* 2000). Exceptions are α4 integrins, which contain an I-like domain within their β chains, but can directly interact with their ligands expressed on the endothelial cell's surface without a previous chemotactic trigger substance (Alon *et al* 1995).

The signaling mediated through the binding of the ligand induces the separation of the cytoplasmic regions of the subunits of the integrin, thus providing its association with the cell's cortical actin cytoskeleton. In particular, the α4 integrins are linked through the focal adhesion protein paxillin, whereas β2 integrins are linked through other focal adhesion molecules, such as talin and filamin (Takala *et al* 2008). Additional integrins are recruited when a ligand is bound to the first integrin in order to increase the tight adhesion of the leukocyte under blood flow stress (Dobereiner *et al* 2006). This process is called the clustering of integrins and is triggered by the release from their tether to the cytoskeleton (regulated by protein kinase C (PKC) and calpain) to enhance the lateral mobility of integrins on the cell membrane (Stewart *et al* 1998). In different types of hematopoietic cells, the role of Rap-1 and its activator CalDAG-GEFI has been reported and also the role of kindlin-3 interacting with talin in the activation process of integrins to provide the firm adhesion of leukocytes (Mory *et al* 2008, Pasvolsky *et al* 2007). The spatial organization of the integrins is critical for their proper function. For example, the nanoclusters of LFA-1, which are not even connected to the ligand, are able to form microclusters efficiently upon ligand binding (Cairo *et al* 2006, Cambi *et al* 2006).

In line with this, many studies show that the flow stress affects the integrins by reinforcing their bonds and even increasing their affinity (Marschel *et al* 2002, Zwartz *et al* 2004). The integration of chemokine-induced and external-force-induced signaling to facilitate transendothelial migration has been presented as a phenomenon called chemorheotaxis (Cinamon *et al* 2001).

How is the crawling of leukocytes altered by integrins?
The signals involved in the integrin-dependent firm adhesion of leukocytes to the endothelial cell lining of blood vessels need to be weakened to allow leukocytes to migrate to the site of the transendothelial migration start. In particular, the β2 integrins are supposed to be necessary for crawling, because the blockade of β2 integrins or their ligands evokes random migration, failure to find the sites for transendothelial migration at the sites of the interendothelial junctions and finally leads to defective diapedesis (Schenkel *et al* 2004). This finding is in line with *in vivo* studies investigating genetically modified mice lacking LFA-1 or Mac-1 to reveal the different underlying mechanisms for each of these β2 integrins. The firm adhesion is triggered by LFA-1, while the crawling is mediated by Mac-1. Together, both processes provide efficient cell migration (Phillipson *et al* 2006). After integrin activation through its ligand, the integrins regulate the signal transduction via different effectors to provide myosin contractility and modulate actin-remodeling GTPases as well as molecules regulating the microtubule network at both the front and the rear (uropod) of the cell. Taken together, both signals generated at the two cellular poles coordinate motility of leukocytes (Vicente-Manzanares and Sánchez-Madrid 2004).

What is the functional role of the endothelial expressed VCAM-1 and ICAM-1 during leukocyte capture?
VCAM-1 and ICAM-1 are members of the immunoglobulin superfamily and are expressed as endothelial adhesion molecules, facilitating binding to the integrins VLA-4 and LFA-1, respectively (Elices *et al* 1990, Marlin and Springer 1987). In more detail, ICAM-1 is slightly expressed on the non-activated endothelial cells, whereas the expression of VCAM-1 and ICAM-1 is enhanced after endothelial cell activation through proinflammatory cytokines such as interleukin-1 (IL-1) and TNF-α (Dustin *et al* 1986, Carlos and Harlan 1994). The binding of VCAM-1 and ICAM-1 to the actin cytoskeleton is mediated by ezrin and moesin linking the membrane to the actin cytoskeleton to provide cortical morphogenesis and adhesion (Barreiro *et al* 2002, Heiska *et al* 1998).

Human umbilical vein endothelial cells (HUVECs) have been used to investigate the dynamics of VCAM-1 and ICAM-1 expression upon activation with TNF during the interaction between leukocytes and the endothelial cell lining of vessels. In particular, after the arrest of the leukocyte at the endothelial cell surface, the interaction of VCAM-1 and ICAM-1 with their ligands leads to the reorganization of the endothelial cell's cortical actin cytoskeleton and subsequently to the formation of a 3D docking structure surrounding the leukocyte. This structure includes a large accumulation of adhesion receptors and activated ezrin and moesin and should protect adhered leukocytes under normal flow from being detached from the endothelial cell lining of blood vessels. This docking structure is supported by the actin cytoskeleton of the endothelial cell and contains actin-bundling proteins such as α-actinin and focal adhesion proteins such as talin, paxillin and vinculin, as well as actin nucleaters. To maintain the docking structure, second messengers such as PI(4,5)P2 and the Rho/160ROCK signal transduction pathway are required (Barreiro *et al* 2002). Even in the case where only one ligand is bound to either ICAM-1 or VCAM-1, both can cluster and form the endothelial docking structure. This kind of clustering occurs independently of the connection to the actin's cytoskeleton and of the assembly of ICAM-1/VCAM-1 heterodimers, because this behavior requires the inclusion of VCAM-1 and ICAM-1 in microdomains that contain many tetraspanins and hence they function as endothelial cell platforms for a special type of adhesion (Barreiro *et al* 2008). In more detail, the tetraspanins are relatively small proteins that are located in the plasma membrane in a special manner, as they cross the membrane four times. In addition, their second extracellular domain interacts laterally with other neighboring integral membrane proteins (Charrin *et al* 2009). Thus, they regulate the function of the membrane by building up multiple protein domains on the cell membrane. Moreover, tetraspanins have been reported to fulfill several cellular functions such as cellular migration, homotypic and heterotypic cell–cell adhesion and the presentation of antigens (Hemler 2005, Gordon-Alonso *et al* 2006, Mittelbrunn *et al* 2002, Yáñez-Mó *et al* 1998, García-López *et al* 2005).

As microscopic analytical methods have improved significantly, the characterization of the diffusive properties of the membrane, the membrane's organization at the nanoscale and certain molecular interactions within the microdomains of the membrane have been investigated in primary human endothelial cells (Barreiro *et al* 2008). Indeed, these precise analyses of the membrane have revealed evidence of certain endothelial adhesion platforms that are different from the so-called lipid raft structures in the cell membrane (Barreiro *et al* 2008). Using scanning electron microscopy to investigate the membranes of endothelial cells that have been treated with a specific peptide blocker of tetraspanins, it has been shown that the nanoclustering (avidity) of VCAM-1 and ICAM-1 facilitated by the endothelial adhesion platforms is a novel molecular organization mechanism regulating the efficient adhesive ability of endothelial adhesion receptors to bind to their integrins' ligands presented on leukocytes (Barreiro *et al* 2008). The functional importance of recruitment of ICAM-1 and VCAM-1 into tetraspanin microdomains on the membrane of primary human endothelial cells has been analyzed using interfering RNAs (RNAi) targeting the tetraspanins CD9 and CD151. Knockdown of CD9 and CD151 revealed that the inclusion of ICAM-1 and VCAM-1 in tetraspanin-containing domains is necessary for their proper function under a highly dynamic microenvironment such as the flow stress within the vessels (Barreiro *et al* 2005). In a second (independent) approach, the alteration of endothelial tetraspanin microdomains by CD9-large extracellular loop (LEL)-glutathione S-transferase (GST) peptides affected ICAM-1 and VCAM-1 functions (Barreiro *et al* 2005). In addition to VCAM-1 and ICAM-1, other adhesion receptors such as CD44, ICAM-2, JAM-A and PECAM-1 can interact with tetraspanin microdomains. Taken together, tetraspanin microdomains seem to be specialized platforms that basically organize certain adhesion receptors in the special membrane regions to provide leukocyte extravasation efficiently (Barreiro *et al* 2008).

Moreover, the endothelial VCAM-1 and ICAM-1 receptors can transmit signals after binding to their ligands. The VCAM-1 receptor is supposed to play a role in the opening of interendothelial junctions to provide the extravasation of leukocytes. Thereby, VCAM-1 activates NADPH oxidase (possibly through NOX2) and produces the reactive oxygen species (ROS) that is regulated by the GTPase Rac activity, with subsequent activation of matrix metalloproteinases and reduction of VE-cadherin dependent cell–cell adhesion, which is mediated by the Pyk-2-induced phosphorylation of β-catenin and supports the extravasation process (Cook-Mills 2002, van Wetering *et al* 2002, Deem and Cook-Mills 2004, van Buul *et al* 2005a, 2005b). By contrast, VCAM-1 and ICAM-1 can mediate a rapid increase in Ca2+ concentration that causes an activation of Src kinase and subsequently the phosphorylation of cortactin (Etienne-Manneville *et al* 2000, Lorenzon *et al* 1998, Yang *et al* 2006a, 2006b). Additionally, ICAM-1 may activate RhoA, causing the formation of F-actin stress fibers and leading to the phosphorylation of focal adhesion kinase (FAK), paxillin and p130Cas, which are in turn involved in signaling cascades involving c-Jun N-terminal kinase (JNK) and p38 (Greenwood *et al* 2002, Hubbard and Rothlein 2000, Thompson *et al* 2002 Wang and Doerschuk 2002, Mierke 2013 FAK). This signaling cascade leads to increased permeability of the endothelial cell lining of blood vessels and is followed by enhanced transmigration of leukocytes through the endothelial cell lining. In addition, the transcription of c-fos and rhoA have been promoted by ICAM-1 (Thompson *et al* 2002). Finally, ICAM-1 is able to mediate its own gene expression and also that of VCAM-1 in order to provide the transendothelial migration of leukocytes in a guided manner (Clayton *et al* 1998).

The role of integrins and their ligands during the transendothelial migration
Endothelial cell–cell junctions are pulled apart during the process of the transmigration of leukocytes to avoid injury to the endothelial monolayer or severe alterations in permeability. During the diapedesis step the leukocytes' membrane and the endothelial cells' are in close contact and thereafter the membranes of the two neighboring endothelial cells interact with each other, forming cell–cell adhesion without the occurrence of any damage due to the transmigration process (Muller 2003, Turowski *et al* 2008, Bamforth *et al* 1997, Carman 2009, Carman and Springer 2004, Yang *et al* 2005).

When leukocytes are at a location on the endothelial membrane site at which the transmigration is likely to occur, such as the preferred intercellular endothelial junctions, they generate pseudopods between the two neighboring endothelial cells, possibly to explore the mechanical properties of the endothelial cells. Then the pseudopods are reformed into a lamella that is able to migrate into the open space within the two adjacent endothelial cells on the monolayer lining of the blood vessels. During this diapedesis step, the LFA-1 molecule plays a major role as it is able to reassemble with high speed to build up a ring-shaped cluster at the contact site between the leukocyte and endothelium, where it still binds to ICAM-1 (Shaw *et al* 2004) or in other cellular models to JAM-A (Woodfin *et al* 2007). After finishing the transmigration process, LFA-1 is still located in the uropod (Sandig *et al* 1997). However, other proteins, such as ICAM-2, JAM-B, JAM-C, PECAM-1 (CD31), ESAM and CD99, are involved in the transendothelial migration process. In particular, these proteins help to provide endothelial cell–cell interaction or they facilitate the leukocyte–endothelial cell interaction by homophilical and heterophilical interactions (Vestweber 2007, Weber *et al* 2007, Woodfin *et al* 2007, Muller 2003).

In addition to the classical diapedsis path through the endothelial cell–cell junctions (paracellular track), there is an alternative route that leukocytes use to walk through the entire individual endothelial cells (transcellular track) without altering the endothelial cell–cell junctions. The latter track may possibly involve use of the endothelial cells' actomyosin cytoskeleton in order to migrate through the cytoplasm of the endothelial cell. However, the question of which track is used at different times is still under discussion. It has been suggested that the mechanical properties of the endothelial cell play a role. In addition, mechanical properties such as protrusive force generation may be of importance for the invasive transmigration behaviors of certain leukocytes. This transcellular track has been seen in the microvasculature, the blood–brain and blood–retinal barriers, and in the high endothelial venules of the secondary lymphoid organs (Engelhardt and Wohburg 2004, Carman and Springer 2004, Carman *et al* 2007). As the mechanism of this transcellular path during transendothelial migration process has been investigated in more detail, the leukocytes can build up Src kinase-dependent and Wiskott–Aldrich syndrome protein (WASP) activity-dependent invasive podosomes to sense the endothelial surface. Finally, these podosomes form a transcellular pore through which the whole cell transmigrates intracellularly through the endothelial cell. In the endothelial monolayer membrane, the fusion of membranes is regulated by calcium and SNARE-containing complexes and new parts of the membrane are delivered by vesicular vacuole organelles (Carman *et al* 2007). In more detail, this refers to the translocation of ICAM-1 into caveolae after the adhesion of the leukocyte to the endothelial cell lining and the assembly of a multivesicular channel, in which ICAM-1 and caveolin-1 are located around the leukocyte's pseudopod and then the leukocyte moves directly through the entire endothelial cell. In particular, ICAM-1 and caveolin line the entire path of the leukocyte through the endothelial cell's interior and move in the direction of the basal endothelial cell's membrane (Millan *et al* 2006). In this special transmigration path, the intermediate filament vimentin is supposed to be important for the transcellular route (Nieminen *et al* 2006). The dome-shaped endothelial structures have been found to facilitate leukocyte's transendothelial migration *in vivo* (Phillipson *et al* 2008). Taken together, the endothelial docking structures are observed to rebuild to dome-like structures that cover the leukocytes on their luminal surface of the endothelial cell lining, through which the basolateral endothelial membrane is ruptured, although this is not leaky and can still provide the endothelial barrier function.

2.2 The dual relationship between inflammation and cancer

Inflammation of tissues is an innate immune response to alterations of tissue homeostasis. Inflammatory processes such as the acute and chronic have the potential to crosstalk with a tumor at all stages of its development and may have a major impact on cancer therapy. As less than 10% (approximately) of all cancer types are evoked by mutations in the germline, the cause of over 90% of cancer may be somatic mutations and alterations in the microenvironment surrounding tumors. The development of tumors is related to many factors, such as chronic infections, dietary influences, obesity, inhaled pollutants (including tobacco and asbestos), high ozone levels and autoimmune diseases, such as rheumatoid arthritis and asthma (Jemal *et al* 2010). The underlying principle of all these processes is that they are caused by a chronic inflammation, which is an indefinite prolonged form of a formerly protective response (acute immune response) to a disturbance of tissue homeostasis (Medzhitov 2008). In line with this, the development of cancer can be regarded in numerous cases as a deregulated formerly protective tissue repair and growth response due to tissue injury. Thus, inflammatory and neoplastic processes can co-develop into so-called wounds that are no longer able to heal (Dvorak 1986). The initial connection between inflammation and tumor growth was observed by Rudolf Virchow in the nineteenth century (Virchow 1881). In his research, Virchow found that leukocytes infiltrate into the middle of primary tumors, which can be seen as the first evidence for a hallmark of cancer described later (Hanahan and Weinberg 2011; see also chapter 1 of this book). These leukocyte infiltrates are regarded as markers of a tumor's immune surveillance and anti-tumor immune responses, whereas, in fact, they support tumor-suppressive and tumor-promoting effects; the mechanisms involved in these divergent effects are not yet well defined.

There are many causes of inflammation-promoted cancer progression. However, infection is a major player in providing inflammation-mediated tumorigenesis, as approximately 20% of all cancer cases worldwide are caused by a microbial infection (Kuper *et al* 2000). It is not currently known at the molecular level what causes inflammation during an infection, but the signals may be different from those from an inflammation caused by no-infection, such as noxious inhalants, obesity and auto-immunity. Although there are basic biological concepts that drive inflammation-associated cancer in various cases, there is a hypothesis that variations in the microbial ecosystems are mirrored in the variability among different inflammatory diseases and among host individuals during the manifestation of inflammation-associated cancer. In particular, common microbial elements have been characterized that are related to inflammation and the development of cancer (Grivennikov *et al* 2012, Goodwin *et al* 2011, Abdulahmir *et al* 2011, Arthur *et al* 2012). Indeed, microbial organisms have been identified as mediating the interaction between inflammation and cancer. In the following the relationship between inflammation and cancer is discussed in more detail.

2.2.1 Inflammation can cause cancer (pro-tumorigenic)

Outside–in: inflammation induces cancer promotion
It is common knowledge that inflammatory responses play decisive roles at the different tumor development stages, including tumor initiation, tumor growth, malignant tumor progression, cellular invasion and the metastasis of the cancer. Within these processes of cancer development and severe progression, inflammation

regulates the immune surveillance and responses to tumor therapy. In more detail, immune cells within primary tumors facilitate extensive and dynamic crosstalk with the broad mass of cancer cells and even molecular events connecting the crosstalk between immune cells and cancer cells have been found. In the following, this crosstalk will be discussed, as well as the principal mechanisms that connect inflammation and immunity with the development of tumors.

The process of carcinogenesis involves a controlled interplay between numerous intrinsic and extrinsic cellular signaling pathways, such as genomic instability, altered proliferation, altered stromal environment and abnormal differentiation of epithelial and mesenchymal states. A feature of inflammation is its ability to support all cellular and molecular mechanisms that promote tumorigenesis (Hanahan and Weinberg 2011). The mechanisms that regulate the neoplastic transformation of cells stimulated by inflammatory cells remain elusive. However, the principles behind inflammation-induced cancer will be discussed below.

Induction of proliferation and cell survival mechanisms
Excessive cancer growth leads to the loss of homeostasis in the tissue's architecture. Within a tissue the architecture is precisely controlled by soluble factors such as growth factors and their receptors, facilitating proliferation, homeostasis and cell survival. Cancer cells are able to utilize growth factor signaling using the following common strategies: they alter the process of endocytosis and the recycling of receptors, abolish negative-feedback mechanisms, reduce growth rates and empower redundant proliferative pathways that act downstream of the receptor tyrosine kinase signaling (Amit *et al* 2007, Mosesson *et al* 2008, Wilson *et al* 2012, Straussman *et al* 2012, Prahallad *et al* 2012). These growth factor receptor tyrosine kinases are known to have a wide range of incoming signals that finally lead to a few prominent signaling cascades inside the cell, such as proliferation, metabolism, survival, differentiation, cellular adhesion and motility (Casaletto and McClatchey *et al* 2012). In particular, these pathways cover the activation of signal transducers and the activators of transcription (STAT) family members such as STAT3, which has been suggested to play a role in tumorigenesis, while it has also been linked to inflammatory processes in gastric, lung, colon, pancreatic and liver cancers (Fukuda *et al* 2011, Lesina *et al* 2011, Bollrath *et al* 2009, Grivennikov *et al* 2009, Bronte-Tinkew *et al* 2009, Gao *et al* 2007). Myeloid cells produce interleukin-6 (IL-6), which is able to activate STAT3. STAT3 expression is a crucial step during the onset tumorigenesis of colitis-associated colorectal cancer (CAC), as it enhances the proliferation of pre-malignant cells and abolishes apoptosis *in vivo* (Bollrath *et al* 2009, Grivennikov *et al* 2009, Waldner *et al* 2012). From a mechanistic point of view, STAT3 induces cell proliferation through the up-regulation of the cell cycle regulators such as cyclin D1, cyclin D2 and cyclin B, as well as the proto-oncogene MYC. Additionally, STAT3 increases cellular survival, as it increases the expression of the anti-apoptotic genes BCL2 and BCL2-like 1 (BCL2L1 encoding BCL-XL) (Bollrath *et al* 2009, Grivennikov *et al* 2009, Yu *et al* 2009).

Another regulatory function of STAT3 has been discovered. The deletion of the sphingosine kinase 2 (Sphk2) in mice has been reported as being an important part of the development of CAC, in which nuclear factor-κB (NF-κB)–IL-6–STAT3 signaling is activated by the production of sphingosine-1-phosphate (S1P). This then leads to an up-regulation of the S1P receptor 1 (S1PR1) and represents an additional amplification step for the support of the chronic inflammation and subsequently of CAC (Liang *et al* 2013). Indeed, excessive STAT3 activation during early neoplasia is supported by immune cell signaling in the local surroundings of a primary tumor. In particular, Kras-driven pancreatic tumor models propose that STAT3 signaling plays an intrinsic role in pancreatic intraepithelial neoplasia development, even in the absence of an inflammation (Fukuda *et al* 2011). This indicates that the inflammation has a supportive effect, but is not essential for tumor development. There are several cases in which epithelial STAT3 signals via the transduction of myeloid cell-derived IL-6 protumorigenic signals, for example, during the transformation of pancreatic intraepithelial neoplasia to pancreatic ductal adenocarcinoma, while it plays no role during its initiation (Lesina *et al* 2011). In line with this, a low grade of inflammation mediated by obesity in the diethylnitrosamine-induced hepatocellular carcinoma (HCC) model activation of tumorigenic STAT3 signaling via IL-6 and through TNFα (Park *et al* 2010). A possible regulatory mechanism during tumor initiation and progression has been reported for immortalized mammary cells that show a transient activation of the oncoprotein Src. This activation of Src is triggered by an epigenetic alteration that leads to a stably transformed state that is facilitated by activated STAT3 (Iliopoulos *et al* 2009). Indeed, Src activation leads to an NF-κB-dependent inflammatory response that is supported by increased IL-6 up-regulation and decreased expression of IL-6's negative regulator, microRNA let-7. Taken together, elevated levels of IL-6 lead to the activation of STAT3 transcription, which is essential for cell trans-formation as well as activation of NF-κB. This behavior can be described as a positive feedback loop that is even present in the absence of the inducer (Iliopoulos *et al* 2009). In summary, an inflammatory process seems to be a supporter of oncogenic activation by increasing the NF-κB–IL-6–STAT3 signaling cascade and subsequently the proliferation of cancer cells.

What role does NF-κB signaling play in tumor development?
NF-κB activation has been reported in most tumors evoked by either inflammatory stimuli or oncogenic mutations leading to gene expressions that mediate inflamma-tion, proliferation and survival of the cells (Grivennikov *et al* 2010, Ben-Neriah and Karin 2011). The first molecular links between inflammation and cancer were discovered in the liver (Pikarsky *et al* 2004) and the colon (Greten *et al* 2004). Additionally this connection emphasizes the importance of the tumor microenviron-ment as a potential source of pro-tumorigenic and inflammatory cytokines, such as NF-κB activators and target genes.

In a mouse model of inflammation-associated HCC (Abcb4–/– mice, Abcb4 encodes the multidrug resistance protein 2 (MDR2)), the secretion of TNFα by surrounding tissue endothelial cells and inflammatory cells induces NF-κB expression in hepatocytes, which is needed for the progression of HCC, whereas the transformation of hepatocytes is not affected (Pikarsky *et al* 2004). In another model, such as the CAC, the inactivation of NF-κB signaling by an epithelial cell-specific inhibitor of NF-κB kinase-β (IKKβ) significantly decreases the occurrence of CAC, whereas the deletion of Ikbkb encoding IKKβ in myeloid cells reduces the primary tumor size (Greten *et al* 2004). In line with the first model, it has been demonstrated that TNFα secretion by leukocytes invading the lamina propria and other submucosal sites of the colon initiates CAC and its progression (Popivanova *et al* 2008). As described above, NF-κB activation induces a tumorigenic autocrine signaling in the transformed cells. In particular, this autocrine loop is necessary for the oncogenic effects of Ras in keratinocytes through stimulation by IL-1α, IL-1R and myeloid differentiation primary response 88 (MYD88), which promotes the activation of NF-κB (Cataisson *et al* 2012). In addition, MYD88 can modulate the colonic macrophage production of cytokines such as IL-6 and transforming growth factor-β (TGFβ), necessary for CAC development (Schiechl *et al* 2011). However, this seems to be an option for a linkage between intrinsic NF-κB and extrinsic tumorigenic factors during an inflammatory process. Moreover, this hypothesis is supported by the role of Toll-like receptor 2 (TLR2) in gastric tumorigenesis through STAT3 activation (Tye *et al* 2012).

As myeloid cell-derived inflammatory cytokines provide proliferative responses during the onset of neoplasia, tissue conditions supporting the release of inflammatory cytokine are of special importance. A response to the damage of tissue is the pronounced production of multiple inflammatory cytokines such as IL-22, which is crucial for the proliferation and survival of epithelial cells (Zenewicz *et al* 2008, Sonnenberg *et al* 2011). Secreted IL-22 by CD11c+ cells facilitates wound-healing responses through the activation of STAT3 in epithelial cells of the intestine after tissue damage and colitis triggered by the dextran sodium sulfate (DSS) treatment (Pickert *et al* 2009). It has been shown that IL-22 leads to CAC development in mice lacking the IL-22 neutralizing receptor IL-22-binding protein (IL-22BP; also called IL22Rα2), indicating that IL-22BP, produced by CD11c+ cells downstream of inflammasome signaling, impedes CAC (Huber *et al* 2012). IL-22 secreted by HCC-infiltrating leukocytes leads to HCC through the activation of STAT3 and subsequently downstream survival genes (Jiang *et al* 2011). Taken together, IL-22 seems to fulfill a direct tumor-promoting function, as indicated by the hepatocyte-specific expression of IL-22 in transgenic mice resulting in HCC survival and proliferation without substantially changing the overall liver inflammation (Park *et al* 2011).

One way to adopt the capabilities of cell proliferation and survival is through inflammatory cytokine signaling and a second way is by dedifferentiation of pre-malignant cells through the achievement of stem-cell-like properties. Therefore a model system of constitutive WNT activation has been used, in which increased NF-κB signaling in epithelial cells increases the activation of WNT–β-catenin and hence triggers the dedifferentiation of intestinal non-stem cells, which then possess tumor-initiating capacity, leading to tumorigenesis in the intestine (Schwitalla *et al* 2013). A third way is the loss of the tumor suppressor adenomatous polyposis coli (APC), which can activate the GTPase Rac1 and lead to hyperproliferation of leucine-rich repeat-containing G protein-coupled receptor 5 (LGR5)-expressing intestinal stem cells, evoked by the Rac1 regulated production of the reactive oxygen species (ROS) and the activation of NF-κB (Myant *et al* 2013). STAT3 has been reported to facilitate cancer stem cells' tumor-promoting properties in certain tissues (Marotta *et al* 2011, Ho *et al* 2012, Zhou *et al* 2007, Scheitz *et al* 2012). Additionally, it has been demonstrated that IL-1β-dependent and IL-6-dependent inflammation may trigger the activation and migration of gastric cardia progenitor cells, leading to the malignant transformation of Barrett's oesophagus and oesophageal adenocarcinoma in a mouse model (Quante *et al* 2012). Overall, inflammation is able to drive preliminary neoplasia toward tumor progression by controlling the expansion of stem cells within the tissue and through acquisition of stem-cell-like properties.

What do we know about the role of inflammation in genomic destabilization?
Besides increased proliferation and the ability to escape apoptosis, cancer cells are characterized by enhanced genomic instability followed by genomic mutagenesis. Inflammation may be a major cause of genomic instability through interference with genomic integrity. Moreover, somatic mutation caused during the inflammatory response may then function with a cell-intrinsic or cell-extrinsic behavior. Firstly, ROS and reactive nitrogen species (RNS) secreted upon inflammatory cytokine stimulation by tissue macrophages and neutrophils, or by pre-malignant cells leading to DNA breaks, point mutations and even more complex DNA lesions (Mantovani *et al* 2008, Campregher *et al* 2008, Mills *et al* 2003). Secondly, several cytokines such as TNFα, IL-1β, IL-4, IL-13 and TGFβ can induce the ectopic expression of activation-induced cytidine deaminase (AID, also called AICDA), which belongs to the DNA and RNA cytosine deaminase family and is able to mutate special cancer-associated genes such as TP53 (encoding p53) and MYC in order to support oncogenesis (Takai *et al* 2009, Okazaki *et al* 2007, Endo *et al* 2008, Komori *et al* 2008). Inflammation can enhance the mutagenesis and indirectly elevate genomic destabilization through the downregulation of DNA repair mechanisms and out-of-order checkpoints of the cell cycle, increasing the random genetic alterations (Campregher *et al* 2008, Colotta *et al* 2009, Schetter *et al* 2010, Singh *et al* 2007). Therefore, mismatch repair (MMR) proteins are useful to prevent genetic instability caused by microsatellite instability and they are targeted during an inflammation, as they are repressed via several mechanisms to elevate the DNA replication errors in the genome. In particular, a multiple inflammatory substance, such as TNFα, IL-1β, prostaglandin E2 (PGE2) and ROS, triggers the hypoxia-inducible factor 1α (HIF1α) to decrease the expression of the MMR genes MSH2 and MSH6 (Colotta *et al* 2009), whereby ROS is able to additionally decrease the MMR proteins' enzymatic activity (Schetter *et al* 2010). Inflammation can also mediate epigenetic silencing of the MMR protein MLH1 and other tumor suppressors by hypermethylation (Schetter *et al* 2010, Hahn *et al* 2008). Moreover, inflammation can inactivate the genomic surveillance evoked by p53 and the overexpression of BCL-2 and MYC, leading to decreased DNA repair and hence more mutations in cancer cells (Mantovani *et al* 2008, Schetter *et al* 2010).

Taken together, it is likely that inflammation can promote genetic instability in several ways, for example, by the ectopic expression of IL-15 inducing large granular lymphocytic leukaemia via MYC and NF-κB and reducing the expression of miR-29b, which can repress the DNA methyltransferase 3B (DNMT3B), leading to DNA hypermethylation and chromosomal instability (Mishra *et al* 2012).

How is the induction of invasion and metastasis regulated by an inflammation?
As a loss of tissue homeostasis has not been associated with high mortality rates, around 90% of cancer mortality is due to cancer metastasis. A major support of metastasis is that myeloid-derived monocytes and macrophages facilitate cancer cell invasion, extravasation and finally metastatic outgrowth (Peinado *et al* 2011, Qian and Pollard 2010), although the underlying mechanisms remain unclear. In particular, IL-4-activated tumor-associated macrophages (TAMs) (Gocheva *et al* 2010, DeNardo *et al* 2009) and CC-chemokine receptor 1 (CCR1)+ immature myeloid cells (Kitamura *et al* 2007) at the tumor periphery play a major role in providing local invasion of cancer cells, as they contain matrix-degrading enzymes such as MMPs, cathepsins and heparanase, which a cancer cell can utilize in order to disseminate and invade the surrounding microenvironment (Talmadge and Fidler 2010, Qian and Pollard 2010, Kalluri and Weinberg 2009, Murdoch *et al* 2008, Lerner *et al* 2011). In addition, the proinflammatory cytokine TNFα increases the vascular permeability, thereby enhancing the cancer cell transendothelial migration events (Grivennikov *et al* 2006), while the cyclooxygenase 2 (COX2)-dependent increase in prostaglandin levels and the increased MMP levels facilitate tissue remodeling (Nguyen *et al* 2009). Taken together, the tissue microenvironment of tumors is a reservoir of multiple inflammatory substances able to induce cellular transformation and tumorigenesis.

What role does acute inflammation play in mediating cancer progression?
As described above, long-lasting and chronic inflammation may lead to cancer, hence, the focus is now on the role of acute inflammation in cancer progression. For example, virally encoded genes such as the oncogenes E6 and E7 of the human papilloma virus (HPV) are able to contribute to cellular transformation (Munger and Howley 2002). However, there are also numerous microbes, such as *Helicobacter pylori*, that are not able to transform cells into malignant cancer cells (Peek and Blaset 2002).

Inflammatory states caused by infection and irritation may alter their micro-environments, promoting genomic lesions and the initiation of a tumor. An effector mechanism is the production of free radicals, such as reactive oxygen intermediated (ROI), hydroxyl radical (OH•), superoxide (O2-•), reactive nitrogen intermediates (RNI), nitric oxide (NO•) and peroxynitrite (ONOO-), through the release of which the host fights against the microbial infection. Although these free radicals are anti-microbial, ROI and RNI also cause oxidative damage and nitration of DNA bases, enhancing the occurrence of DNA mutations (Hussain *et al* 2003). However, cells have mechanisms such as DNA repair, cell cycle arrest, apoptosis and senescence to control unregulated proliferation and thus the accumulation of mutational events in the DNA. If there is little DNA damage or oncogenic activation, the cells will first try to repair the DNA in order to prevent mutations or they will induce programmed death within the mutated cell (apoptosis). In the case of excessive cell death as it occurs in infection or non-infectious tissue after injury, the lost cells need to be replaced by the expansion of other neighboring cells that are normally undifferentiated precursor cells such as tissue stem cells. To fulfill this, there are at least two prerequisites. Firstly, at least some cells must survive during the tissue injury and secondly, they must be expanded to maintain the cell numbers required for a properly functioning tissue. Multiple inflammatory pathways do indeed fulfill these two essential prerequisites for tissue repair (Chen *et al* 2003). However, beyond the physiological role of tissue repair—playing a prominent role in host defense against infection—the inflammatory response clearly provides a strong survival and proliferative signal to initiated cells and thus may as a side-effect facilitate tumor promotion.

Outside regulation of the inside, possibly by the microbiota
What instigates the inflammatory response? What kind of pathogen do we have? What is the cancer-causing stimulus during a microbial infection? In nearly one-fifth of all cancer cases there seems to be direct linkage between cancer and an infectious disease. Our knowledge of oncoviruses triggering and causing cancers is well established, however, the cancer-supporting action of microorganisms such as bacteria and parasites is less well understood, but it has accounted for enhanced tumor development. The role of fungi in possibly causing cancer is not yet clear. In several types of infections, it seems to be the host's inflammatory response to fight against the pathogenic infection that provides the connection between the infection caused by the pathogen and the development of cancer. In some cases, the inflammatory response even leads to the development and progression of cancer that is not directly attributable to an infectious disease; it is caused instead by the tissue damage from the chronic and uncontrolled inflammation leading to the malignant transformation of cells.

There is most evidence for bacterial involvement in inflammation-induced cancer (for example, gastric adenocarcinoma) in respect of *H. pylori*. Specifically, the chronic gastric inflammation facilitated by *H. pylori* evokes aberrant β-catenin signaling in epithelial cells supporting malignant transformation (Franko *et al* 2005). In addition to this and other pathogens, it has been suggested that members of the commensal group of microorganisms are involved in tumor-inducing and tumor-progressive inflammation. In fact, the human body is a host to trillions of viruses, bacteria, fungi and parasites on the skin and in the oral cavity, respiratory tract, urogenital system and gastrointestinal tract (Huttenhower *et al* 2012). However, these commensal microorganisms are necessary for the human physiology under homeostatic conditions and thus they also contribute to abnormalities within inflammatory processes (Gordon 2012). Do these commensal microorganism-dependent inflammatory processes support tumor growth in a similar way to the inflammation caused by a pathogenic infection?

One study reported on commensal microorganism-induced inflammatory tumorigenesis by showing the contribution of the TLR-signaling adaptor protein MYD88 to cancer progression in a ApcMin/+ mouse model for spontaneous intestinal tumorigenesis and in a mouse model in which the animals were treated with repeated injections of azoxymethane (AOM) (Rakoff-Nahoum and Medzhitov 2007).

These results indicate that the innate sensing of microorganisms in the intestine is essential for the appropriate regulation of inflammation and cancer development. In more detail, Erk activates as a downstream target gene MYD88, that leads to the activation of ERK, which then drives intestinal tumorigenesis in ApcMin/+ mice (Lee *et al* 2010). However, a role for MYD88 in intestinal tumorigenesis was reported in colitis-susceptible Il10−/− mice (Uronis *et al* 2009), whereas in another DSS-AOM model of inflammation-induced colorectal cancer, even Myd88−/− mice displayed increased development of intestinal tumors, indicating that MYD88 is not essential for developing an intestinal tumor (Salcedo *et al* 2010). In contrast to the Myd88−/− mouse model, in the Tlr4−/− mouse model the animals showed no DSS-AOM-induced intestinal tumor development (Fukata *et al* 2007), while in another result Tlr2−/− mice showed enhanced intestinal tumorigenesis upon DSS-AOM-induction (Lowe *et al* 2010). How should the contrasting results be dealt with? These contradictory findings can be explained by the differences in the commensial microbiota composition in the innate immune-deficient mice. In particular, these differences in the content of the microbiota lead to different results in promoting or inhibiting the tumorigenic cascade.

As there are contradictory roles for TLR signaling, this may suggest that there is a second alternative pathway of innate immune signaling, which is also dependent on MYD88, but signals through the receptors for IL-1 and IL-18. As suggested, the Il18−/− mice possess a similar susceptibility to the development of intestinal polyps compared with Myd88−/− mice (Salcedo *et al* 2010). Indeed, IL-18 is secreted by cells upon the activation of inflammasomes, which can be seen as an innate immune platform that consists of an upstream NOD-like receptor (NLR), the adaptor protein apoptosis-associated speck-like protein containing a CARD (ASC; also called PYCARD) and caspase 1 (Elinav *et al* 2013). Due to the lack of 'cancer-protecting' IL-18 in mice, including mice deficient in caspase 1 or in the inflammasome-forming NLRs, NLRP3 and NLRP6, these mice show a strongly enhanced susceptibility to colorectal carcinoma (Hu *et al* 2010, Zaki *et al* 2010, Chen *et al* 2011, Normand *et al* 2011). Taken together, these mice possess an altered microbial community with an altered function in the gut system (called dysbiosis), which exerts pro-inflammatory properties on the gut mucosa (Elinav *et al* 2011, Henao-Mejia *et al* 2012). Indeed, the microflora connected with inflammasome deficiency in mice has been reported to be involved in inflammation-mediated colorectal cancer induction through the activation of the inflammatory stimuli-facilitated, epithelial IL-6 signaling. Moreover, this activation of epithelial IL-6 signaling has also been seen in cohabitant wild-type mice (Hu *et al* 2013). These results indicate that alterations in the microbiota composition are a major factor for providing intestinal tumorigenesis upon inflammatory stimulation.

However, as suggested, intestinal dysbiosis has been connected to colorectal cancer. Firstly, for example, in a mouse model, in the Tbet−/− (also called Tbx21−/−) recombination activating gene 2 (Rag2)−/− mouse model for ulcerative colitis (TRUC) (Garrett *et al* 2009); and secondly, subsequently in humans (Sobhani *et al* 2011, Marchesi *et al* 2011). In addition, there is inflammation-mediated colorectal cancer that is connected to the intestinal dysbiosis and is transmissible between mice, as the infectious part of the colorectal cancer that cannot be associated with a pathogenic species. What causes the cancer? It is the aberrant outgrowth and activity of the normally symbiotic (but in this special case 'dysbiosis') species that acts as a pathogenic species ('pathobionts') that is able to elevate the inflammation and induce the neoplastic outgrowth of the tissue cells.

The development of dysbiosis is supposed to be the driving force of the inflammation-associated tumor growth and hence provide a clear example of micro-organism-facilitated inflammation causing a primary tumor, while in turn there are also microorganism-facilitated tumor-driven inflammatory responses. In line with this, the dysbiosis and the enhanced microbial distribution across the intestinal barrier leads to secondary and subsequently to chronic inflammation or even to neoplasia, and hence supports the development of cancer. In more detail, colorectal cancers are characterized by defective mucin production and show altered expression of the proteins regulating the maintenance of the intestinal barrier. The result of the dysregulation is increased microbial translocation to the targeted tissue location of the tumor growth and the up-regulation of IL-23 and IL-17, which supports ongoing local tumor-driven inflammation (Grivennikov *et al* 2012). One example is the model of AOM-treated colitis-susceptible Il10−/− mice, in which the intestinal microbial composition is changed compared to healthy mice, and in particular commensal *Escherichia coli* is able to facilitate an invasive carcinoma that is promoted by the polyketide synthase (pks) genotoxic island. In line with these results, patients with inflammatory bowel disease (IBD) and colorectal cancer show a higher incidence of pks+ *E. coli* compared to healthy controls (Arthur *et al* 2012), which leads to the suggestion that only a single irregularly represented bacterial organism can finally promote increased neoplastic growth. In addition, at least three studies pointed out the importance of microoranisms in supporting cancer progression. In particular, the anaerobe *Fusobacterium nucelatum* has been found in human colorectal carcinoma samples (Castellarin *et al* 2012, Kostic *et al* 2012, Strauss *et al* 2011). It has been reported that *F. nucleatum* increases colorectal neoplasia progression by supporting the recruitment of myeloid cells that are able to infiltrate the primary tumor (Kostic *et al* 2013). On the molecular level, *F. nucleatum* binds to E-cadherin, which is expressed on the cell surface of epithelial cells, and hence activates β-catenin signaling in order to facilitate cell proliferation (Rubinstein *et al* 2013). The enterotoxigenic *Bacteroides fragilis* (ETBF) has been shown to support inflammation-driven colon cancer (Wu *et al* 2009) by inducing spermine oxidase-dependent ROS production (Goodwin *et al* 2011). In summary, these findings propose a major role for bacterial microbiota during inflammatory processes in the induction and progression of cancer development. What roles do viral, fungal and parasitic members of the microbiota play in inflammation-induced cancer? This is not yet fully understood.

The liver is the 'first pass' organ for microbial products from the gastrointestinal tract through the portal vein and enhanced translocation of bacterial products to the liver represent a hallmark of chronic liver disease (Henao-Mejia *et al* 2012). Thus, the intestinal microbiota lead to the development of HCC through TLR4-mediated signaling of non-haematopoietic cells in the liver (Dapito *et al* 2012). In particular, the TLR4-induced hepatomitogen epiregulin supports carcinogenesis in the liver. In addition, obesity can facilitate changes in the gut microbial composition, which increases deoxycholic acid (DCA) and leads to DNA damage. Increased DCA in the enterohepatic circulation enhances cytokine production by hepatic stellate cells and hence supports hepatic carcinogenesis (Yoshimoto *et al* 2013). Taken together, these results demonstrate that the inflammatory-response-driven cancer of commensal microbial elements is not locally restricted to the direct interaction sites of the primary tumor and the microorganisms.

What is the future direction of inflammatory-driven cancer study?
The idea of inflammation-caused cancer has been fully established, whereas the linking mechanisms between inflammatory processes and the various stages of tumor development are still not well understood and require further research. The molecular details coordinating the interaction between cancer cells and the inflammatory response are known, but fundamental questions have still not been answered. Why is it that not every chronic inflammation increases the risk of cancer? Does the side of the inflammation play a role in determining whether or not the inflammatory response is able to induce cancer development?

There is indeed a difference between the organs in which the inflammation occurs and whether this side (for example, the gastrointestinal tract, liver and lung) induces or not changes the risk of cancer in, for example, the joints in rheumatoid arthritis and in the brain. Does this depend on the excessive interaction between the immune cells under inflammatory stimulation and cancer cells? What is the mechanism that commensal microbiota use for inflammation-induced cancer development? Does microorganism-driven inflammation regulate only early neoplasia and tumor progression, or can it also affect cancer metastasis? What is the impact of the mechanical properties of inflamed tissue? Do inflammation-altered mechanical properties lead to increased tumor development?

2.2.2 Inflammation can inhibit cancer (anti-tumorigenic)

Besides its cancer-promoting function, inflammation can also play the opposite role and act as an anti-tumorigenic factor. This second, but no less important role of inflammation in controlling tumor development is not yet fully understood. There is evidence to suggest that inflammatory and immune systems are able to inhibit cancer progression. Currently, there are at least two ways to recognize cancer: the first is tumor immunosurveillance, through which the host can recognize and eliminate transformed cells. The second is the adaptive immune recognition of tumor-associated and specific antigens, through which the immune system can regulate and suppress the development of cancer (Smyth *et al* 2006).

As inflammation is normally suppressed in primary tumors in order to inhibit an immune response and to avoid an inflammatory response switching of the primary tumor development or inhibition of the malignant tumor progression. However, this interaction pathway is not yet well understood. The hypothesis that inflammation inhibits cancer is not new, but not much research effort has been put into this proposal. There is still the question of whether the inflammation changes the tumor micro-environment in such a manner that the primary tumor can no longer grow in that microenvironment and is then just a subject of rejection, apoptosis and clearance. As is discussed above, an inflammation can cause cancer, and there seems to be a precisely regulated balance between tumor-promoting and tumor-antagonizing inflammatory responses and, indeed, there is a dependence of tumor-neutralizing immune responses on the optimal function of key transcription factors and tumor suppressors.

2.2.3 Cancer induces inflammation

The pre-malignant tumors seem to be 'wound-like' (Coussens *et al* 1999), as they are similar to healing or desmoplastic tissue regarding the presence of activated platelets (Dvorak 1986, Mueller and Fusenig 2004). In particular, as reported, tumor growth is thought to be 'biphasic' (Coussens *et al* 1999). In the initial phase, the body handles early tumors as wounds. This first phase involves tumor growth facilitated by the surrounding tumor stroma as an indirect control mechanism caused by the physiological tissue repair during 'wound' healing processes. In more detail, in mouse models for skin and pancreatic carcinogenesis, matrix metalloproteases produced by bone-marrow derived cells such as mast cells can convert VEGF into an active form that is able to facilitate the angiogenic switch to promote further tumor progression (Coussens *et al* 1999, Coussens *et al* 2000, Bergers *et al* 2000). In the later tumor growth state pro-inflammatory factors such as MMPs seem to be directly regulated by the primary tumors (Coussens *et al* 1999). Indeed, this malignant transition has been observed for the regulation of inflammation by early tumors compared to late tumors, leading to spontaneous intestinal tumorigenesis in mice and humans. In particular, in early tumors COX-2 is expressed by stromal cells (Hull *et al* 1999, Sonoshita *et al* 2002), whereas in late tumors COX-2 is expressed by the dysplastic epithelium (Sheehan *et al* 1999). What is the reason for this change of the cell type expressing COX-2? One hypothesis is that there are regulatory mechanisms in tumor-associated stromal cells regulating the upper limit of their expression of tissue repair factors. Thus, these findings are supposed to be a selective pressure for cancer cells that are able to maintain these COX-2 dependent ancillary processes and are then not dependent on a wound-like stroma surrounding the primary tumor. However, the tumor-associated stroma can undergo selective pressure, as there are genetic alterations found in tumor-associated stroma (Moinfar *et al* 2000) and even the loss of p53 has been detected in tumor-associated fibroblasts (Hill *et al* 2005). Moreover, the inflammatory response may not only play a role in promoting tumor growth by increasing the angiogenesis, it may also be involved in cell invasion and metastasis. In more detail, the process of angiogenesis supports the invasion and transmigration of cancer cells. Hence, matrix metalloproteases and MMP inhibitors (TIMPs) play an important role in the process of angiogenesis and help to remodel the extracellular matrix microenvironment (Egeblad and Werb 2002). In order to support cancer progression, infiltrating leukocytes may build a trail through the extracellular matrix tissue (called counter-current invasion theory) (Opdenakker and Van Damme 2004). The countercurrent principle of invasion has been proposed due to observations that cancer cells produce chemokines that regulate cancer progression (Opdenakker and Van Damme 1992,

1999, Balkwill and Mantovani 2001, Mareel and Leroy 2003, Coussens *et al* 2000, Van Coillie *et al* 2001). The term 'countercurrent model' for describing the regulatory role of chemokines in cancer was initially chosen because the fluxes of the primary tumor and host cells are opposite: in particular, chemoattracted leukocytes and growing blood vessels (angiogenesis) are toward the primary tumor, whereas the invasion of cancer cells is away from the primary tumor into the surrounding microenvironment and can be mediated by chemokine-induced proteolysis.

Inside–out signaling: is there an immune modulation by the cancer disease?
Cancer was formerly regarded as a cell-autonomous process. The current concept is based on genetically transformed cells that can develop into malignant neoplasms. As the tumor microenvironment has become the focus of cancer research, tumor progression seems to be regulated by biochemical and mechanical factors of the dense network of interactions between cancer cells and their surrounding tumor stroma, niche-defining cancer subcell populations and the tumor surrounding vasculature. A special feature of cancer cells is that they are able to alter the crosstalk between the tumor and its microenvironment through the regulation of the inflammatory response by soluble mediators. Cancer cells can affect the innate and the adaptive immune response through many pathways, including changing the acquired T cell response from the T helper 1 (TH1) cell subset to the TH2 cell subset, the induction of immunosuppressive T regulatory (TReg) cells, an altered phenotype of macrophages and neutrophils to a type 2 differentiation state, as well as the stimulation of myeloid-derived suppressor cells (MDSCs) (Motz and Coukos 2013). What are the pathways for cancer cells to facilitate the antitumor immune response? Cancer cells can interact at every step of the antitumor inflammatory response by inhibiting the function of immune cells through changing their inflammatory phenotype to a regulatory (immunosuppressive) one.

In more detail, inflammatory responses to tumor growth start with the tumor antigen that is presented by antigen-presenting cells. These immunostimulatory antigens can be of a special type that marks the onset of the primary tumor, and these antigens can also be overexpressed or mutated (called neo-antigens) (Robbins *et al* 2013). Primary tumors alter the function of antigen-presenting cells in many ways, such as through the direct tumor-facilitated regulation of the functions of myeloid cells (Gabrilovich *et al* 1996). In line with this, the tumor-derived vascular endothelial growth factor A (VEGFA) has been shown to suppress the function and maturation of dendritic cells. This VEGFA stimulation of peripheral myeloid cells leads to the expression of programmed cell death 1 ligand 1 (PDL1, also called CD274), which is an inhibitory ligand of the B7 family (Curiel *et al* 2003). Primary tumors can affect the antigen-presenting cell function by release of TGFβ, IL-10 and IL-6 (Geissmann *et al* 1999, Steinbrink *et al* 1999, Menetrier-Caux *et al* 1998). This cytokine stimulation of primary tumors modulates the immune system to an antitumor T cell response. This hypothesis has been confirmed in a mouse model of pancreatic ductal adenocarcinoma (PDA), where tumor-derived granulocyte–macrophage colony-stimulating factor (GM-CSF) induces the development and recruitment of granulocyte differentiation antigen 1 (GR1)+CD11b+ myeloid cells repressing antigen-specific T cells in the local microenvironment of a primary tumor (Bayne *et al* 2012). Thus, regulatory myeloid cells accumulate in the nearby tumor microenvironment and have been described as MDSCs in the peripheral blood of cancer patients. As expected, tumors can secrete CCL2, CXCL5, CXCL12 and stem cell factor (SCF, also called KIT ligand) in order to recruit MDSCs (Murdoch *et al* 2008). At the targeted site around the primary tumor, MDSCs induce their T cell-suppressive function as they secrete IL-10, TGFβ and arginase (Gabrilovich and Nagaraj 2009).

Primary tumors also actively regulate the inflammatory microenvironment through the recruitment of TReg cells. For example, Hodgkin's lymphoma cells and ovarian cancer cells release the chemokine CCL22 that binds to the CCR4 receptor and thereby recruits TReg cells to the tumor stroma (Curiel *et al* 2004). In addition, under special conditions, such as hypoxia, ovarian cancer cells secrete CCL28 that binds to its receptor CCR10, which is expressed predominantly on the cell surface of TReg cells (Facciabene *et al* 2011). Recruited TReg cells are able to convert the local tumor microenvironment to angiogenic mechanisms in order to support oxygen supply (Facciabene *et al* 2012). Moreover, TReg cells are actively targeted toward the primary tumor, as cancer cells secrete TGFβ and adenosine (Zarek *et al* 2008).

In summary, although inflammatory processes support all stages of tumor development, the primary tumor is able to actively vary the immune reactions in the local tumor surrounding microenvironment.

2.2.4 Cancer inhibits inflammation

What role does COX-2 play in the interaction of cancer and inflammation?
A key hypothesis is that constitutive overexpression of cyclooxygenase-2 (COX-2) supports mammary carcinogenesis and, in turn, inhibition of COX-2 promotes breast cancer prevention and cancer therapy. The major results are as follows. Firstly, COX-2 is constitutively expressed in tissue samples during the development of breast cancer and expression is correlated with the cancer disease stage, cancer progression and finally metastasis. Secondly, the essential characteristics of mammary carcinogenesis, such as mutagenesis, mitogenesis, angiogenesis, decreased apoptosis, increased metastasis and immunosuppression, depend on COX-2-driven prostaglandin E2 (PGE-2) biosynthesis. Thirdly, increased levels of COX-2 and PGE-2 expression facilitate the transcription of CYP-19 and aromatase-catalyzed estrogen biosynthesis, promoting increased mitogenesis. Fourthly, extrahepatic CYP-1B1 is detected in mammary adipose tissue and converts paracrine estrogen to carcinogenic quinones supporting mutagenesis. Fifthly, inhibitory substances of COX-2 decrease initial breast cancer occurrence in women and can lower the risk of cancer recurrence and mortality in women carrying breast cancer. Increases in the occurrence of breast cancer and increased breast cancer mortality are mediated by chronic inflammation of mammary adipose tissue and supported by the enhanced expression of COX-2 due to an obesity pandemic. Mammary carcinogenesis is a progressive series of highly specific cellular and molecular alterations evoked by constitutive overexpression of COX-2 and the activation of the prostaglandin cascade during the process of the primary tumor's so-called 'inflammogenesis'.

More than hundred years ago, it had already been proposed that chronic inflammation facilitates cancer development and progression by up-regulation of cellular proliferation (Virchow 1858, Virchow 1863, Balkwill and Mantovani 2001). Several models of carcinogenesis deal with inflammatory stimuli and mediators of wound healing in order to investigate the effect of inflammation on cancer development and progression (Schrieber and Rowley 1999, Coussens and Werb 2002, Philip et al 2004). The inducible COX-2 gene has been shown to connect cancer and inflammation as many models of carcinogenesis supposed an interaction mediated by inflammatory stimuli and COX-2 expression (Koki et al 2002, Jang and Hla 2002, Harris 2002, 2007). Current knowledge indicates that COX-2 driven inflammogenesis represents a good model for breast cancer development. In addition, a general model of inflammogenesis for breast cancer is supposed to be involved in the induction of constitutive COX-2 over-expression and the up-regulation of the prostaglandin cascade. Both models will be presented in the following. There are two primary genes encoding the cyclooxygenase, a constitutive gene (COX-1) and an inducible form (COX-2) (Hla and Neilson 1992, Herschman 1994, 2002). The inducible COX-2 gene represents the switching molecule that induces the inflammatory response via COX-2 activation by an inflammatory stimulus such as alcohol, tobacco, ischemia, trauma, pressure, cytotoxins, foreign objects, bacteria, viruses, fungi and lipopolysaccharides. Moreover, COX-2 induces the biosynthesis of E-prostaglandins such as prostaglandin E2 (PGE-2), which then mediate the inflammatory response. In summary, the basic facts about the connection between inflammation and cancer are listed in table 2.1.

Table 2.1. Inflammation and cancer—basic facts.

1. Chronic inflammation increases the risk of cancer.
2. Subclinical inflammation is important in increasing cancer risk.
3. Multiple types of immune and inflammatory cells are present within tumors.
4. Immune cells affect malignant cells via cytokines, chemokines, growth factors, prostaglandins, reactive oxygen and nitrogen species.
5. Inflammation impacts on each step of tumorigenesis.
6. In growing tumors anti- and pro-tumorigenic immune and inflammatory mechanisms coexist.
7. Signaling pathways providing pro-tumorigenic effects of inflammation are often subject to a feed-forward loop.
8. Special immune and inflammatory substances are dispensable during one stage, but absolutely critical in another stage of tumorigenesis.

Further reading

Bassaganya-Riera J, Viladomiu M, Pedragosa M, De Simone C and Hontecillas R 2012 Immunoregulatory mechanisms underlying prevention of colitis-associated colorectal cancer by probiotic bacteria *PLoS ONE* **7** e34676

Bates R C and Mercurio A M 2003 Tumor necrosis factor-α stimulates the epithelial-to-mesenchymal transition of human colonic organoids *Mol. Biol. Cell* **14** 1790–800

Braumüller H, Wieder T, Brenner E, Aßmann S, Hahn M and Alkhaled M *et al* 2013 T-helper-1-cell cytokines drive cancer into senescence *Nature* **494** 361–5

Campisi J and d'Adda di Fagagna F 2007 Cellular senescence: when bad things happen to good cells *Nature Rev. Mol. Cell Biol.* **8** 729–40

Chen G Y, Liu M, Wang F, Bertin J and Nunez G 2011 A functional role for Nlrp6 in intestinal inflammation and tumorigenesis *J. Immunol.* **186** 7187–94

Chien Y *et al* 2011 Control of the senescence-associated secretory phenotype by NF-κB promotes senescence and enhances chemosensitivity *Genes Dev.* **25** 2125–36

Corthesy B, Gaskins H R and Mercenier A 2007 Cross-talk between probiotic bacteria and the host immune system *J. Nutr.* **137** 781S–90S PMID: 17311975

Del Pozo M A, Sánchez-Mateos P, Nieto M and Sánchez-Madrid F 1995 Chemokines regulate cellular polarization and adhesion receptor redistribution during lymphocyte interaction with endothelium and extracellular matrix. Involvement of cAMP signaling pathway *J. Cell Biol.* **131** 495–508

Engelhardt B and Wolburg H 2004 Mini-review: transendothelial migration of leukocytes: through the front door or around the side of the house? *Eur. J. Immunol.* **34** 2955–63

Gabrilovich D I, Chen H L, Girgis K R, Cunningham H T, Meny G M, Nadaf S, Kavanaugh D and Carbone D P 1996 Production of vascular endothelial growth factor by human tumors inhibits the functional maturation of dendritic cells *Nature Med.* **2** 1096–103

Grivennikov S I and Karin M 2010 Inflammation and oncogenesis: a vicious connection *Curr. Opin. Genet. Dev.* **20** 65–71

Kang T W, Yevsa T, Woller N, Hoenicke L, Wuestefeld T and Dauch D *et al* 2011 Senescence surveillance of pre-malignant hepatocytes limits liver cancer development *Nature* **479** 547–51

Karin M and Greten F R 2005 NF-κB: linking inflammation and immunity to cancer development and progression *Nature Rev. Immunol.* **5** 749–59

Kim Y, Lee D, Kim D, Cho J, Yang J, Chung M, Kim K and Ha N 2008 Inhibition of proliferation in colon cancer cell lines and harmful enzyme activity of colon bacteria by Bifidobacterium adolescentis SPM0212 *Arch. Pharm. Res.* **31** 468–73

Le Leu R K, Brown I L, Hu Y, Bird A R, Jackson M, Esterman A and Young G P 2005 A synbiotic combination of resistant starch and *Bifidobacterium lactis* facilitates apoptotic deletion of carcinogen-damaged cells in rat colon *J. Nutr.* **135** 996–1001 PMID: 15867271

Maroof H, Hassan Z M, Mobarez A M and Mohamadabadi M A 2012 Lactobacillus acidophilus could modulate the immune response against breast cancer in murine model *J. Clin. Immunol.* **32** 1353–9

McDonald B, Spicer J, Giannais B, Fallavollita L, Brodt P and Ferri L E 2009 Systemic inflammation increases cancer cell adhesion to hepatic sinusoids by neutrophil mediated mechanisms *Int. J. Cancer* **125** 1298–305

Nickoloff B J, Ben-Neriah Y and Pikarsky E 2005 Inflammation and cancer: is the link as simple as we think? *J. Invest. Dermatol.* **124** 10–4

Opdenakker G and Van Damme J 1992 Cytokines and proteases in invasive processes: molecular similarities between inflammation and cancer *Cytokine* **4** 251–8

Orlando A, Messa C, Linsalata M, Cavallini A and Russo F 2009 Effects of Lactobacillus rhamnosus GG on proliferation and polyamine metabolism in HGC-27 human gastric and DLD-1 colonic cancer cell lines *Immunopharmacol. Immunotoxicol.* **31** 108–16

Pool-Zobel B L, Neudecker C, Domizlaff I, Ji S, Schillinger U, Rumney C, Moretti M, Vilarini I, Scassellati-Sforzolini R and Rowland I 1996 Lactobacillus- and bifidobacterium-mediated antigenotoxicity in the colon of rats *Nutr. Cancer* **26** 365–80

Pribluda A, Elyada E, Wiener Z, Hamza H, Goldstein R E and Biton M *et al* 2013 A senescence-inflammatory switch from cancer-inhibitory to cancer-promoting mechanism *Cancer Cell* **24** 242–56

Qian B Z, Li J, Zhang H, Kitamura T, Zhang J, Campion L R, Kaiser E A, Snyder L A and Pollard J W 2011 CCL2 recruits inflammatory monocytes to facilitate breast-tumor metastasis *Nature* **475** 222–5

Sanders M E, Guarner F, Guerrant R, Holt P R, Quigley E M, Sartor R B, Sherman P M and Mayer E A 2013 An update on the use and investigation of probiotics in health and disease *Gut* **62** 787–96

Sullivan N J, Sasser A K, Axel A E, Vesuna F, Raman V, Ramirez N, Oberyszyn T M and Hall B M 2009 Interleukin-6 induces an epithelial-mesenchymal transition phenotype in human breast cancer cells *Oncogene* **28** 2940–7

van Wetering S, van Den Berk N, van Buul J D, Mul F P, Lommerse I, Mous R, ten Klooster J P, Zwaginga J J and Hordijk P L 2003 VCAM-1-mediated Rac signaling controls endothelial cell–cell contacts and leukocyte transmigration *Am. J. Physiol. Cell Physiol.* **285** C343–52

Voronov E, Shouval D S, Krelin Y, Cagnano E, Benharroch D, Iwakura Y, Dinarello C A and Apte R N 2003 IL-1 is required for tumor invasiveness and angiogenesis *Proc. Natl Acad. Sci. USA* **100** 2645–50

Wang D, Mann J R and DuBois R N 2005 The role of prostaglandins and other eicosanoids in the gastrointestinal tract *Gastroenterology* **128** 1445–61

Wolf M J et al 2012 Endothelial CCR2 signaling induced by colon carcinoma cells enables extravasation via the JAK2-Stat5 and p38MAPK pathway *Cancer Cell* **22** 91–105

Wu S et al 2009 A human colonic commensal promotes colon tumorigenesis via activation of T helper type 17 T cell responses *Nature Med.* **15** 1016–22

Wyckoff J B, Wang Y, Lin E Y, Li J F, Goswami S, Stanley E R, Segall J E, Pollard J W and Condeelis J 2007 Direct visualization of macrophage-assisted tumor cell intravasation in mammary tumors *Cancer Res.* **67** 2649–56

Xue W, Zender L, Miething C, Dickins R A, Hernando E, Krizhanovsky V, Cordon-Cardo C and Lowe S W 2007 Senescence and tumor clearance is triggered by p53 restoration in murine liver carcinomas *Nature* **445** 656–60

References

Abdulamir A S, Hafidh R R and Abu Bakar F 2011 The association of streptococcus bovis/gallolyticus with colorectal tumors: the nature and the underlying mechanisms of its etiological role *J. Exp. Clin. Cancer Res.* **30** 11

Acevedo A, del Pozo M A, Arroyo A G, Sánchez-Mateos P, González-Amaro R and Sánchez-Madrid F 1993 Distribution of ICAM-3-bearing cells in normal human tissues. Expression of a novel counter-receptor for LFA-1 in epidermal Langerhans cells *Am. J. Pathol.* **143** 774–83 PMID: 8362976

Alon R and Ley K 2008 Cells on the run: shear-regulated integrin activation in leukocyte rolling and arrest on endothelial cells *Curr. Opin. Cell Biol.* **20** 525–32

Alon R, Grabovsky V and Feigelson S 2003 Chemokine induction of integrin adhesiveness on rolling and arrested leukocytes local signaling events or global stepwise activation? *Microcirculation* **10** 297–311

Alon R, Kassner P, Carr M, Finger E, Hemler M and Springer T 1995 The integrin VLA-4 supports tethering and rolling in flow on VCAM-1 *J. Cell Biol.* **128** 1243–53

Amit I et al 2007 A module of negative feedback regulators defines growth factor signaling *Nature Genet.* **39** 503–12

Aplin A E, Howe A, Alahari S K and Juliano R L 1998 Signal transduction and signal modulation by cell adhesion receptors: the role of integrins, cadherins, immunoglobulin-cell adhesion molecules, and selectins *Pharmacol Rev.* **50** 197–263 PMID: 9647866

Arthur J C *et al* 2012 Intestinal inflammation targets cancer-inducing activity of the microbiota *Science* **338** 120–3

Balkwill F and Mantovani A 2001 Inflammation and cancer: back to Virchow? *Lancet* **357** 539–45

Bamforth S D, Lightman S L and Greenwood J 1997 Ultrastructural analysis of interleukin-1 beta-induced leukocyte recruitment to the rat retina *Invest. Ophthalmol. Vis. Sci.* **38** 25–35 PMID: 9008627

Barreiro O, de la Fuente H, Mittelbrunn M and Sánchez-Madrid F 2007 Functional insights on the polarized redistribution of leukocyte integrins and their ligands during leukocyte migration and immune interactions *Immunol. Rev.* **218** 147–64

Barreiro O, Vicente-Manzanares M, Urzainqui A, Yáñez-Mó M and Sánchez-Madrid F 2004 Interactive protrusive structures during leukocyte adhesion and transendothelial migration *Front Biosci.* **9** 1849–63

Barreiro O, Yáñez-Mó M, Sala-Valdes M, Gutiérrez-López M D, Ovalle S, Higginbottom A, Monk P N, Cabañas C and Sánchez-Madrid F 2005 Endothelial tetraspanin microdomains regulate leukocyte firm adhesion during extravasation *Blood* **105** 2852–61

Barreiro O, Yánez-Mó M, Serrador J M, Montoya M C, Vicente-Manzanares M, Tejedor R, Furthmayr H and Sanchez-Madrid F 2002 Dynamic interaction of VCAM-1 and ICAM-1 with moesin and ezrin in a novel endothelial docking structure for adherent leukocytes *J. Cell Biol.* **157** 1233–45

Barreiro O, Zamai M, Yánez-Mó M, Tejera E, López-Romero P, Monk P N, Gratton E, Caiolfa V R and Sánchez-Madrid F 2008 Endothelial adhesion receptors are recruited to adherent leukocytes by inclusion in preformed tetraspanin nanoplatforms *J. Cell Biol.* **183** 527–42

Bayne L J, Beatty G L, Jhala N, Clark C E, Rhim A D, Stanger B Z and Vonderheide R H 2012 Tumor-derived granulocyte-macrophage colony-stimulating factor regulates myeloid inflammation and T cell immunity in pancreatic cancer *Cancer Cell* **21** 822–35

Beglova N, Blacklow S C, Takagi J and Springer T A 2002 Cysteine-rich module structure reveals a fulcrum for integrin rearrangement upon activation *Nat. Struct. Biol.* **9** 282–7

Ben-Neriah Y and Karin M 2011 Inflammation meets cancer, with NF-κB as the matchmaker *Nature Immunol.* **12** 715–23

Bergers G *et al* 2000 Matrix metalloproteinase-9 triggers the angiogenic switch during carcinogenesis *Nat. Cell Biol.* **2** 737–44

Berlin C, Bargatze R, Campbell J, von Andrian U, Szabo M, Hasslen S, Nelson R D, Berg E L, Erlandsen S L and Butcher E C 1995 Alpha 4 integrins mediate lymphocyte attachment and rolling under physiologic flow *Cell* **80** 413–22

Berlin C, Berg E L, Briskin M J, Andrew D P, Kilshaw P J, Holzmann B, Weissman I L, Hamann A and Butcher EC 1993 Alpha 4 beta 7 integrin mediates lymphocyte binding to the mucosal vascular addressin MAdCAM-1 *Cell* **74** 185–95

Bollrath J *et al* 2009 gp130-mediated Stat3 activation in enterocytes regulates cell survival and cell-cycle progression during colitis-associated tumorigenesis *Cancer Cell* **15** 91–102

Bronte-Tinkew D M, Terebiznik M, Franco A, Ang M, Ahn D, Mimuro H, Sasakawa C, Ropeleski M J, Peek R M Jr and Jones N L 2009 Helicobacter pylori cytotoxin-associated gene A activates the signal transducer and activator of transcription 3 pathway *in vitro* and *in vivo Cancer Res.* **69** 632–9

Butcher E C 1991 Leukocyte-endothelial cell recognition: three (or more) steps to specificity and diversity *Cell* **67** 1033–6

Butcher E C and Picker L J 1996 Lymphocyte homing and homeostasis *Science* **272** 60–6

Cabanas C and Hogg N 1993 Ligand intercellular adhesion molecule 1 has a necessary role in activation of integrin lymphocyte function-associated molecule 1 *Proc. Natl Acad. Sci. USA* **90** 5838–42

Cairo C W, Mirchev R and Golan D E 2006 Cytoskeletal regulation couples LFA-1 conformational changes to receptor lateral mobility and clustering *Immunity* **25** 297–308

Cambi A, Joosten B, Koopman M, de Lange F, Beeren I, Torensma R, Fransen J A, Garcia-Parajó M, van Leeuwen F N and Figdor C G 2006 Organization of the integrin LFA-1 in nanoclusters regulates its activity *Mol. Biol. Cell* **17** 4270–81

Campbell J J, Hedrick J, Zlotnik A, Siani M A, Thompson D A and Butcher E C 1998 Chemokines and the arrest of lymphocytes rolling under flow conditions *Science* **279** 381–4

Campregher C, Luciani M G and Gasche C 2008 Activated neutrophils induce an hMSH2-dependent G2/M checkpoint arrest and replication errors at a (CA)13-repeat in colon epithelial cells *Gut* **57** 780–7

Carlos T M and Harlan J M 1994 Leukocyte-endothelial adhesion molecules *Blood* **84** 2068–101 PMID: 7522621

Carman C V 2009 Mechanisms for transcellular diapedesis: probing and pathfinding by 'invadosome-like protrusions' *J. Cell Sci.* **122** 3025–35

Carman C V and Springer T A 2003 Integrin avidity regulation: are changes in affinity and conformation underemphasized? *Curr. Opin. Cell Biol.* **15** 547–56

Carman C V and Springer T A 2004 A transmigratory cup in leukocyte diapedesis both through individual vascular endothelial cells and between them *J. Cell Biol.* **167** 377–88

Carman C V, Sage P T, Sciuto T E, de la Fuente M A, Geha R S, Ochs H D, Dvorak H F, Dvorak A M and Springer T A 2007 Transcellular diapedesis is initiated by invasive podosomes *Immunity* **26** 784–97

Casaletto J B and McClatchey A I 2012 Spatial regulation of receptor tyrosine kinases in development and cancer *Nature Rev. Cancer* **12** 387–400

Castellarin M *et al* 2012 *Fusobacterium nucleatum* infection is prevalent in human colorectal carcinoma *Genome Res.* **22** 299–306

Cataisson C *et al* 2012 IL-1R-MyD88 signaling in keratinocyte transformation and carcinogenesis *J. Exp. Med.* **209** 1689–702

Charrin S, le Naour F, Silvie O, Milhiet P E, Boucheix C and Rubinstein E 2009 Lateral organization of membrane proteins: tetraspanins spin their web *Biochem. J.* **420** 133–54

Chavakis T *et al* 2003 The pattern recognition receptor (RAGE) is a counterreceptor for leukocyte integrins: a novel pathway for inflammatory cell recruitment *J. Exp. Med.* **198** 1507–15

Chen J *et al* 2011 CCL18 from tumor-associated macrophages promotes breast cancer metastasis via PITPNM3 *Cancer Cell* **19** 541–55

Chen L W, Egan L, Li Z W, Greten F R, Kagnoff M F and Karin M 2003 The two faces of IKK and NF-kappaB inhibition: prevention of systemic inflammation but increased local injury following intestinal ischemia-reperfusion *Nat. Med.* **9** 575–81

Cinamon G, Shinder V and Alon R 2001 Shear forces promote lymphocyte migration across vascular endothelium bearing apical chemokines *Nat. Immunol.* **2** 515–22

Clayton A, Evans R A, Pettit E, Hallett M, Williams J D and Steadman R 1998 Cellular activation through the ligation of intercellular adhesion molecule-1 *J. Cell Sci.* **111** 443–53 PMID: 9443894

Colotta F, Allavena P, Sica A, Garlanda C and Mantovani A 2009 Cancer-related inflammation, the seventh hallmark of cancer: links to genetic instability *Carcinogenesis* **30** 1073–81

Constantin G, Majeed M, Giagulli C, Piccio L, Kim J Y, Butcher E C and Laudanna C 2000 Chemokines trigger immediate beta2 integrin affinity and mobility changes: differential regulation and roles in lymphocyte arrest under flow *Immunity* **13** 759–69

Cook-Mills J M 2002 VCAM-1 signals during lymphocyte migration: role of reactive oxygen species *Mol. Immunol.* **39** 499–508

Coussens L M and Werb Z 2002 Inflammation and cancer *Nature* **420** 860–7

Coussens L M, Raymond W W, Bergers G, Laig-Webster M, Behrendtsen O, Werb Z, Caughey G H and Hanahan D 1999 Inflammatory mast cells up-regulate angiogenesis during squamous epithelial carcinogenesis *Genes Dev.* **13** 1382–97

Coussens L M, Tinkle C L, Hanahan D and Werb Z 2000 MMP-9 supplied by bone marrow-derived cells contributes to skin carcinogenesis *Cell* **103** 481–90

Cunningham S A, Rodríguez J M, Arrate M P, Tran T M and Brock T A 2002 JAM2 interacts with alpha4beta1. Facilitation by JAM3 *J. Biol. Chem.* **277** 27589–92

Curiel T J *et al* 2004 Specific recruitment of regulatory T cells in ovarian carcinoma fosters immune privilege and predicts reduced survival *Nature Med.* **10** 942–9

Curiel T J *et al* 2003 Blockade of B7-H1 improves myeloid dendritic cell-mediated antitumor immunity *Nature Med.* **9** 562–7

Dapito D H *et al* 2012 Promotion of hepatocellular carcinoma by the intestinal microbiota and TLR4 *Cancer Cell* **21** 504–16

Deem T L and Cook-Mills J M 2004 Vascular cell adhesion molecule 1 (VCAM-1) activation of endothelial cell matrix metalloproteinases: role of reactive oxygen species *Blood* **104** 2385–93

DeNardo D G, Barreto J B, Andreu P, Vasquez L, Tawfik D, Kolhatkar N and Coussens L M 2009 CD4+ T cells regulate pulmonary metastasis of mammary carcinomas by enhancing protumor properties of macrophages *Cancer Cell* **16** 91–102

Di Lorenzo A, Fernández-Hernando C, Cirino G, William C and Sessa W C 2009 Akt1 is critical for acute inflammation and histamine-mediated vascular leakage *Proc. Natl Acad. Sci. USA* **106** 14552–7

Dobereiner H G, Dubin-Thaler B J, Hofman J M, Xenias H S, Sims T N, Giannone G, Dustin M L, Wiggins C H and Sheetz M P 2006 Lateral membrane waves constitute a universal dynamic pattern of motile cells *Phys. Rev. Lett.* **97** 038102

Dustin M L, Rothlein R, Bhan A K, Dinarello C A and Springer T A 1986 Induction by IL 1 and interferon-gamma: tissue distribution, biochemistry, and function of a natural adherence molecule (ICAM-1) *J. Immunol.* **137** 245–54 PMID: 21505214

Dvorak H F 1986 Tumors: wounds that do not heal. Similarities between tumor stroma generation and wound healing *N. Engl. J. Med.* **315** 1650–9

Dwir O, Kansas G S and Alon R 2001 Cytoplasmic anchorage of L-selectin controls leukocyte capture and rolling by increasing the mechanical stability of the selectin tether *J. Cell Biol.* **155** 145–56

Egeblad M and Werb Z 2002 New functions for the matrix metalloproteinases in cancer progression *Nat. Rev. Cancer* **2** 161–74

Elices M J, Osborn L, Takada Y, Crouse C, Luhowskyj S, Hemler M E and Lobb R R 1990 VCAM-1 on activated endothelium interacts with the leukocyte integrin VLA-4 at a site distinct from the VLA-4/fibronectin binding site *Cell* **60** 577–84

Elinav E, Henao-Mejia J and Flavell R A 2013 Integrative inflammasome activity in the regulation of intestinal mucosal immune responses *Mucosal Immunol.* **6** 4–13

Elinav E *et al* 2011 NLRP6 inflammasome regulates colonic microbial ecology and risk for colitis *Cell* **145** 745–57

Endo Y, Marusawa H, Kou T, Nakase H, Fujii S, Fujimori T, Kinoshita K, Honjo T and Chiba T 2008 Activation-induced cytidine deaminase links between inflammation and the development of colitis-associated colorectal cancers *Gastroenterology* **135** 889–98

Eriksson E E, Xie X, Werr J, Thoren P and Lindbom L 2001 Importance of primary capture and L-selectin-dependent secondary capture in leukocyte accumulation in inflammation and atherosclerosis *in vivo J. Exp. Med.* **194** 205–18

Etienne-Manneville S, Manneville J B, Adamson P, Wilbourn B, Greenwood J and Couraud P O 2000 ICAM-1-coupled cytoskeletal rearrangements and transendothelial lymphocyte migration involve intracellular calcium signaling in brain endothelial cell lines *J. Immunol.* **165** 3375–83

Evans E A and Calderwood D A 2007 Forces and bond dynamics in cell adhesion *Science* **316** 1148–53

Facciabene A, Motz G T and Coukos G 2012 T-regulatory cells: key players in tumor immune escape and angiogenesis *Cancer Res.* **72** 2162–71

Facciabene A *et al* 2011 Tumor hypoxia promotes tolerance and angiogenesis via CCL28 and Treg cells *Nature* **475** 226–30

Franco A T, Israel D A, Washington M K, Krishna U, Fox J G and Rogers A B *et al* 2005 Activation of β-catenin by carcinogenic Helicobacter pylori *Proc. Natl Acad. Sci. USA* **102** 10646–51

Frenette P S and Wagner D D 1996a Adhesion molecules-Part I *N. Engl. J. Med.* **334** 1526–9

Frenette P S and Wagner D D 1996b Adhesion molecules: part II. Blood vessels and blood cells *N. Engl. J. Med.* **335** 43–5

Fukata M *et al* 2007 Toll-like receptor-4 promotes the development of colitis-associated colorectal tumors *Gastroenterology* **133** 1869–81

Fukuda A *et al* 2011 Stat3 and MMP7 contribute to pancreatic ductal adenocarcinoma initiation and progression *Cancer Cell* **19** 441–55

Gabrilovich D I and Nagaraj S 2009 Myeloid-derived suppressor cells as regulators of the immune system *Nature Rev. Immunol.* **9** 162–74

Gahmberg C G, Nortamo P, Kantor C, Autero M, Kotovuori P, Hemiö L, Salcedo R and Patarroyo M 1990 The pivotal role of the Leu-CAM and ICAM molecules in human leukocyte adhesion *Cell Differ. Dev.* **32** 239–45

Gao S P *et al* 2007 Mutations in the EGFR kinase domain mediate STAT3 activation via IL-6 production in human lung adenocarcinomas *J. Clin. Invest.* **117** 3846–56

García-López M A, Barreiro O, García-Díez A, Sánchez-Madrid F and Penas P F 2005 Role of tetraspanins CD9 and CD151 in primary melanocyte motility *J. Invest. Dermatol.* **125** 1001–9

Garrett W S, Punit S, Gallini C A, Michaud M, Zhang D, Sigrist K S, Lord G M, Glickman J N and Glimcher L H 2009 Colitis-associated colorectal cancer driven by T-bet deficiency in dendritic cells *Cancer Cell* **16** 208–19

Geiger B and Bershadsky A 2002 Exploring the neighborhood: adhesion-coupled cell mechano-sensors *Cell* **110** 139–42

Geissmann F, Revy P, Regnault A, Lepelletier Y, Dy M, Brousse N, Amigorena S, Hermine O and Durandy A 1999 TGF-β 1 prevents the noncognate maturation of human dendritic Langerhans cells *J. Immunol.* **162** 4567–75 PMID: 10201996

Gocheva V, Wang H W, Gadea B B, Shree T, Hunter K E, Garfall A L, Berman T and Joyce J A 2010 IL-4 induces cathepsin protease activity in tumor-associated macrophages to promote cancer growth and invasion *Genes Dev.* **24** 241–55

Goodwin A C et al 2011 Polyamine catabolism contributes to enterotoxigenic bacteroides fragilis-induced colon tumorigenesis Proc. Natl Acad. Sci. USA 108 15354-9

Gordon J I 2012 Honor thy gut symbionts redux Science 336 1251-3

Gordon-Alonso M, Yáñez-Mó M, Barreiro O, Álvarez S, Muñoz-Fernández M A, Valenzuela-Fernández A and Sánchez-Madrid F 2006 Tetraspanins CD9 and CD81 modulate HIV-1-induced membrane fusion J. Immunol. 177 5129-37

Grabovsky V et al 2000 Subsecond induction of alpha4integrin clustering by immobilized chemokines stimulates leukocyte tethering and rolling on endothelial vascular cell adhesion molecule 1 under flow conditions J. Exp. Med. 192 495-506

Greenwood J, Etienne-Manneville S, Adamson P and Couraud P O 2002 Lymphocyte migration into the central nervous system: implication of ICAM-1 signalling at the blood–brain barrier Vascul. Pharmacol. 38 315-22

Greten F R, Eckmann L, Greten T F, Park J M, Li Z W, Egan L J, Kagnoff M F and Karin M 2004 IKKβ links inflammation and tumorigenesis in a mouse model of colitis-associated cancer Cell 118 285-96

Grivennikov S I, Greten F R and Karin M 2010 Immunity, inflammation, and cancer Cell 140 883-99

Grivennikov S I, Kuprash D V, Liu Z G and Nedospasov S A 2006 Intracellular signals and events activated by cytokines of the tumor necrosis factor superfamily: From simple paradigms to complex mechanisms Int. Rev. Cytol. 252 129-61

Grivennikov S I, Wang K, Mucida D, Stewart C A, Schnabl B and Jauch D et al 2012 Adenoma-linked barrier defects and microbial products drive IL-23/IL-17-mediated tumor growth Nature 491 254-8 PMID: 23034650

Grivennikov S et al 2009 IL-6 and Stat3 are required for survival of intestinal epithelial cells and development of colitis-associated cancer Cancer Cell 15 103-13

Hahn M A, Hahn T, Lee D H, Esworthy R S, Kim B W, Riggs A D, Chu F F and Pfeifer G P 2008 Methylation of polycomb target genes in intestinal cancer is mediated by inflammation Cancer Res. 68 10280-9

Hanahan D and Weinberg R A 2011 Hallmarks of cancer: the next generation Cell 144 646-74

Harris R E 2002 Cyclooxygenase-2 blockade in cancer prevention and therapy: widening the scope of impact COX-2 Blockade in Cancer Prevention and Therapy ed R E Harris (Totowa, NJ: Humana)

Harris R E 2007 Cyclooxygenase-2 (cox-2) and the inflammo-genesis of cancer Subcell Biochem. 42 93-126

Heiska L, Alfthan K, Gronholm M, Vilja P, Vaheri A and Carpen O 1998 Association of Ezrin with intercellular adhesion molecule-1 and -2 (ICAM-1 and ICAM-2). Regulation by phosphatidylinositol 4,5-bisphosphate J. Biol. Chem. 273 21893-900

Hemler M E 2005 Tetraspanin functions and associated microdomains Nat. Rev. Mol. Cell Biol. 6 801-11

Henao-Mejia J et al 2012 Inflammasome-mediated dysbiosis regulates progression of NAFLD and obesity Nature 482 179-85

Henderson R B, Lim L H, Tessier P A, Gavins F N, Mathies M, Perretti M and Hogg N 2001 The use of lymphocyte function-associated antigen (LFA)-1-deficient mice to determine the role of LFA-1, Mac-1, and alpha4 integrin in the inflammatory response of neutrophils J. Exp. Med. 194 219-26

Herschman H R 1994 Regulation of prostaglandin synthase-1 and prostaglandin synthase-2 Cancer Metastasis Rev. 13 241-56

Herschman H R 2002 Historical aspects of COX-2 *COX-2 Blockade in Cancer Prevention and Therapy* ed R E Harris (Totowa, NJ: Humana) pp 13–32

Hidalgo A, Peired A J, Wild M K, Vestweber D and Frenette P S 2007 Complete identification of E-selectin ligands on neutrophils reveals distinct functions of PSGL-1, ESL-1, and CD44 *Immunity* **26** 477–89

Hill R, Song Y, Cardiff R D and Van Dyke T 2005 Selective evolution of stromal mesenchyme with p53 loss in response to epithelial tumorigenesis *Cell* **123** 1001–11

Hla T and Neilson K 1992 Human cyclooxygenase-2 cDNA *Proc. Natl Acad. Sci. USA* **89** 7384–8

Ho P L, Lay E J, Jian W, Parra D and Chan K S 2012 Stat3 activation in urothelial stem cells leads to direct progression to invasive bladder cancer *Cancer Res.* **72** 3135–42

Hu B, Elinav E, Huber S, Booth C J, Strowig T, Jin C, Eisenbarth S C and Flavell R A 2010 Inflammation-induced tumorigenesis in the colon is regulated by caspase-1 and NLRC4 *Proc. Natl Acad. Sci. USA* **107** 21635–40

Hu B *et al* 2013 Microbiota-induced activation of epithelial IL-6 signaling links inflammasome-driven inflammation with transmissible cancer *Proc. Natl Acad. Sci. USA* **110** 9862–7

Hubbard A K and Rothlein R 2000 Intercellular adhesion molecule-1 (ICAM-1) expression and cell signaling cascades *Free Radic. Biol. Med.* **28** 1379–86

Huber S *et al* 2012 IL-22BP is regulated by the inflammasome and modulates tumorigenesis in the intestine *Nature* **491** 259–63

Hull M A, Booth J K, Tisbury A, Scott N, Bonifer C, Markham A F and Coletta P L 1999 Cyclooxygenase 2 is up-regulated and localized to macrophages in the intestine of Min mice *Br. J. Cancer* **79** 1399–405

Humphries J D, Byron A and Humphries M J 2006 Integrin ligands at a glance *J. Cell Sci.* **119** 3901–3

Hussain S P, Hofseth L J and Harris C C 2003 Radical causes of cancer *Nat. Rev. Cancer* **3** 276–85

Huttenhower C, Gevers D, Knight R, Abubucker S, Badger J H, Asif T and Chinwalla A T 2012 Structure, function and diversity of the healthy human microbiome *Nature* **486** 207–14

Hynes R O 2002 Integrins: bidirectional, allosteric signaling machines *Cell* **110** 673–87

Iliopoulos D, Hirsch H A and Struhl K 2009 An epigenetic switch involving NF-κB, Lin28, Let-7 MicroRNA, and IL6 links inflammation to cell transformation *Cell* **139** 693–706

Ivetic A, Deka J, Ridley A J and Ager A 2002 The cytoplasmic tail of L-selectin interacts with members of the Ezrin–Radixin–Moesin (ERM) family of proteins: cell activation-dependent binding of Moesin but not Ezrin *J. Biol. Chem.* **277** 2321–9

Jacob N and Stohl W 2011 Cytokine disturbances in systemic lupus erythematosus *Arthritis Res. Ther.* **13** 228

Janeway Jr C A and Medzhitov R 2002 Innate immune recognition *Annu. Rev. Immunol.* **20** 197–216

Jang B C and Hla T 2002 Regulation of expression and potential carcinogenic role of cylcooxygenase-2 *COX-2 Blockade in Cancer Prevention and Therapy* ed R E Harris (Totowa, NJ: Humana) pp 171–84

Jemal A, Siegel R, Xu J and Ward E 2010 Cancer statistics *CA Cancer J. Clin.* **60** 277–300

Jiang R, Tan Z, Deng L, Chen Y, Xia Y, Gao Y, Wang X and Sun B 2011 Interleukin-22 promotes human hepatocellular carcinoma by activation of STAT3 *Hepatology* **54** 900–9

Jun C D, Shimaoka M, Carman C V, Takagi J and Springer T A 2001 Dimerization and the effectiveness of ICAM-1 in mediating LFA-1-dependent adhesion *Proc. Natl Acad. Sci. USA* **98** 6830–5

Kadono T, Venturi G M, Steeber D A and Tedder T F 2002 Leukocyte rolling velocities and migration are optimized by cooperative L-selectin and intercellular adhesion molecule-1 functions *J. Immunol.* **169** 4542–50

Kalluri R and Weinberg R A 2009 The basics of epithelial-mesenchymal transition *J. Clin. Invest.* **119** 1420–8

Killock D J, Parsons M, Zarrouk M, Ameer-Beg S M, Ridley A J, Haskard D O, Zvelebil M and Ivetic A 2009 *In vitro* and *in vivo* characterization of molecular interactions between calmodulin, ezrin/radixin/moesin (ERM) and L-selectin *J. Biol. Chem.* **284** 8833–45

Kitamura T *et al* 2007 SMAD4-deficient intestinal tumors recruit CCR1+ myeloid cells that promote invasion *Nature Genet.* **39** 467–75

Koki A T, Leahy K M, Harmon J M and Masferrer J L 2002 Cyclooxygenase-2 and cancer *COX-2 Blockade in Cancer Prevention and Therapy* ed R E Harris (Totowa, NJ: Humana) pp 185–203

Komori J, Marusawa H, Machimoto T, Endo Y, Kinoshita K, Kou T, Haga H, Ikai I, Uemoto S and Chiba T 2008 Activation-induced cytidine deaminase links bile duct inflammation to human cholangiocarcinoma *Hepatology* **47** 888–96

Kostic A D *et al* 2012 Genomic analysis identifies association of *Fusobacterium* with colorectal carcinoma *Genome Res.* **22** 292–8

Kostic A D *et al* 2013 *Fusobacterium nucleatum* potentiates intestinal tumorigenesis and modulates the tumor-immune microenvironment *Cell Host Microbe* **14** 207–15

Kuper H, Adami H O and Trichopoulos D 2000 Infections as a major preventable cause of human cancer *J. Intern. Med.* **248** 171–83

Lamagna C, Meda P, Mandicourt G, Brown J, Gilbert R J, Jones E Y, Kiefer F, Ruga P, Imhof B A and Aurrand-Lions M 2005 Dual interaction of JAM-C with JAM-B and alpha(M)beta2 integrin: function in junctional complexes and leukocyte adhesion *Mol. Biol. Cell.* **16** 4992–5003

Laudanna C 2005 Integrin activation under flow: a local affair *Nat. Immunol.* **6** 429–30

Laudanna C and Alon R 2006 Right on the spot. Chemokine triggering of integrin-mediated arrest of rolling leukocytes *Thromb. Haemost.* **95** 5–11 PMID: 16543955

Lee S H *et al* 2010 ERK activation drives intestinal tumorigenesis in Apcmin/+ mice *Nature Med.* **16** 665–70

Leray E, Yaouanq J, Le Page E, Coustans M, Laplaud D, Oger J and Edan G 2010 Evidence for a two-stage disability progression in multiple sclerosis *Brain* **133** 1900–13

Lerner I *et al* 2011 Heparanase powers a chronic inflammatory circuit that promotes colitis-associated tumorigenesis in mice *J. Clin. Invest.* **121** 1709–21

Lesina M *et al* 2011 Stat3/Socs3 activation by IL-6 transsignaling promotes progression of pancreatic intraepithelial neoplasia and development of pancreatic cancer *Cancer Cell* **19** 456–69

Lévesque J P, Zannettino A C, Pudney M, Niutta S, Haylock D N, Snapp K R, Kansas G S, Berndt M C and Simmons P J 1999 PSGL-1-mediated adhesion of human hematopoietic progenitors to P-selectin results in suppression of hematopoiesis *Immunity* **11** 369–78

Ley K, Laudanna C, Cybulsky M I and Nourshargh S 2007 Getting to the site of inflammation: the leukocyte adhesion cascade updated *Nat. Rev. Immunol.* **7** 678–89

Liang J *et al* 2013 Sphingosine-1-phosphate links persistent STAT3 activation, chronic intestinal inflammation, and development of colitis-associated cancer *Cancer Cell* **23** 107–20

Lorenzon P, Vecile E, Nardon E, Ferrero E, Harlan J M, Tedesco F and Dobrina A 1998 Endothelial cell E- and P-selectin and vascular cell adhesion molecule-1 function as signaling receptors *J. Cell Biol.* **142** 1381–91

Lowe E L, Crother T R, Rabizadeh S, Hu B, Wang H, Chen S, Shimada K, Wong M H, Michelsen K S and Arditi M 2010 Toll-like receptor 2 signaling protects mice from tumor development in a mouse model of colitis-induced cancer *PLoS ONE* **5** e13027

Luster A D 1998 Chemokines—chemotactic cytokines that mediate inflammation *N. Engl. J. Med.* **338** 436–45

Mantovani A, Allavena P, Sica A and Balkwill F 2008 Cancer-related inflammation *Nature* **454** 436–44

Marchesi J R, Dutilh B E, Hall N, Peters W H, Roelofs R, Boleij A and Tjalsma H 2011 Towards the human colorectal cancer microbiome *PLoS ONE* **6** e20447

Mareel M and Leroy A 2003 Clinical, cellular, and molecular aspects of cancer invasion *Physiol. Rev.* **83** 337–76

Marlin S D and Springer T A 1987 Purified intercellular adhesion molecule-1 (ICAM-1) is a ligand for lymphocyte function-associated antigen 1 (LFA-1) *Cell* **51** 813–9

Marotta L L, Almendro V, Marusyk A, Shipitsin M, Schemme J and Walker S R *et al* 2011 The JAK2/STAT3 signaling pathway is required for growth of CD44+CD24− stem-cell-like breast cancer cells in human tumors *J. Clin. Invest.* **121** 2723–35

Marschel P and Schmid-Schonbein G W 2002 Control of fluid shear response in circulating leukocytes by integrins *Ann. Biomed. Eng.* **30** 333–43

Medzhitov R 2008 Origin and physiological roles of inflammation *Nature* **454** 428–35

Mehta P, Cummings R D and McEver R P 1998 Affinity and kinetic analysis of P-selectin binding to P-selectin glycoprotein ligand-1 *J. Biol. Chem.* **273** 32506–13

Menetrier-Caux C, Montmain G, Dieu M C, Bain C, Favrot M C, Caux C and Blay J Y 1998 Inhibition of the differentiation of dendritic cells from CD34+ progenitors by tumor cells: role of interleukin-6 and macrophage colony-stimulating factor *Blood* **92** 4778–91

Mierke C T 2013 The role of focal adhesion kinase in the regulation of cellular mechanical properties *Phys. Biol.* **10** 065005

Millan J, Hewlett L, Glyn M, Toomre D, Clark P and Ridley A J 2006 Lymphocyte transcellular migration occurs through recruitment of endothelial ICAM-1 to caveola- and F-actin-rich domains *Nat. Cell Biol.* **8** 113–23

Mills K D, Ferguson D O and Alt F W 2003 The role of DNA breaks in genomic instability and tumorigenesis *Immunol. Rev.* **194** 77–95

Mishra A, Liu S, Sams G H, Curphey D P, Santhanam R and Rush L J *et al* 2012 Aberrant overexpression of IL-15 initiates large granular lymphocyte leukemia through chromosomal instability and DNA hypermethylation *Cancer Cell* **22** 645–55

Mittelbrunn M, Cabanas C and Sánchez-Madrid F 2006 Integrin alpha4 *UCSD-Nature Molecule Pages* doi:10.1038/mp.a001203.01

Mittelbrunn M, Yáñez-Mó M, Sancho D, Ursa A and Sánchez-Madrid F 2002 Cutting edge: dynamic redistribution of tetraspanin CD81 at the central zone of the immune synapse in both T lymphocytes and APC *J. Immunol.* **169** 6691–5

Moinfar F, Man Y G, Arnould L, Bratthauer G L, Ratschek M and Tavassoli F A 2000 Concurrent NF-kappaB activation by double-strand breaks and independent genetic alterations in the stromal and epithelial cells of mammary carcinoma: implications for tumorigenesis *Cancer Res.* **60** 2562–6 PMID: 10811140

Mory A, Feigelson S W, Yarali N, Kilic S S, Bayhan G I, Gershoni-Baruch R, Etzioni A and Alon R 2008 Kindlin-3: a new gene involved in the pathogenesis of LAD-III *Blood* **112** 2591

Mosesson Y, Mills G B and Yarden Y 2008 Derailed endocytosis: an emerging feature of cancer *Nature Rev. Cancer* **8** 835–50

Motz G T and Coukos G 2013 Deciphering and reversing tumor immune suppression *Immunity* **39** 61–73

Mueller M M and Fusenig N E 2004 Friends or foes—bipolar effects of the tumor stroma in cancer *Nat. Rev. Cancer* **4** 839–49

Muller W A 2003 Leukocyte–endothelial-cell interactions in leukocyte transmigration and the inflammatory response *Trends in Immunology* **24** 327–34

Munger K and Howley P M 2002 Human papillomavirus immortalization and transformation functions *Virus Res.* **89** 213–28

Murdoch C, Muthana M, Coffelt S B and Lewis C E 2008 The role of myeloid cells in the promotion of tumor angiogenesis *Nature Rev. Cancer* **8** 618–31

Myant K B *et al* 2013 ROS production and NF-κB activation triggered by RAC1 facilitate WNT-driven intestinal stem cell proliferation and colorectal cancer initiation *Cell Stem Cell* **12** 761–73

Nagy J A, Dvorak A M and Dvorak H F 2012 Vascular hyperpermeability, angiogenesis and stroma generation *Cold Spring Harb Perspect Med.* **2** a006544

Nguyen D X, Bos P D and Massague J 2009 Metastasis: from dissemination to organ-specific colonization *Nature Rev. Cancer* **9** 274–84

Nicholson M W, Barclay A N, Singer M S, Rosen S D and van der Merwe P A 1998 Affinity and kinetic analysis of L-selectin (CD62L) binding to glycosylation-dependent cell-adhesion molecule-1 *J. Biol. Chem.* **273** 763–70

Nieminen M, Henttinen T, Merinen M, Marttila-Ichihara F, Eriksson J E and Jalkanen S 2006 Vimentin function in lymphocyte adhesion and transcellular migration *Nat. Cell Biol.* **8** 156–62

Nishida N, Xie C, Shimaoka M, Cheng Y, Walz T and Springer T A 2006 Activation of leukocyte beta2 integrins by conversion from bent to extended conformations *Immunity* **25** 583–94

Normand S, Delanoye-Crespin A, Bressenot A, Huot L, Grandjean T, Peyrin-Biroulet L, Lemoine Y, Hot D and Chamaillard M 2011 Nod-like receptor pyrin domain-containing protein 6 (NLRP6) controls epithelial self-renewal and colorectal carcinogenesis upon injury *Proc. Natl Acad. Sci. USA* **108** 9601–6

Okazaki I M, Kotani A and Honjo T 2007 Role of AID in tumorigenesis *Adv. Immunol.* **94** 245–73

Opdenakker G and Van Damme J 1992 Chemotactic factors, passive invasion and metastasis of cancer cells *Immunol. Today* **13** 463–4

Opdenakker G and Van Damme J 1999 Novel monocyte chemoattractants in cancer *Chemokines and Cancer* ed B Rollins (Totowa, NJ: Humana) pp 51–69

Opdenakker G and Van Damme J 2004 The counter-current principle in invasion and metastasis of cancer cells. Recent insights on the roles of chemokines *Int. J. Dev. Biol.* **48** 519–27

Ostermann G, Weber K S, Zernecke A, Schroder A and Weber C 2002 JAM-1 is a ligand of the beta(2) integrin LFA-1 involved in transendothelial migration of leukocytes *Nat. Immunol.* **3** 151–8

Park E J, Lee J H, Yu G Y, He G, Ali S R, Holzer R G, Osterreicher C H, Takahashi H and Karin M 2010 Dietary and genetic obesity promote liver inflammation and tumorigenesis by enhancing IL-6 and TNF expression *Cell* **140** 197–208

Park O *et al* 2011 *In vivo* consequences of liver-specific interleukin-22 expression in mice: implications for human liver disease progression *Hepatology* **54** 252–61

Pasvolsky R *et al* 2007 A LAD-III Syndrome is associated with defective expression of the Rap-1 activator CalDAGGEFI in lymphocytes, neutrophils, and platelets *J. Exp. Med.* **204** 1571–82 PMID: 17576779

Pavalko F M, Walker D M, Graham L, Goheen M, Doerschuk C M and Kansas G S 1995 The cytoplasmic domain of L-selectin interacts with cytoskeletal proteins via alpha-actinin: receptor positioning in microvilli does not require interaction with alpha-actinin *J. Cell Biol.* **129** 1155–64

Peek R M Jr and Blaser M J 2002 *Helicobacter pylori* and gastrointestinal tract adenocarcinomas *Nat. Rev. Cancer* **2** 28–37

Peinado H, Lavotshkin S and Lyden D 2011 The secreted factors responsible for pre-metastatic niche formation: old sayings and new thoughts *Semin. Cancer Biol.* **21** 139–46

Philip M, Rowley D A and Schreiber H 2004 Inflammation as a tumor promoter in cancer induction *Semin. Cancer Biol.* **14** 433–9

Phillipson M, Heit B, Colarusso P, Liu L, Ballantyne C M and Kubes 2006 Intraluminal crawling of neutrophils to emigration sites: a molecularly distinct process from adhesion in the recruitment cascade *J. Exp. Med.* **203** 2569–75

Phillipson M, Kaur J, Colarusso P, Ballantyne C M and Kubes P 2008 Endothelial domes encapsulate adherent neutrophils and minimize increases in vascular permeability in para-cellular and transcellular emigration *PLoS ONE* **3** e1649

Pickert G *et al* 2009 STAT3 links IL-22 signaling in intestinal epithelial cells to mucosal wound healing *J. Exp. Med.* **206** 1465–72

Pikarsky E, Porat R M, Stein I, Abramovitch R, Amit S, Kasem S, Gutkovich-Pyest E, Urieli-Shoval S, Galun E and Ben-Neriah Y 2004 NF-κB functions as a tumor promoter in inflammation-associated cancer *Nature* **431** 461–6

Poon I K H, Hulett M D and Parish C R 2010 Molecular mechanisms of late apoptotic/necrotic cell clearance *Cell Death and Differentiation* **17** 381–97

Popivanova B K, Kitamura K, Wu Y, Kondo T, Kagaya T, Kaneko S, Oshima M, Fujii C and Mukaida N 2008 Blocking TNF-α in mice reduces colorectal carcinogenesis associated with chronic colitis *J. Clin. Invest.* **118** 560–70

Prahallad A, Sun C, Huang S, Di Nicolantonio F, Salazar R, Zecchin D, Beijersbergen R L, Bardelli A and Bernards R 2012 Unresponsiveness of colon cancer to BRAF(V600E) inhibition through feedback activation of EGFR *Nature* **483** 100–3

Qian B Z and Pollard J W 2010 Macrophage diversity enhances tumor progression and metastasis *Cell* **141** 39–51

Quante M *et al* 2012 Bile acid and inflammation activate gastric cardia stem cells in a mouse model of Barrett-like metaplasia *Cancer Cell* **21** 36–51

Rakoff-Nahoum S and Medzhitov R 2007 Regulation of spontaneous intestinal tumorigenesis through the adaptor protein MyD88 *Science* **317** 124–7

Robbins P F *et al* 2013 Mining exomic sequencing data to identify mutated antigens recognized by adoptively transferred tumor-reactive T cells *Nature Med.* **19** 747–52

Rot A and von Andrian U H 2004 Chemokines in innate and adaptive host defense: basic chemokinese grammar for immune cells *Annu. Rev. Immunol.* **22** 891–928

Rubinstein M R, Wang X, Liu W, Hao Y, Cai G and Han Y W 2013 *Fusobacterium nucleatum* promotes colorectal carcinogenesis by modulating E-Cadherin/β-Catenin signaling via its FadA adhesin *Cell Host Microbe* **14** 195–206

Salas A, Shimaoka M, Kogan A N, Harwood C, von Andrian U H and Springer T A 2004 Rolling adhesion through an extended conformation of integrin alphaLbeta2 and relation to alpha I and beta I-like domain interaction *Immunity* **20** 393–406

Salcedo R *et al* 2010 MyD88-mediated signaling prevents development of adenocarcinomas of the colon: role of interleukin 18 *J. Exp. Med.* **207** 1625–36

Sánchez-Madrid F and del Pozo M A 1999 Leukocyte polarization in cell migration and immune interactions *EMBO J.* **18** 501–11

Sandig M, Negrou E and Rogers K A 1997 Changes in the distribution of LFA-1, catenins, and F-actin during transendothelial migration of monocytes in culture *J. Cell Sci.* **110** 807–18

Scheitz C J, Lee T S, McDermitt D J and Tumbar T 2012 Defining a tissue stem cell-driven Runx1/Stat3 signalling axis in epithelial cancer *EMBO J.* **31** 4124–39

Schenkel A R, Mamdouh Z and Muller W A 2004 Locomotion of monocytes on endothelium is a critical step during extravasation *Nat. Immunol.* **5** 393–400

Schetter A J, Heegaard N H and Harris C C 2010 Inflammation and cancer: interweaving microRNA, free radical, cytokine and p53 pathways *Carcinogenesis* **31** 37–49

Schiechl G *et al* 2011 Tumor development in murine ulcerative colitis depends on MyD88 signaling of colonic F4/80+CD11bhighGr1low macrophages *J. Clin. Invest.* **121** 1692–708

Schrieber H and Rowley D A 1999 Inflammation and cancer *Inflammation: Basic Principles and Clinical Correlates* ed J I Gallin and R Snyderman 3rd ed (Philadelphia, PA: Lippincott Williams & Wilkins) pp 1117–29

Schwartz M A 2010 Integrins and extracellular matrix in mechanotransduction *Cold Spring Harb Perspect Biol.* **2** a005066

Schwitalla S *et al* 2013 Intestinal tumorigenesis initiated by dedifferentiation and acquisition of stem-cell-like properties *Cell* **152** 25–38

Shamri R, Grabovsky V, Gauguet J M, Feigelson S, Manevich E, Kolanus W, Robinson M K, Staunton D E, von Andrian U H and Alon R 2005 Lymphocyte arrest requires instantaneous induction of an extended LFA-1 conformation mediated by endothelium-bound chemokines *Nat. Immunol.* **6** 497–506

Shaw S K *et al* 2004 Coordinated redistribution of leukocyte LFA-1 and endothelial cell ICAM-1 accompany neutrophil transmigration *J. Exp. Med.* **200** 1571–80

Sheehan K M, Sheahan K, O'Donoghue D P, MacSweeney F, Conroy R M, Fitzgerald D J and Murray F E 1999 The relationship between cyclooxygenase-2 expression and colorectal cancer *JAMA* **282** 1254–7

Singh B, Vincent L, Berry J A, Multani A S and Lucci A 2007 Cyclooxygenase-2 expression induces genomic instability in MCF10A breast epithelial cells *J. Surg. Res.* **140** 220–6

Smyth M J, Dunn G P and Schreiber R D 2006 Cancer immunosurveillance and immunoediting: the roles of immunity in suppressing tumor development and shaping tumor immunogenicity *Adv. Immunol.* **90** 1–50

Sobhani I, Tap J, Roudot-Thoraval F, Roperch J P, Letulle S, Langella P, Corthier G, Tran Van Nhieu J and Furet J P 2011 Microbial dysbiosis in colorectal cancer (CRC) patients *PLoS ONE* **6** e16393

Sonnenberg G F, Fouser L A and Artis D 2011 Border patrol: regulation of immunity, inflammation and tissue homeostasis at barrier surfaces by IL-22 *Nature Immunol.* **12** 383–90

Sonoshita M, Takaku K, Oshima M, Sugihara K and Taketo M M 2002 Cyclooxygenase-2 expression in fibroblasts and endothelial cells of intestinal polyps *Cancer Res.* **62** 6846–9 PMID: 12460897

Springer T A 1994 Traffic signals for lymphocyte recirculation and leukocyte emigration: the multiple paradigm *Cell* **76** 301–14

Steinbrink K, Jonuleit H, Müller G, Schuler G, Knop J and Enk A H 1999 Interleukin-10-treated human dendritic cells induce a melanoma-antigen-specific energy in CD8+ T cells resulting in a failure to lyse tumor cells *Blood* **93** 1634–42 PMID: 10029592

Stewart M P, McDowall A and Hogg N 1998 LFA-1-mediated adhesion is regulated by cytoskeletal restraint and by a Ca2+-dependent protease, calpain *J. Cell Biol.* **140** 699–707

Strauss J, Kaplan G G, Beck P L, Rioux K, Panaccione R, Devinney R, Lynch T and Allen-Vercoe E 2011 Invasive potential of gut mucosa-derived *Fusobacterium nucleatum* positively correlates with IBD status of the host *Inflamm. Bowel Dis.* **17** 1971–8

Straussman R *et al* 2012 Tumor micro-environment elicits innate resistance to RAF inhibitors through HGF secretion *Nature* **487** 500–4

Takai A *et al* 2009 A novel mouse model of hepatocarcinogenesis triggered by AID causing deleterious p53 mutations *Oncogene* **28** 469–78

Takala H *et al* 2008 Beta2 integrin phosphorylation on Thr758 acts as a molecular switch to regulate 14-3-3 and filamin binding *Blood* **112** 1853–62

Talmadge J E and Fidler I J 2010 AACR centennial series: the biology of cancer metastasis: historical perspective *Cancer Res.* **70** 5649–69

Thompson P W, Randi A M and Ridley A J 2002 Intercellular adhesion molecule (ICAM)-1, but not ICAM-2, activates RhoA and stimulates c-fos and rhoA transcription in endothelial cells *J. Immunol.* **169** 1007–13

Turowski P *et al* 2008 Phosphorylation of vascular endothelial cadherin controls lymphocyte emigration *J. Cell Sci.* **121** 29–37

Tye H *et al* 2012 STAT3-driven up-regulation of TLR2 promotes gastric tumorigenesis independent of tumor inflammation *Cancer Cell* **22** 466–78

Uronis J M, Mühlbauer M, Herfarth H H, Rubinas T C, Jones G S and Jobin C 2009 Modulation of the intestinal microbiota alters colitis-associated colorectal cancer susceptibility *PLoS ONE* **4** e6026

Urzainqui A *et al* 2007 Functional role of P-selectin glycoprotein ligand 1/P-selectin interaction in the generation of tolerogenic dendritic cells *J. Immunol.* **179** 7457–65

Urzainqui A *et al* 2002 ITAM-based interaction of ERM proteins with Syk mediates signaling by the leukocyte adhesion receptor PSGL-1 *Immunity* **17** 401–12

van Buul J D, Anthony E C, Fernández-Borja M, Burridge K and Hordijk P L 2005a Proline-rich tyrosine kinase 2 (Pyk2) mediates vascular endothelial-cadherin-based cell–cell adhesion by regulating beta-catenin tyrosine phosphorylation *J. Biol. Chem.* **280** 21129–36

van Buul J D, Fernández-Borja M, Anthony E C and Hordijk P L 2005b Expression and localization of NOX2 and NOX4 in primary human endothelial cells *Antioxid. Redox. Signal.* **7** 308–17

Van Coillie E, Van Aelst I, Wuyts A, Vercauteren R, Devos R, De Wolf-Peeters C, Van Damme J and Opdenakker G 2001 Tumor angiogenesis induced by granulocyte chemotactic protein-2 as a countercurrent principle *Am. J. Pathol.* **159** 1405–14

van Wetering S, van Buul J D, Quik S, Mul F P, Anthony E C, ten Klooster J P, Collard J G and Hordijk P L 2002 Reactive oxygen species mediate Rac-induced loss of cell–cell adhesion in primary human endothelial cells *J. Cell Sci.* **115** 1837–46 PMID: 11956315

Verfaillie C M 1998 Adhesion receptors as regulators of the hematopoietic process *Blood* **92** 2609–12

Vestweber D 2007 Adhesion and signaling molecules controlling the transmigration of leukocytes through endothelium *Immunol. Rev.* **218** 178–96

Viatte S, Darren Plant D and Raychaudhuri S 2013 Genetics and epigenetics of rheumatoid arthritis *Nat. Rev. Rheumatol.* **9** 141–53

Vicente-Manzanares M and Sánchez-Madrid F 2004 Role of the cytoskeleton during leukocyte responses *Nat. Rev. Immunol.* **4** 110–22

Viegas M S, Estronca L M B B and Vieira O V 2012 Comparison of the kinetics of maturation of phagosomes containing apoptotic cells and IgG-opsonized particles *PLoS One* **7** e48391

Virchow R 1858 Reizung and Reizbarkeit *Arch. Pathol. Anat. Klin. Med.* **14** 1–63

Virchow R 1863 Aetiologie der neoplastischen Geschwulst/Pathogenie der neoplastischen Geschwulste *Die Krankhaften Geschwulste* (Berlin: August Hirschwald) pp 57–101

Virchow R 1881 An address on the value of pathological experiments *Br. Med. J.* **2** 198–203

von Andrian U H and Mackay C R 2000 T-cell function and migration. Two sides of the same coin *N. Engl. J. Med.* **343** 1020–34

von Andrian U H, Hasslen S R, Nelson R D, Erlandsen S L and Butcher E C 1995 A central role for microvillous receptor presentation in leukocyte adhesion under flow *Cell* **82** 989–99

von Hundelshausen P, Koenen R R and Weber C 2009 Platelet-mediated enhancement of leukocyte adhesion *Microcirculation* **16** 84–96

Waldner M J, Foersch S and Neurath M F 2012 Interleukin-6—a key regulator of colorectal cancer development *Int. J. Biol. Sci.* **8** 1248–53

Walzog B and Gaehtgens P 2000 Adhesion molecules: the path to a new understanding of acute inflammation *Physiology* **15** 107–13 PMID: 11390891

Wang Q and Doerschuk C M 2002 The signaling pathways induced by neutrophil-endothelial cell adhesion *Antioxid. Redox. Signal.* **4** 39–47

Weber C, Fraemohs L and Dejana E 2007 The role of junctional adhesion molecules in vascular inflammation *Nat. Rev. Immunol.* **7** 467–77

Wilson T R *et al* 2012 Widespread potential for growth-factor-driven resistance to anticancer kinase inhibitors *Nature* **487** 505–9

Woodfin A, Reichel C A, Khandoga A, Corada M, Voisin M B, Scheiermann C, Haskard D O, Dejana E, Krombach F and Nourshargh S 2007 JAM-A mediates neutrophil transmigration in a stimulus-specific manner *in vivo*: evidence for sequential roles for JAM-A and PECAM-1 in neutrophil transmigration *Blood* **110** 1848–56

Woodfin A, Voisin M B and Nourshargh S 2007 PECAM-1: a multi-functional molecule in inflammation and vascular biology *Arterioscler. Thromb. Vasc. Biol.* **27** 2514–23

Wu Y, Deng J, Rychahou P G, Qiu S, Evers B M and Zhou B P 2009 Stabilization of snail by NF-κB is required for inflammation-induced cell migration and invasion *Cancer Cell* **15** 416–28

Yáñez-Mó M, Alfranca A, Cabanas C, Marazuela M, Tejedor R, Ursa M A, Ashman L K, de Landázuri M O and Sánchez-Madrid F 1998 Regulation of endothelial cell motility by complexes of tetraspan molecules CD81/TAPA-1 and CD151/PETA-3 with alpha3 beta1 integrin localized at endothelial lateral junctions *J. Cell Biol.* **141** 791–804

Yang L, Froio R M, Sciuto T E, Dvorak A M, Alon R and Luscinskas F W 2005 ICAM-1 regulates neutrophil adhesion and transcellular migration of TNF-alpha-activated vascular endothelium under flow *Blood* **106** 584–92

Yang L, Kowalski J R, Zhan X, Thomas S M and Luscinskas F W 2006a Endothelial cell cortactin phosphorylation by Src contributes to polymorphonuclear leukocyte transmigration *in vitro Circ. Res.* **98** 394–402

Yang L, Kowalski J R, Yacono P, Bajmoczi M, Shaw S K, Froio R M, Golan D E, Thomas S M and Luscinskas F W 2006b Endothelial cell cortactin coordinates intercellular adhesion molecule-1 clustering and actin cytoskeleton remodeling during polymorphonuclear leukocyte adhesion and transmigration *J. Immunol.* **177** 6440–9

Yoshimoto S *et al* 2013 Obesity-induced gut microbial metabolite promotes liver cancer through senescence secretome *Nature* **499** 97–101

Yu H, Pardoll D and Jove R 2009 STATs in cancer inflammation and immunity: a leading role for STAT3 *Nature Rev. Cancer* **9** 798–809

Zaki M H, Vogel P, Body-Malapel M, Lamkanfi M and Kanneganti T D 2010 IL-18 production downstream of the Nlrp3 inflammasome confers protection against colorectal tumor formation *J. Immunol.* **185** 4912–20

Zarbock A, Abram C L, Hundt M, Altman A, Lowell C A and Ley K 2008 PSGL-1 engagement by E-selectin signals through Src kinase Fgr and ITAM adapters DAP12 and FcR gamma to induce slow leukocyte rolling *J. Exp. Med.* **205** 2339–47

Zarbock A, Lowell C A and Ley K 2007 Spleen tyrosine kinase Syk is necessary for E-selectin-induced alpha(L)beta(2) integrin-mediated rolling on intercellular adhesion molecule-1 *Immunity* **26** 773–83

Zarek P E, Huang C T, Lutz E R, Kowalski J, Horton M R, Linden J, Drake C G and Powell J D 2008 A2A receptor signaling promotes peripheral tolerance by inducing T-cell energy and the generation of adaptive regulatory T cells *Blood* **111** 251–9

Zenewicz L A, Yancopoulos G D, Valenzuela D M, Murphy A J, Stevens S and Flavell R A 2008 Innate and adaptive interleukin-22 protects mice from inflammatory bowel disease *Immunity* **29** 947–57

Zhou J, Wulfkuhle J, Zhang H, Gu P, Yang Y, Deng J, Margolick J B, Liotta L A, Petricoin E 3rd and Zhang Y 2007 Activation of the PTEN/mTOR/STAT3 pathway in breast cancer stem-like cells is required for viability and maintenance *Proc. Natl Acad. Sci. USA* **104** 16158–63

Zwartz G J, Chigaev A, Dwyer D C, Foutz T D, Edwards B S and Sklar L A 2004 Real-time analysis of very late antigen-4 affinity modulation by shear *J. Biol. Chem.* **279** 38277–86

IOP Publishing

Physics of Cancer

Claudia Tanja Mierke

Chapter 3

Cellular stiffness and deformability

Summary

As single-molecule force spectroscopy has emerged as a powerful tool to investigate the forces and motions necessary for the malignant progression of cancer, this chapter focuses on them and discusses each method, such as magnetic and optical tweezers, atomic force microscopy and traction force microscopy, in a critical way by highlighting their limitations and elucidating their developmental potential in the light of physics-driven cancer research.

How can cellular stiffness and the deformability of cells be measured?

Forces play a fundamental role in the process of cancer development and are facilitated by molecular scale forces. Over the last two decades, the ability to study forces has been revolutionized by the development of biophysical techniques that allow measurement of the force generated by single molecules, ranging from cells to proteins, for example through analysis of the displacement of marker beads. The use of single-molecule manipulation techniques, such as optical tweezers, magnetic tweezers, AFM, micro-needle manipulation (Cluzel *et al* 1996, Kishino and Yanagida 1988), biomembrane force probing (Evans *et al* 1995) and flow induced stretching (Smith *et al* 1992, Kim *et al* 2007), to measure the forces exerted by cells is currently still expanding and undergoing steady refinement. As the first three methods are still the most important and most commonly used biophysical methods, this chapter focuses on them and additionally highlights the optical cell stretcher and traction force microscopy.

The current state-of-the-art single-molecule manipulation techniques span six orders of magnitude in length $(10^{-10}–10^{-4}\,\text{m})$ and force $(10^{-14}–10^{-8}\,\text{N})$. These methods can manipulate cells $(\sim 100\,\mu\text{m})$, measure RNA polymerase advancing a single base pair $(0.34\,\text{nm})$ along DNA (Abbondanzieri *et al* 2005) and the mechanical rupture forces of covalent bonds (Grandbois *et al* 1999) (nN), as well as the kinetics of nucleic acid folding processes (Hohng *et al* 2007) $(\sim 0.1\,\text{pN})$. The importance of these methods comes from their individual power and the breadth that is evoked by the wide variety of applications and systems in which these methods can be used. Indeed, single cells can be probed to investigate the strength and location of receptor binding interactions (Litvinov *et al* 2002) and adhesion, or to measure traction and adhesion forces (Prass *et al* 2006). In particular, viscoelastic properties can be measured on short length scales and in small volumes (MacKintosh and Schmidt 1999), or even within living cells using a live-cell imaging approach combined with an incubation chamber (Daniels *et al* 2006). Single-molecule force and displacement measurements started with the characterization of classical motor proteins such as kinesins (Block *et al* 1990, Svoboda and Block 1994) and myosins (Simmons *et al* 1993). The step size, stall-force and processivity of these motors were determined and led to the investigation of fundamental questions concerning the connection and regulation of chemical and mechanical cycles (Tinoco and Bustamante 2002). The application of external forces allows the selective deregulation of the single steps in a biochemical reaction cycle that drives cellular migration through connective tissue. Detailed analysis of the force extension relationship (for example, the elasticity of individual polymers such as nucleic acids) has been reported (Cluzel *et al* 1996, Smith *et al* 1992). As this was possible, the next step was to investigate unconventional nucleic-acid molecular motors, which are able to translocate or alter DNA or RNA in a special manner. At the same time, techniques were invented or further developed to mechanically disrupt molecular bonds (Tskhovrebova *et al* 1997, Kellermayer *et al* 1997, Rief *et al* 1997). The determination of rupture forces or even full force spectra enables us to measure whole bond energies, lifetimes and even entire energy landscapes (Hummer and Szabo 2005, Woodside *et al* 2006). Force spectroscopy is an established tool to measure ligand and antibody binding strength (Lim *et al* 2006) and has been used to investigate the multi-state unfolding of single proteins as well as nucleic acid structures (Zhuang and Rief 2003). Indeed, force spectroscopy gained attention from theoretical approaches that allow us to extract detailed equilibrium thermodynamic parameters from naturally non-equilibrium pulling experiments (Hummer and Szabo 2005, Evans *et al* 2001, Jarzynski *et al* 1997).

In their ability to apply external forces and measure their displacement, single-molecule techniques exhibit the following advantages. Firstly, the measurement of single molecules avoids the problems associated with population averaging. Secondly, rare or transient phenomena can now be detected, as the required resolution is available and the events can be measured often enough to ensure that this is no artifact. Thirdly, it is possible to measure multi-state or multi-species distributions, together with static and even dynamic enzymatic heterogeneity (Lu *et al* 1998). Fourthly, it is even possible to obtain kinetic rates from single-molecule measurements or by investigation of the event time distributions (Xie and Lu 1999, Hua *et al* 1997, Schnitzer and Block 1997). Fifthly, the processivity of enzymes can be analyzed by intrinsically synchronizing single-molecule measurements.

3.1 Magnetic tweezers

In the last two decades, the biological and medical sciences have reached a completely new phase of measurement techniques. Until now, genomics, proteomics and metabolomics have revealed numerous components, physical interactions and chemical reactions that occur in the living cells under certain conditions or at a specific disease or developmental step. As we have detected all the components, we need integrate them into networks of molecular signaling pathways, which are suggested to regulate fundamental processes. Among these processes are the orchestration of gene expression, intermediary metabolism, outside–in or inside–out signal transduction and cell cycle control, all of which are precisely controlled in both time and space. These regulatory interacting cycles are the basis of life on a molecular scale and they are needed to reveal the pathological state of cells such as tumorigenesis in order to be able to design drugs and develop appropriate therapies.

Indeed, a wide variety of methods have been established to analyze the localization, dynamics and interactions of molecules in living cells. The use of the green fluorescent protein together with live cell microscopy, fluorescence recovery after photobleaching and fluorescence resonance energy transfer has delivered a major breakthrough in the field of cancer research (Wouters et al 2001). In particular, these methods provide insights into molecular processes in living cells, but their limitation is that they can only passively reveal these processes in time and space.

A number of analysis methods have been invented in order to nanomanipulate biological systems and perform single molecule research. The AFM (Viani et al 1999), optical tweezers (Svoboda et al 1993), optical cell stretcher (Guck et al 2000, 2005) and magnetic tweezers (Strick et al 1996, Fabry et al 2001) are commonly used to study the biophysical behavior of individual macromolecules or living cells. In particular, molecules are usually connected to micron-sized latex and/or magnetic beads, allowing the exact movement in the nanometer-range and positioning of the molecule as well as the measurement. Moreover, the application of forces on these molecules or on cells that bind these molecules via their cell surface receptor can be performed in the physiologically relevant picoNewton (pN) range. In more detail, the dynamics of biomolecular systems such as the movement of a single RNA-polymerase along a DNA double helix (Davenport et al 2000), the movement of kinesin on microtubules in order to transport a vesicle (Howard et al 1989) and even the analysis of chromatin structure, can be investigated (Pope et al 2002). However, these techniques are still limited, as they have many problems regarding the measurement of molecular systems inside of living cells or even tissues. There are special requirements for these methods to be adopted for the nanomanipulation of objects located in the cytoplasm of living cells. First, the force objects used must be small (1 mm); secondly, the exerted forces should be in the relevant physiological range of a few pN or even nN; and thirdly, it would be good to control the forces in their amplitude and direction in order to move the force probe to the targeted site.

A good improvement is the use of magnetic forces in order to meet the prerequisites for the proper measurements. The state-of-the-art magnetic tweezers to analyze living cells rely mainly on microrheology (Bausch et al 1998, 1999). Indeed, it has been described that large forces can be exerted on magnetic beads that are located inside living cells (Hosu et al 2003). In more detail, these magnetic tweezers are constructed of one or two poles and thus the manipulation of magnetic beads in different directions and even the change of one direction into another are possible. A major limitation of this method is that it is technically very difficult to establish and the required speed and accuracy for the repositioning of the pole(s) is needed for exact manipulation. Instead, there are more practical methods, consisting of multiple magnetic poles, with 2D or 3D manipulation therefore possible, while only low forces can be applied (Amblard *et al* 1996, Gosse and Croquette 2002, Huang *et al* 2002, Sacconi *et al* 2001).

However, a new magnetic tweezers type has been reported that uses micron scale magnetic poles, which can be in close neighborhood to living cells (de Vries *et al* 2005). This technique allows us to combine high forces with the ability of a 2D manipulation of the magnetic beads and subsequently of molecular systems inside living cells. In comparison with magnetic tweezers, the technique of optical tweezers has the major advantage of a possible 3D manipulation in order to stimulate beads inside cells or on cell surfaces for microrheology (Caspi *et al* 2002, Laurent *et al* 2002) or molecular manipulation (Peters *et al* 1999). By contrast, the advantages of magnetic tweezers are seen when intracellular properties are explored. The restriction of optical tweezers is that they exert forces on microscopic objects that have a higher or lower refractive index compared to the local surroundings. Another disadvantage is that there is a myriad of objects inside a cell and thus optical tweezers cannot act selectively in the intracellular microenvironment. Moreover, the relatively high optical intensities needed for the proper measurement damage the living cells (Neumann *et al* 1999).

Microrheology

The physical properties such as the viscoelasticity of the cell's cytoplasm are crucial for numerous cellular processes such as the intracellular vesicle or compartment transport and cellular motility. However, the measurement of this viscoleastic property is affected by the high degree of diversity among the cytoplasms of cells in a population due to different cell-cycle phases or different degrees of stimulation and hence large forces (in the nano-Newton range) are required to determine the viscoelastic moduli of the cells. Indeed, a highly promising technique is microrheology involving magnetic beads bound to the cells, which was invented roughly 40 years ago (Amblard *et al* 1996, Crick and Hughes 1949, Ziemann *et al* 1994). In former days, due to technical difficulties in generating high local magnetic fields and magnetic field gradients, as well as the poor commercially availability of large spherical magnetic beads (10 μm in diameter), the applicability of quantitative measurements was restricted to large cells such as sea urchin eggs (Hiramoto 1969a, 1996b). However, the viscoelastic properties of smaller cells such as macrophages were measured by the torsional deformations inside of living cells that are evoked by magnetic twisting of magnetic beads (Valberg and Butler 1987, Valberg and Feldman 1987, Möller *et al* 1997, Mijailovich *et al* 2002). This technique is called magnetic twisting cytometry and has delivered the apparent viscosities of the cell's cytoplasm as well as non-linear viscoelastic behavior. So far only a few measurements of forces exerted inside moving cells have begun and thus they are not yet known and further effort is needed to obtain them. The first measurement of internal forces was reported in 1995 (Guilford *et al* 1995). In this report, they were able to use magnetic beads with a diameter of 6 µm, which where then phagocytized by macrophages. Another microrheometer measurement setup is capable of generating forces in the nano-Newton range on superparamagnetic as well as ferromagnetic beads that both have diameters smaller than 5 µm. Using conventional and exact particle tracking techniques, the bead deflection was analyzed in order to determine the viscoelastic response and draw the recovery curves with a spatial resolution of 10 nm and a time resolution of $t = 0.04$ s. Moreover, this biophysical technique has been applied to analyze the viscoelastic moduli of fibroblasts' plasma membranes (Bausch *et al* 1998). When comparing this particular method with others that enable us to determine local measurements of viscoelasticity from outside the cells (Daily *et al* 1984, Radmacher *et al* 1996), the biophysical technique described above even allows local measurements of the inside and outside of single living cells.

In another report, the first local analysis of the viscoelastic moduli of the cytoplasm of macrophages was performed using viscoelastic creep experiments, in which solid and filled ferromagnetic beads with a diameter of 1.3 µm are used (Bausch *et al* 1999). In addition to this, an attempt was made to measure the displacement field generated by local forces by the single observation of the induced motion of non-magnetic colloidal objects within cells (Bausch *et al* 1998). It has been reported that transgressing elastic fields exist in cells (Maniotis *et al* 1997), whereas there has been no long-range elastic coupling within the cells. Further, by analyzing the magnetophoretic motion of the beads, to which a constant magnetic force has been applied, local active opposing forces generated by a cell's cytoplasm can be detected. Taken together, micromanipulation and microrheology methods such as AFM, micropipettes, optical tweezers, magnetic tweezers and microneedles still have important applications in biophysics, cell biology and cancer research. Historically, the use of magnetic forces in biology was reported by Heilbronn, who inserted magnetic particles into protoplasts and analyzed their movement in a magnetic field gradient (Heilbronn 1922). In line with this, Freundlich and Seifriz investigated the movement of particles in a magnetic field gradient in echinoderm eggs (Freundlich and Seifriz 1923). In addition, Crick and Hughes analyzed the viscoelastic response of chick fibroblasts (Crick and Hughes 1949) and then both Yagi (Yagi 1961) and Hiramoto (Hiramoto 1969a, 1996b) further improved the magnetic particle tracking method in order to provide a reliable tool for investigating the viscoelasticity of living cells. Numerous researchers have used magnetic tweezers and even further improved and refined different selected variants of this biophysical method. Finally, the magnetic tweezers technique has become an important tool for mechanically stimulating biomolecules in soft materials such as living cells. For example, multidimensional magnetic tweezers consisting of several magnetic poles have been developed, which are used to move magnetic beads or objects in certain directions (Amblard *et al* 1996, de Vries *et al* 2005, Fisher *et al* 2005). A disadvantage of this method is that the magnitude of force that can be applied is strictly limited. An improvement has been the exchangeable poles of the magnetic tweezers, in which the pole geometry is altered depending on the desired experimental setup (Fisher *et al* 2006). In addition to the multidimensional magnetic tweezer, two-pole experimental setups can be built to apply high and even alternating forces between two opposing magnets (Sato *et al* 1983, Guilford and Gore 1992, Ziemann *et al* 1994, Barbic *et al* 2001). As a variation of this setup, Guilford and Gore (Guilford and Gore 1992) used the optical tracking and current feedback approach in their two-pole magnetic trap, in which they were able to exert external forces of up to 80 nN on spherical metal beads of 10 µm diameter. Indeed, this amount of force is needed and is sufficient for performing tissue studies, whereas it is not suited to single cell experiments.

Another option of the setup is to use vertical magnetic tweezers and hence the vertical gradient between two magnetic poles placed above the sample object's surface. This special geometry has been used for DNA unfolding experimental approaches and adhesion experiments, however, while it is able to generate a homogenous gradient field, it is rather limited regarding the small forces, such as 0.2 nN, that can be applied to the magnetic objects (Barbic *et al* 2001, Assi *et al* 2002, Gosse and Croquette 2002, Walter *et al* 2006). It is known that the highest forces that can be exerted are generally obtained using one-pole microneedle geometric approaches. For these one-pole setups two kinds of magnetism can be implemented. A permanent magnet can be installed (Matthews *et al* 2004) or electromagnets with soft iron cores can be integrated (Bausch *et al* 1998, Alenghat *et al* 2000, Huang *et al* 2005, Overby *et al* 2005). Another fundamentally different setup is the use of homogenous magnetic fields to rotate permanently magnetized particles (magnetic twisting), as provided in magnetic twisting cytometry (Valberg and Albertini 1985, Wang *et al* 1993, Fabry *et al* 2001). A major weakness of this method is the small forces that can be applied, which are in the range of 500 pN (Mierke 2011).

A major breakthrough in the field of magnetic-based biophysical methods has been the improvement of the magnetic tweezers method by increasing the magnitude of the exerted forces by an easily reproducible setup for magnetic tweezers, which can be constructed easily in all kinds of research laboratories. In more detail, this setup contains a single solenoid with a high-permeability soft iron core, called µ-metal, that is attached to a remote-controlled commercially available micromanipulator. Now, high forces of up to 100 nN can be applied to 5 µm superparamagnetic beads by reaching distances between the core metal and superparamagnetic bead that are in the range of 10 to 30 µm. A problem may be that the tweezers needle is probably not so far away from the cell with the attached magnetic bead and may come into contact with the cell, which leads to a mechanical stimulation of the cell of interest. The bead position has to be monitored by tracking the bead on the cell's surface, without any external force application. When the displacement of the bead follows a straight line, which is parallel to the *x*-axis in a certain time interval (10 s), the measurement can then be started by applying the external force on the superparamagnetic bead (Mierke *et al* 2008a, 2008b). During and after force application the bead displacement will be monitored and plotted over time (classical creep experiment). An inherent problem of this setup for magnetic tweezers is the poor control of force, as there is a highly non-linear force–distance relationship. However, this problem can be solved by constantly tracking the distance between superparamagnetic bead and the magnetic core of the electromagnet. If there is an alteration in the distance, this change can be compensated for by adapting either the solenoid current or moving the core position using the moveable micromanipulator. Sometimes, both compensations revealed the best result for the measurement, as the significant movement of the tweezers needle disturbs the medium and may induce local disturbance and high currents may head up the tweezers needle, which may increase the temperature of the cell setup within the culture dish. In more detail, the force is calibrated using the classical viscous drag approach (Yagi 1961). An empirical fit can be applied to describe the non-linear relationship between the distance, the current and the exerted force using an analytical formula. However, this approach provides exact timing and force accuracy opportunities and can even be mounted on all commonly available inverted microscopes that carry a charge coupled device (CCD) camera.

There are at least two major advantages to this magnetic tweezers technique. The first is that the forces can be applied locally to certain structures of interest and targeted to certain receptors or molecules displayed on the cell's surface. The second is that the cells can be measured while cultured on a 2D surface. However, whether the cell can also be measured within 3D networks near the surface remains elusive. It seems to be likely that this application can be adapted to the current setups by applying higher forces toward the superparamagnetic beads and increasing the distance between the magnetic tweezers core and the superparamagnetic bead.

3.1.1 Adhesion forces

In order to investigate adhesion forces, the detachment of receptor-bound beads can be measured by step-wise increasing the external force applied to the superparamagnetic beads using a magnetic tweezers device. Before the measurements start, it is important to decide whether single cells, cell clusters or even cell monolayers will be analyzed, as the external applied forces propagate throughout the cell cluster or cell monolayer, indicating that the cytoskeletons of the individual cells are connected between neighboring cells (Raupach et al 2007). For these measurements the superparamagnetic beads are coated with a ligand for a special cell surface receptor or with an antibody directed toward a certain target on the cell surface. In order to investigate the detachment of beads bound to a certain cell surface receptor, the cells are seeded into culture dishes, which may or may not be surface coated with a special ligand for a cell-matrix receptor. After one or two days of cell culture at 37°C, 5% CO_2, and 95% humidity, the cells are then incubated with special coated beads for 30 min under the same conditions. In more detail, the detachment of the coated superparamagnetic beads from the cells is measured during external force application ranging from 0.5 to 10 nN using the magnetic tweezers technique. The percentage of detached beads in relation to the force required for the bead detachment is used to quantify the bead binding strength to the targeted receptor on the cell surface. The specificity of the binding and hence of the detachment forces can be determined by using an inhibitory antibody directed against the receptor or by using an excess of soluble ligand to block the binding of bead-bound ligand, applied 30 min before the coated beads are added to the cells. However, an alternative way of determining the detachment forces will be discussed below, in the AFM section.

3.1.2 Overall cellular stiffness and fluidity

The magnetic tweezers setup is extremely well suited for creep measurements, where, for example, a staircase-like sequence of step forces ranging from 0.5 to 10 nN is applied to superparamagnetic epoxylated beads of a diameter of 4.5 μm, which are coated with the extracellular matrix protein fibronectin (FN) (100 μg ml^{-1}) in order to investigate the interaction between the integrins and their ligand fibronectin (Mierke et al 2008a, 2008b, McNulty et al 2009). For the proper measurement it is important that the coated beads do not cluster after addition to the cells, otherwise they cannot be tracked accurately. The best option is to have just one bead attached to one cell in order to apply the external force locally and to calculate the exact amount of force toward this cell. Thus, only those cells carrying one bead will be chosen for the measurement. Because of the bead concentration and the amount of bound protein on the bead's surface it is possible to have most of the beads attached just to one cell without selecting a minority of cells for the magnetic tweezers measurement.

When an external force step ΔF is applied to a cell-bound bead, the bead moves and thus a displacement d(t) toward the tip of the tweezers needle can be detected. The ratio of d(t)/ΔF defines a creep response $J(t)$. The creep response $J(t)$ of the cells follows a power law with time, hence, the equation is $J(t) = a(t/t_0)^b$, where both factors a and b are force-dependent and the reference time t_0 is set to 1 s. The factors a and b are then determined by a least squares fit (Mierke et al 2008a). The prefactor a (in units of μm nN^{-1}) characterizes the elastic properties of the cell and corresponds to a compliance, which is indeed the reciprocal of the stiffness (Mierke et al 2008a). The power law exponent b reflects the dynamics of the force-bearing structures of the cell that are directly connected to the bead, such as the acto-myosin cytoskeleton (Fabry et al 2001). In more detail, if the power law exponent b is zero, then this is indicative of a purely elastic solid. However, if b is one, then this behavior is indicative of a purely viscous fluid (Fabry et al 2001). As is true for quite a few soft matter materials, such as cells, the power law exponent b usually falls in a range between 0.1 and 0.5, whereas higher values of b are associated with a higher turnover rate of cytoskeletal structures and proteins. In addition, higher values of b are often associated with a reduced cellular stiffness, indicating that the protein turnover or the de- or reassembly of the structure affect the mechanical properties of the cell (Fabry et al 2001).

Magnetic tweezers have been used to investigate the functional role of vinculin in providing mechanical properties such as cellular stiffness and in influencing cytoskeletal remodeling dynamics. How does vinculin affect cytoskeletal dynamics and cellular stiffness in order to facilitate cellular motility? In order to investigate whether altered 3D motility of vinculin-expressing wildtype cells is regulated through altered cellular stiffness and remodeling dynamics of the cytoskeleton, microrheology measurements are performed. In particular, using the magnetic tweezers technique, forces of up to 10 nN can be applied to FN-coated super-paramagnetic beads that are bound to certain integrin receptors exposed on the cell surface. The bead displacement after the first force application or after a further step of an increased force application, normalized to the force magnitude, defines the creep response $J(t) = a(t/t_0)^b$. This creep response consists of an elastic response (cell elasticity or stiffness, $1/a$) and a frictional response (cytoskeletal fluidity, b) (Mierke et al 2008b, Fabry and Frederg 2003). Vinculin knockout embryonic fibroblasts (MEFvin−/− cells) possess a higher cytoskeletal fluidity and a lower stiffness compared with Vinculin wildtype embryonic fibroblasts (MEFvinwt/wt cells), indicating that MEFvin−/− cells are more deformable. Hence this increased deformability may help them to squeeze more easily through a dense extracellular matrix compared to the stiffer MEFvinwt/wt cells. However, these MEFvin−/− cells are less invasive into 3D extracellular matrices compared to the MEFvinwt/wt cells. Thus, the higher cytoskeletal fluidity and increased deformability of MEFvin−/− cells cannot explain their decreased invasiveness (Mierke et al 2010).

In addition, the magnetic tweezers experiments revealed that in both cell lines, the cytoskeletal fluidity and stiffness increased with elevated external forces on the cell-connected beads. However, this behavior indicates that both MEFs respond to external applied forces by showing stress stiffening and, surprisingly, increased fluidization of the actomyosin cytoskeleton, which seems to be independent of vinculin's mechano-coupling function. How does this behavior depend on the contractility of the cells? Therefore, the inhibition of contractile forces after the addition of the myosin light chain kinase inhibitor ML-7 has been determined. The finding was that both cell lines displayed increased cellular fluidity. This finding suggests that the cellular fluidity depends mainly on the cell contractility. Interestingly, after the addition of ML-7 the cellular stiffness of both MEFs decreased to similar levels, indicating that the higher initial stiffness of the MEFvinwt/wt cells prior to ML-7 addition is provided by higher actomyosin-mediated contractile forces, which have also been measured independently using traction force microscopy (Mierke et al 2010, Mierke et al 2008a). As the MEFvinwt/wt cells are able to generate and transmit higher contractile forces to their microenvironment compared to MEFvin−/− cells, this result seems to be a good explanation for the increased invasiveness into dense 3D extracellular matrices, where cells need to transmit contractile forces in order to move through the matrix network.

3.2 Optical cell stretcher

The optical stretcher (usually called the optical cell stretcher) has become an important and useful device for studying the mechanical properties of cells that are in suspension at the measurement time point. The important limitation for the measurement of adherent cells is now overcome by a detailed and precise measurement protocol and the surprisingly suitable measurement of rounded single cells in suspension; these cells display differences in their cellular mechanics, even when they are rounded up. However, there might be adherent cell types, where differences in cellular mechanics are detected, but this could be confirmed using an optical cell stretcher device. So, a major criticism of the optical cell stretcher method has been removed for many cell types and genetically altered or even stimulated cell types. There seems to be even more potential for this method if it is combined with a commercially available cell sorter unit, or if fluorescent molecules lined to the protein of interest during the (very short) cell stretching process are used. Hopefully, it will even be possible to measure whole tissue samples in a native manner without any additional and hence artificial cell culture.

A short introduction to the historical development of the optical stretcher
The development of the laser traps technique in the light of biomedical research such as cancer research began at least three decades ago, as these laser traps have been implemented to manipulate small soft matter objects ranging in size from atoms to cells (Ashkin 1970, Chu 1991, Svoboda and Block 1994). The fundamental general principle underlying laser traps is the momentum transfer from the laser light to the sample. Due to Newton's second law, this momentum transfer exerts a force on the sample. Before the optical cell stretcher was established, these optical forces evoked by the momentum transfer were only used to trap a sample. The best-known optical trap is the one laser beam gradient trap, called optical tweezers (Ashkin *et al* 1986). Optical tweezers have developed as a promising tool for cell and molecular biological research. In particular, they have been used for trapping cells (Ashkin *et al* 1987, Ashkin and Dziedzic 1987), measuring the forces exerted by molecular motors such as myosin or kinesin (Block *et al* 1990, Shepherd *et al* 1990, Kuo and Sheetz 1993, Simmons *et al* 1993, Svoboda *et al* 1993), determining the swimming forces of sperm (Tadir *et al* 1990, Colon *et al* 1992) and studying the polymeric properties of single DNA strands (Chu 1991).

The optical cell stretcher was developed later, and in contrast to the optical tweezers technique, it is based on a double laser beam optical trap (Ashkin 1970, Constable *et al* 1993). In this setup, two opposed, but slightly divergent and identical laser beams with a Gaussian intensity profile, can trap a sample such as a living cell flowing in a liquid object by holding it in the middle of the two laser beams (figure 3.1). This trapping of a sample is stable when the total net force on the sample is zero and restoring. This special condition is provided exactly when the refractive index of the sample is higher than that of the surrounding fluid and when the beam sizes are larger than the trapped sample. However, in extended samples such as living cells, the momentum transfer primarily occurs at the cell's surface. Indeed, the total force acting on the center of gravity is zero because the geometry of the two-beam optical trap is symmetric and all the resulting surface forces cancel each other out. Nevertheless, when the sample is sufficiently elastic, as with living cells, the forces acting on the cell's surface stretch the sample along the beam axis (figure 3.1) (Guck *et al* 2000). At first glance, this optical stretching may seem to be counterintuitive. However, there is a simple explanation. It is current knowledge that light, including laser light, carries momentum, and when a ray of light is reflected or refracted at an interface between two media with different refractive indices, the light will change the direction or velocity and subsequently its momentum is altered. Due to the law of momentum conservation, some momentum is transferred from the laser light to the sample's interface and in order not to disobey Newton's second law a force is exerted on the sample's surface.

To illustrate what happens in the sample when the two laser beams are applied to it, we can consider that a ray of light passes through a cube of optically denser

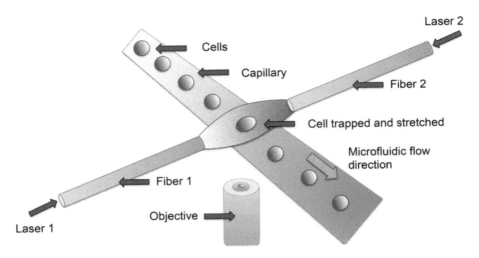

Figure 3.1. Optical cell stretcher principle.

material. What happens? As the light enters the dielectric object, it gains momentum so that the surface gains momentum in the opposite direction. In addition, the light also loses momentum when it leaves the dielectric object: at this point, the opposite surface gains momentum in the direction of the laser light propagation. Moreover, the reflection of the laser light on both surfaces of the sample also leads to a momentum transfer on both surfaces in the direction of the laser light propagation. This contribution to the surface forces is smaller than the contribution that arises from the increase of the laser light's momentum inside the cube. In summary, the two resulting surface forces at the front and back sides of the sample are in opposite directions and thus stretch the sample (Guck *et al* 2000). However, the asymmetry between the surface forces of the sample leads to a total force that acts on the center of the cube. If there is a second, identical ray of laser light, as in the optical cell stretcher approach, which passes through the cube from the opposite side, there is no total force on the cube, but the forces on the surface generated by the two rays of laser light are then additive. In contrast to asymmetric trapping geometries, where the total force is equal to the trapping force (called optical traps), the optical cell stretcher exploits surface forces to stretch the samples. High laser light power, in the range 800 to 1200 mW, can be used in each beam, which leads to surface forces of up to hundreds of picoNewton acting on the sample's surface. However, there is no excessive radiation damage to the cells observed by the laser beams, as these beams are not focused when using the optical cell stretcher, which reduces the light flux through the analyzed cells compared to other optical traps such as optical tweezers. In several reports (Guck *et al* 2000, 2005 and Schulze *et al* 2012), the optical cell stretcher has been used to demonstrate cellular deformability due to altered gene expression, altered cell age or altered aggressiveness of cancer cells.

Human erythrocytes such as red blood cells (RBCs) have served as a simple model system and have thus been analyzed. These RBCs have some advantages for use as a model system for optical cell stretching, because they have no internal organelles, hemoglobin is homogeneously distributed within their cytoplasm and in addition they have a nearly spherical shape when they are osmotically swollen. Taken together, they model very closely an isotropic, soft, dielectric sphere without any internal structures that they can be used for stress profile calculations. Moreover, they are soft cells and hence their deformations due to optical stretching can be easily analyzed. Another advantage of the use of RBCs is that their elastic properties are well known (Bennett 1985, 1990, Mohandas and Evans 1994). In particular, the only elastic component of RBCs is a thin biomembrane that is composed of a phospholipid bilayer. On the inner side of the membrane is a triagonal network composed of spectrin filaments and on the outside there is a glycocalyx network (Mohandas and Evans 1994). In more detail, the ratio between cell radius r and the thickness h of the membrane is $r/h = 100$. The bending energy is very small compared to the stretching energy and hence can be omitted. Linear membrane theory predicts the deformations of RBCs upon exertion of surface stresses using the optical cell stretcher technique.

A much better sample are the BALB 3T3 fibroblasts that, unlike RBCs, have an extensive 3D network (called the cytoskeleton) of protein filaments throughout the whole cytoplasm with elastic properties (Lodish et al 1995). This cytoskeleton contains semiflexible actin microfilaments, rod-like microtubules and flexible intermediate filaments that build up a complicated 3D scaffold network with the help of accessory proteins (Adelman et al 1968; Pollard 1984, Elson 1988, Janmey 1991). It has been shown that the classical concepts used in polymer physics fail to explain how these three types of filaments provide the mechanical stability of the cells (MacKintosh et al 1995), but it is known that in most cells microfilaments are a principal determinant of mechanical strength and stability (Stossel 1984, Janmey et al 1986, Sato et al 1987, Elson 1988). However, over the last decade this hypothesis has changed and there is evidence that intermediate filaments also contribute to the mechanical properties of cells (for example, deformability) in order to provide cellular functions such as motility (Seltmann et al 2013).

The actin cortex is a thick ($r/h = 10$) homogeneous layer located underneath the cell membrane. When the cells adhere to an extracellular matrix substrate or cell culture plastic, they form additional bundles of actin microfilaments (called F-actin stress fibers) that then lead to the formation of focal adhesion plaques (called focal adhesions) spanning over the entire cytoplasm of the cell. The dynamic remodeling of the F-actin cytoskeleton facilitates cellular functions such as motility and cytoplasmic cleavage, which occurs as the last step of the mitosis process (Pollard 1986, Carlier 1998, Stossel et al 1999).

In order to show that the actin cytoskeleton is responsible for the mechanical properties of cells, they are pronouncedly softened upon addition of actin-disrupting drugs such as cytochalasins (Petersen et al 1982, Pasternak and Elson 1985) and gelsolin (Cooper et al 1987), indicating the importance of actin in providing the mechanical properties of cells. Frequency-dependent AFM-based microrheology revealed that fibroblasts exhibit the same viscoelastic response as homogeneous artificial actin networks in vitro (Mahaffy et al 2000). Another experimental approach using rat embryo fibroblasts showed that actin plays an important role in providing the mechanical responses upon the deformation caused by glass needles (Heidemann *et al* 1999). In particular, actin and microtubules were tagged with green fluorescent protein and the role of these two cytoskeletal components in determining the cell's shape during mechanical deformation was visualized directly using a fluorescent microscope. Indeed, actin has been reported to be almost exclusively responsible for the cell's elastic response, whereas the microtubules displayed solely fluid-like behavior.

In non-mitotic (interphase) cells, the microtubules are located radially outward from the microtubule-organizing center in close neighborhood to the cell's nucleus (Lodish *et al* 1995). Microtubules can be seen as tracks for motor proteins such as dynein and kinesin, which are able to transport vesicles through the cell's cytoplasm. Another function of microtubules is the separation of chromosomes during mitosis (Mitchison *et al* 1986, Mitchison 1992). The third type of the main cytoskeletal components are intermediate filaments, which are flexible polymers and are specific to certain differentiated cell types such as keratinocytes (Herrmann and Aebi 1998, Janmey *et al* 1998). Vimentin is an intermediate filament that is expressed in mesenchymal cells such as fibroblasts. In more detail, vimentin fibers terminate at the cell's nuclear membrane and at hemidesmosomes (cell–matrix adhesion) or desmosomes (cell–cell adhesion) on the cell's membrane. Another type of inter-mediate filament is lamin, which builds up the nuclear lamina (a polymer cortex underneath the nuclear membrane) (Aebi *et al* 1986). Intermediate filaments have been reported to be co-localized with microtubules, suggesting an interaction between these two cytoskeletal components. In the past, both microtubules and intermediate filaments have been reported to be less important in providing the elastic strength and structural response of cells upon which external stress is being applied (Petersen *et al* 1982, Pasternak and Elson 1985, Heidemann *et al* 1999, Rotsch and Radmacher 2000). It is suggested that intermediate filaments are important in large deformations that cannot be achieved with deforming stresses in the range of several Pascals. In more detail, intermediate filaments are shown to be more important for cellular elasticity in adherent cells compared to suspended cells, where the initially fully extended filaments become slack (Janmey *et al* 1991, Wang and Stamenovic 2000). However, this is still under investigation and needs to be further confirmed by additional experiments.

A quantitative description of the cytoskeletal contribution to a cell's viscoelasticity is still elusive. The optical cell stretcher can measure the viscoelastic properties of the entire cell's cytoskeleton and is hence able to shed light on the regulation of cellular elasticity. The entire cytoskeleton is an intricate polymer network and a structural framework that provides the shape of a cell and its mechanical rigidity (Elson 1988, Lodish *et al* 2000). It has been shown that enormous external pressures can be endured by cells with their integrity still being preserved (Janmey *et al* 1991). The momentum carried by light and the forces it can exert on material objects such as cells is fast and even not noticeable on macroscopic length scales. The relatively small forces exerted by light are ideally suited to the deformation of microscopic small objects such as living cells and may additionally serve as a detection marker (prognostic marker) for cytoskeleton-altering diseases such as cancer (Guck *et al* 2005).

The capability of cells to withstand external deforming stresses is essential for tissue integrity and homeostasis and has led to the development of many techniques to probe cellular elasticity. In particular, AFM (Radmacher *et al* 1996), manipulation with microneedles (Felder and Elson 1990), microplate manipulation (Thoumine and Ott 1997) and cell poking (Dailey *et al* 1984) are not able to detect small variations in cellular elasticity because these biophysical devices have a high spring constant compared to the elastic modulus of the sample material. However, the AFM technique has been significantly improved for the measurement of cellular elasticity by attaching micron-sized beads to the cantilever's scanning tip in order to decrease the pressure that is applied to the cell sample (Mahaffy *et al* 2000). There is a minor problem with the micropipette aspiration of cellular segments (Discher *et al* 1994) and the displacement of surface-attached microspheres (Wang *et al* 1993), as they can lead to inaccurate measurements if the cellular membrane detaches from the cytoskeleton and hence uncouples the membrane's stiffness from the overall cellular stiffness during the deformation step. In addition, a major disadvantage is that all of these techniques are very laborious and can only analyze the elasticity over a small part of a cell's surface. The whole cellular elasticity can instead be measured indirectly by determining the compression and the shear moduli of densely packed cell pellets (Elson 1988, Eichinger *et al* 1996) or by using microarray assays (Carlson *et al* 1997). Microarray assays can be used to dissect the impact of various interacting factors on cellular functions and phenotypes by simultaneously synthesizing more than 1,000 unique microenvironments with automated nanoliter liquid-dispensing technology (Ranga *et al* 2014). Using this novel 3D microarray technique, the combined effects of matrix elasticity, proteolytic degradability and three distinct classes of signaling proteins on cells (such as mouse embryonic stem cells) can be observed simultaneously to unveil a comprehensive map of the interactions responsible for self-renewal. This novel approach can be used to reveal multifactorial 3D cell–matrix interactions, depending on the microenvironmental properties, such as matrix stiffness (Ranga *et al* 2014). A limitation of this technique is that these measurements only represent an average value rather than a reliable single-cell measurement and depend on adhesion forces such as cell–cell and cell–matrix adhesion. Alternatively, the optical cell stretcher is a technique that is able to solve most of the problems of the other biophysical techniques described above, but in addition it allows us to measure large numbers of individual cells by the incorporation of an automated flow chamber fabricated using modern soft lithography techniques in order to guide the detached cells through the laser beams.

Using optical cell stretching, the cytoskeletal properties of the cells can be measured, providing not only the mechanical rigidity, but also many important cellular functions, such as cellular motility and transendothelial migration behavior (Lodish *et al* 2000). The cytoskeleton composed of actin, microtubules and intermediate filaments works together with accessory proteins regulating cellular motility, ribosomal and vesicle transport, mitosis and mechano-sensing and mechano-transduction processes (Wang *et al* 1993, Wirtz and Dobbs 1990). All of these cellular functions are precisely regulated by cytoskeletal elements that are finely tuned and even synchronized with the overall cellular function. Subsequently, alterations of the cellular function during differentiation (Olins *et al* 2000) or because of disease are manifested in changes to the cytoskeleton. For example, cytoskeletal alterations can evoke capillary clogs in circulatory problems (Worthen *et al* 1989) and lead to various blood diseases, such as sickle-cell anemia, hereditary spherocytosis or immune hemolytic anemia (Bosch *et al* 1994, Williamson *et al* 1985), while genetic disorders of intermediate filaments and their cytoskeletal networks can cause problems with the skin, hair, liver and colon, in addition to motor neuron diseases such as amyotrophic lateral sclerosis (Fuchs and Cleveland 1998, Kirfel *et al* 2003).

However, the most common example of the malignant transformation of cells is cancer, during which the morphology of cells changes due to changes of the cytoskeleton. These changes are diagnostic for cancer. During the cell's malignant transformation from a fully mature and postmitotic state to a replicating, highly motile and immortal cancer cell, the cytoskeleton undergoes changes from an ordered and rigid structure to a highly irregular and compliant state. In particular, the alterations include a decreased quantity of constituent polymers and accessory proteins, as well as a restructuring of the cellular network (Cunningham *et al* 1992, Katsantonis *et al* 1994, Moustakas and Stournaras 1999, Rao and Cohen 1991). As expected, these cytoskeletal alterations serve as markers for the malignancy of cells mirrored in their increased replication and enhanced motility, both of which are inconsistent with a less dynamic cytoskeleton. In summary, these alterations in the cytoskeletal components and their concentrations as well as the cytoskeletal structure seem to be manifested in the overall mechanical properties of a cell. Hence, measuring a cell's rigidity may provide information about its malignant state and can be used as a novel marker of cellular malignancy.

There are currently only a few techniques capable of assessing cellular mechanical properties, but they all suggest a correlation between cellular rigidity and the cellular malignant stage. From a historical point of view, the state-of-the-art technique has been the micropipette aspiration method (Hochmuth 2000). Indeed, using this micropipette aspiration technique, a 50% reduction in the elasticity of malignantly transformed fibroblasts as compared to their healthy controls has been observed (Ward *et al* 1991). As an alternative method, AFM has been used for the investigation of cellular rigidity (Mahaffy *et al* 2000, Rotsch *et al* 1999). In particular, normal human bladder epithelial cell lines and complimentary cancerous cell lines have been analyzed using AFM (Lekka *et al* 1999). They found that the rigidity of cancerous cells and their healthy counterparts differed by an order of magnitude. It has been reported that for small indentations (up to 500–1000 nm) the actin filaments are mainly responsible for providing the mechanical properties of the cells, whereas the disruption of microtubules has no pronounced effect on the mechanical properties of cells measured by AFM (Rao and Cohen 1991, Yamazaki *et al* 2005). Cells possess *in vitro* Young's modulus values in the range of 1 to 100 kPa (Radmacher 1997, Dulińska *et al* 2006), which includes vastly different types of cells such as vascular smooth muscle cells, fibroblasts, bladder cells, RBCs and epithelial cells. As such different cell types are being measured, the large variation in the Young's modulus is fully justified.

However, the Young's modulus varied even within human lung carcinoma cells from 13 kPa to 150 kPa (Weisenhorn *et al* 1993). Additional studies reported, for example, that the vinculin-deficient F9 mouse embryonic carcinoma cells possess a slightly lower Young's modulus (2.5 ± 1.5 kPa) compared to the wild-type cells (3.8 ± 1.1 kPa). Perhaps this was due to an altered cytoskeletal organization of actin (Goldmann and Ezzell 1996), indicating a prominent functional role for the focal adhesion protein vinculin as an essential part of the cytoskeletal matrix scaffold. Later, a comparative study of the cellular mechanical properties of malignant and non-malignant cells was performed (Lekka *et al* 1999). The Young's modulus of three human cancerous bladder cell lines was one order of magnitude lower than for healthy control cells. Additionally, these measurements were confirmed in 2005 using differently aggressive human breast cancer cell lines (Guck *et al* 2005, Park *et al* 2005). The cellular deformability was shown to be sensitive enough to monitor alterations during the malignant progression of mouse fibroblasts and human breast epithelial cells from normal to cancerous cells. The distribution of the Young's modulus of normal fibroblasts was much wider than for the malignantly transformed cells.

Taken together, cellular stiffness determined via AFM is dependent on the local point of the cantilever position and varies over the whole cellular surface, which may lead to a discrepancy in the Young's modulus, not only if measured at different parts on a single cell, but also when measured for a population of cells. In addition, the cell–cell interaction can also impact on the mechanical measurement of cells and they can behave differently than single cells. Hence, experimental errors increase alongside the large degree of cellular heterogeneity and they may also be time-dependent. Comparison between different cell types is still possible, and it is even possible to distinguish between different cancer cell types displaying different malignancy states, because of their altered mechanical properties, evoked by the cell's cytoskeleton.

Other techniques, such as magnetic bead rheology (Wang *et al* 1993), microneedle probes (Zahalak *et al* 1990), microplate manipulation (Thoumine and Ott 1997), acoustic microscopes (Kundu *et al* 2000), sorting in microfabricated sieves (Carlson *et al* 1997) and the manipulation of beads attached to cells with optical tweezers (Sleep *et al* 1999), can also be used to determine the mechanical properties of cells. The main and most common result is that malignant cells respond to the external stresses applied by becoming either less elastic (softer and more deformable) or less viscous (less resistant to flow), depending on the measurement method used. However, metastatic cancer cells have been reported to display an even lower resistance to deformation compared to healthy cells (Raz and Geiger 1982, Ward *et al* 1991). In line with this, when using an optical stretcher it has been found that malignant cancer cells are more deformable than healthy controls (Guck *et al* 2005). Metastatic cancer cells must be capable of being highly deformable as they have to be able to squeeze through the surrounding tissue matrix as they enter the circulatory systems by transendothelial migration, after which they travel to targeted distant organs in order to build up secondary tumors (Wyckoff *et al* 2000).

Finally, these results indicate that cellular elasticity is suitable for use as a cell marker and even a diagnostic factor for determining the malignant state of cancer cells and thus their aggressive potential. However, all these techniques (excluding the optical cell stretcher) have to deal with low cell throughput measurements that are not good for statistics. In addition, the mechanical contact of the probe with the device may also lead to adhesion and active cellular response and moreover special cell preparation as well as non-physiological probe handling causes measurement artefacts. By contrast, an optical stretcher (Guck *et al* 2001) combined with a microfluidic chamber is able to deform individual suspended cells by optically induced surface forces at rates that are in the flow through rate of commercially available flow cytometers and cell sorters. Thus, the high throughput rate is suitable for the statistics and leads to reliable results. In addition, it may also be possible to combine an optical cell stretcher with a cell-sorting unit, in order to separate the stimulated and stretched cells from the non-stretched cells. Taken together, the deformability of cells measured with a microfluidic optical cell stretcher is suitable as an inherent cell marker for the malignancy of cancer cells.

Does optical cell stretching affect the viability of stretched cells?
The viability of the cells analyzed with an optical cell stretcher is important, as dead cells are known to possess an altered cytoskeleton. Thus, radiation damage should be avoided or reduced during optical cell stretcher measurements by selecting a wavelength of 785 nm with low absorption. In particular, the mechanical properties of BALB 3T3 cells are significantly different when they are not alive compared to their healthy counterparts. In phase contrast microscopy living cells display a characteristic bright rim around their cell edge, while dead cells show no sharp contour and appear diffuse.

However, it is not obvious that two 800 mW laser beams can be used for the deformation of cells without causing any radiation damage. It is extremely important to choose the right wavelength for the cell deformation measurements in order to avoid radiation damage. If we consider using a short wavelength, this might be desirable for optical tweezers, because it may result in higher gradients and better trapping efficiencies due to the small spot size of a focused laser beam, which is about half the wavelength of the light used before. However, the main problem with short wavelengths is that they are not appropriate for avoiding cell damage from radiation, because the absorption by the chromophores in cells is high with decreasing wavelength compared to increasing wavelength in the infrared region (Svoboda and Block 1994). It has been reported that the absorption peaks of proteins are in the ultraviolet region of the electromagnetic spectrum. Therefore, optical cell stretchers with a long wavelength of 1064 nm of an Nd-YAG laser have been implemented to improve this method and achieve better results (Ashkin *et al* 1987). At first glance, this choice of wavelength seems less than optimal because 70% of a cell's weight is water (Alberts *et al* 1994), which absorbs light more strongly with increasing wavelength. Another problem is that most cells trapped with optical tweezers do not survive light powers greater than 20–250 mW, which depends on the particular cell type and the laser wavelength (Ashkin *et al* 1987, Ashkin and Dziedzic 1987, Kuo and Sheetz 1992). Optical tweezers have been reported to induce local heating of water, which is a limiting factor in optical trapping experiments in the near infrared region. In more detail, theoretical calculations predict a temperature increase of less than 3 K per

100 mW in the wavelength range of 650 to 1050 nm for durations of up to 10 s (Schönle and Hell 1998). Thus, the duration of the laser exposure to the cells is reduced to 2 s in optical cell stretcher experiments. These theoretical results have been confirmed by a 0.3°C increase in temperature per 100 mW at powers of up to 400 mW by using Chinese hamster ovary cells (CHO cells) trapped in optical tweezers at a wavelength of 1064 nm (Liu *et al* 1996).

In addition to the effects of heating due to water absorption, the cells can also be damaged by radiation through metabolic change and cellular viability when trapped in optical tweezers. Further to this, microfluorometric measurements on CHO cells (Liu *et al* 1996) revealed that even up to 400 mW of 1064 nm cw laser light causes either no alterations to the DNA structure or the cellular pH. However, the right choice of laser wavelength is important, as photodamage has been reported for rotating *E. coli* assays with maxima at 870 and 930 nm and minima at 830 and 970 nm (Neuman *et al* 1999). The presence of oxygen seems to be responsible for the damage, but the origin of the damage is still elusive. The damage at 785 nm is at least as small as at 1064 nm, with the latter being the most commonly used wavelength for biological trapping experiments (Neumann *et al* 1999). In addition, the sensitivity to light is linearly correlated to the intensity, which cancels out multi-photon processes. For an optical cell stretcher device such damage has not yet been seen. An explanation for this is that the laser beams are not focused and the power densities are lower compared to the optical tweezers device by about two orders of magnitude for the same light power. In summary, in order to apply higher forces much higher light powers can be used without the risk of cellular damage.

Cell cytoskeletons in suspension

This part discusses the fact that the cells measured using the optical cell stretcher are in suspension and not adhered to a substrate. For cells in the bloodstream such as leukocytes, which are generally non-adherent, this presents no problem when measuring them. There is the concern, however, that adherent eukaryotic cells may dissolve their acto-myosin cytoskeleton when they are in suspension and thus are non-adherent. The actin cytoskeleton of BALB 3T3 fibroblasts in suspension has been analyzed through TRITC-phalloidin staining using fluorescence microscopy. The result did not confirm this hypothesis, as suspended cells possess an extensive actin network throughout the whole cytoplasm. In particular, the peripheral cortical actin is visible underneath the cell membrane. The only features of the cell's cytoskeleton not observed in suspension are stress fibers, due to the lack of focal adhesion plaques in suspended cells. However, even in the absence of stress fibers BALB 3T3 possesses a large resistance to deformation induced by the optical cell stretcher. Stress fibers are mainly reported in cells adhered to a planar 2D substrate, whereas they are less pronounced in cells directly embedded in a connective tissue matrix. Thus, the presence of many stress fibers may even be non-physiological. Indeed, several studies have found that the influence of the cytoskeleton on the deformability of normally adherent cells, such as bovine aortic endothelial cells (Sato *et al* 1987), chick embryo fibroblasts (Thoumine and Ott 1997) and rat embryo fibroblasts, (Heidemann *et al* 1999) can be determined in suspension by using, for example, an optical cell stretcher.

These results are further supported by frequency-dependent AFM microrheology experiments, comparing *in vitro* actin gels and NIH3T3 fibroblasts (Mahaffy *et al* 2000). In more detail, the viscoelastic properties of the cells are similar to those of homogeneous actin gels. Thus, the elastic strength of cells seems to be stored within the actin cortex that is still assembled in non-adherent cells. As most polymer theoretic models describe isotropic actin networks *in vitro*, they can be adapted to non-adherent cells. Moreover, it seems to be likely that these theoretic models will be compared with data obtained from living cells that only possess the actin cortex and not anisotropic structures such as F-actin stress fibers.

The optical deformability of dielectric matter has not yet been the focus of biophysical research. However, the optical stretcher serves as a non-destructive optical tool for the high-throughput and quantitative deformation of cells. Indeed, the forces exerted by light on the cells due to the momentum transferred are adequate to hold as well as move the objects and even they are high enough to deform the samples. In more detail, the momentum is predominantly transferred to the surface of the sample. The total force, such as the gradient and scattering force, that traps the sample arises from the asymmetry of the resulting surface stresses when the sample is not located at its equilibrium position and thus acts on the sample's center of gravity. These trapping forces are significantly smaller than the forces on the sample's surface because the surface forces cancel nearly completely upon integration. This phenomenon is most obvious in a two-beam trap as used for the optical cell stretcher. In the case when the dielectric object (such as a cell) is located in the center of the optical trap, the total force is zero, and the forces acting on the cell's surface can be several hundred picoNewton. The forces used for cell elasticity measurements with the optical cell stretcher are in the range of those for optical tweezers and AFMs.

At first glance, the result that the surface forces pull on the surface rather than compressing it seems surprising. This result can be explained through a ray optics (RO) approach by the increase of the light's momentum as it enters a denser medium, such as the interior of a cell, and the resulting stresses on the cell's surface. Another approach is guided by the minimization of energy. In particular, it is energetically favorable for a dielectric object such as a cell to have most of its volume in the area with the highest intensity along the laser beam axis. This leads to the result that the cell is pulled to the axis. This seems to be conceptually correct, but it would be difficult to calculate.

In more detail, the magnitude of the induced forces scales linearly with the incident laser light power and with $(n - 1)$, where $n = n_{cell}/n_{medium}$ is the relative refractive index. For example, with 1 W of power from each laser beam the forces are in the range of 200 to 500 pN, even when the relative refractive indices of biological materials are typically relatively low, such as $n = 1.02–1.05$.

The optical deformability of cells is a cell-type specific property and hence serves as a marker for a particular cell type, for example red blood cells (RBCs) and BALB 3T3 fibroblasts. Trapped cells show no radiation damage, even when stretched with 800 mW of light in each beam, and hence show a representative cytoskeleton in suspension.

Biomedical application of the optical cell stretcher
There seems to be a medical application for the optical cell stretcher technique. Using it, one cell per second can be measured to determine cellular elasticity. As the lifespan of a living cell *in vitro* is limited, measurement frequency is crucial to obtain good statistics. In particular, the optical stretcher seems to be suitable for applications in the research and diagnosis of diseases such as cancer, which are evoked by alterations of the cytoskeleton. Current cancer-detection methods are based on markers and optical inspection (Sidransky 1996). An increasing disorder of the actin cytoskeleton may serve as a marker for malignancy (Koffer *et al* 1985, Takahashi *et al* 1986). Alterations in the total amount of actin and in the relative ratio of the several actin isoforms (Wang and Goldberg 1976, Goldstein *et al* 1985, Leavitt *et al* 1986, Takahashi *et al* 1986, Taniguchi *et al* 1986), an overexpression of gelsolin in breast cancer cells (Chaponnier and Gabbiani 1989) and the lack of filamin in human malignant melanoma cells may all cause actin cytoskeletal alterations (Cunningham *et al* 1992). However, all of these alterations may evoke an altered viscoelastic response of the cells that can be detected with the optical cell stretcher. Models of actin networks (MacKintosh *et al* 1995) reported that the shear modulus increases with the actin concentration raised to 2.2. In particular, a slight decrease in the actin concentration leads to a detectable decrease in the cell's elasticity. The optical cell stretcher may serve as an advanced diagnostic tool in medical laboratories. Moreover, this novel analytical tool requires minimal tissue samples, which can be obtained using simple cytobrushes on the surfaces of the lung, esophagus and stomach, as well as in the cervix or through fine-needle aspiration (Dunphy and Ramos 1997, Fajardo and DeAngelis 1997).

The optical deformability of mouse fibroblasts
In order to demonstrate the connection between cellular function and the optical deformability of cells, the optical deformability of BALB/3T3 and SV-T2 fibroblasts was analyzed using the microfluidic optical cell stretcher. BALB/3T3 is a fibroblast cell line established from disaggregated BALB/c mouse embryos, whereas SV-T2 cells are, through a SV40 DNA virus, oncogenic, transformed cells derived from BALB/3T3 (Aaronson and Todaro 1968). These two cell lines are good model systems for studying the malignant transformation of cells (Aaronson and Todaro 1968; Thoumine and Ott 1997). Indeed, the optical deformability of the SV-T2 cells is increased compared to that of BALB/3T3 cells.

The optical deformability is time-dependent due to the viscoelastic property of cells. Thus, the cells were stretched for one second, as this timescale exploits both elastic and viscous contributions to the cellular deformability (Wottawah *et al* 2005). This difference in deformability has been obtained after analyzing only 30 cells for each cell type. In order to analyze these cells further using proteomics slightly more cells may be required. In the future, it may be possible that microfluidic cell stretchers reach the same sample sizes and measurement rates as fluorescence-activated cell sorter (FACS) devices. As the refractive indexes are not distinguishable between the two cells, the difference in optical deformability can only be attributed to their different mechanical properties. Moreover, the optical deformability of cells can be equated with cellular compliance.

In a theoretic model, for isotropic networks of semiflexible polymers there is a strong dependence of the shear modulus on the filament concentration (Gardel *et al* 2003, Janmey *et al* 1991, Wilhelm and Frey 2003). The relative amounts of F-actin in BALB/3T3 and SV-T2 cells using Alexa-532 Phalloidin staining, and measurements of their integrated intensity distributions, have shown that the total amount of filamentous actin in malignantly transformed cells is reduced by approximately 40%. Moreover, due to the reduced cell size of the SV-T2 cells, the concentration of F-actin is reduced by approximately 50%. The reduction in the absolute amount of F-actin is followed by a remodeling of the whole actin cytoskeleton of the malignantly transformed SV-T2 cells compared to the normal BALB/3T3 cells. In summary, the decreased quantity and the altered structures of both actin cytoskeletons provide an explanation for the enhanced optical deformability of the cancerous SV-T2 cells. In line with this critically important role for F-actin, latrunculin-treated cells (inhibited actin polymerization) soften by about half.

The optical deformability of human breast carcinoma cells
The optical deformability of differently aggressive and invasive cancer cells of epithelial origin can be measured using an optical cell stretcher. It has been established that MCF and MDA-MB-231 cells are model cell lines for investigating breast cancer (Johnson *et al* 1999, Soule *et al* 1973, Tait *et al* 1990, Wang *et al* 2001). In particular, MCF-10 is a non-tumorigenic human epithelial cell line isolated from the benign breast tissue of a fibrocystic disease. Although these cells are immortal, they are as normal as non-cancerous mammary epithelial cells. Another breast cancer cell line (adenocarcinoma) is MCF-7, which is isolated from the pleural effusion. In particular, these cells are non-motile and non-metastatic epithelial cancer cells. When phorbol ester tetradecanoylphorbol-acetate (TPA) is added to MCF-7 cells (modMCF-7 cells) these are then stimulated and display a dramatic increase (18-fold) in their invasiveness and ability to metastasize (Johnson *et al* 1999). In addition, TPA induces the secretion of MMPs in MCF-7 cells, which are able to degrade the surrounding extracellular matrix of tissues and are associated with cancer metastasis and malignant progression *in vivo* (Himelstein *et al* 1994). Using an optical cell stretcher, the cancerous MCF-7 cells showed an increased deformability compared to normal MCF-10 cells and non-metastatic MCF-7 cells.

In addition, the optical deformability of MDA-MB-231 breast cancer cells has been analyzed in the presence and absence of all-trans-retinoic acid (modMDA-MB-231), which leads to less aggressive MDA-MB-231 cells (Wang *et al* 2001). As expected, the optical deformability reduced metastasizing capacity. The differences in the refractive indexes of all five cancer cell lines lay at around 1.365 and hence cannot explain the differences in optical deformability, which instead seem to be correlated with a reduced cytoskeletal resistance to cellular deformation in the cancerous cells compared to normal cells. The reduction in structural strength is consistent with their capability to migrate through connective tissue, to intravasate into the blood and lymph vessels, to circulate through these microvascular systems and possibly extravasate to migrate and build up metastases in targeted organs. Although a major cytoskeletal element in epithelial cells is the intermediate filament keratin (Kirfel *et al* 2003), it is hypothesized that the contribution of keratins to the mechanical properties is important at strains larger than those applied in these optical cell stretcher experiments (Janmey *et al* 1991, Wang and Stamenovic 2000). Thus, the differences in optical deformability are evoked by the decrease of F-actin stress fibers during the malignant transformation by approximately 30% (Katsantonis 1994). By contrast, in another study it has been shown that the keratins play a role in providing optical cell deformability (Seltmann *et al* 2013).

Alterations in optical deformability are detectable due to subtle changes in cytoskeletal composition and structure. Indeed, this dependence confirms results from polymer physics where even minute variations in the concentration of cytoskeletal filaments lead to a non-linearly enhanced overall elasticity (Gardel *et al* 2003, Janmey *et al* 1991, Wilhelm and Frey 2003). Optical deformability can in general serve as a unique and useful biophysical cell marker, which is able to detect even small cytoskeletal changes through the altered optical properties of the cells.

The results obtained with differently invasive and aggressive cancer cell lines can be adapted to primary cells in order to obtain diagnostic markers for the invasive potential of cancer cells. Indeed, the microfluidic optical cell stretcher permits the investigation of samples of individual cells and determines their metastatic potential (Caraway *et al* 1993, Epstein *et al* 2002, Sherman and Kurman 1996). Subsequently, pure samples of isolated cancer cells can be investigated without artificial corruption using established genomic or proteomic techniques, which cannot detect small changes. Another possible field of application for an optical cell stretcher is the isolation and separation of therapeutic stem cells (Olins *et al* 2000), which can be identified and sorted with a microfluidic optical cell stretcher without knowing a special set of markers able to identify unique a stem cell.

The mechanical parameter, the optical deformability of cells, is an inherent cellular marker that provides a sensitive cellular mechanical-based alternative to current proteomic techniques and thus may lead to novel promising findings regarding all cellular processes that involve the mechanical properties of the cytoskeleton.

Further reading

Allen P G and Janmey P A 1994 Gelsolin displaces phalloidin from actin filaments. A new fluorescence method shows that both Ca2+ and Mg2+ affect the rate at which gelsolin severs F-actin *J. Biol. Chem.* **269** 32916–23 PMID: 7806519

Ashkin A and Dziedzic J M 1973 Radiation pressure on a free liquid surface *Phys. Rev. Lett.* **30** 139–42

Barer R and Joseph S 1954 Refractometry of living cells, part I. Basic principles *Q. J. Microsc. Sci.* **95** 399–423

Barer R and Joseph S 1955a Refractometry of living cells, part II. The immersion medium *Q. J. Microsc. Sci.* **96** 1–26

Barer R and Joseph S 1955b Refractometry of living cells, part III. Technical and optical methods *Q. J. Microsc. Sci.* **96** 423–47

Brevik I 1979 Experiments in phenomenological electrodynamics and the electromagnetic energy-momentum tensor *Phys. Rep.* **52** 133–201

Evans E and Fung Y C 1972 Improved measurements of the erythrocyte geometry *Microvasc. Res.* **4** 335–47

Henon S, Lenormand G, Richert A and Gallet F 1999 A new determination of the shear modulus of the human erythrocyte membrane using optical tweezers *Biophys. J.* **76** 1145–51

Hochmuth R M 1993 Measuring the mechanical properties of individual human blood cells *J. Biomech. Eng.* **115** 515–9

Jackson J D 1975 *Classical Electrodynamics* (New York: Wiley) pp 281–2

Käs J, Strey H, Tang J X, Finger D, Ezzell R, Sackmann E and Janmey P A 1996 F-actin, a model polymer for semiflexible chains in dilute, semidilute, and liquid crystalline solutions *Biophys. J.* **70** 609–25

Mazurkiewicz Z E and Nagorski R T 1991 *Shells of Revolution* (New York: Elsevier) pp 360–9

Roosen G 1977 A theoretical and experimental study of the stable equilibrium positions of spheres levitated by two horizontal laser beams *Opt. Commun.* **21** 189–95

Strey H, Peterson M and Sackmann E 1995 Measurements of erythrocyte membrane elasticity by flicker eigenmode decomposition *Biophys. J.* **69** 478–88

Ugural A C 1999 *Stresses in Plates and Shells* (New York: McGraw-Hill) pp 339–409

van de Hulst H C 1957 *Light Scattering by Small Particles* (New York: Wiley) pp 172–6

References

Aaronson S A and Tadaro G J 1968 Development of 3T3-like lines from Balb/c mouse embryo cultures: transformation susceptibility to SV40 *J. Cell Physiol.* **72** 141–8

Abbondanzieri E A, Greenleaf W J, Shaevitz J W, Landick R and Block S M 2005 Direct observation of base-pair stepping by RNA polymerase *Nature* **438** 460–5

Adelman M R, Borisy G G, Shelanski M L, Weisenberg R C and Taylor E W 1968 Cytoplasmic filaments and tubules *Fed. Proc.* **27** 1186–93

Aebi U, Cohn J, Buhle L and Gerace L 1986 The nuclear lamina is a meshwork of intermediate-type filaments *Nature* **323** 560–4

Alberts B, Bray D, Lewis J, Raff M, Roberts K and Watson J D 1994 *Molecular Biology of the Cell* (New York: Garland) pp 786–861

Alenghat F J, Fabry B, Tsai K Y, Goldmann W H and Ingber D E 2000 Analysis of cell mechanics in single vinculin-deficient cells using a magnetic tweezer *Biochem. Biophys. Res. Commun.* **277** 93–9

Amblard F, Yurke B, Pargellis A and Leibler S 1996 A magnetic manipulator for studying local rheology and micromechanical properties of biological systems *Rev. Sci. Instr.* **67** 818–27

Ashkin A 1970 Acceleration and trapping of particles by radiation pressure *Phys. Rev. Lett.* **24** 156–9

Ashkin A and Dziedzic J M 1987 Optical trapping and manipulation of viruses and bacteria *Science* **235** 1517–20

Ashkin A, Dziedzic J M and Yamane T 1987 Optical trapping and manipulation of single cells using infrared laser beams *Nature* **330** 769–71

Ashkin A, Dziedzic J M, Bjorkholm J E and Chu S 1986 Observation of a single-beam gradient force optical trap for dielectric particles *Opt. Lett.* **11** 288–90

Assi F, Jenks R, Yang J, Love C and Prentiss M 2002 Massively parallel adhesion and reactivity measurements using simple and inexpensive magnetic tweezers *J. Appl. Phys.* **92** 5584

Barbic M, Mock J J, Gray A P and Schultz S 2001 Electromagnetic micromotor for microfluidics applications *Appl. Phys. Lett.* **79** 1399–1401

Bausch A R, Möller W and Sackmann E 1999 Measurement of local viscoelasticity and forces in living cells by magnetic tweezers *Biophys. J.* **76** 573–9

Bausch A R, Ziemann F, Boulbitch A A, Jacobson K and Sackmann E 1998 Local measurements of viscoelastic parameters of adherent cell membranes by magnetic bead microrheometry *Biophys. J.* **75** 2038–49

Bennett V 1985 The membrane skeleton of human erythrocytes and its implication for more complex cells *Annu. Rev. Biochem.* **54** 273–304

Bennett V 1990 Spectrin-based membrane skeleton—a multipotential adapter between plasma membrane and cytoplasm *Physiol. Rev.* **70** 1029–60 PMID: 2271059

Block S M, Goldstein L S and Schnapp B J 1990 Bead movement by single kinesin molecules studied with optical tweezers *Nature* **348** 348–52

Bosch F H, Werre J M, Schipper L, Roerdinkholder-Stoelwinder B, Huls T, Willekens F L, Wichers G and Halie M R 1994 Determinants of red blood cell deformability in relation to cell age *Eur. J. Haematol.* **52** 35–41

Caraway N P, Fanning C V, Wojcik E M, Staerkel G A, Benjamin R S and Ordonez N G 1993 Cytology of malignant melanoma of soft parts: fine-needle aspirates and exfoliative specimens *Diagn. Cytopathol.* **9** 632–8

Carlier M F 1998 Control of actin dynamics *Curr. Opin. Cell Biol.* **10** 45–51

Carlson R H, Gabel C V, Chan S S, Austin R H, Brody J P and Winkelman J W 1997 Self-sorting of white blood cells in a lattice *Phys. Rev. Lett.* **79** 2149–52

Caspi A, Granek R and Elbaum M 2002 Diffusion and directed motion in cellular transport *Phys. Rev. E* **66** 11916-1–12

Chaponnier C and Gabbiani G 1989 Gelsolin modulation in epithelial and stromal cells of mammary carcinoma *Am. J. Pathol.* **134** 597–603 PMID: 2538057

Chu S 1991 Laser manipulation of atoms and particles *Science* **253** 861–6

Cluzel P, Lebrun A, Heller C, Lavery R, Viovy J L, Chatenay D and Caron F 1996 DNA: An extensible molecule *Science* **271** 792–4

Colon J M, Sarosi P G, McGovern P G, Ashkin A and Dziedzic J M 1992 Controlled micromanipulation of human sperm in three dimensions with an infrared laser optical trap: effect on sperm velocity *Fertil. Steril.* **57** 695–8

Constable A, Kim J, Mervis J, Zarinetchi F and Prentiss M 1993 Demonstration of a fiber-optical light-force trap *Opt. Lett.* **18** 1867–9

Cooper J A, Bryan J, Schwab B, Frieden C, Loftus D J and Elson E L 1987 Microinjection of gelsolin into living cells *J. Cell Biol.* **104** 491–501

Crick F H C and Hughes A F W 1949 The physical properties of cytoplasm: a study by means of the magnetic particle method *Exp. Cell Res.* **1** 37–80

Cunningham C C, Gorlin J B, Kwiatkowski D J, Hartwig J H, Janmey P A, Byers H R and Stossel T P 1992 Actin-binding protein requirement for cortical stability and efficient locomotion *Science* **255** 325–7

Daily B, Elson E L and Zahalak G I 1984 Cell poking: determination of the elastic area compressibility modulus of the erythrocyte membrane *Biophys. J.* **45** 661–82

Daniels B R, Masi B C and Wirtz D 2006 Probing single-cell micromechanics *in vivo*: the microrheology of C-elegans developing embryos *Biophys. J.* **90** 4712–19

Davenport R J, Wuite G J L, Landick R and Bustamante C 2000 Single-molecule study of transcriptional pausing and arrest by *E. coli* RNA polymerase *Science* **287** 2497–500

de Vries A H E, Krenn B E, van Driel R and Kanger J S 2005 Micro magnetic tweezers for nanomanipulation inside live cells *Biophys. J.* **88** 2137–44

Discher D E, Mohandas N and Evans E A 1994 Molecular maps of red cell deformation: hidden elasticity and *in situ* connectivity *Science* **266** 1032–5

Dulińska I, Targosz M, Strojny W, Lekka M, Czuba P, Balwierz W and Szymoński M 2006 Stiffness of normal and pathological erythrocytes studied by means of atomic force microscopy *J. Biochem. Biophys. Methods* **6** 1–11

Dunphy C and Ramos R 1997 Combining fine-needle aspiration and flow cytometric immunophenotyping in evaluation of nodal and extranodal sites for possible lymphoma: a retrospective review *Diagn. Cytopathol.* **16** 200–6

Eichinger L, Köppel B, Noegel A A, Schleicher M, Schliwa M, Weijer K, Wittke W and Janmey P A 1996 Mechanical perturbation elicits a phenotypic difference between dictyostelium wild-type cells and cytoskeletal mutants *Biophys. J.* **70** 1054–60

Elson E L 1988 Cellular mechanics as an indicator of cytoskeletal structure and function *Annu. Rev. Biophys. Biophys. Chem.* **17** 397–430

Epstein J B, Zhang L and Rosin M 2002 Advances in the diagnosis of oral premalignant and malignant lesions *J. Can. Dent. Assoc.* **68** 617–21

Evans E 2001 Probing the relation between force–lifetime–and chemistry in single molecular bonds *Annu. Rev. Biophys. Biomol. Struct.* **30** 105–28

Evans E, Ritchie K and Merkel R 1995 Sensitive force technique to probe molecular adhesion and structural linkages at biological interfaces *Biophys. J.* **68** 2580–7

Fabry B and Fredberg J J 2003 Remodeling of the airway smooth muscle cell: are we built of glass? *Respir. Physiol. Neurobiol.* **137** 109–24

Fabry B, Maksym G N, Butler J P, Glogauer M, Navajas D and Fredberg J J 2001 Scaling the microrheology of living cells *Phys. Rev. Lett.* **87** 148102–16

Fajardo L L and DeAngelis G A 1997 The role of stereotactic biopsy in abnormal mammograms *Surg. Oncol. Clin. North Am.* **6** 285–99 PMID: 9115496

Felder S and Elson E L 1990 Mechanics of fibroblast locomotion: quantitative analysis of forces and motions at the leading lamellas of fibroblasts *J. Cell Biol.* **111** 2513–26

Fisher J K *et al* 2006 Thin-foil magnetic force system for high-numerical-aperture microscopy *Rev. Sci. Instrum.* **77** 023702

Fisher J K *et al* 2005 Three-dimensional force microscope: a nanometric optical tracking and magnetic manipulation system for the biomedical sciences *Rev. Sci. Instrum.* **76** 053711

Freundlich H and Seifriz W 1923 *Z. Phys. Chem. Stoechiom. Verwandtschaftsl.* **104** 233

Fuchs E and Cleveland D W 1998 A structural scaffolding of intermediate filaments in health and disease *Science* **279** 514–9

Gardel M L, Valentine M T, Crocker J C, Bausch A R and Weitz D A 2003 Microrheology of entangled F-actin solutions *Phys. Rev. Lett.* **91** 158302

Goldmann W H and Ezzell R M 1996 Viscoelasticity in wild-type and vinculin-deficient (5.51) mouse F9 embryonic carcinoma cells examined by atomic force microscopy and rheology *Exp. Cell Res.* **226** 234–7

Goldstein D, Djeu J, Latter G, Burbeck S and Leavitt J 1985 Abundant synthesis of the transformation-induced protein of neoplastic human fibroblasts, plastin, in normal lymphocytes *Cancer Res.* **45** 5643–7 PMID: 4053036

Gosse C and Croquette V 2002 Magnetic tweezers: micromanipulation and force measurement at the molecular level *Biophys. J.* **82** 3314–29

Grandbois M, Beyer M, Rief M, Clausen-Schaumann H and Gaub H E 1999 How strong is a covalent bond? *Science* **283** 1727–30

Guck J, Ananthakrishnan R, Mahmood H, Moon T J, Cunningham C C and Käs J 2001 The optical stretcher: a novel laser tool to micromanipulate cells *Biophys. J.* **81** 767–84

Guck J, Ananthakrishnan R, Moon T J, Cunningham C C and Käs J 2000 Optical deformability of soft biological dielectrics *Phys. Rev. Lett.* **84** 5451–4

Guck J *et al* 2005 Optical deformability as an inherent cell marker for testing malignant transformation and metastatic competence *Biophys. J.* **88** 3689–98

Guilford W H and Gore R W 1992 A novel remote-sensing isometric force transducer for micromechanics studies *Am. J. Physiol.* **263** C700–7 PMID: 1415519

Guilford W H, Lantz R C and Gore R W 1995 Locomotive forces produced by single leukocytes *in vivo* and *in vitro Am. J. Physiol.* **268** C1308–12 PMID: 7762625

Heilbronn A 1922 *Jahrbuch für wiss. Botanik* **61** 284

Heidemann S R, Kaech S, Buxbaum R E and Matus A 1999 Direct observations of the mechanical behaviors of the cytoskeleton in living fibroblasts *J. Cell Biol.* **145** 109–22

Herrmann H and Aebi U 1998 Structure, assembly, and dynamics of intermediate filaments *Subcell. Biochem.* **31** 319–62 PMID: 9932497

Himelstein B P, Canete-Soler R, Bernhard E J, Dilks D W and Muschel R J 1994 Metalloproteinases in tumor progression: the contribution of MMP-9 *Invasion Metastasis* **14** 246–58

Hiramoto Y 1969a Mechanical properties of the protoplasm of the sea urchin egg. I. Unfertilized egg *Exp. Cell Res.* **56** 201–8

Hiramoto Y 1969b Mechanical properties of the protoplasm of the sea urchin egg. II. Fertilized egg *Exp. Cell Res.* **56** 209–18

Hochmuth R M 2000 Micropipette aspiration of living cells *J. Biomech.* **33** 15–22

Hohng S, Zhou R, Nahas M K, Yu J, Schulten K, Lilley D M and Ha T 2007 Fluorescence-force spectroscopy maps two-dimensional reaction landscape of the holliday junction *Science* **318** 279–83

Hosu B G, Jakab K, Banki P, Toth F I and Forgacs G 2003 Magnetic tweezers for intracellular applications *Rev. Sci. Instrum.* **74** 4158–63

Howard J, Hudspeth A J and Vale R D 1989 Movement of microtubules by single kinesin molecules *Nature* **342** 154–8

Hua W, Young E C, Fleming M L and Gelles J 1997 Coupling of kinesin steps to ATP hydrolysis *Nature* **388** 390–3

Huang H, Dong C Y, Kwon H-S, Sutin J D, Kamm R D and So P T C 2002 Three-dimensional cellular deformation analysis with a two-photon magnetic manipulator workstation *Biophys. J.* **82** 2211–23

Huang H, Sylvan J, Jonas M, Barresi R, So P T C, Campbell K P and Lee R T 2005 Cell stiffness and receptors: evidence for cytoskeletal subnetworks *Am. J. Physiol. Cell Physiol.* **288** C72–80

Hummer G and Szabo A 2005 Free energy surfaces from single-molecule force spectroscopy *Acc. Chem. Res.* **38** 504–13

Janmey P A 1991 Mechanical properties of cytoskeletal polymers *Curr. Opin. Cell Biol.* **3** 4–11

Janmey P A, Euteneuer V, Traub P and Schliwa M 1991 Viscoelastic properties of vimentin compared with other filamentous biopolymer networks *J. Cell Biol.* **113** 155–60

Janmey P A, Peetermans J, Zaner K S, Stossel T P and Tanaka T 1986 Structure and mobility of actin filaments as measured by quasielastic light scattering, viscometry, and electron microscopy *J. Biol. Chem.* **261** 8357–62 PMID: 3013849

Janmey P A, Shah J V, Janssen K P and Schliwa M 1998 Viscoelasticity of intermediate filament networks *Subcell. Biochem.* **31** 381–97 PMID: 9932499

Jarzynski C 1997 Nonequilibrium equality for free energy differences *Phys. Rev. Lett.* **78** 2690–3

Johnson M D, Torri J A, Lippman M E and Dickson R B 1999 Regulation of motility and protease expression in PKC-mediated induction of MCF-7 breast cancer cell invasiveness *Exp. Cell Res.* **247** 105–13

Katsantonis J, Tosca A, Koukouritaki S B, Theodoropoulos P A, Gravanis A and Stournaras C 1994 Differences in the G/total actin ratio and microfilament stability between normal and malignant human keratinocytes *Cell Biochem. Funct.* **12** 267–74

Kellermayer M S Z, Smith S B, Granzier H L and Bustamante C 1997 Folding-unfolding transitions in single titin molecules characterized with laser tweezers *Science* **276** 1112–6

Kim S J, Blainey P C, Schroeder C M and Xie X S 2007 Multiplexed single-molecule assay for enzymatic activity on flow-stretched DNA *Nat. Methods.* **4** 397–9

Kirfel J, Magin T M and Reichelt J 2003 Keratins: a structural scaffold with emerging functions *Cell. Mol. Life Sci.* **60** 56–71

Kishino A and Yanagida T 1988 Force measurements by micromanipulation of a single actin filament by glass needles *Nature* **334** 74–6

Koffer A, Daridan M and Clarke G 1985 Regulation of the microfilament system in normal and polyoma virus transformed cultured (BHK) cells *Tissue Cell* **17** 147–59

Kundu T, Bereiter-Hahn J and Karl I 2000 Cell property determination from the acoustic microscope generated voltage versus frequency curves *Biophys. J.* **78** 2270–9

Kuo S C and Sheetz M P 1992 Optical tweezers in cell biology *Trends Cell Biol.* **2** 116–8

Kuo S C and Sheetz M P 1993 Force of single kinesin molecules measured with optical tweezers *Science* **260** 232–4

Laurent V M, Henon S, Planus E, Fodil R, Balland M, Isabey D and Gallet F 2002 Assessment of mechanical properties of adherent living cells by bead micromanipulation: comparison of magnetic twisting cytometry vs optical tweezers *J. Biomech. Eng.* **124** 408–21

Leavitt J, Latter G, Lutomski L, Goldstein D and Burbeck S 1986 Tropomyosin isoform switching in tumorigenic human fibroblasts *Mol. Cell Biol.* **6** 2721–6

Lekka M, Laidler P, Gil D, Lekki J, Stachura Z and Hrynkiewicz A Z 1999 Elasticity of normal and cancerous human bladder cells studied by scanning force microscopy *Eur. Biophys. J.* **28** 312–6

Lim C T, Zhou E H, Li A, Vedula S R K and Fu H X 2006 Experimental techniques for single cell and single molecule biomechanics *Mater Sci. Eng. C-Biomimetic Supramol Syst.* **26** 1278–88

Litvinov R I, Shuman H, Bennett J S and Weisel J W 2002 Binding strength and activation state of single fibrinogen-integrin pairs on living cells *Proc. Natl. Acad. Sci. USA* **99** 7426–31

Liu Y, Sonek G J, Berns M W and Tromberg B J 1996 Physiological monitoring of optically trapped cells: assessing the effects of confinement by 1064-nm laser tweezers using micro-fluorometry *Biophys. J.* **71** 2158–67

Lodish H, Baltimore D, Berk A, Zipurski S L, Matsudaira P and Darnell J 1995 *Mol. Cell Biol.* (New York: Scientific American Books) pp 1051–9

Lodish H B, Berk A, Zipursky S L, Matsudaira P, Baltimore D and Darnell J E 2000 *Molecular Cell Biology* (New York: Freeman)

Lu H P, Xun L and Xie X S 1998 Single-molecule enzymatic dynamics *Science* **282** 1877–82

MacKintosh F C and Schmidt C F 1999 Microrheology *Curr. Opin. Colloid Interface Sci.* **4** 300–7

MacKintosh F C, Käs J and Janmey P A 1995 Elasticity of semiflexible biopolymer networks *Phys. Rev. Lett.* **75** 4425–8

Mahaffy R E, Shih C K, MacKintosh F C and Käs J 2000 Scanning probe-based frequency-dependent microrheology of polymer gels and biological cells *Phys. Rev. Lett.* **85** 880–3

Maniotis A J, Chen C S and Ingber D E 1997 Demonstration of mechanical connections between integrins, cytoskeletal filaments, and nucleoplasm that stabilize nuclear structure *Proc. Natl Acad. Sci. USA* **94** 849–54

Matthews B D, Overby D R, Alenghat F J, Karavitis J, Numaguchi Y, Allen P G and Ingber D E 2004 Mechanical properties of individual focal adhesions probed with a magnetic micro-needle *Biochem. Biophys. Res. Commun.* **313** 758–64

McNulty A L, Weinberg J B and Guilak F 2009 Inhibition of matrix metalloproteinases enhances *in vitro* repair of the meniscus *Clin. Orthop. Relat. Res.* **467** 1557–67

Mierke C T 2011 Cancer cells regulate biomechanical properties of human microvascular endothelial cells *J. Biol. Chem.* **286** 40025–37

Mierke C T, Kollmannsberger P, Zitterbart D P, Diez G, Koch T M, Marg S, Ziegler W H, Goldmann W H and Fabry B 2010 Vinculin facilitates cell invasion into three-dimensional collagen matrices *J. Biol. Chem.* **285** 13121–30

Mierke C T, Kollmannsberger P, Zitterbart D P, Smith J, Fabry B and Goldmann W H 2008a mechano-coupling and regulation of contractility by the vinculin tail domain *Biophys. J.* **94** 661–70

Mierke C T, Zitterbart D P, Kollmannsberger P, Raupach C, Schlötzer-Schrehardt U, Goecke T W, Behrens J and Fabry B 2008b Breakdown of the endothelial barrier function in tumor cell transmigration *Biophys. J.* **94** 2832–46

Mijailovich S M, Kojic M, Zivkovic M, Fabry B and Fredberg J J 2002 A finite element model of cell deformation during magnetic bead twisting *J. Appl. Physiol.* **93** 1429–36

Mitchison T J 1992 Compare and contrast actin filaments and microtubules *Mol. Biol. Cell* **3** 1309–15

Mitchison T, Evans L, Schulze E and Kirschner M 1986 Sites of microtubule assembly and disassembly in the mitotic spindle *Cell* **45** 515–27

Mohandas N and Evans E 1994 Mechanical properties of the red cell membrane in relation to molecular structure and genetic defects *Annu. Rev. Biophys. Biomol. Struct.* **23** 787–818

Moustakas A and Stournaras C 1999 Regulation of actin organisation by TGF-beta in H-ras-transformed fibroblasts *J. Cell Sci.* **112** 1169–79

Möller W, Takenaka S, Rust M, Stahlhofen W and Heyder J 1997 Probing mechanical properties of living cells by magnetopneumography *J. Aerosol Med.* **10** 173–86

Neuman K C, Chadd E H, Liou G F, Bergman K and Block S M 1999 Characterization of photodamage to Escherichia coli in optical traps *Biophys. J.* **77** 2856–63

Olins A L, Herrmann H, Lichter P and Olins D E 2000 Retinoic acid differentiation of HL-60 cells promotes cytoskeletal polarization *Exp. Cell Res.* **254** 130–42

Overby D R, Matthews B D, Alsberg E and Ingber D E 2005 Novel dynamic rheological behavior of individual focal adhesions measured within single cells using electromagnetic pulling cytometry *Acta Biomaterialia* **1** 295–303

Park S, Koch D, Cardenas R, Käs J and Shih C K 2005 Cell motility and local viscoelasticity of fibroblasts *Biophys. J.* **89**(6) 4330–42

Pasternak C and Elson E L 1985 Lymphocyte mechanical response triggered by cross-linking surface receptors *J. Cell Biol.* **100** 860–72

Peters I M, van Kooyk Y, van Vliet S J, de Grooth B G, Figdor C G and Greve J 1999 3D single-particle tracking and optical trap measurements on adhesion proteins *Cytometry* **36** 189–94

Petersen N O, McConnaughey W B and Elson E L 1982 Dependence of locally measured cellular deformability on position on the cell, temperature, and cytochalasin B *Proc. Natl Acad. Sci. USA* **79** 5327–31

Pollard T D 1984 Molecular architecture of the cytoplasmic matrix *Kroc Found. Ser.* **16** 75–86

Pollard T D 1986 Assembly and dynamics of the actin filament system in nonmuscle cells *J. Cell Biochem.* **31** 87–95

Pope L H, Bennink M L and Greve J 2002 Optical tweezers stretching of chromatin *J. Muscle Res. Cell Motil.* **23** 397–407

Prass M, Jacobson K, Mogilner A and Radmacher M 2006 Direct measurement of the lamellipodial protrusive force in a migrating cell *J. Cell Biol.* **174** 767–72

Radmacher M 1997 Measuring the elastic properties of biological samples with the AFM *IEEE Med. Eng. Biol.* **16** 47–57

Radmacher M, Fritz M, Kacher C M, Cleveland J P and Hansma P K 1996 Measuring the viscoelastic properties of human platelets with the atomic force microscope *Biophys. J.* **70** 556–67

Ranga A, Gobaa S, Okawa Y, Mosiewicz K, Negro A and Lutolf M P 2014 3D niche microarrays for systems-level analyses of cell fate *Nature Commun.* **5** 4324

Rao K M and Cohen H J 1991 Actin cytoskeletal network in aging and cancer *Mutation Res.* **256** 139–48

Raupach C, Zitterbart D P, Mierke C T, Metzner C, Müller F A and Fabry B 2007 Stress fluctuations and motion of cytoskeletal-bound markers *Phys. Rev. E Stat Nonlin. Soft Matter Phys.* **76** 011918

Raz A and Geiger B 1982 Altered organization of cell-substrate contacts and membrane-associated cytoskeleton in tumor cell variants exhibiting different metastatic capabilities *Cancer Res.* **42** 5183–90

Rief M, Gautel M, Oesterhelt F, Fernandez J M and Gaub H 1997 Reversible unfolding of individual titin immunoglobulin domains by AFM *Science* **276** 1109–12

Rotsch C, Jacobson K and Radmacher M 1999 Dimensional and mechanical dynamics of active and stable edges in motile fibroblasts investigated by using atomic force microscopy *Proc. Natl. Acad. Sci. USA* **96** 921–6

Rotsch C and Radmacher M 2000 Drug-induced changes of cytoskeletal structure and mechanics in fibroblasts: an atomic force microscopy study *Biophys. J.* **78** 520–35

Sacconi L, Romano G, Ballerini R, Capitanio M, De Pas M, Dunlap D, Giuntini M, Finzi L and Pavone F S 2001 Three-dimensional magneto-optic trap for micro-object manipulation *Opt. Lett.* **26** 1359–61

Sato M, Levesque M J and Nerem R M 1987 An application of the micropipette technique to the measurement of the mechanical properties of cultured bovine aortic endothelial cells *J. Biomech. Eng.* **109** 27–34

Sato M, Wong T Z and Allen R D 1983 Rheological properties of living cytoplasm: endoplasm of Physarum plasmodium *J. Cell Biol.* **97** 1089–97

Schnitzer M J and Block S M 1997 Kinesin hydrolyses one ATP per 8-nm step *Nature* **388** 386–90

Schönle A and Hell S W 1998 Heating by absorption in the focus of an objective lens *Opt. Lett.* **23** 325–7

Schulze C, Wetzel F, Kueper T, Malsen A, Muhr G, Jaspers S, Blatt T, Wittern K P, Wenck H and Käs J A 2012 Stiffening of human skin fibroblasts with age *Clin. Plast. Surg.* **39** 9–20

Seltmann K, Fritsch A W, Käs J A and Magin T M 2013 Keratins significantly contribute to cell stiffness and impact invasive behavior *Proc. Natl Acad. Sci. USA* **110** 18507–12

Shepherd G M, Corey D P and Block S M 1990 Actin cores of hair-cell stereocilia support myosin motility *Proc. Natl Acad. Sci. USA* **87** 8627–31

Sherman M E and Kurman R J 1996 The role of exfoliative cytology and histopathology in screening and triage *Obstet. Gynecol. Clin. North Am.* **23** 641–55

Sidransky D 1996 Advances in cancer detection *Sci. Am.* **275** 104–9

Simmons R M, Finer J T, Warrick H M, Kralik B, Chu S and Spudich J A 1993 Force on single actin filaments in a motility assay measured with an optical trap *Adv. Exp. Med. Biol.* **332** 331–6

Sleep J, Wilson D, Simmons R and Gratzer W 1999 Elasticity of the red cell membrane and its relation to hemolytic disorders: an optical tweezers study *Biophys. J.* **77** 3085–95

Smith S B, Finzi L and Bustamante C 1992 Direct mechanical measurements of the elasticity of single DNA molecules by using magnetic beads *Science* **258** 1122–6

Soule H D, Vazquez J, Long A, Albert S and Brennan M 1973 A human cell line from a pleural effusion derived from a breast carcinoma *J. Natl. Cancer Inst.* **51** 1409–16

Stossel T P 1984 Contribution of actin to the structure of the cytoplasmic matrix *J. Cell Biol.* **99** 15s–21s

Stossel T P, Hartwig J H, Janmey P A and Kwiatkowski D J 1999 Cell crawling two decades after Abercrombie *Biochem. Soc. Symp.* **65** 267–80 PMID: 10320944

Strick T R, Allemand J F, Bensimon D, Bensimon A and Croquette V 1996 The elasticity of a single supercoiled DNA molecule *Science* **276** 1835–7

Svoboda K and Block S 1994 Force and velocity measured for single kinesin molecules *Cell* **77** 773–84

Svoboda K, Schmidt C F, Schnapp B J and Block S M 1993 Direct observation of kinesin stepping by optical trapping interferometry *Nature* **365** 721–7

Tadir Y, Wright W H, Vafa O, Ord T, Asch R H and Berns W M 1990 Force generated by human sperm correlated to velocity and determined using a laser generated optical trap *Fertil. Steril.* **53** 944–7

Tait L, Soule H D and Russo J 1990 Ultrastructural and immunocytochemical characterization of an immortalized human breast epithelial cell line, MCF-10 *Cancer Res.* **50** 6087–94

Takahashi K, Hein V I, Junker J L, Colburn N H and Rice J M 1986 Role of cytoskeleton changes and expression of the H-ras oncogene during promotion of neoplastic transformation in mouse epidermal JB6 cells *Cancer Res.* **46** 5923–32 PMID: 3093072

Taniguchi S, Kawano T, Kakunaga T and Baba T 1986 Differences in expression of a variant actin between low and high metastatic B16 melanoma *J. Biol. Chem.* **261** 6100–6 PMID: 3700386

Thoumine O and Ott A 1997 Time scale dependent viscoelastic and contractile regimes in fibroblasts probed by microplate manipulation *J. Cell Sci.* **110** 2109–16 PMID: 9378761

Tinoco I and Bustamante C 2002 The effect of force on thermodynamics and kinetics of single molecule reactions *Biophys. Chem.* **101–102** 513–33

Tskhovrebova L, Trinick J, Sleep J A and Simmons R M 1997 Elasticity and unfolding of single molecules of the giant muscle protein titin *Nature* **387** 308–12

Valberg P A and Albertini D F 1985 Cytoplasmic motions, rheology, and structure probed by a novel magnetic particle method *J. Cell Biol.* **101** 130–40

Valberg P A and Butler J P 1987 Magnetic particle motions within living cells Physical theory and techniques *Biophys. J.* **52** 537–50

Valberg P A and Feldman H A 1987 Magnetic particle motions within living cells Measurement of cytoplasmic viscosity and motile activity *Biophys. J.* **52** 551–61

Viani M B, Schaffer T E, Chand A, Rief M, Gaub H E and Hansma P K 1999 Small cantilevers for force spectroscopy of single molecules *J. Appl. Phys.* **86** 2258–62

Walter N, Selhuber C, Kessler H and Spatz J P 2006 Cellular unbinding forces of initial adhesion processes on nanopatterned surfaces probed with magnetic tweezers *Nano Lett.* **6** 398–402

Wang E and Goldberg A R 1976 Changes in microfilament organization and surface topography upon transformation of chick embryo fibroblasts with Rous sarcoma virus *Proc. Natl Acad. Sci. USA* **73** 4065–9

Wang N and Stamenovic D 2000 Contribution of intermediate filaments to cell stiffness, stiffening, and growth *Am. J. Physiol. Cell Physiol.* **279** C188–94 PMID: 10898730

Wang N, Butler J P and Ingberg D E 1993 Mechanotransduction across the cell surface and through the cytoskeleton *Science* **260** 1124–7

Wang Q, Lee D, Sysounthone V, Chandraratna R A S, Christakos S, Korah R and Wieder R 2001 1,25-dihydroxyvitamin D3 and retonic acid analogues induce differentiation in breast cancer cells with function- and cell-specific additive effects *Breast Cancer Res. Treat.* **67** 157–68

Ward K A, Li W I, Zimmer S and Davis T 1991 Viscoelastic properties of transformed cells: role in tumor cell progression and metastasis formation *Biorheology* **28** 301–13

Weisenhorn A L, Khorsandi M, Kasas S, Gotzos V and Butt H J 1993 Functional Imaging of Early Markers of Disease *Nanotechnology* **4** 106–13

Wilhelm J and Frey E 2003 Elasticity of stiff polymer networks *Phys. Rev. Lett.* **91** 108103

Williamson J R, Gardner R A, Boylan C W, Carroll G L, Chang K, Marvel J S, Gonen B, Kilo C, Tran-Son-Tay R and Sutera S P 1985 Microrheologic investigation of erythrocyte deform-ability in diabetes mellitus *Blood* **65** 283–8

Wirtz H R and Dobbs L G 1990 Calcium mobilization and exocytosis after one mechanical stretch of lung epithelial cells *Science* **250** 1266–9

Woodside M T, Behnke-Parks W M, Larizadeh K, Travers K, Herschlag D and Block S M 2006 Nanomechanical measurements of the sequence-dependent folding landscapes of single nucleic acid hairpins *Proc. Natl Acad. Sci. USA* **103** 6190–5

Worthen G S, Schwab B 3rd, Elson E L and Downey G P 1989 Mechanics of stimulated neutrophils: cell stiffening induces retention in capillaries *Science* **245** 183–6

Wottawah F, Schinkinger S, Lincoln B, Ananthakrishnan R, Romeyke M, Guck J and Käs J 2005 Optical rheology of biological cells *Phys. Rev. Lett.* **94** 098103

Wouters F S, Vermeer P J and Bastiaens P I H 2001 Imaging biochemistry inside cells *Tr. Cell Biol.* **11** 203–11

Wyckoff J B, Jones J G, Condeelis J S and Segall J E 2000 A critical step in metastasis: *in vivo* analysis of intravasation at the primary tumor *Cancer Res.* **60** 2504–11

Xie X S and Lu H P 1999 Single-molecule enzymology *J. Biol. Chem.* **274** 15967–70

Yagi K 1961 The mechanical and colloidal properties of Amoeba protoplasm and their relations to the mechanism of amoeboid movement *Comp. Biochem. Physiol.* **3** 73–80

Yamazaki D, Kurisu S and Takenawa T 2005 Regulation of cancer cell motility through actin reorganization *Cancer Sci.* **7** 379–86

Zahalak G I, McConnaughey W B and Elson E L 1990 Determination of cellular mechanical properties by cell poking, with an application to leukocytes *J. Biomech. Eng.* **112** 283–94

Zhuang X W and Rief M 2003 Single-molecule folding *Curr. Opin. Struct. Biol.* **13** 88–97

Ziemann F, Rädler J and Sackmann E 1994 Local measurements of viscoelastic moduli of entangled actin networks using an oscillating magnetic bead micro-rheometer *Biophys. J.* **66** 2210–6

IOP Publishing

Physics of Cancer

Claudia Tanja Mierke

Chapter 4

Cell–cell and cell–matrix adhesion strength, local cell stiffness and forces

Summary

Cell–cell and cell–matrix adhesion strength is important for many cellular functions, including motility and adhesion. Moreover, it influences the mechanical properties of cells, such as their cellular stiffness and contractile force generation. It is still under investigation which of the three main cytoskeletal structural components—actin microfilaments, keratin intermediate filaments or microtubules—contributes the most to the mechanical properties of cancer cells. Is it possible to dissect the contribution of each to the overall mechanical properties? The answer to this question may be difficult to give, as the microfilaments can interact with intermediate filaments and microtubules, and this raises the question of whether one cytoskeletal structural component can replace another. In the following, this is described and how adhesion strength, cellular stiffness and forces can be measured with state-of-the art biophysical methods is discussed.

4.1 Atomic force microscopy

AFM is basically a biophysical technique for acquiring images and other information from a wide variety of objects such as living cells, at extremely high resolution in the nanometer range. The technique works by scanning a very sharp probe (end radius approximately 10 nm) along the object's surface. This is done very carefully in order to keep the force between the cantilever carrying the probe and surface of the object of interest at a set, low level. In more detail, the probe is a silicon (Si) or silicon nitride (Si3N4) cantilever with a sharp integrated tip, and the vertical bending (called deflection) of the cantilever due to the forces acting on the tip is detected by a laser that is focused on the back of the cantilever. Subsequently, the laser is reflected by the cantilever and detected by a photodetector. The movement of the laser beam on the photodetector is a measure of the movement of the probe. This kind of set-up is called an optical lever. During the measurement procedure, the probe is moved over the sample by a scanner (typically a piezoelectric element), which is able to make extremely precise movements. The extremely high resolution of the AFM is achieved because of the combination of the sharp tip, the very sensitive optical lever and the very precise scanner movements, which is also combined with precise control of probe–sample forces.

AFM is a very high-resolution type of scanning probe microscopy with a resolution in the range of a nanometer and it is thus more than 1000 times better than the optical diffraction limit (Hirschfeld and Huber 2010, Grazioso *et al* 2010, Lang *et al* 2004, Capella and Dietler 1999). The precursor to the AFM was the scanning tunneling microscope, developed by Gerd Binnig and Heinrich Rohrer in the early 1980s (earning them the Nobel Prize for physics). The AFM was invented by Binnig and then the first experimental implementation was performed by Binnig, Quate and Gerber in 1986 (Binnig *et al* 1986). Three years later, the AFM became commercially available. It is one of the best tools for imaging, measuring and manipulating matter such as cells, tissues and proteins at the nanoscale. In particular, the surface of the sample is scanned by the mechanical probe (the cantilever). The very precise scanning of the sample's surface is mediated by piezoelectric elements that facilitate tiny, accurate and precise movements on (electronic) command. In some special variations, electric potentials can be scanned using conducting cantilevers. In advanced types of AFM, the current can be passed through the tip to probe the electrical conductivity or transport of the underlying surface of the sample, but this is much more challenging, with few research groups reporting consistent data (Binnig and Quate 1986). The electron micrograph of a used AFM cantilever has an image width of approximately 100 μm and 30 μm.

An AFM can operate in one of three modes: the non-contact, contact and tapping modes, in line with the cantilever's tip motion.

1. The non-contact mode or frequency modulation AFM (dynamic mode) (called NC-AFM, close contact AFM or FM-AFM).
2. The contact (static) mode (called C-AFM or CMAFM).
3. The tapping (dynamic) mode, also intermittent contact (AC mode, vibrating mode or amplitude modulation AFM) (called TMAFM, IC-AFM or AM-AFM).

The contact mode is often called a static mode. The tapping and non-contact modes are, by contrast, described as dynamic, because the cantilever is oscillated in them. This is done by adding an extra piezoelectric element, which oscillates up and down at somewhere between 5–400 kHz to the cantilever holder. There is still a little confusion over terminology, because the tapping mode was trademarked by one AFM manufacturer. Thus this mode has many other names, such as the intermittent contact (IC-AFM), AC or vibrating mode, which are all the same. The modes of the AFM can also be named by their detection mechanisms: the tapping mode is then called amplitude modulation AFM (AM-AFM) and the non-contact mode is called frequency modulation AFM (FM-AFM). The main difference between the tapping mode and non-contact modes (NC-AFM) is that in the former the tip of the probe of the cantilever touches the sample and moves completely away from it in each oscillation cycle. In NC-AFM, the cantilever stays close to the sample during the whole measurement and has a pronouncedly smaller oscillation amplitude. The NC-AFM is more sensitive to small oscillations of the cantilever and can be used in close contact (almost touching). Compared to other AFM techniques, the NC-AFM has a true atomic resolution. The AM-AFM is typically used for the tapping mode, in which the cantilever tip taps the sample during each oscillation cycle. This has turned out to be the most stable mode for use in air and hence it is currently the most commonly used mode for most applications.

Non-topographic modes
Unlike the topographic modes that collect images, many modes to measure other properties of the sample have been reported, such as the magnetic field or the potential difference between the sample and the cantilever probe.
- Magnetic force microscopy (MFM) measures the distribution of the magnetic field within the sample.
- Kelvin probe microscopy (KPM) measures the contact potential difference across the sample.
- Force spectroscopy measures individual molecular interactions.
- The nanoindentation method measures the hardness or softness of the sample.
- Thermal modes determine thermal parameters, such as thermal conductivity on the nanoscale level.

The AFM consists of a cantilever with a sharp tip (called the probe) at its end that is used to scan the surface of the sample. The cantilever consists of Si or Si3N4 and a tip radius of curvature on the order of nanometers. When the cantilever's tip is brought in close to a sample's surface, the forces between the tip and the sample deflect off the cantilever, according to Hooke's law. The forces measured in AFM include mechanical contact forces, van der Waals forces, capillary forces, chemical bonding, electrostatic forces, magnetic forces (MFM), Casimir forces and solvation forces. Together with force, additional quantities can be measured simultaneously through the use of specialized types of probes using scanning thermal microscopy, scanning joule expansion microscopy and photothermal microspectroscopy. The deflection of the cantilever carrying the probe is usually measured using a laser spot reflected from the top surface of the cantilever into an array of photodiodes. Other useful methods use optical interferometry, capacitive sensing or piezoresistive AFM cantilevers. These cantilevers are fabricated with piezoresistive elements acting as a strain gauge. For example, a Wheatstone bridge can be used to detect strain in the AFM cantilever due to deflection, but this method is not as sensitive as laser deflection or interferometry.

An AFM topographical scan of a glass surface can be performed. Thus, the features of the glass on the microscale and nanoscale can be determined and they represent the roughness of the glass surface. An image space in 3D (x, y, z) can be detected and the size is 20 μm × 20 μm × 420 nm. However, if the cantilever's tip is scanned at a constant height, there is a risk that the tip will stretch and damage the surface. In order to avoid this, typically a feedback mechanism is incorporated to adjust the tip-to-sample distance to have a constant force between the tip and the sample.

The tip or sample is mounted on a 'tripod' of three piezo crystals, with each being responsible for scanning in the x, y and z directions. In 1986, as the AFM was invented, a new piezoelectric scanner (called the tube scanner) was developed for use in STM. Then, tube scanners were used for AFMs. In particular, this tube scanner is able to scan the sample in the x, y and z directions using a single tube piezo consisting of a single interior contact and four external contacts. A major advantage of the tube scanner is the better vibrational isolation that results on the one hand from the higher resonant frequency of the single-crystal construction and on the other hand from a low resonant frequency isolation stage. However, a disadvantage is that the x–y motion may lead to unwanted z motion, resulting in distortion (Binnig and Smith 1986).

The AFM can be operated in a number of modes. There are two major groups: static ('contact') modes and a variety of dynamic ('non-contact' or 'tapping') modes, in which the cantilever is vibrated.

What role does the cantilever's probe play?
The AFM probe is a widely distributed and well-established measuring device with a sharp tip on the free swinging end of a cantilever, which protrudes from a holder plate. In more detail, the dimensions of the cantilever are in the range of micrometers and the radius of the tip is in the range of a few nanometers. The holder plate (named the holder chip) often has a size of 1.6 mm by 3.4 mm, supporting the AFM probe with tweezers and hence it fits into the corresponding holder clips on the scanning head of the atomic microscope. This device is most commonly called the AFM probe, but other names include the AFM tip and cantilever (using the name of a single part as the name for the whole device) (figure 4.1). Taken together, an AFM probe is a particular type of scanning probe microscopy (SPM) device.

Most of the AFM probes used consist of Si, while borosilicate glass and Si3N4 can also be used. During the AFM measurement the tip is very close to the surface of the object (sample) and then the cantilever is deflected by the interaction between the tip and the surface. Then a spatial map of the interaction between the cantilever and the sample is recorded by measuring the deflection at many points of a 2D surface.

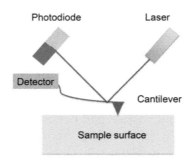

Indeed, several types of interaction can be detected. Depending on the interaction to be measured, the surface of the tip of the AFM probe needs to be modified with a coating such as extracellular matrix proteins. Gold is one coating, and serves for covalent bonding of biological molecules and the detection of their interaction with a surface. In addition, diamond coatings can be used for increased wear resistance and magnetic coatings for detecting the magnetic properties of the investigated surface. Moreover, the surface of the cantilevers can also be altered. These coatings are used to increase the reflectance of the cantilever and subsequently to improve the deflection signal (Bryant *et al* 1988).

The non-contact mode AFM (also called modulation AFM)

This is also called the non-contact mode (NC-AFM), the tip of the cantilever does not touch the surface of the sample. In addition, this non-contact mode can also be named after the detection mode, which is then called frequency modulation AFM. In this mode, the cantilever can be oscillated at its resonant frequency, which is called the frequency modulation mode, or it can be oscillated above the resonant frequency, where the amplitude of oscillation is typically a few nanometers (less than 10 nm) down to a few picometers, which is called the amplitude modulation mode.

In particular, the van der Waals forces are strongest at a distance of 1 nm to 10 nm above the surface of the sample and they decrease as any other long-range force extends the resonance frequency of the cantilever above the surface of the sample. Moreover, this decrease in the resonant frequency is restored by using a feedback loop system in order to maintain a constant oscillation amplitude or frequency through adjusting the distance between the cantilever's tip and the sample's surface. In addition, measuring the tip-to-sample distance at each 2D data point (in x and y dimensions) enables the scanning software to construct a topographic image of the sample's surface.

An advantage of the non-contact mode AFM is that neither tip nor sample degradation effects occur; these have been observed in some cases after taking numerous scans with contact AFM. Therefore, non-contact AFM is preferable to contact AFM when soft samples such as living cells are measured. Although for soft samples the non-contact mode AFM is preferred, in the case of rigid samples there seems to be no difference between contact and non-contact images. If a few monolayers of an adsorbed fluid are located on the surface of a rigid sample, the images of contact and non-contact AFM are substantially different. In the contact mode the cantilever's tip will penetrate the liquid layer in order to measure the underlying surface, whereas in the non-contact mode the cantilever's tip will oscillate above the adsorbed fluid layer in order to image the fluid and the rigid surface.

The dynamic mode operation is based on frequency modulation, in which a phase-locked loop is utilized to track the cantilever's resonance frequency, whereas in the more common amplitude modulation a servo loop acts to keep the cantilever excitation to a defined amplitude.

In frequency modulation, alterations in the oscillation frequency deliver information about the interactions between the tip and the sample. The frequency can be detected with very high sensitivity and hence the frequency modulation mode allows the use of very stiff cantilevers. Another advantage is that stiff cantilevers provide stability, even very close to the surface, and hence this was the first AFM technique that led to data with true atomic resolution under ultra-high vacuum conditions.

In amplitude modulation, alterations in the oscillation amplitude or phase deliver provide the feedback signal for imaging. In particular, in amplitude modulation alterations in the phase of oscillation are used to distinguish between different types of surface materials. The amplitude modulation can be performed in the non-contact as well as in the intermittent contact regime (tapping mode). In the dynamic contact mode, the cantilever is oscillated in such a manner that the distance between the cantilever's tip and the sample's surface is altered. In addition, the amplitude modulation can be performed in the non-contact regime in order to image with atomic resolution by using pronouncedly stiff cantilevers and small amplitudes in an ultra-high vacuum environment.

The AFM beam deflection method
The beam deflection method is the most prominent one for cantilever deflection measurements. In more detail, laser light from a solid-state diode hits the back of the cantilever, is reflected off the back of the cantilever and is collected by a position-sensitive detector (PSD) that contains two closely spaced photodiodes. The output signals are collected with a differential amplifier. An angular displacement of the cantilever leads to an unequal light distribution between the two photodiodes, because one collects more light than the other. This produces an output signal that is based on the difference between the photodiode signals normalized by their sum, which is proportional to the cantilever deflection. In particular, it can detect cantilever deflections of less than 10 nm until the thermal noise provides the limitation. However, a long beam path of several centimeters can easily amplify alterations in the beam angle.

Other deflection measurement methods
Several other methods for measuring beam deflection have been established. They are described briefly in the following.

Piezoelectric detection. The cantilevers are made from quartz or another piezo-electric material and they are able to detect deflection directly as an electrical signal. Indeed, cantilever oscillations of up to 10 pm have been reported using this method.

Laser Doppler vibrometry. A laser Doppler vibrometer is used to obtain highly accurate deflection measurements for an oscillating cantilever; this is only performed in the non-contact mode. However, this method is expensive and hence rarely used.

The scanning tunneling microscope (STM). The STM is based on quantum tunneling. When a conducting tip is placed near to the sample's surface, a bias (voltage difference) applied between the two will enable electrons to tunnel through the vacuum between them. In particular, the resulting tunneling current is a function of tip position, the applied voltage and the local density of states of the sample. The current is detected as the tip's position scans across the sample's surface and is hence usually displayed as an image. The first atomic microscope used a complete STM system containing its own feedback mechanism to detect deflection. In particular, this method is difficult to implement and is also slow to respond to deflection alterations compared to the modern methods.

Optical interferometry. Optical interferometry is used to measure the deflection of the cantilever. Due to the nanometer scale deflections measured in AFM, the interferometer measures in the sub-fringe regime, hence, any drift even in the laser power or wavelength has a strong impact on the measurement and therefore must be omitted. For these reasons, optical interferometer measurements are performed with great care using index matching fluids between optical fiber junctions and very stable lasers. Thus, optical interferometry is rarely used as it is difficult to perform and expensive.

Capacitive detection. Metal-coated cantilevers can build a capacitor with another contact located behind the cantilever. In particular, deflection alters the distance between the contacts and can be measured as a change in capacitance.

Piezoresistive detection. This technique is similar to capacitive detection, but it uses piezoresistive cantilevers in order to record the detection. This is not often used as the piezoresistive detection dissipates energy from the system, affecting the quality factor Q of the resonance that characterizes how underdamped the resonator is (Roiter and Minko 2005).

Force spectroscopy. In addition to imaging, another major application of AFM is force spectroscopy, which is the direct measurement of tip–sample interaction forces as a function of the distance between the tip and sample. In this method, the AFM tip is extended towards and retracted from the sample's surface and in parallel the deflection of the cantilever is monitored as a function of piezoelectric displacement. Thus, a force–distance curve is recorded. These measurements have been used successfully to measure nanoscale contacts, atomic bonding, van der Waals forces and Casimir forces, dissolution forces in liquids and single molecule stretching and also rupture forces. In addition, AFM can be used to measure in an aqueous microenvironment. For instance, the dispersion force due to a polymer adsorbed on the substrate can be determined. Indeed, forces of the order of a few piconewtons can be measured with a vertical distance resolution of better than 0.1 nm. Moreover, force spectroscopy can be performed with either static or dynamic modes. In the dynamic modes, information about the cantilever vibration is monitored in addition to the static deflection (Geisse 2009). The problems with this technique are the absence of direct measurement of the tip–sample separation and the requirement for low-stiffness cantilevers, which tend to snap and stick to the sample's surface. These problems are solvable, as an AFM has been developed that directly measures the tip–sample separation. Hence, the snap-in can be reduced by measuring in liquids or by using stiffer cantilevers, whereas in the latter case a more sensitive deflection sensor is needed to monitor even small alterations in the cantilever deflection. In addition, by applying a small dither to the tip, the stiffness (in the form of a force gradient) of the bond can then be measured.

Atomic force microscopy applications. Force spectroscopy is used in biophysics to analyze the mechanical properties of living materials, such as tissues and cells.

Can even individual surface atoms be identified? Indeed, the AFM can be used to image and manipulate atoms as well as structures on many surfaces. The atom at the apex of the tip can sense individual atoms on the underlying surface by forming incipient chemical bonds with each atom. As these chemical interactions subtly alter the tip's vibration frequency, they can be detected and hence mapped. This principle has been used to distinguish between atoms of Si, tin and lead on an alloy surface by comparing these small atomic values to values obtained from large-scale density functional theory (DFT) simulations. The principle is to first measure these forces precisely for each type of atom expected to be located in the sample and then to compare these values with the forces obtained by DFT simulations. As expected, it has been found that the tip interacted most strongly with Si atoms and interacted 23% and 41% less strongly with tin and lead atoms, respectively. Thus, each different type of atom can be identified in the matrix when the cantilever's tip is moved across the surface (Gross *et al* 2009).

Contact mode
In this mode, the tip is moved (or dragged) across the surface of the sample and the contours of the surface are measured using either the deflection of the cantilever directly or, more commonly, the feedback signal, which is required to keep the cantilever at a constant position.

As the measurement of a static signal is subject to noise and drift, low stiffness cantilevers are used to increase the deflection signal. However, in close proximity to the sample's surface the attractive forces are strong and hence lead to a 'snap-in' of the cantilever to the sample's surface. The contact mode AFM is usually performed at a depth where the overall force is repulsive, which is the case in firm contact with the solid surface below any adsorbed layers.

The tapping mode
Single polymer chains of a thickness of 0.4 nm can be recorded in a tapping mode under aqueous media with different pH values.

In ambient conditions, most samples develop a liquid meniscus layer. Thus, keeping the probe tip close enough to the sample in order to detect short-range forces while the tip is prevented from sticking to the surface represents a major problem for the non-contact dynamic mode in ambient conditions. Thus, the dynamic contact mode (also called the intermittent contact, AC or tapping mode) has been developed to solve this problem (Willemsen *et al* 2000).

In the tapping mode, the cantilever is driven to oscillate up and down close to its resonance frequency by a small piezoelectric element mounted in the AFM tip holder, similar to the non-contact mode. However, the amplitude of this oscillation is even larger than 10 nm and is typically between 100 and 200 nm. The force interactions acting on the cantilever when the tip approaches close to the surface include van der Waals forces, dipole–dipole interactions and electrostatic forces, which all cause the amplitude of this oscillation to decrease as the tip comes into close proximity with the sample. However, the height of the cantilever above the sample is precisely controlled by an electronic servo that uses the piezoelectric actuator. The servo adjusts the height to maintain the oscillation amplitude of the cantilever precisely when it is scanned over the sample. A tapping AFM image is hence gained by imaging the force of the intermittent contacts of the tip with the sample surface. This tapping method decreases the damage to the surface and the tip compared to the amount caused in contact mode. Hence, the tapping mode is gentle enough even for the visualization of supported lipid bilayers or adsorbed single polymer molecules, such as 0.4 nm thick chains of synthetic polyelectrolytes under liquid medium. With defined scanning parameters, the conformation of single molecules can remain unaltered for hours.

The first AFM

An AFM's usefulness has certain limitations, as with other biophysical devices. When determining whether analyzing a sample with an AFM is appropriate, there are various advantages and disadvantages that have to be considered.

Advantages. The AFM has several advantages over the SEM. Compared to the electron microscope, which provides a 2D projection or a 2D image of a sample, the AFM provides a 3D surface profile. In addition, samples analyzed by AFM do not require any special treatment, such as metal/carbon coatings, which would irreversibly alter or even damage the sample, and they do not typically suffer from charging artifacts in the final image.

While an electron microscope needs an expensive vacuum environment for proper operation, most AFM modes can work perfectly well in ambient air or even a liquid environment, which allows the investigation of biological macromolecules and even living organisms. In principle, AFM can provide higher resolution than SEM. It has been shown to provide true atomic resolution in ultra-high vacuum and in liquid environments. Thus, high resolution AFM is comparable in resolution to STM and transmission electron microscopy. In addition, AFM can be combined with a variety of optical microscopic techniques such as fluorescent microscopy, which increases its applicability to a wide field of experiments. Indeed, combined AFM–optical instruments have been applied primarily in the biological sciences, but have also found a niche in some materials applications, such as those involving photovoltaics research (Nishida *et al* 2008).

Disadvantages. A disadvantage of the AFM compared with the SEM is the single scan image size. In particular, in one pass the SEM is able to image an area on the order of square millimeters with a depth of field on the order of millimeters, whereas the AFM can only image a maximum height on the order of 10–20 μm and a maximum scanning area of about 150×150 μm. In order to improve the scanned area size for AFM is the usage of parallel probes in a fashion similar to that of millipede data storage.

In addition, the scanning speed of an AFM is another limitation. Traditionally, an AFM cannot scan images as fast as a SEM. An AFM requires several minutes for a typical scan, whereas a SEM is capable of scanning at near real time, although at relatively low quality. The relatively slow rate of scanning during AFM imaging mostly leads to thermal drift in the image, making the AFM less well suited for measuring accurate distances between topographical features on the image.

However, several fast-acting designs have been suggested to increase microscope scanning productivity, such as video AFM, which means that reasonable quality images are being obtained with video AFM at video rate. This video rate is even faster than the average SEM. In order to eliminate image distortions induced by thermal drift, several methods have become available.

AFM images may also be affected by non-linearity, hysteresis and creep of the piezoelectric material and cross-talk between the x, y, z axes, which seems to require software enhancement and filtering. However, filtering may flatten out real topographical features. Special AFMs utilize real-time correction software or closed-loop scanners, which practically eliminate these problems. Some AFMs also use separated orthogonal scanners (as opposed to a single tube), which additionally serve to eliminate some of the cross-talk problems. In addition, an AFM artifact arises from a tip with a high radius of curvature with respect to the special feature that is to be visualized.

Similar to any other imaging technique, there is the possibility of image artifacts, which could be induced by an unsuitable tip, a poor operating environment or even by the sample itself. These image artifacts are usually unavoidable. However, their occurrence and effect on the results can be reduced through various methods. Artifacts resulting from a too coarse tip can be caused, for instance, by inappropriate handling or through collisions with the sample by either scanning too fast or having an unreasonably rough surface, which may cause wearing of the tip.

One AFM artifact is steep sample topography. Due to the nature of AFM probes, they cannot normally measure steep walls or overhangs. However, specially produced cantilevers and AFMs can be used to modulate the probe sideways as well as up and down (as with dynamic contact and non-contact modes) to measure sidewalls, at the cost of more expensive cantilevers, lower lateral resolution and additional artifacts.

Piezoelectric scanners. AFM scanners are indeed made from piezoelectric material, which expands and contracts proportionally to an applied voltage. Whether they elongate or contract depends upon the polarity of the voltage applied. The scanner is constructed by combining independently operated piezo electrodes for X, Y and Z in a single tube, forming a scanner that can manipulate samples and probes with extreme precision in 3D. Moreover, independent stacks of piezos can be used instead of a tube, resulting in decoupled X, Y and Z movement.

Scanners are characterized by their sensitivity, which is the ratio of piezo movement to piezo voltage, for example, by how much the piezo material extends or contracts per applied volt. Due to differences in material or size, the sensitivity varies from scanner to scanner. In more detail, the sensitivity varies non-linearly with respect to scan size. Piezo scanners exhibit more sensitivity at the end than at the beginning of a scan. In particular, this causes the forward and reverse scans to behave differently and display hysteresis between the two scan directions. However, this can be corrected by applying a non-linear voltage to the piezo electrodes, which leads to a linear scanner movement and calibrates the scanner accordingly. One disadvantage of this approach is that it requires recalibration because the precise non-linear voltage needed to correct non-linear movement will alter as the piezo ages. However, this problem is easily solved by adding a linear sensor to the sample's or the piezo's stage, which is then able to detect the true movement of the piezo element. Thus, deviations from the ideal movement can be detected by the sensor and corrections applied to the piezo drive signal to correct for non-linear piezo movement. This design is known as a closed loop AFM. In turn, non-sensored piezo AFMs are referred to as open loop AFMs.

It is common knowledge that the sensitivity of piezoelectric materials decreases exponentially with time. However, this causes most of the alterations in sensitivity to occur in the initial stages of the scanner's life. In particular, piezoelectric scanners are run for approximately 48 h before they are shipped from the factory so that they are past the point where they may have large alterations in sensitivity. As the scanner ages, the sensitivity will alter less with time and the scanner needs to be recalibrated only in rare cases, although various manufacturer manuals recommend monthly to semi-monthly calibration of open loop AFMs (Lapshin 1998).

Receptor–ligand interactions are known to play a crucial role in biological systems and their measurement forms an important part of modern pharmaceutical development. Numerous assay formats are available that can be used to screen and quantify receptor ligands. An overview for both radioactive and non-radioactive assay technologies will be presented, with special emphasis on the latter. While radioreceptor assays are fast, easy to use and reproducible, their major disadvantage is that they are hazardous to human health, produce radioactive waste, require special laboratory conditions and are thus rather expensive on a large scale. This has promoted the development of non-radioactive assays based on optical methods like fluorescence polarization, fluorescence resonance energy transfer or surface plasmon resonance. In light of their application in high-throughput screening environments, there has been special emphasis on so called 'mix-and-measure' assays, which do not require separation of bound from free ligand.

The advent of recombinant production of receptors has contributed to the increased availability of specific assays and some aspects of the expression of recombinant receptors will be covered briefly. Applications of receptor–ligand binding assays relate to screening and the quantification of pharmaceuticals in biological matrices. Drug discovery is a highly complex and costly process, which demands integrated efforts in several relevant aspects, involving innovation, knowledge, information, technologies and expertise. The shift from traditional to genomics- and proteomics-based drug research has fundamentally transformed the key strategies in the pharmaceutical industry addressed to the design of new chemical entities for drug candidates against a variety of biological targets. Thus, drug discovery has moved toward more rational strategies based on increased understanding of the fundamental principles of protein–ligand interactions. The combination of the knowledge of several 3D protein structures and of hundreds to thousands of small molecules has attracted attention for the application of structure- and ligand-based drug design approaches. In line with this, virtual screening technologies have substantially enhanced the impact of computational methods applied to chemistry and biology and the goal of applying such methods is to reduce large compound databases and to select a limited number of promising candidates for drug design. A novel concept for a protein–ligand docking simulator using virtual reality (VR) technologies, in particular tactile sense technology, has been designed and a prototype developed. Most conventional docking simulators are based on numerical differential calculations of the total energy between a protein and a ligand.

However, the basic concept of this method differs from that of conventional simulators. This particular design utilizes the force between a ligand and a protein instead of the total energy. The most characteristic function of the system is its ability to touch and sense the electrostatic potential field of a protein molecule. The surface of a protein of interest can be scanned using a globular probe, which is given an electrostatic charge and controlled through a force feedback device. The electrostatic force between the protein and the probe is calculated in real time and immediately fed back into the force feedback device. The operator can easily search interactively for positions where the probe is strongly attracted to the force field. These positions can be regarded as candidate sites where functional groups of ligands corresponding to the probe can bind to the target protein. Certain limitations remain, for example only twenty protein atoms can be used to generate the electrostatic field. Moreover, the system can only use globular probes, excluding drug molecules or small chemical groups from being simulated.

4.1.1 Cellular stiffness

It is well established that cell functions are essentially determined by structure. At different hierarchy levels the structural organization of cells is characterized by special mechanical properties. It is obvious that the cell structure should be different for a variety of physiological processes, such as cell differentiation, growth and adhesion, and under pathogenesis, such as oxidative stress and attacks by viruses or parasites. For a long time there have been two approaches for the investigation of cellular mechanical properties: (i) the cell mechanical properties are studied integrally, with the cell considered as a single whole object, and (ii) the mechanical properties of the cellular structural components are studied in detail, using isolated lipid bilayers, biomembranes and cytosolic proteins. However, it has become possible to probe the micro- and nanomechanical properties of cell structures and to investigate the spatial distribution of the mechanical properties of special cellular structures within a single cell using the AFM technique (Hansma 2001). AFM is a novel method for high-resolution imaging of any surface, including those of living and fixed cells (Bischoff and Hein 2003). This powerful technique can also be used for the characterization of the mechanical, electrical and magnetic characteristics of samples to be studied both qualitatively and quantitatively. In particular, AFM operation is based on the detection of repulsive and/or attractive surface forces. The interaction between the sample surface and a tip of the cantilever located very close to it corresponds to the forces between the atoms of the sample and those of the tip that scans its surface. The contrast of the image is provided by monitoring the forces of interaction between the tip and the surface. The tip is fabricated under a flexible cantilever responsible for the signal transduction. The interaction between the sample and the tip causes bending or twisting of the cantilever in a manner proportional to the interaction force. A small laser, which is focused on the cantilever, detects any bending or twisting of it. In more detail, the reflection of the laser beam is focused on a photodiode detector. Then the interaction of the sample with the tip is measured by the variation in the reflected beam's point of incidence on the photodiode. Deflection of the cantilever by interaction with features on the sample surface is monitored during scanning and is translated into a 3D image of the surface (Mozafari *et al* 2005).

An AFM can be operated in a number of diverse imaging modes, depending on the nature of the interaction between the tip and sample surface. The mechanical properties of cell surface and subsurface layers can be determined either by contact mode AFM techniques, such as force modulation, lateral force microscopy and force-curve analysis, or by phase imaging in tapping mode AFM, such as the intermittent or semi-contact modes.

Probing of the cell surface by AFM techniques can detect heterogeneities of the mechanical properties of the surface at the nanolevel and subsurface layers of cells. In particular, the resolution of AFM in air at the vertical direction is 0.1–0.5 nm and at the horizontal direction it is approximately 1–5 nm, depending on the rigidity of the sample. A horizontal resolution can be reached for living cells in an aqueous medium even at a range of several tens of nanometers due to the softness of the cell membrane. The thickness of cellular membranes is known to be approximately 5–10 nm. When analyzing the cell's heterogeneities using AFM techniques, it is possible to draw a picture of the cellular structure of certain specific regions within a single cell. However, the sensitivity and resolution of the AFM method depends on tip and cantilever characteristics such as radius, shape and material type (Alessandrini and Facci 2005).

The basic AFM technique for the quantitative study of the mechanical characteristics of cells and tissues is force spectroscopy (called force-curve analysis). By recording the force value and the vertical deflection of the cantilever, the probe approaches the surface under investigation at the fixed point and usually performs force-curve analysis. Then, the force value versus the distance between the probe and the surface can be plotted for this case. The force curve contains information about long- and short-range interactions and forms the basis for estimating the sample Young's modulus. There is a serious problem in estimating the absolute value of a cellular Young's modulus using AFM force-curve analysis because of the question of which mechanical model is appropriate and should be chosen.

To date, the Hertz model has been used in the majority of studies devoted to the evaluation of the Young's modulus of cells. In principle, the Hertz model describes the simple case of the elastic deformation of two perfectly homogeneous smooth bodies touching each other under a specific load (Hertz 1881, Johnson 1985). The two important assumptions of the Hertz model are: (i) the indenter must have a parabolic shape and (ii) the indented sample such as a cell is assumed to be extremely thick in comparison to the indentation depth. The first assumption remains a valid one, if the spherical tip radius is much larger than the indentation depth ($h < 0.3\,R$) (Mahaffy *et al* 2000).

If the tip of an AFM nanoscope is approximated by a sphere with the radius R, then the force on cantilever $F(h)$ is given by equation (4.1):

$$F(h) = \frac{4\sqrt{R}}{3} E^* h^{\frac{3}{2}},$$

where h is the depth of the indentation and E^* the effective modulus of a system tip sample, which is calculated from equation (4.2):

$$\frac{1}{E^*} = \frac{1 - v_{tip}^2}{E_{tip}} + \frac{1 - v_{sample}^2}{E_{sample}},$$

in which E_{tip}, v_{tip} and E_{sample}, v_{sample} are the Young's modules and the Poisson ratios for the materials of the tip and sample, respectively. However, if the material of the tip is considerably harder than the sample, equation (4.3) is used (Vinckier and Semenza 1998):

$$E^* \approx \frac{E_{sample}}{1 - v_{sample}^2}.$$

The Sneddon's variation of the Hertz model (equation (4.4)) is used for the case of cone tip of an AFM cantilever (Laurent et al 2005):

$$F(h) = \frac{2}{\pi} \tan\alpha \frac{E_{sample}}{1 - v_{sample}^2} h^2,$$

where α is the half-opening angle of the AFM tip.

Although the indentation depth in the case of AFM probing of the cell is in the range of hundreds of nanometers, which is too high to be an appropriate depth for the Hertz model, it has been demonstrated in many studies that the Hertz model still describes the experimental results sufficiently well. The original Hertz theory did not allow for adhesion of the indenter to the material. Thus, the Hertz theory has been improved for that special case (Johnson et al 1971). The Hertz model was mainly used for an estimation of the static Young's modulus of cells, whereas the dynamic Young's modulus was sometimes used for the characterization of cells' elastic properties (Mahaffy et al 2004). As the cell surface is heterogeneous, consisting of a network of cell membrane and submembrane structures, the cellular Young's modulus evaluation using the Hertz model assumes an error.

The second model used for investigating cell elastic properties is based on force spectroscopy data, but it is founded on the theory of elastic shells (ES) (A-Hassan et al 1998, Scheffer et al 2001, Timoshenko and Woinowsky-Krieger 1970). This theory considers cells as shells filled with liquid. In such an approach, the effective Young's modulus can be evaluated from the relationship between the effective Young's modulus, shell thickness and the bending modulus. A serious problem for such an evaluation procedure is determining the particular boundary conditions for the calculation of any constants involved in the main relation and defining the tip–sample contact radius. This has not yet been fully solved.

Among other theoretical models used in AFM-based evaluation of cells' mechanical features, the finite element model should be mentioned; it is the most popular model for analysis of elasticity in engineering (Ohashi *et al* 2002).

The elasticity parameter values calculated using various models differ from each other. For example, the mechanical properties of endothelial cells exposed to shear stress have revealed different Young's modulus values in respect of various models (Ohashi *et al* 2002). The modulus values calculated using a finite element model appeared to be significantly higher: from 12.2 to 18.7 kPa with exposure to shear stress. However, the modulus value calculated using the Hertz model reflects the same tendency, but has different means (0.87 and 1.75 kPa for control and sheared endothelial cells, respectively).

The AFM experimental approach known as force integration to equal limits (FIEL) mapping, used to produce quantitative maps of relative cell elasticity, was developed in 1998 (A-Hassan *et al* 1998). The FIEL theory assumes a simple relationship between the values of the work done by the AFM cantilever during an indentation and the elastic constants at different surface positions.

In more detail, the collection of force curves over a certain area ensures the development of the elasticity map of the cell's surface. The surface elasticity map can also be performed using either force modulation (static mode) or phase imaging (tapping mode) techniques. The characteristics of cantilever oscillation, such as the amplitude and phase shift, carry information about local elastic and friction properties of the sample in both modes. The image of the changes to the oscillation characteristics represents a map of the relative mechanical properties of the cell surface.

However, indirect information regarding the elastic properties of the cell surface can be obtained by recording the lateral force map of the surface of interest. The AFM cantilever's lateral deflections, such as torsion, arise because of either alterations to the surface slope or the heterogeneity of the surface's frictional properties. The lateral force map is analyzed simultaneously with the sample's surface topography to elaborate the specificity of the elastic property map. In particular, the AFM is used to probe the elasticity of mammalian cells in order to investigate their temporal and spatial structural dynamics under physiological and pathological processes. A reliable parameter to compare different cell types under different experimental conditions, the Young's modulus has been determined using AFM measurements.

The Young's modulus and its AFM measurement in living cells
The elastic modulus value of living cells varies widely. It is evident that this reflects the real variability of the parameter and also the imperfection of AFM measurement methods and the numerical estimation of cell elasticity. However, almost a decade of progress in this area allows the generalization of some methodological factors that have a significant influence on the Young's modulus value. The problems connected with cell specificity will be discussed below. Under a change of the external conditions, the elasticity of cell membranes alters much more strongly than the morphology of the cell.

The first factor is a question of AFM sample handling. Erythrocytes illustrate progress in this acute area well. The AFM method has been used to investigate both living (Nowakowski *et al* 2001, Kamruzzahan *et al* 2004) and fixed erythrocytes in either air (Gould *et al* 1990) or in a buffer (Butt *et al* 1990). The living cells are rather soft and delicate for AFM probing under physiological conditions. Their drying, freezing and fixing with chemical agents can even improve the AFM images and AFM indentation results. However, these procedures alter the cell structure, viability and elasticity. The Young's modulus values for erythrocytes treated with 5% formalin solution are increased ten-fold (119.5 kPa) over those for viable (native) erythrocytes (16.05 kPa) (Mozhanova *et al* 2003). The transverse stiffness of cardiomyocytes is also increased by a factor of 16 after fixing with formalin (Shroff *et al* 1995). Comparing a variety of methods for preparing erythrocyte ghosts for AFM studies showed that air drying is not suitable even after fixation in glutaraldehyde (Takeuchi *et al* 1998). On the other hand, fixation may enhance the images of cell structures such as the cytoskeleton (Shroff *et al* 1995, Hofmann *et al* 1997). Moreover, the highest resolution for cells (such as 10 nm) may only be achieved in air, which assumes cell fixation before AFM probing. The high mobility of the erythrocyte shape in buffer solutions leads to smearing of the AFM image and the maximal spatial resolution of living erythrocytes becomes poor at about 200 nm (Mozhanova *et al* 2003).

The standard AFM technique for cell elasticity measurement is based on indentation of the cells firmly attached to the substrate. For reliable indentation results, firm substrate–cell contact is indeed required, which is a problem for non-adhered cells in solution. A good approach for the immobilization of native erythrocytes in liquid is attachment to a glass surface previously modified with poly-L-lysine solution. Poly-L-lysine provides accurate localization of red blood cells on the glass surface due to the electrostatic interaction between the negatively charged cell surface and the positively charged poly-L-lysine layer. However, poly-L-lysine can promote membrane rearrangement, with the formation of a specific membrane deformation pattern within the contact area (Dulinska *et al* 2006). A novel method of indentation for leukemia cells placed at special microwells has been presented (Rosenbluth *et al* 2006). This method provides the mechanical immobilization of cells, whereas it also has an influence on the estimated Young's modulus value. In this case, the cell deformation is described precisely by the elastic model based on Hertzian mechanics.

The second factor is related to the heterogeneity of the mechanical properties of cells. However, there are pronounced variations of the elastic modulus values at different cell regions. It has been reported that the elastic modulus value of the human umbilical vein endothelial cells is 7.22 kPa directly over the nucleus, dropping to only 2.97 kPa over the cell body in close proximity to the nucleus and 1.27 kPa on the cell body near the edge (Mathur *et al* 2000). Moreover, it has been found that the cell body of bovine pulmonary artery endothelial cells is two- to three-fold softer compared to the cell periphery (Costa and Yin 1999). A corresponding study on cardiomyocytes also revealed that cells are softer at the nuclear region and become stiffer toward the periphery (Shroff *et al* 1995). Mapping of the

Young's modulus across the living chicken cardiocytes indicated that the stress fibers are characterized by the presence of areas with a stiffness of 100–200 kPa embedded in softer parts of the cell, with elastic modulus values between 5 and 30 kPa (Hofmann et al 1997). The elasticity map images of living astrocytes such as the glial cells of nerve tissue showed that the cell membrane above the nucleus is softer (2–3 kPa) than the surroundings and that the cell membrane above ridge-like structures reflecting F-actin is stiffer (10–20 kPa) than the surroundings. In the elasticity map images of fixed astrocytes, the elasticity value of cells was found to be relatively uniform (200–700 kPa), irrespective of the inner structure of cells (Yamane et al 2000). The variation of the elastic modulus value within a single cell is often a subject of study in itself.

The third factor influencing elastic modulus measurement is connected with cell thickness. However, substrate contributions can be neglected if the AFM tip is never indented by more than 10% of the cell thickness (Mathur et al 2001). If the cell compartment under the investigation is very thin (<1000 nm), as in the case of lamellipodium, it is necessary to negotiate the special challenges for accurate measurement of its viscoelastic behavior. Thus, the reported AFM-based micro-rheology method ensures the estimation of the viscoelastic constants of thin parts of cell (<1000 nm) and those of thick areas, applying two different models, one for well-adhered regions and another model for non-adhered regions (Mahaffy et al 2004).

Cellular elasticity and the functional mechanics of endothelial cells
Although the Young's modulus determined by AFM techniques must be assessed carefully as absolute values, it is also very useful as a relative parameter in special experiments. Thus, the Young's modulus can be used successfully in the study of a variety of cellular functions.

Vascular endothelial cells are an interesting system for analyzing cell mechanics and cytoskeletal dynamics. These cells are located in mechanically active micro-environments and are necessary to withstand shear stress, blood pressure and any alteration in pressure due to breathing cycles. The earliest AFM work on living aortic endothelial cells studied the effects of shear stress on cellular organization and other factors that may have an impact on the mechanical response of cells to flow (Barbee et al 1994, Barbee 1995, Sato et al 2000, Ohashi et al 2002). It has been reported that the local elastic parameters of aortic endothelial cells increases significantly (from 0.87–1.75 kPa) with exposure to shear stress (Ohashi et al 2002). The average elastic modulus values of bovine pulmonary artery endothelial cells (BPAECs) are in a similar range of 0.2–2 kPa (Pesen and Hoh 2005). However, a difference in the mechanical properties of rabbit endothelium has been demonstrated (Miyazaki and Hayashi 1999). Moreover, cells are stiffer in the medial wall of aortic bifurcation compared to the endothelium located in the lateral wall. Using AFM and human umbilical vein endothelial cells (HUVECs), it has been shown that the cell responds globally to the localized applied force over the cell edge and the nucleus (Mathur et al 2000). Thus it has been concluded that the nuclear region of the cell appears to be stiffer than the rest of the cell body, although the nucleus appears to be offset from the basal surface. By contrast, the focal adhesion movement upon the apical cell surface perturbation suggests a link between the nucleus and the focal adhesions through the cytoskeleton (Mathur *et al* 2000).

4.1.2 Adhesion forces between cells

Historically, cell–cell adhesion has been studied through fluorescent microscopy and through molecular biological approaches and biochemical methods (Behrens *et al* 1989). However, quantification of cell–cell adhesion forces was not yet possible. A classical micropipette aspiration method has now been used to address the problem of the force quantification of cell–cell adhesions (Chu *et al* 2014). The micro aspiration method has delivered quantitative data sets, but this system has been restricted to a limited number of cells and to sophisticated biophysical laboratories, and hence it is not suitable as a routine device and thus the experimental throughput has been restricted to small numbers of repeated experiments.

In order to investigate cell–cell adhesions the AFM technique can be used. This AFM method was developed in 1986 (Binnig and Quate 1986). Its advantage is that structures at atomic resolution and the forces between them can be measured. A major advantage it has over the STM is that insulators can also be analyzed (Binnig and Quate 1986). However, cell–cell or cell–matrix adhesion strength measurement with an AFM is restricted to the cantilever kind used and the forces that can be used to rupture the cell–cell or cell–adhesion connections. The cell–cell adhesion technique uses a cell glued to the cantilever tip by binding to fibronectin that has been coated to the cantilever tip prior cell adhesion to the cantilever. This cell attached to the cantilever serves as a probe for the adherent cells in the culture dish. Then the cantilever is brought to an adherent cell and pressed against it with a certain amount of force, for example 0.5 nN. and incubated for 20 s or even much longer in order to establish cell–cell adhesion.

Quantitative analysis of cellular interactions with the extracellular microenviron-ment is necessary to reveal how cells regulate adhesion during the development and maintenance of multicellular organisms and how alterations in cell adhesion may cause diseases. Indeed, a practical guide to quantify the adhesive strength of living animal cells to various substrates using AFM-based single-cell force spectroscopy (SCFS) has been developed (Friedrichs *et al* 2013). How are we to control cell states and attachment to the AFM cantilever? How are we to functionalize support for SCFS measurements? How are we to conduct cell adhesion measurements? How are we to analyze and interpret the recorded SCFS data? This section will answer all these questions and is intended to assist researchers to perform reliable AFM-based SCFS measurements.

The adhesive interactions of cells with their microenvironment trigger signal trans-duction pathways, which regulate important cellular processes such as cell migration, gene expression, cell survival, tissue organization and differentiation (Legate *et al* 2009, Weber *et al* 2011). Accordingly, mutations in genes encoding adhesion receptors (Fässler *et al* 1996, Sheppard 2000) or adhesion-associated components (Monkley *et al* 2000, Montanez *et al* 2008, Dowling *et al* 2008) may lead to developmental disorders and diseases. Consequently, methods that enable the characterization of cell adhesion are important for cell biological, clinical, pharmaceutical, biophysical and biomaterials research, such as tissue engineering and regeneration.

Due to the central importance of these processes, a variety of assays have been developed and established in order to characterize cell adhesion. Among these, the SCFS methods are most suitable for directly quantifying cell adhesion forces at the cellular level and even down to the contribution of single molecules (Benoit *et al* 2000, Helenius *et al* 2008, Taubenberger *et al* 2007). A brief overview of the most common methods applied to characterize cell adhesion is presented, which focuses mainly on AFM-based SCFS (AFM–SCFS). In particular, the technical basis of AFM–SCFS, with an emphasis on clear descriptions of experimental procedures and pitfalls, is discussed and moreover, how AFM–SCFS data can be evaluated and analyzed is decribed in detail. This background information will help understanding of the principles underlying AFM–SCFS, as well as the possibilities and limitations of the method for the quantification of cellular interactions.

Methods used to characterize cell adhesion
Insight into the mechanical interactions between cells and their microenvironment can be obtained using different artificial cell culture substrates. Evidence has been provided that migrating cells exert compressive forces on culture substrates, for example through the finding that fibroblasts introduce wrinkles into thin Si rubber film substrates (Harris *et al* 1980). With knowledge of the substrate's stiffness and the length of the microscopic wrinkles, the forces exerted by the cells can be estimated on the order of nanonewtons (nN) (Beningo *et al* 2002). In order to reveal even local deformations, the spatial resolution of this wrinkling substrate approach was further improved by embedding micrometer-sized beads into the substrate (Oliver *et al* 1998). In line with this, other studies showed that dynamic deformations of various other substrates were the result of contractile and adhesive cellular forces (Burton and Taylor 1997, Lee *et al* 1994). Although these substrate-based methods were revealed to be important for understanding the mechanical interactions between cells and their microenvironment, they do not directly provide values for the adhesion strength of the cells and their substrates.

Methods for examining cell adhesion strength essentially focus on measuring the capability of cells to remain attached when exposed to an external detachment force. The most common adhesion assay, the plate-and-wash assay, relies on seeding cells onto substrates of interest, washing off 'non-adherent' cells with physiological buffers after a certain adhesion time interval and counting the remaining adherent cells (Klebe 1974). Plate-and-wash assays have identified key adhesion components and generated valuable insights into the mechanisms regulating adhesion (Amano *et al* 1997, Ridley and Hall 1992, Sieg *et al* 2000). However, these specific assays provide no information on adhesion strength and hence report only the initial rate of attachment of cells to the substrate, as the formation of more than ten receptor–ligand bonds is sufficient to prevent their removal from the plate (Boettiger and Wehrle-Haller 2010).

However, several semi-quantitative adhesion assays have been developed to apply controlled shear stress to adherent cells. For instance, in special flow chamber assays, shear stress is exerted on cells by a homogenous buffer flow (Kaplanski *et al* 1993). In spinning-disc assays, both controlled centrifugal forces and shear flow generated by rotation are applied to adherent cells (Garcia *et al* 1997). Both assays are reported to provide reproducible and controllable results. Nevertheless, these techniques have limitations, as the resistance of cells to detachment by flow and centrifugal forces depends not only on the number, distribution and strength of the adhesion bonds present, but also on the spreading area and the surface topography of these cells. Thus the adhesive strength of the cells to the substrate can only be estimated and not directly measured.

In order to quantitatively analyze the interaction forces of cells with given substrates, sensitive SCFS assays can be performed. In brief, SCFS assays allow adhesive interaction forces and binding kinetics to be measured under physiologically relevant conditions on the cellular level as well as on the single molecule level. In micropipette-based manipulation assays a single cell, which is caught through the suction pressure at the tip of a micropipette, is brought into contact with an adhesive surface and finally retracted to measure the adhesive forces, which have been established during the adhesion time. Several micropipette-based experimental techniques that operate at both cellular and molecular levels have been developed, including the step pressure technique (Sung et al 1986), the biomembrane force probe (Evans et al 1995) and the micropipette aspiration technique (Evans et al 1976). All these methods were applied to investigate surface receptor expression, membrane tether formation from single cells and single molecule or bond dynamics (Shao and Hochmuth 1996). However, the disadvantage of these micropipette-based techniques is either low force resolution (nN range; step pressure technique) or low detectable forces (from approximately 10 pN to approximately 1 nN; biomembrane force probe). In addition, optical tweezers, which trap nano- or micrometer-sized particles in the center of a laser focus, can in principle be employed to analyze cell–substrate interactions (Thoumine et al 2000, Choquet et al 1997). However, this method is restricted because of the difficulty of measuring forces higher than 100 to 200 pN. Among all of these assays, AFM-based SCFS is currently the most versatile method to analyze adhesive interactions of cells with other cells, proteins and surfaces. This is because SCFS offers a large range of detectable forces, from 10 pN to approximately 100 nN and offers precise spatial (1 nm to 100 μm) and temporal (0.1 s to more than 10 min) control over the adhesion experiment and the experimental parameters (Helenius et al 2008).

The development and applications of SCFS

Initially, AFM-based SMFS was used to investigate the interaction of isolated receptors attached to the AFM cantilever, with ligand-decorated surfaces (Florin et al 1994), and also to measure the interaction of cantilever-linked ligands with receptors bound to a supporting surface (Lehenkari and Horton 1999). For these SMFS experiments the tip of the AFM is modified with a ligand (or receptor) of interest, then the tip is brought forward until the ligand (or receptor) on the tip binds to the receptor (or ligand) attached to the supporting surface. After a certain contact time the tip is withdrawn in order to rupture the receptor–ligand bond. In more detail, the rupture force is detected by the deflection of the AFM cantilever to which the tip is attached. The same principle of measurement is used to quantify local receptor–ligand interaction forces on a cellular surface (Grandbois et al 2000).

Based on this design principle, a SCFS setup has been developed where the overall adhesion of the cantilever's tip to an immobilized and adherent cell is measured (Lehenkari and Horton 1999). As expected, this setup has certain restrictions. In order to overcome these limitations, an inverted SCFS setup has been developed (Benoit *et al* 2000). In more detail, this SCFS setup is simply inverted, as a living cell is attached to an AFM cantilever, thus converting the cell into a 'probe'. Using this inverted setup the interactions of this 'cellular probe' with a sample of interest can be quantified (Thie *et al* 1998, Li *et al* 2003).

The SCFS was designed a long time ago as a tool to quantify cell adhesion (Benoit *et al* 2000, Helenius *et al* 2008, Benoit and Gaub 2000, Krieg *et al* 2008). Pioneering experiments have revealed the adhesion strength between two cells of *Dictyostelium discoideum* to single-molecule resolution (Benoit *et al* 2000). Then this was done for important proteins regulating the differential, adhesive behavior of zebra fish mesendodermal progenitor cells to fibronectin, which led to an insight into germ layer formation and separation (Puech *et al* 2005). Subsequently, the adhesiveness between gastrulating zebra fish cells derived from different germ layers was quantified (Krieg *et al* 2008). In this manner, the mechanisms underlying cellular sorting during gastrulation can be analyzed and, in particular, the contribution of diverse cell adhesion strengths and cell cortex tensions in germ layer organization can be exhibited. In another series of experiments, the α2β1 integrin facilitated adhesion of Chinese hamster ovary (CHO) cells to nanoscopically structured collagen type I matrices was investigated using recombinant knock-out cells and blocking substrates (Taubenberger *et al* 2007). In still another set of experiments, the role of the integrin activator TPA in strengthening integrin–cytoskeleton connections and enhancing α2β1-integrin avidity was determined (Tulla *et al* 2008). These examples and another example (Helenius *et al* 2008) indicated the applicability of SCFS to reveal the dynamic strengthening of cell adhesive bonds. Characterizing how cells form cell adhesions during the initial binding of individual cell adhesion molecules, their clustering on the cell surface and adhesion strengthening is essential to understand the establishment and regulation of cell adhesion (Taubenberger *et al* 2007, Cuerrier *et al* 2009, Friedrichs *et al* 2007, 2008, Selhuber-Unkel *et al* 2008). However, the spatial assembly of, for instance, extracellular matrix molecules also plays a role in cell adhesion, spreading and differentiation (Trappmann *et al* 2012). In this respect, the importance of molecular spacing has been investigated in modulating adhesion strength by measuring the force required to harvest an adhering cell from a nanopatterned surface by an adhesive cantilever (Selhuber-Unkel *et al* 2010). Taken together, these studies clearly demonstrate the versatility of SCFS in investigating different aspects of cell adhesion.

Experimental configurations
As mentioned earlier, there are two basic configurations for SCFS measurements, which are distinguishable in the relationship between the cell and cantilever.
  (1) A cell adhered to a substrate is touched by a functionalized tipless AFM cantilever (alternatively a pointed tip or a micrometer sized bead may be

used) for a defined contact time with a defined force and is then retracted from the cell (Grandbois *et al* 2000, Munoz Javier *et al* 2006).

(2) A cell adhered to an AFM cantilever is brought into contact with functionalized substrates for a defined contact time and is then retracted from the substrate. This configuration can also be used to determine the cell–cell adhesions between two cells, when the cell-carrying cantilever is placed onto an adherent cell for a certain time interval.

In more detail, configuration (1) has the advantage of testing one type of cantilever functionalization on several cells and thus decreases the influence of outlier cells. However, there are still various disadvantages to this approach. First, cellular contacts often leave behind debris, which may contaminate the functional-ized cantilever tip and thus reduce the specificity of the probed interaction (Culp *et al* 1979, Cohen *et al* 2003). Consequently, for long-term cell adhesion contacts this method provides just one reliable force measurement. In the case of highly structured cell surface topographies, such as filopodia, microvilli, microridges and lipid rafts, the matching of the geometry of the cantilever's tip surface (pyramid, cone or sphere) with the local geometry of the cell's surface strongly affects the effective surface area interacting between the cantilever's tip and cell and subse-quently the determined adhesive forces. Finally, adherent cells show diverse behavior in the spreading area and cellular polarization, which can have pronounced effects on the local density and identity of the cell surface receptors. However, seeding cells onto defined micropatterned substrates may reduce the variability in cell spreading and polarization, whereas the risk of cantilever contamination by debris still remains. For these reasons, this configuration may not increase the number of reliable data points, which is very important for measuring long-term adhesion contacts. Nevertheless, for short and weak contacts this approach can assist in mapping local adhesion spots on the cells (Grandbois *et al* 2000). A special application of this configuration is the harvesting of already adherent cells by addressing a cell with a cantilever, which has an even greater affinity for the cell compared to the cell's underlying adhesive substrate (Selhuber-Unkel *et al* 2010). This special format supports the investigation of cell adhesion over time points ranging from minutes to hours prior to the detachment measurement.

Configuration (2) has the advantage that the same cell can be probed at different surfaces and even to different local spots of the same surface in order to overcome the problem of contamination by cell debris left from prior adhesion events. In more detail, the cell attached to the cantilever is not allowed to spread on it, because this flat configuration of the cell would then also eliminate differences in cell area and polarization. The disadvantages of this configuration are the time-consuming and difficult additional procedure of immobilizing a living cell to the cantilever and the possibility of influencing the overall state of the cell through the adhesive contact to the functionalized AFM cantilever, which indeed occupies adhesion receptors (Friedrichs *et al* 2010). Moreover, single cells can be in different states regarding cell cycle and differentiation and thus display distinct adhesive properties that may be difficult to compare. This fact requires that the state of the cells must be controlled as tightly as possible by ensuring a homogeneous preparation protocol for each cell type and state, and also that many measurements from numerous different cells should be collected in order to obtain an 'average' adhesion response.

A special application of configuration (2) is the investigation of cell–cell adhesion by approaching a second cell or even a whole cell layer as a substrate with a cantilever carrying an attached cell. The advantages of cell–cell measurements are the perfectly prepared surfaces, regarding the orientation, functionality and natural microenvironment of the interacting molecules. However, cells usually interact through many molecules of various kinds, thus masking the signal arising from the interaction of interest through a mixture of intercellular cell–cell bonds. Moreover, variable responses from even two individual cells, rather than only one cell, contribute to enhanced variability in the collected data. Nevertheless, for some experiments these exact cellular contributions are the subject of investigation.

4.1.3 Adhesion forces between a cell and the extracellular matrix

Cell adhesion is the central mechanism that ensures the structural integrity of tissues. It is usually studied in cell culture, where the most prominent cell–matrix adhesion structures are the so-called focal adhesions or focal contacts (Gumbier 1996). Focal contacts consist of large patches of transmembrane adhesion receptors such as the integrin family (Critchley 2000). These integrin patches within the cell membrane can have lateral sizes of several micrometers. On the extracellular side, an integrin binds to ligands, such as the extracellular matrix protein fibronectin (Ruoslahti 1996). On the intracellular side, the receptors are linked to the actin cytoskeleton via a cytoplasmic plaque composed of many diverse proteins, such as talin, vinculin, paxillin and α-actinin (Zamir and Geiger 2001). This connection of the extracellular matrix to the cytoskeleton, which is often organized in the form of stress fibers, allows the transmission of forces between the cells and the extracellular matrix through focal contacts. A key player for force generation and transmission is vinculin (Mierke et al 2008, 2010, Mierke 2009).

Cells use focal adhesions or focal contacts to integrate biochemical and mechanical information about their microenvironment and regulate the organization of their cytoskeleton. In more detail, cell behavior is largely affected by the mechanical properties of the extracellular matrix and extracellular mechanical signals. For instance, the stiffness of elastic substrates provides their focal contact formation, cell elasticity, cell migration and cell differentiation (Discher et al 2005, Engler et al 2004a, 2004b, 2004c, Pelham and Wang 1997, Solon et al 2007). When intracellular force generation is inhibited by the myosin II inhibitor blebbistatin, these effects are eliminated (Discher et al 2005). On the other hand, stimulating cells by an external shear force can initiate a substantial growth of focal contacts (Paul et al 2008, Riveline et al 2001). These observations demonstrate the mechanosensitive nature of focal contacts and their importance for cellular decision-making. Mechanosensing through focal contacts seems to be based on a mechanobiochemical feedback loop involving force-induced signal transduction through small GTPases from the Rho family and activation of force generation in the actin cytoskeleton (Geiger and

Bershadsky 2002). However, the molecular basis of the mechanosensing process is still elusive. It has been suggested that the integrin–fibronectin bond as well as talin, one of the integrin binding partners in the cytoplasmic plaque, undergoes conformational alterations on stretching (del Rio et al 2009, Friedland et al 2009). Thus, investigating the mechanical properties of focal contacts is a major step to fully reveal the underlying mechanism of the mechanosensing process.

Micro- and nanopatterned substrates with a controlled distribution and density of binding sites for integrin receptors have emerged as a suitable tool for determining the impact of the extracellular matrix network structure on cell behavior. For instance, the forces that cells actively exert at focal contact sites (Balaban et al 2001, Tan et al 2003) and the dependence of cell spreading, polarization, motility, proliferation and focal contact formation on the extracellular matrix pattern (Lehnert et al 2004, Singhvi et al 1994) as well as on the nano-scale density (Maheshwari et al 2000, Massia and Hubbell 1991) of integrin binding sites has been analyzed. In more detail, it has been revealed that it is not only the average density of integrin binding sites underneath the cell body, but also the details of their lateral spacing at the nano-scale that seem to control focal contact assembly and cell behavior at the microscale. For an integrin binding site spacing larger or equal to 73 nm, focal contact assembly is inhibited and hence the cells do not spread. However, for a spacing larger or equal to 58 nm, focal contacts and actin stress fibers are assembled and indeed cells adopt a well-spread, pancake-like shape (Arnold et al 2004, Cavalcanti-Adam et al 2007). On surfaces with a nano-scale gradient of binding site spacing, the cells are able to sense effective differences in binding site spacing down to 1 nm over the cell diameter (Arnold et al 2008). These results indicate that cell adhesion is extremely sensitive to the organization of molecular binding sites at the nano-scale level.

In order to investigate the role of the nano-scale spacing of individual integrin binding sites for the detachment strength of adhered cells and the stability of focal contacts under load, nanopatterned substrates can be used. These substrates are prepared using diblock copolymer nanolithography, functionalized with the cyclic RGD peptide c[RGDfK(Ahx-Mpa)], which can bind even to the $\alpha v \beta 3$ integrin with high affinity (Kantlehner et al 2000, Pfaff et al 1994). In this way, the minimal spacing of individual $\alpha v \beta 3$-integrin molecules bound to the surface is precisely controlled. This, in turn, controls the structure and formation of focal contacts and subsequently the cell spreading. In order to quantify cell adhesion and focal contact strength, an AFM in conjunction with phase contrast and fluorescence microscopy is used, making it possible to investigate the force-induced detachment of spread cells and focal contacts. Indeed, the AFM has become a standard tool to investigate the adhesion strength of specific receptor–ligand bonds, both for single molecules and for multiple bonds in the context of single cells (Ludwig et al 2008, Müller et al 2009). However, in most studies on cell adhesion the contact time between cell and substrate is restricted to only a few minutes (Benoit et al 2000, Puech et al 2005, Taubenberger et al 2007). Studies on long-term adhesion strength with contact times of several hours have been carried out primarily with techniques such as the spinning disk apparatus (Garcia et al 1997) and centrifugation assays (Thoumine et al 1996), which can only provide information about the average adhesion strength in cell ensembles. Thus, the AFM-based method for studying long-term adhesion here significantly extends the timescale and precision for probing the adhesion strength of individual cells. Moreover, it allows analysis of the stability of subcellular components such as focal contacts, which only develop after long adhesion time.

Previous studies have shown the cooperative effect of integrin spacing for cell adhesion strength during the initial cell–substrate contact (Selhuber-Unkel *et al* 2008), which means before mature focal contacts are assembled. Using this novel method, the role of integrin spacing has been investigated for the long-term stability of cell adhesion and the elasticity of the cell body. Indeed, it has been found that focal contact formation plays the key role in controlling local cell adhesion strength and cell stiffness. Interestingly, these experiments reveal that focal contacts preferentially break at their intracellular side. This suggests that the weak link between the cell and extracellular matrix is the one between integrin receptors and cytoskeleton rather than the cell–extracellular matrix connection. It has been further investigated whether there is a dependence of the rupture force of focal contacts on the loading rate. Applying a theoretical model for the rupture of parallel bonds under shared load, two different scaling regimes for the dependence of cluster stability on the loading rate can be identified. The quantitative information obtained from analyzing focal contact stability on substrates with different integrin binding site spacing indicates that there seems to be a maximum packing density for the proteins in the cytoplasmic adhesion plaque. Indeed, the intermolecular spacing of integrins at the nanometer scale plays an essential role for cell adhesion strength and elasticity.

RGD-functionalized nanopatterns providing binding sites for individual integrin molecules have been used to investigate the role of integrin binding site spacing for cell elasticity, cell adhesion strength and focal contact stability in mature cell adhesion. However, these long-term adhesions require very different experiments compared to initial adhesion, as the cells are spread out on the substrates and have formed stable, micrometer-sized adhesion clusters. Due to the large adhesion strength at this timescale, standard protocols that are normally used in AFM-based single-cell force microscopy experiments (Benoit *et al* 2000, Puech *et al* 2005, Taubenberger *et al* 2007, Selhuber-Unkel *et al* 2008, Wojcikiewicz *et al* 2004) have to be adapted to this special situation.

Thus, a method that is capable of investigating cell adhesion forces at adhesion timescales of 5–7 h using AFM in conjunction with optical microscopy has been developed (Selhuber-Unkel *et al* 2010). In particular, the AFM cantilever is functionalized with fibronectin so that a cytoskeletal connection between the adhesion points at the cantilever and at the surface can be established. Then the cytoskeleton can transmit forces from the dorsal side of the cell to the adhesion zones at the surface. As fibronectin also binds $\alpha v \beta 3$ integrins, similar to the RGD on the substrates, the cell–cantilever contact can diminish the cell–substrate binding strength by a depletion of the integrin contacts. However, this effect is limited because the cantilever touches the dorsal side of the cell for no longer than 10–15 min before cell detachment. However, fibronectin also binds to other integrins in the cell membrane (Hynes 2002) and hence a large excess of fibronectin-binding molecules is available in the membrane (Akiyma and Yamada 1985).

Observing the fluorescence of YFP-marked paxillin molecules in the cytoplasmic plaque during detachment of cells, it has been found that only a part of the paxillin is removed. This result indicates that focal contacts predominantly break on their intracellular side, which means the link between receptors and the acto-myosin cytoskeleton is breaking instead of the receptor–extracellular matrix bonds. This result is reminiscent of studies on cell migration, where an integrin release at the rear end of the cell has been revealed (Palecek *et al* 1998, Zimmermann *et al* 1999, 2001). An intracellular breakage of focal contacts has also been observed through de-roofing cells with short ultrasonic bursts (Franz and Müller 2005). In more detail, the remaining parts of the focal contacts on the surface still contain paxillin and F-actin and the structure of the focal contact remains intact even in the rare event of actin stress fiber rupture. However, the inhomogeneous distribution of actin in a focal contact (Zaidel-Bar *et al* 2004) can be an explanation for the partial detachment of paxillin.

A well-known response of focal contacts stimulated by an external force with a tangential component is that they grow in the direction of the applied force (Paul *et al* 2008, Riveline *et al* 2001, Zaidel-Bar *et al* 2005). In these experiments, a lateral growth of focal contacts has never been observed. There may be several reasons for this. First, the detachment process was probably too fast for the cell to pronouncedly alter the focal adhesion area, in particular as stress gradually built up so that the effective time where high stress was applied to the cells was less then 1 min. Second, the force was applied to the cell in the vertical direction so that the tangential component was possibly too small to induce a lateral growth of the focal contact. However, it would be interesting to further reveal the mechanisms of focal contact growth and rupture as a function of the direction and magnitude of applied load.

It is known that cells adapt their stiffness to the microenvironment, mainly through reorganization of the actin cytoskeleton (Solon *et al* 2007). Indeed, the stiffness of the cell body before each detachment step can be determined by measuring the slope of the force-extension curve. On average, it has been shown that this stiffness is increased by a factor of four due to focal contact formation and the binding site density is supposed to play a minor role in determining this parameter. However, this result seems to be plausible, as the state of the actin cytoskeleton is tightly coupled to the focal contact formation. Without the presence of focal contacts, the actin monomers form a loose network of fibers in the intracellular space, which tends to condensate close to focal contacts to form stress fibers (Arnold *et al* 2004, Gardel *et al* 2008).

The experimental approach was complemented by a theoretical investigation of focal contact stability. A mean field description was used to determine the rupture strength of the adhesion clusters. Although the determination of the loading rate on a focal contact was based on a number of assumptions, the experimental results indeed reproduced the two theoretically predicted scaling regimes very well and gave reasonable values for the parameters. However, the focal contact rupture force density (F_0/A_0) does not increase in direct proportion to the binding site density. As the focal contact rupture forces are measurements of the cohesion strength of the focal contact plaque rather than a measurement of the adhesion strength to the extracellular connection, this observation suggests the existence of a maximum packing density of the adhesion plaque. This maximum packing density is reached on substrates with an integrin spacing of 28 nm. This observation should be further investigated, for example by analyzing the fluorescence intensity of different focal contact proteins as a function of binding site density. However, the hierarchical organization of the focal contact plaque suggests a well-defined spacing as well as precise orientation of its proteins and hence the existence of a maximum packing density is not surprising.

Taken together, it has been shown that the intermolecular spacing of integrins at the nanometer scale plays a crucial role for cell adhesion strength, cell elasticity and the stability of focal contact and that it is possible to investigate the detachment forces of well-adhered individual cells. These results, together with the method presented, are a good basis for further experiments investigating related questions concerning the mechanical properties of focal contacts, the role and mechanism of focal contact rupture and growth, and the influence of intermolecular spacing at different timescales.

4.2 Traction forces

Cell adhesion and migration are reported to depend on the transmission and generation of actomyosin-dependent contractile forces through focal adhesions to the extracellular matrix of connective tissue. Indeed, during the last two decades, experimental and computational advances have improved the resolution and reliability of 2D and 3D traction force microscopy. It has been suggested that forces exerted by cells drive the migration and invasion of cancer cells through the extracellular matrix or through an endothelial barrier when entering or leaving blood or lymphoid vessels (Mierke *et al* 2008, 2011a, 2011b).

4.2.1 2D forces on planar substrates

First, the use of two differently colored nanobeads as fiducial markers in poly-acrylamide gels is discussed and how the displacement field can be computationally extracted from the fluorescence data is explained. Second, different improvements regarding standard methods for force reconstruction from the displacement field are presented, such as the boundary element method, Fourier-transform traction cytometry and traction reconstruction with point forces. Using extensive data simulation, it can be demonstrated that the spatial resolution of the boundary element method can be improved considerably by splitting the elastic field into near, intermediate and far field. Moreover, Fourier-transform traction cytometry requires considerably less computer time, whereas it can achieve a comparable resolution only when combined with Wiener filtering or appropriate regularization schemes. However, both methods seem to underestimate forces, in particular at small adhesion sites. The traction reconstruction with point forces does not suffer from this restriction, but is only applicable to stationary and well-developed adhesion sites. Third, both advantages are combined and used for the reconstruction of fibroblast traction with a spatial resolution of approximately 1 μm.

It has been hypothesized that physical force plays a crucial role as a regulator of many cellular processes, including cell adhesion and migration (Geiger and Bershadsky 2002, Discher *et al* 2005, Orr *et al* 2006, Vogel and Sheetz 2006, Schwarz 2007). In particular, the dynamics of establishing actomyosin-generated traction force and its transmission in an elastic microenvironment seems to be essential for the way cells sense and respond to the mechanical properties of their microenvironment. Thus, measuring the cellular traction forces exerted on elastic substrates upon cell–matrix adhesion is an essential tool for investigating and understanding the regulation of cell adhesion and migration in a quantitative way (Beningo and Wang 2002, Roy *et al* 2002).

The traction force microscopy method was pioneered at least 30 years ago using thin flexible Si sheets as a first wrinkling assay to provide a qualitative measure for the mechanical activity of cells (Harris 1980). By contrast, in order to perform quantitative studies it is essential to suppress wrinkling, which has been achieved by using films under prestress (Lee *et al* 1994, Dembo *et al* 1996) and later it has also been achieved for thick films attached to a cover slide (Dembo and Wang 1999, Pelham and Wang 1997). Now the state-of-the-art method is to investigate traction forces on thick elastic substrates, which ensures the reconstruction of cellular traction forces. In more detail, the films are usually prepared from polyacrylamide (PAA) or polydimethylsiloxane (PDMS) coated with adhesive ligands such as fibronectin or collagen, where fibronectin is preferable to collagen as its coating procedure works better. RGD peptide can also be used for the coating, and it is even better suited than fibronectin. The use of PAA has the advantage that its stiffness can be tuned easily over the physiologically relevant range from 100 Pa to 100 kPa (Engler *et al* 2004a, 2004b, 2004c, Yeung *et al* 2005). PDMS, which is hard to prepare with a bulk stiffness below 10 kPa, has the advantage that it can easily be micropatterned (Balaban *et al* 2001, Goffin *et al* 2006, Cesa *et al* 2007). An alternative to flat elastic substrates is the pillar assay, which represents an array of microfabricated PDMS-pillars deforming easily under cellular traction due to their small diameter (Tan *et al* 2003, du Roure *et al* 2005, Cai *et al* 2006, Saez *et al* 2007). As each pillar represents a localized force sensor, the pillar assay ensures a simple, albeit spatially constrained, readout of the forces. For cell adhesion, which is not spatially confined, however, flat elastic substrates are the method of choice.

A setup for traction force microscopy on flat elastic substrates has to be combined with different experimental and computational techniques. In more detail, a cell adheres to a flat substrate and thereby exerts force through its sites of adhesion. In particular, the resulting deformations in the substrate are tracked by monitoring the individual movement of embedded marker beads. For PAA gels, fluorescent microbeads are usually used, and these are embedded near the surface of the gel. In some applications they are even collected in the uppermost layer of the gels through centriguation of the inverted PAA gels before the polymerization. For PDMS, micropatterning can be performed to create a pattern that is easily detected under phase contrast microscopy. The displacement field has to be extracted from a pair of images, one image showing the substrate as it has deformed under cell traction and one reference image showing the undeformed substrate without any cell traction acting on the gel. In general, there are two ways to perform the image processing: either one directly tracks the movement of the fiducial markers (called particle tracking velocimetry (PTV)) or one uses a cross-correlation function to obtain the local motion statistically (called particle image velocimetry (PIV)). The next step is to reconstruct the cellular traction pattern from the displacement field. For synthetic and substrates such as PAA or PDMS, elastic behavior can be assumed, which is indeed a homogeneous, isotropic and linear material. In more detail, during cell adhesion in tissue culture the cell is rather flat and hence force in the normal direction can be neglected. However, this is a crude assumption that still needs to be proven. Then both the displacement field $u(x)$ and the traction stress field $f(x)$ are 2D in the plane of the substrate ($x = x1$, $x2$). They are related by the following integral equation (4.5):

$$u_i(x) = \int \sum_j G_{ij}(x - x')f_j(x')\mathrm{d}x'.$$

Given the experimental displacement $u(x)$ and the relevant Green's function $G_{ij}(x)$, one simply needs to invert equation (4.5) to obtain the desired traction field $f(x)$. For thick films, the substrate can be approximated by an elastic half-space and hence one can use the Boussinesq Green's function as in equation (4.6) (Landau and Lifshitz 1970):

$$
\begin{aligned}
G_{ij}(x) &= \frac{(1 + v)}{\pi E}\left[(1 + v)\frac{\delta_{ij}}{r} + v\frac{x_i x_j}{r^3}\right] \\
&= \frac{(1 + v)}{\pi E r^3}\begin{pmatrix} (1 - v)r^2 + vx^2 & vxy \\ vxy & (1 - v)r^2 + vy^2 \end{pmatrix},
\end{aligned}
$$

where v and E represent the Poisson ratio and Young modulus, respectively, and $r = |x|$. For clarity, the Green's tensor has been written here in both the index notation and in full form. As is typical for a 3D elastic Green's function, it scales as $\sim 1\ r^{-1}$ with distance, thus it seems to be long-ranged and has a singularity at the origin. Equation (4.5) is a Fredholm integral equation of the first kind with a weakly singular kernel. The long-ranged nature of the kernel implies that the direct problem corresponds to a smoothing operation. This suggests that the inverse problem may be very sensitive to small differences in the displacement field, depending on the exact nature of the experimental data. In more detail, noise in the experimental data can result from different sources, such as elastic inhomogeneities in the substrate, insufficient coupling between marker beads and polymer matrix, deficiencies in the optical setup and lack of accuracy in the tracking routines.

Three standard methods have been established to calculate the force from displacement. Both the boundary element method (BEM) (Dembo *et al* 1996, Dembo and Wang 1999) and Fourier transform traction cytometry (FTTC) (Butler *et al* 2002) approximate the integral on a grid (discretized methods). While BEM effectively corresponds to inverting a large system of linear equations in real space, FTTC uses the fact that the relevant system of linear equations is much smaller in Fourier space, thus facilitating inversion. Traction reconstruction with point forces

(TRPF) (Balaban *et al* 2001, Cesa *et al* 2007, Schwarz *et al* 2002) uses additional experimental knowledge about the precise location of the adhesion sites, which can, for instance, be obtained by fluorescence data on proteins localizing to focal adhesions such as vinculin or paxillin (Balaban *et al* 2001, Schwarz *et al* 2002) or by reflection interference contrast microscopy (Cesa *et al* 2007). Then the integral in equation (4.5) converts into a simple sum. In the past, there has been disagreement about the advantages and disadvantages of these diverse methods. In particular, which of the two discretized methods is better in respect to resolution and reliability remains an open quesiton. Moreover, different solutions have been proposed to deal with the issue of experimental noise, whereas a systematic and detailed analysis of these approaches is still missing.

Different experimental and computational advances are presented here, which together provide much higher resolution and greater reliability in traction force microscopy than was possible formerly. A new method to track the deformations of a PAA substrate has been utilized experimentally in which two differently colored nanobeads are used simultaneously as fiducial markers. In order to extract the corresponding displacement field, a new image processing method has been adapted in which particle tracking velocimetry (PTV) and particle image velocimetry (PIV) are combined. Regarding the computational reconstruction of the traction field, variants of all three standard techniques have been implemented, including BEM, FTTC and TRPF, in order to compare their performance systematically using extensive data simulation. In particular, an improved version of BEM has been adapted and compared with new variants of FTTC. Analysis of simulated data revealed the importance of filtering for FTTC and indicated that certain variants of FTTC can perform almost as well as BEM while being much more efficient in terms of computer time required. However, as a disadvantage, both discretized methods are found to be strongly biased regarding small adhesion sites. TRPF does not suffer from this limitation, but its underlying assumption of accurate knowledge of adhesion site location restricts its applicability. Finally, it has been demonstrated with fibroblasts that the overall setup results in a spatial resolution that constitutes a 5–10-fold improvement over earlier methods (Selhuber-Unkel *et al* 2010).

Two distinct kinds of fluorescent marker beads improves the measured displacement field
The spatial resolution of the reconstructed traction field depends directly on the spatial resolution with which the displacement field is sampled. The use of a multispectral confocal spinning disk microscope enables the acquisition of images from different fluorescence channels at high resolution. In order to achieve the highest possible information content, two differently colored beads can be densely embedded in the PAA gel. Indeed, the number of features is much higher if a combination of both channels is used. As the quality of images usually differs from channel to channel, it is impossible to interlace both extracted vector fields when tracking each channel separately using correlation-based PTV.

According to the Nyquist–Shannon sampling theorem, the optimal spatial resolution that can be achieved is determined by half the sampling frequency.

For one- and two-channel tracking, this should be roughly 1.5 μm and 1 μm, respectively. To check whether this limit can be reached, the capability of the different computational methods to distinguish very small traction sources is analyzed. In particular, the resolution is defined by the Rayleigh criterion, as the minimum distance between two point forces at which they can still be separated. The results of Fourier-transform traction cytometry (FTTC) after four different noise treatments, such as Gaussian filtering, Wiener filtering, 0th, and second-order regularization, showed after visual inspection that very weak traction is not reproduced. The reconstructions in several cases revealed that nearby forces can no longer be separated. Indeed, all the methods provide a clear representation of the overall traction pattern and strong point forces can in any case be distinguished clearly. The differences in resolution between BEM and FTTC are mainly due to mesh geometry, which is a slight advantage for the BEM. Moreover, the choice of the filter procedure strongly influences the point-force resolution. Indeed, the new method using two colors results in a much denser and more precise sampling of the displacement field compared to the traditional approach with one color.

To determine the detection limit of the traction reconstruction quantitatively, a ten-point force traction pattern is generated and measured for a high substrate stiffness of $E = 10$ kPa and the reconstructed traction field is determined outside the sites of adhesion. These values represent a background in the traction pattern, which potentially masks small cellular traction forces. An increase in the noise level only leads to a slight increase in the detection limit. However, increasing the absolute level of the displacement field by increasing the original traction also leads to a strong increase in the traction background. For instance, for a maximum displacement of approximately eight pixels the detection limit is around 200 Pa, corresponding to a force of 0.2 nN μm^{-2}. Indeed, for the same displacement, but for a doubled substrate stiffness of $E = 20$ kPa, the traction background increases by a factor of two (such as 400 Pa), which corresponds to a force of 0.4 nN μm^{-2}. Thus small forces cannot be resolved when the overall magnitude of traction is high, regardless of the noise level.

BEM and FTTC produce comparable results

In order to obtain a more general quantitative comparison between the different variants of traction force microscopy, three different scores for circular adhesions are measured with different noise levels in the displacement field. For all three scores used—DTM, DTMS and DTA—it is the case that the closer its value to zero, the better the performance. In the absence of noise, filtering or regularization cannot improve the results of FTTC. Indeed, FTTC works comparably well compared to BEM with 0th order regularization, whereas BEM with first-order regularization works relatively poorly. Moreover, this result persists in the presence of noise, thus in general BEM is best together with 0th order regularization. For FTTC, it has been shown that Wiener filtering is always better than Gaussian filtering and that 0th order regularization is always better than second-order regularization. Thus these three methods—Wiener filtering, 0th order regularization and BEM—are equally suitable. However, there are dramatic differences in the required computation times.

On a standard desktop computer BEM requires several hours and in addition needs large storage resources, whereas FTTC only requires seconds. In addition, the programming effort is pronouncedly reduced for FTTC, which can be encoded in a few pages, whereas the BEM is normally spread out over several subroutines. Taken together, FTTC with Wiener filtering or 0th order regularization is reliable and the most efficient method.

The traction magnitude of small adhesion sites is underestimated with discretized methods

The deviation of traction magnitude (DTM) is always negative for both BEM and FTTC, hence the significant underestimation of traction is a persistent challenge with the discretized methods. The causes of systematic and random underestimation are manifold, but depend strongly on the sampling density of the displacement field, as suggested by the Nyquist criterion. Indeed, the analysis of adhesion sites smaller than two mesh sizes is really problematic. The difference between reconstructing traction force using BEM with 0th order regularization and FTTC with Wiener filtering on an irregular versus regular grid is clearly visible: in both cases, the overall traction pattern is nicely reproduced. In more detail, the DTM is plotted as a function of adhesion size (measured in units of mesh size) for noise levels of 0 and 10%. Below a critical size, there is an inverse relation between DTM and adhesion size, suggesting that force reconstruction is only reliable for large adhesions. For 0% noise, the critical value is approximately two mesh sizes, whereas it shifts to larger values for higher noise levels. In the presence of 10% noise, only adhesions with a size approximately four times larger than the mesh size are reconstructed in a reliable manner. For smaller adhesions, the DTM becomes strongly negative, thus the traction at these adhesions (and hence the overall strain energies) is severely underestimated. In conclusion, it is important to start with a dense sampling of the displacement field, as this leads to small mesh sizes and thus reliable force reconstruction for small adhesions and even better estimates for the strain energy.

The traction reconstruction with point forces is precise, but depends on correct localization of focal adhesions

To improve the force reconstruction at small adhesions, traction reconstruction with point forces (TRPF) is used, which starts with the assumption of a highly localized force (Schwarz *et al* 2002). However, the DTM depends much less on the size of the adhesion, which leads to a reliable force reconstruction over a large range of adhesion sizes. The critical point in this method is that the localization of the adhesions has to be very precise. If the adhesions are not placed well, the quantitative analysis using DTM and DTA shows directional errors and strong magnitude fluctuations. Finally, this finding restricts the applicability of the method to cells with few and distinct focal adhesions.

High resolution reconstruction for fibroblast traction measurements

How can the whole setup be used to determine fibroblast tractions? In order to obtain the required spatial resolution for the displacement field, it is proposed to image only a limited region of the cell. Thus a region of interest has been selected, which does not have any additional nearby adhesions that might alter the traction pattern inside the region of interest. Indeed, the mesh size of approximately 500 nm results in a spatial resolution of around 1 µm, because the computational reconstruction is combined with adequate filtering. For the substrate stiffness $E = 15.6$ kPa and the given fibroblast strength, the traction background is approximately 500 Pa, corresponding to a force of 0.5 nN $µm^{-2}$. In the reconstruction forces of up to 10 kPa are found frequently, corresponding to a force of 10 nN $µm^{-2}$, which is 20-fold above the detection limit. For the BEM the computational mesh is an irregular grid, confined by the wedge-like region including the cell contour. For FTTC, force vectors are reconstructed on a square lattice covering the whole image. Both the BEM and FTTC methods provide very similar results. In this case (TRPF), one point is first been selected for each adhesion, whereas for very large adhesions two points are selected. The resulting forces are in visual agreement with the results from BEM and FTTC. Direct comparison between the three methods for the small region indicates the difference between them: one force vector for TRPF, an irregular pattern for BEM and a regular pattern for FTTC. It shows that the spatial resolution of BEM and FTTC is indeed around 1 µm.

Established and newly developed procedures to reconstruct cellular traction force on flat elastic substrates have been discussed. Furthermore, the approach that guarantees an optimal spatial resolution and magnitude reconstruction for diverse experimental conditions has been indicated. As an initial step, the known experimental traction force protocols have been implemented so that they can be used for very small and differently colored fluorescent beads under a confocal microscope. The use of two kinds of distinctly colored beads densely embedded in the gel ensures the extraction of displacement fields with an average mesh size of 500 nm, setting the basic scale for the spatial resolution of the force reconstruction as 1 µm. This resolution is a major improvement (by a factor of ~5–10) over earlier work.

It is common knowledge that cellular adhesion sizes vary considerably, ranging from a few hundred nanometers in nascent adhesions of migrating cells to tens of microns for supermature focal adhesions in myofibroblasts. For small adhesions, the mesh size of the displacement data is larger than the adhesion feature size and thus the Nyquist frequency determining the theoretical upper restriction for the resolution of traction forces is typically too low to capture all the details of the exerted traction field. There is still an open question as to whether there is a special computational method to provide the best results in reconstructing cellular traction fields. The main methods to investigate cellular traction fields are BEM (Dembo *et al* 1996, Dembo and Wang 1999) and FTTC (Butler *et al* 2002). For both methods, different improvements have been proposed, such as analytical integration procedures and adaptive mesh generation for BEM, and different filtering and regularization schemes for both BEM and FTTC.

An intrinsic feature of both methods is that solving the inverse stress–strain problem is equivalent to multiplying each Fourier component of the displacement field by its respective wave number. Indeed, high frequency noise will thus be amplified and has to be removed while leaving as much as possible of the original signal untouched. Thus, different strategies can be used. On the one hand, it seems reasonable to impose a smoothness constraint on the displacement field and filter it before the calculation of traction stress is performed. This technique only works with the Fourier transform methods. On the other hand, it is better to choose a good solution *a posteriori* by inspection of its properties. However, regularization constrains the resulting traction field and one can fine-tune it by compensating the above-mentioned effect of noise amplification to a desired level in the solution.

FTTC is shown to work best with Wiener filtering and 0th order regularization in simulated data, whereas the analysis of real data leads to the suggestion that regularization is a more robust approach. As expected, BEM mostly works well with 0th order regularization. An overall comparison of both approaches leads to the conclusion that FTTC, when combined with a proper filtering procedure, is in large parts comparable with BEM regarding resolution. The initial advantages of the boundary element approach, resulting from the exact incorporation of irregular data, are lost in the presence of noise. Aliasing and boundary effects are not very prominent with boundary element methods, but can also limit the performance of FTTC. However, the great advantage of FTTC is the very small run time compared with the BEM.

Both discretized methods suffer from a systematic underestimation of traction forces at small adhesion sites. It would be interesting to know whether traction force reconstruction with point forces (TRPF) can measure force magnitudes properly independent of adhesion size. The concept involved, of *a priori* localized traction sources, does, in fact, serve to avoid the above bias. However, this concept is only applicable when the main assumption of this method, a reasonable localization of all traction sources, can be ensured. In general, traction measurement systems which do not permit the presumably force-facilitated, yet temporary, assembly of very small adhesion sites seem to be a slightly ill-fated choice when the whole spectrum of forces and morphologies of cellular adhesion sites is being investigated in migrating cells.

Keeping in mind that several mechanisms may contribute to force development at focal adhesions, a high spatial resolution is required for a quantitative understanding of traction forces. From the work just presented, it seems to be clear that dense displacement fields are the most important strategy to achieve this goal. However, a major concern seems to be that incorporating many marker beads in the gel contradicts the assumption of linear elasticity in the substrate, as the beads may alter the mechanical properties of the matrix. In more detail, inclusions in an elastic material are known to perturb the Young's modulus and Poisson ratio only in a small surrounding shell of the scale of the bead diameter (Jasiuk *et al* 1997). Thus this effect only seems to be a problem when large beads such as microbeads are used.

In summary, our work shows that high-resolution traction force microscopy has been reported to rely on the combination of advances in substrate preparation, image processing and computational force reconstruction. The systematic and quantitative comparison shows that, depending on the specific situation and the resources available, different approaches can be useful. For example, BEM with 0th order regularization seems to be the best choice to obtain high-resolution traction patterns for migrating cells possessing many small adhesions and with a small noise level in the displacement data. If computer time is restricted, FTTC with 0th order regularization is an equivalent alternative, especially at higher noise levels. For stationary cells in which focal adhesions can be precisely localized, TRPF is both computationally cheap and reliable.

4.2.2 3D forces within a 3D collagen matrix scaffold

Quantitative measurements of cell-generated forces have been developed and can be performed easily when the cells are cultured on 2D substrates (Mierke *et al* 2008). Indeed, a technique has been developed to quantitatively measure the 3D traction forces exerted by cells that are fully encapsulated within well-defined elastic hydrogel matrices. This approach has been used to measure tractions from a variety of cell types and diverse contexts and thus reveal patterns of force generation that are attributable to morphologically distinct regions of cells as they migrate into the surrounding extracellular matrix.

It has been known for a long time that cells are constantly probing, pushing and pulling on the surrounding extracellular matrix of connective tissues, for instance to migrate and invade through it. Therefore the cells need to apply forces to the extracellular microenvironment in order to sense the matrix's mechanical properties. In particular, these cell-generated forces facilitate cell migration as well as tissue morphogenesis, and maintain the intrinsic mechanical balance of tissues (Dembo and Wang 1999, Keller *et al* 2003). These forces not only guide mechanical and structural events, but also facilitate signal transduction pathways promoting functions ranging from cell proliferation to stem cell differentiation (Huang *et al* 1998, McBeath *et al* 2004). Hence, precise measurement of the spatial and temporal nature of these forces is needed in order to reveal when and where mechanical events play a prominent role in physiological and pathological settings.

Indeed, methods employing planar 2D elastic surfaces or arrays of flexible cantilevers have mapped with subcellular resolution and the forces that cells generate to their substrates have been determined (Balaban *et al* 2001, Butler *et al* 2002, Tan *et al* 2003, Mierke *et al* 2008, 2011). However, many processes are altered when cells are removed from native 3D microenvironments and are then cultured on 2D substrates. At a structural level, cells encapsulated within a 3D matrix display dramatically different morphology, cytoskeletal organization and focal adhesion structure compared to those cultured on 2D substrates (Cukiermann *et al* 2001). However, even the initial requirements for cells to adhere and spread against a 2D substrate are different from the invasive process essential for cells to extend inside a 3D matrix. In more detail, these differences suggest that dimensionality alone may significantly impact how cellular forces are generated and transduced into biochemical or structural alterations. Although the mechanical properties of 3D extracellular matrices and the cellular forces generated within a 3D microenvironment have been shown to regulate many cellular functions (Pampaloni *et al* 2007), the quantitative measurement of cellular forces within a 3D context has to be demonstrated.

In particular, the traction stresses (force per area), hereafter tractions, exerted by cells embedded within a hydrogel matrix can be determined quantitatively (Legant *et al* 2011, 2013). GFP-expressing fibroblasts were encapsulated within mechanically well-defined polyethylene glycol (PEG) hydrogels that incorporate proteolytically degradable domains in the polymer backbone and pendant adhesive ligands (Miller *et al* 2010). In more detail, the incorporation of adhesive and degradable domains enables the cells to invade, spread and adopt physiologically relevant cellular morphologies. These hydrogels had a Young's modulus of 600 to 1,000 Pa, a range similar to those of commonly used extracellular matrices, such as reconstituted collagen or Matrigel and to *in vivo* tissues such as mammary and brain tissue (Paszek *et al* 2005, Discher *et al* 2005). As expected, cells in 3D PEG gels deformed the surrounding nearby matrix, which was visualized by tracking the displacements of 60,000–80,000 fluorescent beads in the close neighborhood of each cell. In particular, bead displacements were determined relative to a reference, stress-free, state of the gel after lysing the cell with a detergent. Deformations of 20–30% peak principal strain have been measured in much of the cell surrounding hydrogel. The largest strains (even up to 50%) occurred in close vicinity to long impressive extensions, which is indeed consistent with the observations of strong forces exerted by these regions on 2D substrates (Chan *et al* 2008). As the mechanics of the PEG hydrogels showed no substantial dependence on strain or frequency, a linear elasticity theory and the finite element method was used to determine the cellular tractions that would give rise to the measured bead displacements. In brief, a finite element mesh of the hydrogel is generated that surrounds the cell from confocal images. In particular, a discretized Green's function was constructed by applying unit tractions to each facet on the surface of the cell mesh and solving the finite element equations in order to calculate the induced bead displacements. As usual, the standard regularization methods for ill-posed, over-determined linear systems of equations were then used to compute the tractions exerted by the cell. The time needed to calculate a single data set is approximately 4.5 h. However, this calculation time can be reduced dramatically by using a simplified finite element mesh of the cell and hydrogel. Indeed, these lower resolution datasets still capture the fundamental character of higher-resolution measurements.

Simulated traction fields are used to validate the approach and to characterize its spatial resolution. The experimental noise for the bead displacements was determined by measuring the displacement of the beads in cell-free regions of the hydrogel before and after detergent treatment. In addition, the surface discretization noise was measured from multiple discretizations of the same cells. These datasets then were superimposed onto the displacements generated by simulated loadings prior to the traction reconstruction. In this setting, the percentage of traction recovered was proportional to the magnitude and characteristic length of the simulated loadings, defined as the average period of spatial oscillation. Indeed, for all cases, the presence of noise reduced recovery accuracy by approximately 20–30%. Despite these limitations, the recovered tractions can still capture the essential periodic features of the most spatially complex simulated loadings with characteristic lengths of spatial variation down to only 10 µm.

In the next step, the tractions from living cells encapsulated within 3D hydrogels are measured. Indeed, it has been found that cells have exerted tractions in the range of 100–5000 Pa, with strong forces located predominantly near the tips of long extensions. For all measurements, the forces were in static equilibrium, with a typical error of approximately 1–5% of the total force applied by the cell. Further analysis revealed that these tractions were minimally altered by possible variations in local hydrogel mechanics or by uncertainty in the recorded bead displacements. Measurements of cellular forces on 2D surfaces have generally been limited to shear loadings, although some studies have measured small forces exerted normal to the planar surface (Maskarinec *et al* 2009, Hur *et al* 2009). However, it is still elusive whether these relationships might be altered for cells inside 3D extracellular matrices. Cells encapsulated within a 3D matrix have been reported to exert predominantly shear tractions, although small normal tractions were additionally present in close proximity to the cell body. In order to investigate whether patterns of force are associated with specific regions of cells, the magnitude and angle of tractions is quantified with respect to the center of mass of the cell. In general, tractions increased as a function of distance from the center of mass. Interestingly, cells encapsulated in hydrogels with a Young's modulus of approximately 1,000 Pa generated stronger tractions than those in approximately 600 Pa hydrogels, indicating that cells adapt to the mechanical properties of their microenvironment. However, the observed differences in tractions were not due to an overall increase in total cellular contractility, as measured by the net contractile moment, whereas they were most prominent in strong inward tractions near the tips of long extensions. This reveals a local and non-linear reinforcement of cellular contractility due to the substrate rigidity and suggests that such regions may be hubs for force-facilitated mechanotransduction in 3D settings. The cell bodies showed no bias in the traction angle. However, strong tractions became progressively aligned back toward the center of mass in more well spread regions of the cell, for instance, near the tips of long extensions. These patterns of force were typically reflected in multiple cell types, but could be altered by cell–cell proximity or culture as a multicellular aggregate. The neighboring 3T3 cells preferentially extended away from each other, whereas proliferating multicellular tumor spheroids exerted outward normal tractions on the extracellular matrix.

In more detail, a subset of extensions was observed that displayed strong tractions several microns behind the leading tip, whereas the tractions at the tip itself were substantially lower. As such traction profiles are similar to those determined behind the leading edge of a lamellipodia for a migrating cell on a 2D substrate, it can be hypothesized that such regions represent invading or growing cellular extensions in 3D. To analyze this possibility, the tractions from time-lapse images of cells as they invaded the surrounding hydrogel have been determined. Indeed, tractions at the tips of growing extensions were pronouncedly lower compared to the strong tractions exerted by proximal regions of the same extension. However, normal forces pushing into the extracellular matrix have not been detected in these extensions, which suggests that a local inhibition of myosin-dependent contractility allows tip encouragement. Strong tractions have been detected from small extensions on the cell face opposite to the invading extensions. These stable extensions exhibited very different force distributions compared to the growing extensions and often lack the characteristic drop in force near the leading edge and hence may correspond to an anterior–posterior polarity axis build-up within the cell.

Taken together, these data suggest that cells in 3D matrices probe their surrounding extracellular matrix primarily through strong inward tractions near the tips of long slender extensions such as protrusions. Importantly, it has been demonstrated that this 3D force measurement technique is applicable to many different cell types, cell–cell interactions and even to multicellular tumor structures where both tumor growth and invasion have previously been revealed to be mechanoresponsive (Paszek *et al* 2005). Because synthetic hydrogels possess a similar elastic moduli compared to *in vivo* tissues (Paszek *et al* 2005, Discher *et al* 2005) and can hence support a wide range of cellular functions (Lutolf and Hubbell 2005), this approach will enable investigations into the role of cellular forces in a variety of biological settings.

Further reading

Besser A and Schwarz U S 2007 Coupling biochemistry and mechanics in cell adhesion: a model for inhomogeneous stress fiber contraction *New J. Phys.* **9** 425

Bell G I 1987 Models for the specific adhesion of cells to cells *Science* **200** 618–627

Berdyyeva T K, Woodworth C D and Sokolov I 2005 Human epithelial cells increase their rigidity with ageing *in vitro*: direct measurements *Phys. Med. Biol.* **50** 81–92

Braet F, de Zanger R, Seynaeve C, Baekeland M and Wisse E 2001 A comparative atomic force microscopy study on living skin fibroblasts and liver endothelial cells *J. Electron Microsc. (Tokyo)* **50** 283–90

Bushell G R, Cahill C, Clarke F M, Gibson C T, Myhra S and Watson G S 1999 Imaging and force–distance analysis of human fibroblasts *in vitro* by atomic force microscopy *Cytometry* **36** 254–64

Cavalcanti-Adam E A, Volberg T, Micoulet A, Kessler H, Geiger B and Spatz J P 2007 Cell spreading and focal adhesion dynamics are regulated by spacing of integrin ligands *Biophys. J.* **92** 2964–74

Chung K-H and Ki D-E 2007 Wear characteristics of diamond-coated atomic force microscope probe *Ultramicroscopy* **108** 1–10

Collinsworth A M, Zhang S, Kraus W E and Truskey G A 2002 Apparent elastic modulus and hysteresis of skeletal muscle cells throughout differentiation *Am. J. Physiol. Cell Physiol.* **283** C1219–27

Domke J, Dannohl S, Parak W J, Muller O, Aicher W K and Radmacher M 2000 Substrate dependent differences in morphology and elasticity of living osteoblasts investigated by atomic force microscopy *Colloids Surf. B: Biointerfaces* **19** 367–79

Erdmann T and Schwarz U S 2004 Adhesion clusters under shared linear loading: a stochastic analysis *Europhys. Lett.* **66** 603–9

Erdmann T and Schwarz U S 2004a Stability of adhesion clusters under constant force *Phys. Rev. Lett.* **92** 108102

Evans E and Ritchie K 1997 Dynamic strength of molecular adhesion bonds *Biophys. J.* **72** 1541–55

Fritz M, Radmacher M and Gaub H E 1993 *In vitro* activation of human platelets triggered and probed by atomic force microscopy *Exp. Cell Res.* **205** 187–90

Fritz M, Radmacher M and Gaub H E 1994 Granula motion and membrane spreading during activation of human platelets imaged by atomic force microscopy *Biophys. J.* **66** 1328–34

Geiger B and Bershadsky A 2002 Exploring the neighborhood: adhesion-coupled cell mechano-sensors *Cell* **110** 139–42

García A J, Ducheyne P and Boettiger D 1997 Quantification of cell adhesion using a spinning disc device and application to surface-reactive materials *Biomaterials* **18** 1091–8

Gladilin E, Micoulet A, Hosseini B, Rohr K, Spatz J and Eils R 2007 3D finite element analysis of uniaxial cell stretching: from image to insight *Phys. Biol.* **4** 104–13

Haga H, Nagayama M, Kawabata K, Ito E, Ushiki T and Sambongi T 2000 Time-lapse viscoelastic imaging of living fibroblasts using force modulation mode in AFM *J. Electron Microsc. (Tokyo)* **49** 473–81

Koo L Y, Irvine D J, Mayes A M, Lauffenburger D A and Griffith L G 2002 Co-regulation of cell adhesion by nanoscale RGD organization and mechanical stimulus *J. Cell Sci.* **115** 1423–33 PMID: 11896190

Lekka M, Laidler P, Gil D, Lekki J, Stachura Z and Hrynkiewicz A Z 1999 Elasticity of normal and cancerous human bladder cells studied by scanning force microscopy *Eur. Biophys. J.* **28**(4) 312–6

Lekka M, Laidler P, Ignacak J, Labedz M, Lekki J, Struszczyk H, Stachura Z and Hrynkiewicz A Z 2001 The effect of chitosan on stiffness and glycolytic activity of human bladder cells *Biochim. Biophys. Acta* **1540**(2) 127–36

Liang X, Mao G and Simon Ng K Y 2004 Probing small unilamellar EggPC vesicles on mica surface by atomic force microscopy *Colloids Surf. B: Biointerfaces* **34** 41–51

Lieber S C, Aubry N, Pain J, Diaz G, Kim S-J and Vatner S F 2004 Aging increases stiffness of cardiac myocytes measured by atomic force microscopy nanoindentation *Am. J. Physiol. Heart Circ. Physiol.* **287** H645–51

Lussi J W, Falconnet D, Hubbell J A, Textor M and Csucs G 2006 Pattern stability under cell culture conditions—a comparative study of patterning methods based on PLL-g-PEG background passivation *Biomaterials* **27** 2534–41

Merkel R, Nassoy P, Leung A, Ritchie K and Evans E 1999 Energy landscapes of receptor-ligand bonds explored with dynamic force spectroscopy *Nature* **397** 50–3

Micoulet A, Spatz J P and Ott A 2005 Mechanical response analysis and power generation by single-cell stretching *Chem. Phys. Chem.* **6** 663–70

Murakoshi M, Yoshida N, Iida K, Kumano S, Kobayashi T and Wada H 2006 Local mechanical properties of mouse outer hair cells: atomic force microscopic study *Auris Nasus Larynx* **33** 149–57

Nagayama M, Haga H and Kawabata K 2001 Drastic change of local stiffness distribution correlating to cell migration in living fibroblasts *Cell Motil. Cytoskel* **50** 173–9

Oliver T, Jacobson K and Dembo M 1998 Design and use of substrata to measure traction forces exerted by cultured cells *Methods Enzymol.* **298** 497–521

Pasche S, Textor M, Meagher L, Spencer N D and Griesser H J 2005 Relationship between interfacial forces measured by colloid-probe atomic force microscopy and protein resistance of poly(ethylene glycol)-grafted poly(L-lysine) adlayers on niobia surfaces *Langmuir* **21** 6508–20

Pelham R J and Wang Y-L 1999 High resolution detection of mechanical forces exerted by locomoting fibroblasts on the substrate *Mol. Biol. Cell* **10** 935–45

Radmacher M, Fritz M, Kacher C M, Cleveland J P and Hansma P K 1996 Measuring the viscoelastic properties of human platelets with the atomic force microscope *Biophys. J.* **70** 556–67

Rotsch C and Radmacher R 2000 Drug-induced changes of cytoskeletal structure and mechanics in fibroblasts: an atomic force microscopy study *Biophys. J.* **78** 520–35

Rotsch C, Braet F, Wisse E and Radmacher M 1997 AFM imaging and elasticity measurements on living rat liver macrophages *Cell Biol. Int.* **21** 685–96

Rotsch C, Jacobson K and Radmacher R 1999 Dimensional and mechanical dynamics of active and stable edges in motile fibroblasts investigated by using atomic force microscopy *Proc. Natl Acad. Sci. USA* **96** 921–6

Sasaki S, Morimoto M, Haga H, Kawabata K, Ito E, Ushiki T, Abe K and Sambongi T 1998 Elastic properties of living fibroblasts as imaged using force modulation mode in atomic force microscopy *Arch. Histol. Cytol.* **61** 57–63

Sato H, Kataoka N, Kajiya F, Katano M, Takigawa T and Masuda T 2004 Kinetic study on the elastic change of vascular endothelial cells on collagen matrices by atomic force microscopy *Colloids Surf. B: Biointerfaces* **34** 141–6

Schrot S, Weidenfeller C, Schäffer T E, Robenek H and Galla H J 2005 Influence of hydrocortisone on the mechanical properties of the cerebral endothelium *in vitro Biophys. J.* **89** 3904–10

Simon A, Cohen-Bouhacina T, Porte V C, Aime J P, Amedee J, Bareille R and Baquey C 2003 Characterization of dynamic cellular adhesion of osteoblasts using atomic force microscopy *Cytometry A.* **54** 36–47

Spatz J P, Mößmer S, Hartmann C, Möller M, Herzog T, Krieger M, Boyen H-G, Ziemann P and Kabius B 2000 Ordered deposition of inorganic clusters from micellar block copolymer films *Langmuir* **16** 407–15

Städler B, Falconnet D, Pfeiffer I, Höök F and Vörös J 2004 Micropatterning of DNA-tagged vesicles *Langmuir* **20** 11348–54

Sugawara M, Ishida Y and Wada H 2002 Local mechanical properties of guinea pig outer hair cells measured by atomic force microscopy *Hear. Res.* **174** 222–9

Sugawara M, Ishida Y and Wada H 2004 Mechanical properties of sensory and supporting cells in the organ of Corti of the guinea pig cochlea—study by atomic force microscopy *Hear. Res.* **192** 57–64

Takai E, Costa K D, Shaheen A, Hung C T and Guo X E 2005 Osteoblast elastic modulus measured by atomic force microscopy is substrate dependent *Ann. Biomed. Eng.* **33** 963–71

Tolomeo J F, Steele C R and Holley M C 1996 Mechanical properties of the lateral cortex of mammalian auditory outer hair cells *Biophys. J.* **71** 421–9

Wada H, Usukura H, Katori Y, Kakehata S, Ikeda K and Kobayashi T 2003 Relationship between the local stiffness of the outer hair cell along the cell axis and its ultrastructure observed by atomic force microscopy *Hear. Res.* **177** 61–70

Wu H W, Kuhn T and Moy V N 1998 Mechanical properties of L929 cells measured by atomic force microscopy: effects of anticytoskeletal drugs and membrane crosslinking *Scanning* **20** 389–97

Xiong J P, Stehle T, Diefenbach B, Zhang R, Dunker R, Scott D L, Joachimiak A, Goodman S L and Arnaout M A 2001 Crystal structure of the extracellular segment of integrin alpha Vbeta3 *Science* **294** 339–45

Xu X and Raman A 2007 Comparative dynamics of magnetically, acoustically and Brownian motion driven microcantilevers in liquids *J. Appl. Phys.* **102** 034303

Yoshikawa Y, Yasuike T, Yagi A and Yamada T 1999 Transverse elasticity of myofibrils of rabbit skeletal muscle studied by atomic force microscopy *Biochem. Biophys. Res. Comm.* **256** 13–9

Zamir E, Katz BZ, Aota S, Yamada K M, Geiger B and Kam Z 1999 Molecular diversity of cell–matrix adhesions *J. Cell Sci.* **112** 1655–69 PMID: 10318759

Zhang S, Kraus W E and Truskey G A 2004 Stretch-induced nitric oxide modulates mechanical properties of skeletal muscle cells *Am. J. Physiol. Cell Physiol.* **287** C292–9

References

A-Hassan E, Heinz W F, Antonik M D, D'Costa N P, Nageswaran S, Schoenenberger C-A and Hoh J N 1998 Relative microelastic mapping of living cells by atomic force microscopy *Biophys. J.* **74** 1564–78

Akiyama S K and Yamada K M 1985 The interaction of plasma fibronectin with fibroblastic cells in suspension *J. Biol. Chem.* **260** 4492–500 PMID: 2932436

Alessandrini A and Facci P 2005 AFM: a versatile tool in biophysics *Meas. Sci. Technol.* **16** R65–92

Amano M, Chihara K, Kimura K, Fukata Y, Nakamura N, Matsuura Y and Kaibuchi K 1997 *Science* **275** 1308–11

Arnold M, Cavalcanti-Adam E A, Glass R, Blümmel J, Eck W, Kantlehner M, Kessler H and Spatz J P 2004 Activation of integrin function by nanopatterned adhesive interfaces *Chem. Phys. Chem.* **5** 383–8

Arnold M *et al* 2008 Induction of cell polarization and migration by a gradient of nanoscale variations in adhesive ligand spacing *Nano Lett.* **8** 2063–9

Balaban N Q *et al* 2001 Force and focal adhesion assembly: a close relationship studied using elastic micropatterned substrates *Nat. Cell Biol.* **3** 466–72

Barbee K A 1995 Changes in surface topography in endothelial monolayers with time at confluence: influence on subcellular shear stress distribution due to flow *Biochem. Cell Biol.* **73** 501–5

Barbee K A, Davies P F and Lal R 1994 Shear stress-induced reorganization of the surface topography of living endothelial cells imaged by atomic force miroscopy *Circ. Res.* **74** 163–71

Behrens J, Mareel M M, Van Roy F M and Birchmeier W 1989 Dissecting tumor cell invasion: epithelial cells acquire invasive properties after the loss of uvomorulin- mediated cell–cell adhesion *J. Cell Biol.* **108** 2435–47

Beningo K A and Wang Y-L 2002 Flexible substrata for the detection of cellular traction forces *Trends Cell Biol.* **12** 79–84

Beningo K A, Lo C-M and Wang Y-L 2002 Flexible polyacrylamide substrata for the analysis of mechanical interactions at cell-substratum adhesions *Methods Cell Biol.* **69** 325–39

Benoit M, Gabriel D, Gerisch G and Gaub H E 2000 Discrete interactions in cell adhesion measured by single-molecule force spectroscopy *Nat. Cell Biol.* **2** 313–7

Benoit M and Gaub H 2000 Measuring cell adhesion forces with the atomic force microscope at the molecular level *Cells Tissues Organs* **172** 174–89

Binnig G and Quate C F 1986 Atomic Force Microscope *Phys. Rev. Lett.* **56** 930–4

Binnig G and Smith D P E 1986 Single-tube three-dimensional scanner for scanning tunneling microscopy *Rev. Sci. Instrum.* **57** 1688

Binnig G, Quate C F and Gerber C 1986 Atomic force microscope *Phys. Rev. Lett.* **56** 930–3

Bischoff G and Hein H-J (ed) 2003 *Micro- and Nanostructures of Biological Systems* (Aachen: Shaker)

Boettiger D and Wehrle-Haller B 2010 Integrin and glycocalyx mediated contributions to cell adhesion identified by single cell force spectroscopy *J. Phys. Condens. Matter* **22** 194101

Bryant P J, Miller R G and Yang R 1988 Scanning tunneling and atomic force microscopy combined *Appl. Phys. Lett.* **52** 2233–5

Burton K and Taylor D L 1997 Traction forces of cytokinesis measured with optically modified elastic substrata *Nature* **385** 450–4

Butler J P, Tolic-Norrelykke I M, Fabry B and Fredberg J J 2002 Traction fields, moments, and strain energy that cells exert on their surroundings *Am. J. Physiol. Cell Physiol.* **282** C595–605

Butt H-J, Wolff E K, Gould S A C, Northern B D, Peterson C M and Hansma P K 1990 Imaging cells with the atomic force microscope *J. Struct. Biol.* **105** 54–61

Cai Y *et al* 2006 Nonmuscle myosin IIa-dependent force inhibits cell spreading and drives F-actin flow *Biophys. J.* **91** 3907–20

Cappella B and Dietler G 1999 Force-distance curves by atomic force microscopy *Surf. Sci. Rep.* **34** 1–104

Cesa C, Kirchgessner N, Mayer D, Schwarz U S, Hoffmann B and Merkel R 2007 Micropatterned silicone elastomer substrates for high resolution analysis of cellular force patterns *Rev. Sci. Instrum.* **78** 034301

Chan C E and Odde D J 2008 Traction dynamics of filopodia on compliant substrates *Science* **322** 1687–91

Choquet D, Felsenfeld D P and Sheetz M P 1997 Extracellular matrix rigidity causes strengthening of integrin-cytoskeleton linkages *Cell* **88** 39–48

Chu Y-S, Thomas W A, Eder O, Pincet F, Perez E, Thiery J P and Dufour S 2014 Force measurements in E-cadherin-mediated cell doublets reveal rapid adhesion strengthened by actin cytoskeleton remodeling through Rac and Cdc42 *J. Cell Biol.* **167**(6) 1183–94

Cohen M, Klein E, Geiger B and Addadi L 2003 Organization and adhesive properties of the hyaluronan pericellular coat of chondrocytes and epithelial cells *Biophys. J.* **85** 1996–2005

Costa K D and Yin F C P 1999 Analysis of indentation: implications for measuring mechanical properties with atomic force microscopy *J. Biomech. Eng.* **121** 462–71

Critchley D R 2000 Focal adhesions—the cytoskeletal connection *Curr. Opin. Cell Biol.* **12** 133–9

Cuerrier C M, Benoit M, Guillemette G, Gobeil F and Grandbois M 2009 Real-time monitoring of angiotensin II-induced contractile response and cytoskeleton remodeling in individual cells by atomic force microscopy *Pflugers Arch.* **457** 1361–72

Cukierman E, Pankov R, Stevens D R and Yamada K M 2001 Taking cell–matrix adhesions to the third dimension *Science* **294** 1708–12

Culp L A, Murray B A and Rollins J B 1979 Fibronectin and proteoglycans as determinants of cell-substratum adhesion *Supramol. Struct.* **11** 401–27

del Rio A, Perez-Jimenez R, Liu R, Roca-Cusachs P, Fernandez J M and Sheetz M P 2009 Stretching single talin rod molecules activates vinculin binding *Science* **323** 638–41

Dembo M and Wang Y-L 1999 Stresses at the cell-to-substrate interface during locomotion of fibroblasts *Biophys. J.* **76** 2307–16

Dembo M, Oliver T, Ishihara A and Jacobson K 1996 Imaging the traction stresses exerted by locomoting cells with the elastic substratum method *Biophys. J.* **70** 2008–22

Discher D E, Janmey P and Wang Y-L 2005 Tissue cells feel and respond to the stiffness of their substrate *Science* **310** 1139–43

Dowling J J, Gibbs E, Russell M, Goldman D, Minarcik J, Golden J A and Feldman E L 2008 Kindlin-2 is an essential component of intercalated discs and is required for vertebrate cardiac structure and function *Circ. Res.* **102** 423–31

du Roure O, Saez A, Buguin A, Austin R H, Chavrier P, Silberzan P and Ladoux B 2005 Force mapping in epithelial cell migration *Proc. Natl Acad. Sci. USA* **102** 2390–05

Dulinska I, Targosz M, Strojny W, Lekka M, Czuba P, Balwierz W and Szymonski M 2006 Stiffness of normal and pathological erythrocytes studied by means of atomic force microscopy *J. Biochem. Biophys. Meth.* **66** 1–11

Engler A J, Griffin M A, Sen S, Bönnemann C G, Sweeney H L and Discher D E 2004a Myotubes differentiate optimally on substrates with tissue-like stiffness: pathological implications for soft or stiff microenvironments *J. Cell Biol.* **166** 877–87

Engler A J, Richert L, Wong J Y, Picart C and Discher D E 2004b Surface probe measurements of the elasticity of sectioned tissue, thin gels and polyelectrolyte multilayer films: correlations between substrate stiffness and cell adhesion *Surf. Sci.* **570** 142–54

Engler A, Bacakova L, Newman C, Hategan A, Griffin M and Discher D 2004c Substrate compliance versus ligand density in cell on gel response *Biophys. J.* **86** 617–28

Evans E A, Waugh R and Melnik L 1976 Elastic area compressibility modulus of red cell membrane *Biophys. J.* **16** 585–95

Evans E, Ritchie K and Merkel R 1995 Sensitive force technique to probe molecular adhesion and structural linkages at biological interfaces *Biophys. J.* **68** 2580–7

Fässler R, Georges-Labouesse E and Hirsch E 1996 Genetic analyses of integrin function in mice *Curr. Opin. Cell Biol.* **8** 641–6

Florin E L, Moy V T and Gaub H E 1994 Adhesion forces between individual ligand–receptor pairs *Science* **264** 415–7

Franz C M and Müller D J 2005 Analyzing focal adhesion structure by atomic force microscopy *J. Cell Sci.* **118** 5315–23

Friedland J C, Lee M H and Boettiger D 2009 Mechanically activated integrin switch controls alpha5beta1 function *Science* **323** 642–4

Friedrichs J, Manninen A, Müller DJ and Helenius J 2008 Galectin-3 regulates integrin alpha2beta1-mediated adhesion to collagen-I and -IV *J. Biol. Chem.* **283** 32264–72

Friedrichs J, Helenius J and Müller D J 2010 Stimulated single-cell force spectroscopy to quantify cell adhesion receptor crosstalk *Proteomics* **10** 1455–62

Friedrichs J, Torkko J M, Helenius J, Teräväinen T P, Füllekrug J, Müller D J, Simons K and Manninen A 2007 Contributions of galectin-3 and -9 to epithelial cell adhesion analyzed by single cell force spectroscopy *J. Biol. Chem.* **282** 29375–83

Friedrichs J, Legate K R, Schubert R, Bharadwaj M, Werner C, Müller DJ and Benoit M 2013 A practical guide to quantify cell adhesion using single-cell force spectroscopy *Methods* **60** 169–78

Garcia C R S, Takeuschi M, Yoshioka K and Miyamoto H 1997 Imaging Plasmodium falciparum-infected ghosts and parasite by atomic force microscopy *J. Struct. Biol.* **119** 92–8

Gardel M L, Sabass B, Ji L, Danuser G, Schwarz U S and Waterman C M 2008 Traction stress in focal adhesions correlates biphasically with actin retrograde flow speed *J. Cell Biol.* **183** 999–1005

Geiger B and Bershadsky A 2002 Exploring the neighborhood: adhesion-coupled cell mechano-sensors *Cell* **110** 139–42

Geisse N A 2009 AFM and Combined Optical Techniques *Mater. Today* **12** 40–5

Goffin J M, Pittet P, Csucs G, Lussi J, Meister J J and Hinz B 2006 Focal adhesion size controls tension-dependent recruitment of α-smooth muscle actin to stress fibers *J. Cell Biol.* **172** 259–68

Gould S A C *et al* 1990 From atoms to integrated circuit chips, blood cells, and bacteria with the atomic force microscope *J. Vac. Sci. Technol.* **A8** 369–73

Grandbois M, Dettmann W, Benoit M and Gaub H E J 2000 Affinity imaging of red blood cells using an atomic force microscope *Histochem. Cytochem.* **48** 719–24

Grazioso F, Patton B R and Smith J M 2010 A high stability beam-scanning confocal optical microscope for low temperature operation *Rev. Sci. Instrum.* **81** 093705–4

Gross L, Mohn F, Moll N, Liljeroth P and Meyer G 2009 The chemical structure of a molecule resolved by atomic force microscopy *Science* **325** 1110–4

Gumbier B M 1996 Cell adhesion: the molecular basis of tissue architecture and morphogenesis *Cell* **84** 345–57

Hansma H G 2001 Surface biology of DNA by atomic force microscopy *Annu. Rev. Phys. Chem.* **52** 71–92

Harris A K, Wild P and Stopak D 1980 Silicone rubber substrata: a new wrinkle in the study of cell locomotion *Science* **208** 177–9

Helenius J, Heisenberg C-P, Gaub H E and Muller D J 2008 Single-cell force spectroscopy *J. Cell Sci.* **121** 1785–91

Hertz H 1881 Ueber den Kontakt elastischer Koerper *J. fuer die Reine Angewandte Mathematik* **92** 156

Hirschfeld V and Hübner C G 2010 A sensitive and versatile laser scanning confocal optical microscope for single-molecule fluorescence at 77 K *Rev. Sci. Instrum.* **81** 113705

Hofmann U G, Rotsch C, Parak W J and Radmacher M 1997 Investigating the cytoskeleton of chicken cardiocytes with the atomic force microscope *J. Struct. Biol.* **119** 84–91

Huang S, Chen C S and Ingber D E 1998 Control of cyclin D1, p27(Kip1) and cell cycle progression in human capillary endothelial cells by cell shape and cytoskeletal tension *Mol. Biol. Cell.* **9** 3179–93

Hur S S, Zhao Y, Li Y S, Botvinick E and Chien S 2009 Live cells exert 3-dimensional traction forces on their substrata *Cell Mol. Bioeng.* **2** 425–36

Hynes R O 2002 Integrins: bidirectional, allosteric signaling machines *Cell* **110** 673–87

Jasiuk I, Sheng P and Tsuchida E 1997 A spherical inclusion in an elastic half-space under shear *J. Appl. Mech.* **64** 471–9

Johnson K L 1985 *Contact Mechanics* (Cambridge: Cambridge University Press)

Johnson K L, Kendall K and Roberts A D 1971 Surface energy and the contact of elastic solids *Proc. R. Soc. Lond. Ser. A.* **324** 301–21

Kamruzzahan A S, Kienberger F, Stron C M, Berg J, Huss R, Ebner A, Zhu R, Rankl C, Gruber H J and Hinterdorfer P 2004 Imaging morphological details and pathological differences of red blood cells using tapping-mode AFM *Biol. Chem.* **385**(10) 955–60

Kantlehner M, Schaffner P, Finsinger D, Meyer J, Jonczyk A, Diefenbach B, Nies B, Hölzemann G, Goodman S L and Kessler H 2000 Surface coating with cyclic RGD peptides stimulates osteoblast adhesion and proliferation as well as bone formation *Chem. Bio. Chem.* **1** 107–14

Kaplanski G, Farnarier C, Tissot O, Pierres A, Benoliel A M, Alessi M C, Kaplanski S and Bongrand P 1993 Granulocyte-endothelium initial adhesion. Analysis of transient binding events mediated by E-selectin in a laminar shear flow *Biophys. J.* **64** 1922–33

Keller R, Davidson L A and Shook D R 2003 How we are shaped: the biomechanics of gastrulation *Differentiation* **71** 171–205

Klebe R J 1974 Isolation of a collagen-dependent cell attachment factor *Nature* **250** 248–51

Krieg M, Arboleda-Estudillo Y, Puech P-H, Käfer J, Graner F, Müller D J and Heisenberg C-P 2008 Tensile forces govern germ-layer organization in zebrafish *Nat. Cell Biol.* **10** 429–36

Landau L D and Lifshitz E M 1970 Theory of elasticity *Course of Theoretical Physics* vol 7 2nd ed (Oxford: Pergamon)

Lang K M, Hite D A, Simmonds R W, McDermott R, Pappas D P and Martinis J M 2004 Conducting atomic force microscopy for nanoscale tunnel barrier characterization *Rev. Sci. Instrum.* **75** 2726–31

Lapshin R V 1998 Automatic lateral calibration of tunneling microscope scanners *Rev. Sci. Instrum.* **69** 3268–76

Laurent V M, Kasas S, Yersin A, Schäffer T E, Catsicas S, Dietler G, Verkhovsky A B and Meister J-J 2005 Gradient of rigidity in the lamellipodia of migrating cells revealed by atomic force microscopy *Biophys. J.* **89** 667–75

Lee J, Leonard M, Oliver T, Ishihara A and Jacobson K 1994 Traction forces generated by locomoting keratocytes *J. Cell Biol.* **127** 1957–64

Legant W R, Choi C K, Miller J S, Shao L, Gao L, Betzig E and Chen C S 2013 Multidimensional traction force microscopy reveals out-of-plane rotational moments about focal adhesions *Proc. Natl Acad. Sci. USA* **110** 881–6

Legant W R, Miller J S, Blakely B L, Cohen D M, Genin G M and Chen C S 2011 Measurement of mechanical tractions exerted by cells within three-dimensional matrices *Nat. Methods* **7** 969–71

Legate K R, Wickström S A and Fässler R 2009 Genetic and cell biological analysis of integrin outside–in signaling *Genes Dev.* **23** 397–418

Lehenkari P P and Horton M A 1999 Single integrin molecule adhesion forces in intact cells measured by atomic force microscopy *Biochem. Biophys. Res. Commun.* **259** 645–50

Lehnert D, Wehrle-Haller B, David C, Weiland U, Ballestrem C, Imhof B A and Bastmeyer M 2004 Cell behavior on micropatterned substrata: limits of extracellular matrix geometry for spreading and adhesion *J. Cell Sci.* **117** 41–52

Li F, Redick S D, Erickson H P and Moy V T 2003 Force measurements of the alpha5beta1 integrin-fibronectin interaction *Biophys. J.* **84** 1252–62

Ludwig T, Kirmse R, Poole K and Schwarz U S 2008 Probing cellular microenvironments and tissue remodeling by atomic force microscopy *Pflug. Arch.* **456** 29–49

Lutolf M P and Hubbell J A 2005 Synthetic biomaterials as instructive extracellular micro-environments for morphogenesis in tissue engineering *Nat. Biotechnol.* **23** 47–55

Mahaffy R E, Park S, Gerde E, Käs J and Shih S K 2004 Quantitative analysis of the viscoelastic properties of thin regions of fibroblasts using atomic force microscopy *Biophys. J.* **86** 1777–93

Mahaffy R E, Shih C K, MacKintosh F C and Käs J 2000 Scanning probe-based frequency-dependent microrheology of polymer gels and biological cells *Phys. Rev. Lett.* **85** 880–3

Maheshwari G, Brown G, Lauffenburger D A, Wells A and Griffith L G 2000 Cell adhesion and motility depend on nanoscale RGD clustering *J. Cell Sci.* **113** 1677–86

Maskarinec S A, Franck C, Tirrell D A and Ravichandran G 2009 Quantifying cellular traction forces in three dimensions *Proc. Natl Acad. Sci. USA* **106** 22108–13

Massia S P and Hubbell J A 1991 An RGD spacing of 440 nm is sufficient for integrin alpha V beta 3-mediated fibroblast spreading and 140 nm for focal contact and stress fiber formation *J. Cell Biol.* **114** 1089–100

Mathur A B, Collinsworth A M, Reichert W M, Kraus W E and Truskey G A 2001 Endothelial, cardiac muscle and skeletal muscle exhibit different viscous and elastic properties as determined by atomic force microscopy *J. Biomech.* **34** 1545–53

Mathur A B, Truskey G A and Reichert W M 2000 Atomic force and total internal reflection fluorescence microscopy for the study of force transmission in endothelial cells *Biophys. J.* **78** 1725–35

McBeath R, Pirone D M, Nelson C M, Bhadriraju K and Chen C S 2004 Cell shape, cytoskeletal tension, and RhoA regulate stem cell lineage commitment *Dev. Cell* **6** 483–95

Mierke C T 2009 The role of vinculin in the regulation of the mechanical properties of cells *Cell Biochem. Biophys.* **53** 115–26

Mierke C T 2011a Cancer cells regulate biomechanical properties of human microvascular endothelial cells *J. Biol. Chem.* **286** 40025–37

Mierke C T, Frey B, Fellner M, Herrmann M and Fabry B 2011b Integrin facilitates cancer cell invasion through enhanced contractile force *J. Cell Sci.* **124** 369–83

Mierke C T, Kollmannsberger P, Paranhos-Zitterbart D, Smith J, Fabry B and Goldmann W H 2008 Mechano-coupling and regulation of contractility by the vinculin tail domain *Biophys. J.* **94** 661–70

Mierke C T, Kollmannsberger P, Zitterbart D P, Diez G, Koch T M, Marg S, Ziegler W H, Goldmann W H and Fabry B 2010 Vinculin facilitates cell invasion into three-dimensional collagen matrices *J. Biol. Chem.* **285** 13121–30

Miller J S, Shen C J, Legant W R, Baranski J D, Blakely B L and Chen C S 2010 Bioactive hydrogels made from step-growth derived PEG-peptide macromers *Biomaterials* **31** 3736–43

Miyazaki H and Hayashi K 1999 Atomic force microscopic measurement of the mechanical properties of intact endothelial cells in fresh arteries *Med. Biol. Eng. Comput.* **37**(4) 530–6

Monkley S J, Zhou X H, Kinston S J, Giblett S M, Hemmings L, Priddle H, Brown J E, Pritchard C A, Critchley D R and Fässler R 2000 Disruption of the talin gene arrests mouse development at the gastrulation stage *Dev. Dyn.* **219** 560–74

Montanez E, Ussar S, Schifferer M, Bösl M, Zent R, Moser M and Fässler R 2008 Kindlin-2 controls bidirectional signaling of integrins *Genes Dev.* **22** 1325–30

Mozafari M R, Reed C J, Rostron C and Hasirci V 2005 A review of scanning probe microscopy investigations of liposome–DNA complexes *J. Liposome Res.* **15** 93–107

Mozhanova A A, Nurgazizov N I and Bukharaev A A 2003 Local elastic properties of biological materials studied by SFM *Proceedings Nizhni Novgorod* 266–7

Muñoz Javier A *et al* 2006 Combined atomic force microscopy and optical microscopy measurements as a method to investigate particle uptake by cells *Small* **2** 394–400 PMID: 17193058

Müller D J, Helenius J, Alsteens D and Dufrêne Y F 2009 Force probing surfaces of living cells to molecular resolution *Nat. Chem. Biol.* **5** 383–90

Nishida S, Kobayashi D, Sakurada T, Nakazawa T, Hoshi Y and Kawakatsu H 2008 Photothermal excitation and laser Doppler velocimetry of higher cantilever vibration modes for dynamic atomic force microscopy in liquid *Rev. Sci. Instrum.* **79** 123703

Nowakowski R, Luckham P and Winlove P 2001 Imaging erythrocytes under physiological conditions by atomic force microscopy *Biochim. Biophys. Acta* **1514** 170–6

Ohashi T, Ishii Y, Ishikawa Y, Matsumoto T and Sato M 2002 Experimental and numerical analyses of local mechanical properties measured by atomic force microscopy for sheared endothelial cells *BioMed. Mater. Eng.* **12** 319–27 PMID: 12446947

Orr A W, Helmke B P, Blackman B R and Schwartz M A 2006 Mechanisms of mechano-transduction *Dev. Cell* **10** 11–20

Palecek S P, Huttenlocher A, Horwitz A F and Lauffenburger D A 1998 Physical and biochemical regulation of integrin release during rear detachment of migrating cells *J. Cell Sci.* **111** 929–40 PMID: 9490637

Pampaloni F, Reynaud E G and Stelzer E H 2007 The third dimension bridges the gap between cell culture and live tissue *Nat. Rev. Mol. Cell Biol.* **8** 839–45

Paszek M J *et al* 2005 Tensional homeostasis and the malignant phenotype *Cancer Cell* **8** 241–54

Paul R, Heil P, Spatz J P and Schwarz U S 2008 Propagation of mechanical stress through the actin cytoskeleton toward focal adhesions: model and experiment *Biophys. J.* **94** 1470–82

Pelham R J Jr and Wang Y 1997 Cell locomotion and focal adhesions are regulated by substrate flexibility *Proc. Natl Acad. Sci. USA* **94** 13661–5

Pesen D and Hoh J H 2005 Micromechanical architecture of the endothelial cell cortex *Biophys. J.* **88** 670–9

Pfaff M, Tangemann K, Müller B, Gurrath M, Müller G, Kessler H, Timpl R and Engel J 1994 Selective recognition of cyclic RGD peptides of NMR defined conformation by alpha IIb beta 3, alpha V beta 3, and alpha 5 beta 1 integrins *J. Biol. Chem.* **269** 20233–8 PMID: 8051114

Puech P-H *et al* 2005 *J. Cell Sci.* **118** 4199–206

Puech P-H, Taubenberger A, Ulrich F, Krieg U M, Müller D J and Heisenberg C-P 2005 Measuring cell adhesion forces of primary gastrulating cells from zebra fish using atomic force microscopy *J. Cell Sci.* **118** 4199–206

Ridley A J and Hall A 1992 The small GTP-binding protein rho regulates the assembly of focal adhesions and actin stress fibers in response to growth factors *Cell* **70** 389–99

Riveline D, Zamir E, Balaban N Q, Schwarz U S, Ishizaki T, Narumiya S, Kam Z, Geiger B and Bershadsky A D 2001 Focal contacts as mechanosensors: externally applied local mechanical force induces growth of focal contacts by an mDia1-dependent and ROCK-independent mechanism *J. Cell Biol.* **153** 1175–86

Roiter Y and Minko S 2005 AFM single molecule experiments at the solid-liquid interface: *in situ* conformation of adsorbed flexible polyelectrolyte chains *J. Am. Chem. Soc.* **127** 15688–9

Rosenbluth M J, Lam W A and Fletcher D A 2006 Force microscopy of nonadherent cells: a comparison of leukemia cell deformability *Biophys. J.* **90** 2994–3003

Roy P, Rajfur Z, Pomorski P and Jacobson K 2002 Microscope-based techniques to study cell adhesion and migration *Nat. Cell Biol.* **4** E91–6

Ruoslahti E 1996 RGD and other recognition sequences for integrins *Annu. Rev. Cell Dev. Biol.* **12** 697–715

Saez A, Ghibaudo M, Buguin A, Silberzan P and Ladoux B 2007 Rigidity-driven growth and migration of epithelial cells on microstructured anisotropic substrates *Proc. Natl Acad. Sci. USA* **104** 8281–6

Sato H, Nagayama K, Kataoka N, Sasaki M and Hane K 2000 Local mechanical properties measured by atomic force microscopy for cultured bovine endothelial cells exposed to shear stress *J. Biomech.* **33** 127–35

Scheffer L, Bitler A, Ben-Jacob E and Korenstein R 2001 Atomic force pulling: probing the local elasticity of the cell membrane *Eur. Biophys. J.* **30** 83–90

Schwarz U S 2007 Soft matters in cell adhesion: rigidity sensing on soft elastic substrates *Soft Matter* **3** 263–6

Schwarz U S, Balaban N Q, Riveline D, Bershadsky A, Geiger G and Safran S A 2002 Calculation of forces at focal adhesions from elastic substrate data: the effect of localized force and the need for regularization *Biophys. J.* **83** 1380–94

Selhuber-Unkel C, López-García M, Kessler H and Spatz J P 2008 Cooperativity in adhesion cluster formation during initial cell adhesion *Biophys. J.* **95** 5424–31

Selhuber-Unkel C, Erdmann T, López-García M, Kessler H, Schwarz U S and Spatz J P 2010 Cell adhesion strength is controlled by intermolecular spacing of adhesion receptors *Biophys. J.* **98** 543–51

Shao J Y and Hochmuth R M 1996 Micropipette suction for measuring piconewton forces of adhesion and tether formation from neutrophil membranes *Biophys. J.* **71** 2892–901

Sheppard D 2000 *In vivo* functions of integrins: lessons from null mutations in mice *Matrix Biol.* **19** 203–9

Shroff S G, Saner D R and Lal R 1995 Dynamic micromechanical properties of cultured rat atrial myocytes measured by atomic force microscopy *Am. J. Physiol. Cell Physiol.* **269** C286–92

Sieg D J, Hauck C R, Ilic D, Klingbeil C K, Schaefer E, Damsky C H and Schlaepfer D D 2000 FAK integrates growth-factor and integrin signals to promote cell migration *Nat. Cell Biol.* **2** 249–56

Singhvi R, Kumar A, Lopez G P, Stephanopoulos G N, Wang D I, Whitesides G M and Ingber D E 1994 Engineering cell shape and function *Science* **264** 696–8

Solon J *et al* 2007 Fibroblast adaptation and stiffness matching to soft elastic substrates *Biophys. J.* **93** 4453–61

Sung K L, Sung L A, Crimmins M, Burakoff S J and Chien S 1986 Determination of junction avidity of cytolytic T cell and target cell *Science* **234** 1405–8

Takeuchi M, Miyamoto H, Sako Y, Komizu H and Kuzumi A 1998 Structure of the erythrocyte membrane skeleton as observed by atomic force microscopy *Biophys. J.* **74** 2171–83

Tan J L, Tien J, Pirone D M, Gray D S, Bhadriraju K and Chen C S 2003 Cells lying on a bed of microneedles: an approach to isolate mechanical force *Proc. Natl Acad. Sci. USA* **100** 1484–9

Taubenberger A, Cisneros D A, Friedrichs J, Puech P H, Muller D J and Franz C M 2007 Revealing early steps of alpha2beta1 integrin-mediated adhesion to collagen type I by using single-cell force spectroscopy *Mol. Biol. Cell* **18** 1634–44

Thie M, Röspel R, Dettmann W, Benoit M, Ludwig M, Gaub H E and Denker H W 1998 Interactions between trophoblast and uterine epithelium: monitoring of adhesive forces *Hum. Reprod.* **13** 3211–9

Thoumine O, Ott A and Louvard D 1996 Critical centrifugal forces induce adhesion rupture or structural reorganization in cultured cells *Cell Motil. Cytoskeleton* **33** 276–87

Thoumine O, Kocian P, Kottelat A and Meister JJ 2000 Short-term binding of fibroblasts to fibronectin: optical tweezers experiments and probabilistic analysis *Eur. Biophys. J.* **29** 398–408

Timoshenko S P and Woinowsky-Krieger S 1970 *Theory of Plates and Shells* (New York: McGraw-Hill)

Trappmann B *et al* 2012 Extracellular-matrix tethering regulates stem-cell fate *Nat. Mater.* **11** 642–9

Tulla M, Helenius J, Jokinen J, Taubenberger A, Müller D J and Heino J 2008 TPA primes alpha2beta1 integrins for cell adhesion *FEBS Lett.* **582** 3520–4

Vinckier A and Semenza G 1998 Measuring elasticity of biological materials by atomic force microscopy *FEBS Lett.* **430** 12–6

Vogel V and Sheetz M 2006 Local force and geometry sensing regulate cell functions *Nat. Rev. Mol. Cell Biol.* **7** 265–75

Weber G F, Bjerke M A and DeSimone D W 2011 Integrins and cadherins join forces to form adhesive networks *J. Cell Sci.* **124** 1183–93

Willemsen O H, Snel M M E, Cambi A, Greve J, De Grooth B G and Carl G 2000 Figdor Biomolecular Interactions Measured by Atomic Force Microscopy *Biophys. J.* **79** 3267–81

Wojcikiewicz E P, Zhang X and Moy V T 2004 Force and compliance measurements on living cells using atomic force microscopy (AFM) *Biol. Proceed. Online* **6** 1–9

Yamane Y, Shiga H, Haga H, Kawabata K, Abe K and Ito E 2000 Quantitative analyses of topography and elasticity of living and fixed astrocytes *J. Electron Microsc. (Tokyo)* **49** 463–71

Yeung T, Georges P C, Flanagan L A, Marg B, Ortiz M, Funaki M, Zahir N, Ming W, Weaver V and Janmey V A 2005 Effects of substrate stiffness on cell morphology, cytoskeletal structure, and adhesion *Cell Motil. Cytoskeleton* **60** 24–34

Zaidel-Bar R, Kam Z and Geiger B 2005 Polarized downregulation of the paxillin-p130CAS-Rac1 pathway induced by shear flow *J. Cell Sci.* **118** 3997–4007

Zaidel-Bar R, Cohen M, Addadi L and Geiger B 2004 Hierarchical assembly of cell–matrix adhesion complexes *Biochem. Soc. Trans.* **32** 416–20

Zamir E and Geiger B 2001 Molecular complexity and dynamics of cell–matrix adhesions *J. Cell Sci.* **114** 3583–90 PMID: 11707510

Zimmermann E, Richter C, Reichle I, Westphal P, Geggier U, Rehn S, Rogaschewski W and Bleiss G R 2001 FuhrMammalian cell traces—morphology, molecular composition, artificial guidance and biotechnological relevance as a new type of 'bionanotube' *Appl. Phys.* **73** 11–26

Zimmermann H, Hagedorn R, Richter E and Fuhr G 1999 Topography of cell traces studied by atomic force microscopy *Eur. Biophys. J.* **28** 516–25

IOP Publishing

Physics of Cancer

Claudia Tanja Mierke

Chapter 5

Cytoskeletal remodeling dynamics

Summary

Cytoskeletal remodeling dynamics are involved in many cellular processes, such as cell motility, transendothelial migration and cell–matrix or cell–cell adhesion receptor recycling. In particular, the connection between the cell's cytoskeleton and the surrounding microenvironment is crucial for the proper function of the cells, including in respect of the mechanical properties regulating cellular motility.

5.1 Nano-scale particle tracking

Nano-scale single-particle tracking (NSPT) is an effective method to analyze the dynamics of molecules in living cells. NSPT exists in two dimensions as well as three, with 3D-NSPT being more sophisticated than 2D-NSPT. 2D-NSPT has been used in numerous biological and medical research fields, for example in the analysis of membrane dynamics (Kusumi *et al* 2005a, Murase *et al* 2004, Fujiwara *et al* 2002), to reveal the virus infection mechanism (Lakadamyali *et al* 2003, Liu *et al* 2011a, 2012, Joo *et al* 2011) and to discover the intracellular or intercellular transport dynamics of molecules (Wang *et al* 2012, Liu *et al* 2011b, He *et al* 2010). For 2D-NSPT, the particle center of mass can be determined with a precision of several nanometers by combining it with a 2D Gaussian fitting method (Kural *et al* 2005, Liang *et al* 2007). The main limitation of the 2D-NSPT method is that the particle can only be tracked within a thin focal plane, otherwise the tracking mechanism will fail. The nano-scale tracking of beads inside cells is of great interest as it is important for determining the cytoskeletal remodeling dynamics within cells that migrate through a dense 3D extracellular matrix microenvironment. Thus, 3D NSPT tracking that captures the full spatiotemporal behavior of the beads embedded by the living cells is becoming more and more important for investigating the cellular environment. Several microscopy techniques have been developed for 3D localization in the SPT field (Katayama *et al* 2009, Wells *et al* 2010, Dupont and Lamb 2011, Juette *et al* 2013, van den Broek 2013), resulting in many single-particle applications (Han *et al* 2012, Quirin *et al* 2012, Planchon *et al* 2011, Jones *et al* 2011, Ram *et al* 2008). There are two main types of technique for tracking individual particles in 3D: (i) 3D tracking by post-analyzing the recorded image data and (ii) real-time 3D tracking via feedback approaches (Dupont and Lamb 2011). The first approach is useful for tracking multiple particles simultaneously. Most versions require modifications to a standard microscope and use of a 2D Gaussian fitting algorithm to achieve super-resolution localization precision, including astigmatic imaging (Holtzer *et al* 2007, Huang *et al* 2008), multifocal plane (Ram *et al* 2008), biplane detection (Juette *et al* 2008) and double-helix point spread function (Pavani *et al* 2009, Thompson *et al* 2010). However, with the development of sensitive cameras and bright fluorescent tags, a high number of 2D images at different Z-positions can be recorded, which seems to be the most straightforward method for 3D single-particle imaging and tracking in living cells.

Although several corresponding algorithms have been developed for locating the particle center from a series of 3D image z-stacks (Kubitscheck *et al* 1996, Patwardhan 2003, Ruthardt *et al* 2011), the frequently used centroid and Gaussian non-linear least-squares fitting methods are suitable for 3D SPT. In more detail, the high-speed centroid method has very low localization precision and high edge sensitivity. Although the iterative Gaussian fitting methods are considered to be the most accurate algorithm, they are not suitable for 3D SPT because the large data volumes require too much time to analyze. Additionally, the initial values should be set for fitting parameters before calculation, which is a very tedious process. Thus, it has been considered whether an efficient algorithm could be developed for 3D SPT with high computation speed and high precision.

There is a localization algorithm based on a radial symmetry method that can solve the challenges of 3D SPT mentioned above. Although the whole-pixel and sub-pixel localizations using the radial symmetry method in 2D have been described in previous reports (Loy and Zelinsky 2003, Parthasarathy 2012), such a method for 3D and sub-pixel localization has not been reported so far. The localization accuracy and computation speed of the method is primarily assessed based on two applications from theoretical and practical perspectives: (i) localization of simulated 3D particle images by wide-field and confocal microscopy and (ii) 3D SPT of a quantum dots (Qdots)-labeled influenza virus in living host cells. However, this novel method determines the 3D particle center with low edge sensitivity and sub-pixel precision, which is similar to the Gaussian non-linear least-squares fitting method. Without any iterative or numerical fitting steps, the computation speed of this method is about two orders of magnitude higher than that of the Gaussian fitting method and thus similar to that of the centroid method. Moreover, these features make this algorithm a very competitive method for 3D SPT applications.

The 3D single-particle localization method
The 3D point spread function (PSF) of the Born–Wolf model (Born and Wolf 1980) is introduced to simulate the 3D charge-coupled device (CCD) images of individual particles. In particular, the axial intensity distribution of a 3D PSF has the shape of 'butterfly' in typical wide-field microscopy. Moreover, the intensity of the 'wings' is

dramatically reduced for confocal microscopy, whereas the 'body' is an approximate ellipsoid, which can be characterized by the ratio of axial resolution (R_z) to lateral resolution (R_{xy})

$$R_z/R_{xy} = 3.28n/NA,$$

where n is the refractive index of the objective medium and the particular NA is the numerical aperture of the objective. Based on this approximation and by scaling of the 3D images in the axial direction according to the ratio, images with the intensity distribution of an approximate sphere are obtained. This 3D radial symmetry localization method is based on the fact that the gradient line at each pixel in a 3D sphere would intersect theoretically with the true center of the 3D sphere, which has a distance of 0 to all gradient lines. Considering any optical noise and deformation error, the center is estimated using the point which has the least total distance to all gradient lines. Moreover, a displacement-weighted method is implemented to reduce the influence of the intensity of the wings, leading to a more accurate approach regarding the high-intensity areas.

Accuracy for wide-field microscopy
The performance of this method is evaluated for wide-field microscopy using 1000 3D CCD images, which are simulated with a signal-to-noise (S/N) ratio of 20 and with the true centers distributed randomly between ±0.5 pixels in x, y and z directions. By tracking 1000 images with a centered $7 \times 7 \times 11$ pixel region (pixel size: $100 \times 100 \times 150$ nm), the 3D radial symmetry method leads to a mean lateral error of 0.011 pixels and an axial error of 0.044 pixels, much smaller than that of the centroid method (lateral error: 0.077 pixels, axial error: 0.163 pixels) and in a similar range to that of the Gaussian fitting method (lateral error: 0.011 pixels, axial error: 0.045 pixels). The 3D scatter plots of the errors in 3D further illustrate that the error distribution of this method is similar to that of the Gaussian fitting method and more concentrated compared to that of the centroid method. Meanwhile, the accuracy of the 3D radial symmetry method has been investigated without scaling the 3D image in the axis direction by using simulated 3D CCD images with an S/N ratio of 20. The accuracy of the 3D radial symmetry method (lateral error: 0.018 pixels, axial error: 0.089 pixels) is lower than that of the Gaussian fitting method, indicating that the scaling process is very important for improving the precision of the 3D radial symmetry method.

The intensity distribution of the scaled image of a 3D particle shows that the intensity in the axial direction is not only the shape of a sphere, but it also contains halos (the wings of the 3D PSF image). In order to eliminate the influence of the wings of the 3D particle image on the localization accuracy, the centers have been determined with different lateral sizes. The total error plots show that the accuracy decreases with the shrinking of the lateral size. Moreover, the accuracy is similar to that of the Gaussian fitting method for large lateral size and even better for small size, indicating that this approach can reduce the surrounding interference.

The accuracy of this method is further determined using simulated 3D images with S/N ratios of 3–100. The error plots indicate that the lateral accuracy is slightly lower than that of the Gaussian fitting method for low S/N ratios, whereas it is higher for high S/N ratios and the axial accuracy has the same trend. The total error for three localization methods has been revealed. The total error is 7.06 nm for the radial symmetry method, 7.14 nm for the Gaussian fitting method and 28.44 nm for the centroid method at the S/N ratio of 20. The total error plots indicate that the accuracy of the 3D radial symmetry method is much better than that of the centroid method and seems to be as good as that of the Gaussian fitting method over almost the entire range and even better for high S/N ratios. Moreover, the average computation time is 0.39 ms for the radial symmetry method, 0.16 ms for centroid method and 39.12 ms for the Gaussian fitting method, which is about 100 times longer than that of the radial symmetry method. Indeed, this is a key attribute of this method for processing a mass of 3D images in the SPT field.

The accuracy of the confocal microscopy
Taking the z-stacks by confocal microscopy is a widely used tool for 3D particle imaging and tracking. Thus, the accuracy of different localization methods is obtained by using simulated 3D CCD images for confocal microscopy. Hence, the 3D radial symmetry method (lateral error: 0.009 pixels, axial error: 0.036 pixels, total error: 5.58 nm) is still as accurate as the Gaussian fitting method (lateral error: 0.008 pixels, axial error: 0.038 pixels, total error: 5.59 nm) and much better compared to the centroid method (lateral error: 0.064 pixels, axial error: 0.097 pixels, total error: 18.59 nm) for 1000 CCD images with a S/N ratio of 20. As shown above, a series of simulated images with a wide range of S/N values has been examined. Thus it has been found that the accuracy of this method is slightly lower than that of the Gaussian fitting method for low S/N ratios, but even higher in the axial direction for high S/N ratios. Indeed, these results demonstrate that the localization accuracy for three methods is improved by using the 3D confocal CCD images and thus the 3D radial symmetry method is applicable for confocal microscopy. With the decrease of axial size, the accuracy decreases similarly for the three localization methods. Taken together, this method is as accurate as the Gaussian fitting method.

The accuracy estimated by actual experiments
In order to investigate the locating ability under particular distortions or imperfections, the algorithm using 3D CCD images from actual experiments has been refined. In particular, the positions of multiple fluorescent beads have been detected in 3D images using different methods. However, the true centers could not be obtained in actual experiments, but the positions located by the radial symmetry method are consistent with those obtained with the Gaussian fitting method. In addition, the 3D SPT technique has been utilized in order to investigate the behavior of a Qdot-labeled influenza virus in live Madin–Darby canine kidney cells and analyze the trajectory of a single virus precisely. It has been observed that the trajectory of radial symmetry-based tracking is very similar to that of the Gaussian fitting method. It is has to be pointed out that the 3D radial symmetry method took only 0.76 s to calculate the particle positions of the trajectory (273 frames), whereas

47.77 s was needed by the Gaussian fitting method. The relationship of mean square displacement (MSD) with time interval is highly important for analyzing the motion mode and calculating the relative motional parameters, such as the diffusion coefficient and fitting velocity (Liu *et al* 2012, Saxton and Jacobson 1997). Indeed, it has been found that the MSD–time plot for radial symmetry-based tracking overlaps very significantly with that of the Gaussian fitting method. In addition, the results obtained further indicate that the radial-symmetry based method is as accurate as the Gaussian fitting method in the actual experiments, and has a much higher computation speed.

Although z-stacks microscopy has been widely used for 3D single-particle tracking, the relatively rapid algorithms for high-resolution single-particle localization are only rarely provided. A non-iterative localization algorithm has been developed that first utilizes the scaling of 3D image in the axial direction and then focuses on evaluating the radial symmetry center of the scaled image to achieve the desired single-particle localization. In the first step, the deformation of a 3D image of single particles is accomplished by scaling the image in the axial direction based on the ratio of axial resolution to lateral resolution. By evaluating the accuracy using simulated 3D particle images, it can be confirmed that the scaling step is essential for improving the accuracy of this localization method. In practice, the ratio of the axial resolution to the lateral resolution may alter from microscope to microscope. However, the localization method can be adjusted slightly with different microscopes for higher precision. Localizing the particle position in the second step is based on the particle center that has the least total distance to all gradient lines at each pixel in the scaled 3D image. In addition, a displacement-weighted method is applied to improve the localization precision pronouncedly, which can reduce the influence of the edge intensity. Finally, the combination of these two steps achieves 3D single-particle localization with a sub-pixel localization accuracy and sub-millisecond computation time.

Compared with the frequently used centroid and Gaussian fitting methods (Kubitschek *et al* 1996, Patwardhan 2003, Ruthardt *et al* 2011), this approach seems to be more precise than the centroid method in its localization precision and it has the same precision as the 3D Gaussian fitting method, although the computation speed is two orders of magnitude higher compared than for the Gaussian fitting method. High computation speed and precise localization ability are the main contributions of the 3D radial symmetry method. Meanwhile, although the radial symmetry method has been used in whole-pixel and sub-pixel localization in 2D (Loy and Zelinsky 2003, Parthasarathy 2012), the present 3D radial symmetry method is a significant expansion of the radial symmetry-based algorithm in 3D and is expected to be exploited in some new algorithms. Moreover, by tracking a Qdot-labeled influenza virus in living cells, it was further confirmed that the 3D radial symmetry method is suitable to rapidly localize the position of single viruses with a high precision in the 3D SPT field. As a main advantage, this approach can considerably reduce the time and costs for processing the large volume data of 3D images, which makes it especially suited for 3D high-precision single-particle tracking, 3D single-molecule imaging and new microscopy techniques.

Single particle tracking in 3D in a living cell microenvironment holds the promise of revealing important new biological insights. However, conventional microscopy-based imaging techniques are not well suited for fast 3D tracking of single particles in cells. An imaging modality multifocal plane microscope (MUM) has been developed to image fast intracellular dynamics in 3D in living cells. Thus, an algorithm is introduced, the MUM localization algorithm (MUMLA), in order to determine the 3D position of a point source that is imaged using MUM. MUMLA has been validated through simulated and experimental data and indeed showed that the 3D position of quantum dots can be determined over a wide spatial range. Moreover, it has been shown that MUMLA provides the best possible accuracy for determining the 3D position. This analysis demonstrates that MUM overcomes the poor depth discrimination of the conventional microscope and thus paves the way for high-accuracy tracking of nanoparticles in a living cell microenvironment. Using MUM and MUMLA the full 3D trajectories of Qdot-labeled antibody molecules undergoing endocytosis in living cells from the cell membrane to the sorting endosome deep inside the cell have been reported.

Fluorescence microscopy of living cells represents a major tool for analyzing intracellular trafficking events. Using state-of-the-art microscopy techniques, only a single focal plane can be imaged at a particular time. Membrane protein dynamics can be investigated in one focal plane and the significant advances can be regarded as an advantage of fluorescence microscopy (Kwik et al 2003, Zenisek et al 2000). However, cells are 3D objects and intracellular trafficking pathways are typically not constrained to one focal plane. If the dynamics are not restricted to one focal plane, the currently available technology is inadequate for the detailed investigation of fast intracellular dynamics (Ehrlich et al 2004, Hua et al 2006, Ober et al 2004a, Oheim 2004, Rutter and Hill 2006). For instance, significant advances have been made in the investigation of events that precede endocytosis at the cell membrane (Conner and Schmid 2003, Maxfield and McGraw 2004, Pelkmans and Zervial 2005).

However, the dynamic events postendocytosis typically cannot be imaged, as they occur outside the focal plane that is set to image the cell membrane. Classical approaches based on altering the focal plane are often ineffective in such situations, as the focusing devices are relatively slow in comparison to many of the intracellular dynamics (Schütz et al 2001, Arhel et al 2006, Thomann et al 2002). In addition, the focal plane may frequently fail to be in the right place at the right time, thus missing important aspects of the dynamic events.

However, modern microscopy techniques have generated significant interest in investigating the intracellular trafficking pathways at the single-molecule level (Ober et al 2004b, Sako et al 2000). In particular, single molecule experiments overcome the averaging effects and thus provide information that is not accessible using conventional bulk studies. However, the 3D tracking of single molecules poses several challenges. In addition to whether or not images of the single molecule can be captured while it undergoes potentially highly complex 3D dynamics (Moerner 2007), the question arises of whether the 3D location of the single molecule can be determined and how accurately this can be performed.

Several imaging techniques have been proposed to determine the Z-position of a single molecule/particle. Approaches (Speidel *et al* 2003, Toprak *et al* 2007) that use out-of-focus rings of the 3D PSF to infer the Z-position are not capable of tracking Qdots (Toprak *et al* 2007) and pose several challenges, especially for live-cell imaging applications, since the out-of-focus rings can be detected only when the particle is at certain depths. Moreover, a large number of photons must be collected so that the out-of-focus rings can be detected above the background, which severely compromises the temporal resolution. Similar problems are also encountered with the approach that infers the Z-position from out-of-focus images acquired in a conventional fluorescence microscope (Aguet *et al* 2005). Moreover, this approach is applicable only at certain depths and is problematic, for instance, when the point source is close to the plane of focus. The technique based on encoding the 3D position using a cylindrical lens (Kao and Verkman 1994, Holtzer *et al* 2007, Huang *et al* 2008) is somewhat limited in its spatial range to 1 μm in the *z*-direction (Holtzer *et al* 2007). Moreover, this technique uses epi-illumination and thus poses the same problems as conventional epifluorescence microscopy in tracking events that fall outside even one focal plane. The approach based on *z*-stack imaging in order to determine the 3D position of a point source (Schütz *et al* 2001, Watanabe and Higuchi 2007) has limitations in terms of the acquisition speed and the accuracy achievable for the location estimates, and thus poses problems for investigating fast and highly complex 3D dynamics. It should be pointed out that the above-mentioned techniques have not been able to image the cellular microenvironment with which the point sources interact. This is especially important for gaining useful biological information, such as identification of the final destination of the single molecules of interest. Confocal/two-photon particle tracking approaches that can scan the sample in 3D can only track one or very few particles within the cell and require high photon emission rates of the bead.

One of the key requirements for 3D tracking of single molecules within a cellular microenvironment is that the targeted molecule must be continuously tracked for extended periods of time at high spatial and temporal precision. However, conventional fluorophores such as organic dyes and fluorescent proteins typically have a limited fluorescent on-time (typically 1–10 s), after which they are irreversibly photobleached, thus severely limiting the duration over which the tagged molecule can be tracked. However, the use of Qdots, which are extremely bright and photostable fluorescent labels when compared to conventional fluorophores, enables continuous tracking of single molecules for extended periods of time (several minutes or even hours). There have been several reports on single Qdot tracking within a cellular microenvironment, for example on the cell membrane (Dahan *et al* 2003, Crane and Verkman 2008) or inside the cells (Nan *et al* 2005, Courty *et al* 2006, Cui *et al* 2007). All of these reports have focused on Qdot tracking in 2D. However, the 3D tracking of Qdots in cells has been problematic due to the above-mentioned challenges that relate to imaging fast 3D dynamics with conventional microscopy-based techniques.

Rapid progress has been made recently in developing localization-based super-resolution imaging techniques (Lidke *et al* 2005, Rust *et al* 2006, Betzig *et al* 2006,

Hess *et al* 2006, Egner *et al* 2007). In particular, these techniques typically use photoactivated fluorescent labels and exploit the fact that the location of a point source can be determined with a very high (nanometer) level of accuracy (Ober *et al* 2004b, Thompson *et al* 2002). This is in conjunction with the working assumption that during photoactivation sparsely distributed (thus spatially well separated) labels are turned on, enabling the retrieval of nanoscale positional and distance information for the point sources well below Rayleigh's resolution limit.

Originally demonstrated in 2D fixed cell samples, these techniques have also been extended to 3D imaging of non-cellular/fixed-cell samples (Huang *et al* 2008, Folling *et al* 2007, Juette *et al* 2008) and more recently have even been used to track single molecules in 2D in living cells (Hess *et al* 2007, Manley *et al* 2008, Shroff *et al* 2008). However, single molecules were tracked only for a short period of time because of the use of conventional fluorophores, which are susceptible to rapid photobleaching. Moreover, live-cell imaging was carried out using conventional microscopy-based imaging approaches, with their 3D tracking problems in terms of imaging events that fall outside the plane of focus. Thus, these techniques do not support the long-term, continuous (time-lapse) 3D imaging of fluorophores, which limits their applicability to 3D tracking in living cells.

An imaging modality, multifocal plane microscopy (MUM) has been developed to ensure 3D subcellular tracking within a living cell microenvironment (Prabhat *et al* 2004, 2007). In MUM, the sample is simultaneously imaged at distinct focal planes. This is achieved by placing detectors at specific distances in the microscope's emission-light path. The sample can be concurrently illuminated in the epi-fluorescence mode and also in the total internal reflection fluorescence (TIRF) mode. In MUM, the temporal resolution is determined by the frame rate of the camera that images the corresponding focal plane, which does not produce a realistic limitation, given current camera technology. Indeed, MUM has been performed to investigate the exocytic pathway of immunoglobulin G molecules from the sorting endosome to exocytosis on the cell membrane (Prabhat *et al* 2007), as facilitated by the Fc receptor FcRn (Ghetie and Ward 2000). Prior results addressed the problem of providing qualitative data, for instance, the imaging of the dynamic events at different focal planes within a cell. A major obstacle to high accuracy in 3D location estimation is the poor depth discrimination of a conventional microscope. This means that the Z-position, which means the position of the point source along the optical axis, is difficult to determine, and this is particularly the case when the point source is close to being in focus. Aside from this, the accuracy with which the 3D location of the point source can be determined is of fundamental importance. The latter is especially relevant in live-cell imaging applications, where the S/N ratio is typically very poor.

Here we present a methodology for determining the 3D coordinates of single fluorescent point sources imaged using MUM in living cells. The specifics of MUM acquisition exploit the fact that for each point in time more than one image of the point source is available, each at a different focal level. By appropriately exploiting this data structure it can be demonstrated that estimates can be obtained and that they are significantly more accurate than could be obtained by classical approaches, especially when the point source is near the focus in one of the focal planes. Moreover, simulations and experimental data have revealed that the proposed MUM localization algorithm (MUMLA) is applicable over a wide spatial range (approximately 2.5 µm depth) and produces estimates whose standard deviations are very close to the theoretically best possible level. This analysis shows that MUM overcomes the poor depth discrimination of the conventional microscope and thus paves the way for high-accuracy tracking of nanoparticles in a living cell microenvironment.

It should be pointed out that MUM supports multicolor imaging. This has enabled us to image Qdots in 3D and also to image, at the same time, the cellular microenvironment with which the Qdot-labeled molecules interact. The latter was realized by labeling the cellular structures with spectrally distinct fluorescent fusion proteins. As is shown here, this has enabled us to track the fate of Qdot-labeled antibody molecules from endocytosis at the cell membrane to its delivery into the sorting endosome inside the cell in another focal plane.

The impressive developments in optical microscopy have made this technology increasingly compatible with other biological studies. Fluorescence microscopy in particular has contributed to investigation of the dynamic behaviors of living specimens and it can resolve objects with nanometer precision and resolution due to super-resolution imaging. Additionally, single particle tracking provides information on the dynamics of individual proteins at the nanometer scale both *in vitro* and in cells. The development of fluorescent probes has complemented advances in microscopy technologies. The quantum dot, a semi-conductor fluorescent nanoparticle, is particularly suitable for single particle tracking and super-resolution imaging.

Fluorescence microscopy has become standard for investigating the dynamic behavior of biological phenomena such as the expression, movement and localization of proteins and other molecules (Ellinger 1940, Lichtman and Conchello 2005, Drummen 2012, Miyawaki 2013, Peter *et al* 2014). However, optical diffraction, limits the spatial resolution to several 100 nm, thus no information is obtained on many details for these phenomena (Abbe 1873). Two technologies have overcome this limitation and permit the observation of even smaller nano-scale dynamics: single particle tracking (Ritchie and Kusumi 2003, Saxton 2009, Chenouard *et al* 2014) and super-resolution microscopy (Schermelleh *et al* 2010, Galbraith and Galbraith 2011, Leung and Chou 2011). Single particle tracking ensures the observation of the precise position of single fluorescent probes conjugated to separate target proteins over a 2D plane. In addition, super-resolution microscopy provides highly resolved optical images beyond the aforementioned spatial resolution.

In order to conduct the above imaging techniques, it is often required to label the target protein with a fluorescent probe. Fluorescent proteins are most popular for this purpose because of their simple and easy labeling procedure in living cells (Shimomura and Johnson 1692, Tsien 1998, Nifosí *et al* 2007). In more detail, organic dyes are also common because of their wide application (Wombacher and Cornish 2011, Wysocki and Lavis 2011, Terai and Nagano 2013). Another group of probes gaining attention is inorganic nanoparticles made of semiconductors, metals and silicon (Ruedas-Rama *et al* 2012, Chinnathambi *et al* 2014, Cupaioli *et al* 2014). Although usually larger than fluorescent proteins and organic dyes, inorganic nanoparticles generally have stronger and more stable fluorescence profiles, which makes them suitable and even applicable not only to basic research, but also to clinical studies (Byers and Hitchman 2010, Choi and Frangioni 2010, Saadeh *et al* 2014, Wang and Wang 2014). In more detail, these same properties make them well suited for single particle tracking methods (Chang *et al* 2008, Saxton 2008, Barroso 2011, Bruchez 2011, Clausen and Lagerholm 2011, Ruthardt *et al* 2011, Pierobon and Cappello 2012, Kairdolf *et al* 2013, Petryayeva *et al* 2013).

This section is focused on advanced microscopy using Qdots, perhaps the most studied inorganic nanoparticles for biological applications (Pilla *et al* 2012). Single particle tracking using Qdots has reached 3D (X, Y, Z) (Genovesio *et al* 2006, Holtzer *et al* 2007, Watanabe and Higuchi 2007, Watanabe *et al* 2007, Ram *et al* 2008, 2012, Wells *et al* 2008, 2010, Yajima *et al* 2008) and more recently has even reached four dimensions (X, Y, Z, θ) (Ohmachi *et al* 2012, Watanabe *et al* 2013). For all their benefits, Qdots do have major drawbacks, such as high blinking (Nirmal *et al* 1996, van Sark *et al* 2001, Schlegel *et al* 2002, Hohng and Ha 2004, Ko *et al* 2011) and a spectral blue-shift during observation (Nirmal *et al* 1996, van Sark *et al* 2002, Hoyer *et al* 2011), which complicate the continuous tracking of the single particle and emerge due to photo-oxidation while under high-power illumination. These limitations have stimulated research into new super-resolution microscopy methods (Lidke *et al* 2005, Dertinger *et al* 2009, Watanabe *et al* 2010, Chien *et al* 2011, Hoyer *et al* 2011, Deng *et al* 2014).

Qdots as fluorescent labeling probes
A Qdot is a semiconductor nanocrystal with electronic characteristics that depend on its size and shape (Rossetti *et al* 1980, Ekimov and Onushchenko 1981). Due to their unique characteristics and ease of synthesis, Qdots have not only been applied to biomedical research, but also to engineering- and industry-related fields, such as transistors, solar cells, LEDs and diode lasers (Pilla *et al* 2012). Qdots used in biological studies have a core-shell structure; the most famous is the cadmium selenide (CdSe) core and zinc sulfide (ZnS) shell (Dabbousi *et al* 1997, Bruchez *et al* 1998, Chan and Nie 1998, Pilla *et al* 2012). As a major advantage, this structure results in Qdots having narrow emission spectra, but wide absorption spectra. There are two important criteria when applying Qdots to biological studies: solubility and conjugating capability (Li *et al* 2010). Highly fluorescent Qdots are usually synthesized in organic solvents in coordination with compounds such as tri-n-octylphosphine oxide (TOPO) or alkylamine. These compounds coat the Qdot, in order to make it too hydrophobic to be dissolved in water. Thus, further surface coating or exchange with hydrophilic compounds is needed for use in biological assays. Furthermore, upon becoming water soluble, the surface of the Qdot must have reactive groups such as amino and carboxyl chains in order for the Qdot to conjugate with the target biological sample. The surface coating not only contributes to the water solubilization, but also to the stabilization of the fluorescence of the

Qdot in water because the photophysical properties are affected by the surface coating (Kuno *et al* 1997, Kloepfer *et al* 2005). Some surface coating methods even suppress the blinking that is a major drawback of Qdots (Hohng and Ha 2004, Fomenko and Nesbitt 2008, Mandal and Tamai 2011, Zhang *et al* 2013).

There are two main ways to prepare water-soluble Qdots (Erathodiyil and Ying 2011, Zhang and Clapp 2011). The first is to encapsulate a hydrophobic Qdot with an amphiphilic polymer or phospholipid (Dubertret *et al* 2002, Gao *et al* 2005, Li *et al* 2010, Tomczak *et al* 2013). The second is a ligand-exchange method in which the capping hydrophobic ligands are exchanged with hydrophilic ligands (Gerion *et al* 2001, Guo *et al* 2003, Pinaud *et al* 2004, Kim *et al* 2005, Nann 2005, Jiang *et al* 2006, Dubois *et al* 2007). While the water-solubilized Qdot obtained by the first method is more stable and suitable for commercialization, its size increases to about 20–40 nm, which increases the risk of steric hindrance against the function of the target protein (Li *et al* 2010). The ligand-exchange method is inferior in stability, but it has a simpler synthesis process and produces a smaller Qdot. The thin coating layer is another advantage of this method, as it reduces the risk of steric effects that could compromise the function of the protein upon conjugation with the Qdot.

Many coating reagents exist for the ligand-exchange method, including mer-captocarboxylic acid (Jiang *et al* 2006), carbon disulfide (Dubois *et al* 2007), thiosilanol (Gerion *et al* 2001), dendrimer (Guo *et al* 2003), peptide (Pinaud *et al* 2004), phosphine oxide (Kim *et al* 2005), and polyethylenimine (Nann 2005). The coating reagents can also functionalize Qdots for specific purposes. Examples of this include β-cyclodextrin for ion-sensing (Palaniappan *et al* 2004), cyclodextrin for redox-active substrates (Palaniappan *et al* 2006) and cyclodextrin thiol for pH sensing (Cao *et al* 2006). Usually glutathione can be used as the coating compound because of its relatively easy preparation, which only requires the mixing of hydrophobic Qdots with an aqueous glutathione solution (Jin *et al* 2008, Tiwari *et al* 2009). In particular, glutathione-coated Qdots have two reactive groups (amino and carboxyl) that ensure easy conjugation with the target protein and show no cytotoxy (Tiwari *et al* 2009).

Fluorescence microscopy for nano-scale measurements
The microscopy introduced requires a regular wide-field fluorescence microscope and no complicated optical principles or special devices. However, because nano-scale measurements require a high S/N ratio, a highly photon-sensitive camera, such as an electron multiplying charge coupled device (EMCCD) one, is necessary. More recently, complementary metal–oxide–semiconductor (sCMOS) cameras have become available as alternatives (Huang *et al* 2011, Long *et al* 2012, Ma *et al* 2013). The vibration and/or stage drift of the microscope should of course be considered, as these can lead to artifacts in the measurement by obscuring the behavior and the structure of the target. Consequently, the microscope should be set on a vibration-isolation table and built with minimum height and maximum rigidity to reduce any vibration. As thermal expansion of the metals composing the microscope causes drifts in the stage and focus position, microscopes made of metals with lower thermal expansion such as invar are generally preferred. In

addition, the drifts can be further suppressed by setting the microscopic system in a room with constant temperature and humidity.

One strategy for reducing vibrations is discussed in the following. In particular, the transition images of a silica bead with 1 μm diameter absorbed on a coverslip surface were acquired with excess illumination so that the camera gain could be set to zero. The frame rate was 2.0 ms, the images were acquired for 1.0 s, and the precise position (X, Y) of the bead was calculated by image analysis. Moreover, in the usual initial setup, the position of the bead was kept stable within 0.7 nm in the X-axis and 0.4 nm in the Y-axis. When a screw to fix the CCD camera was loosened, the vibration increased to 0.8 nm in both axes. Normally, a mono-objective revolver is used, but when instead a commercially available six-position revolver was used, the vibration was enhanced in the Y-axis to 2 nm. Thus, rigid construction of the microscope is paramount for nano-scale measurements and observations.

Single particle tracking using Qdots

Single particle tracking is well adapted for investigation of motor and membrane proteins, because resolving nano-scale movements is necessary for understanding the protein function (Ritchie and Kusumi 2003, Park *et al* 2007, Toprak and Selvin 2007, Saxton 2009). Although the resolution of conventional fluorescence microscopes is constrained by the diffraction limit, the 2D position of a single particle can be determined by calculating the weight center of the image of the fluorescent spot. The fluorescence emitted from a fluorescent probe forms a point PSF that can be fitted with a Gaussian distribution where I0 and (x0, y0) are the fluorescence intensity and the position of the fluorescing center, respectively. Moreover, σ is the radial standard deviation of the Gaussian function and C is the background fluorescence. Indeed, this analysis can be used to measure the center position of the image (Kubitscheck *et al* 2000, Cheezum *et al* 2001, Thompson *et al* 2002, Small and Stahlheber 2014). Although there are other common methods for determining the center, including cross-correlation, sum-absolute difference and the simple centroid, Gaussian fitting has the highest robustness at low S/N ratios, which is why it is commonly used (Thompson *et al* 2002). In this case, the actual fitting computation is performed using the Levenberg–Marquardt method (Levenberg 1944).

The calculation precision from Gaussian fitting depends strongly on the photon number that the detection device receives from the emission of the fluorescent probe and can be as small as a few nanometers (Deschout *et al* 2014, Small and Stahlheber 2014). The method described above is called fluorescence imaging with one-nanometer accuracy (FIONA) and has become a standard method (Yildiz *et al* 2003, Yildiz and Selvin 2005, Park *et al* 2007, Hoffman *et al* 2011). However, the number of photons emitted by single organic dyes and fluorescent protein molecules before photobleaching, about 110,000 (Kubitscheck *et al* 2000), is too low for the observation of protein movement over a long time. As the cause of photobleaching is thought to be oxygen collisions with the dye molecule in its excited state, it can be reduced by the addition of oxygen scavengers (Sambongi *et al* 1999, Adachi *et al* 2000). Thus, the photon number from a single dye molecule can be increased to 1.4 million photons before photobleaching (Yildiz and Selvin 2005). Meanwhile,

Qdots show little photobleaching and strong fluorescence even in the absence of scavengers (Bruchez *et al* 1998). Although non-fluorescent nano-particles such as gold nano-particles are becoming increasingly popular for precise and long-term tracking using absorption (Kusumi *et al* 2005b, Lasne *et al* 2006) or scattering (Nishikawa *et al* 2010), the Qdot is still preferred in biological studies because of its wider color spectrum.

The relationship between the tracking precision and the average number of photons emitted from a Qdot has been investigated. The tracking precision was defined as the standard deviation of 100 data obtained with a Qdot immobilized on a glass surface in our case. However, the experimental accuracy was a little lower compared to the theoretical expectation because of high blinking, thus it was still 2 nm when the photon number from a Qdot was 15,000 per exposure. In order to demonstrate the potential of single particle tracking as a biological tool, the movement of kinesin, a microtubule-mediated motor protein was analyzed. In more detail, the motor domain of the kinesin was fused with biotin career protein (BCCP) and conjugated with a Qdot via biotin–avidin affinity. The Qdot-labeled kinesin were then bound to microtubules adsorbed onto a cover slip. Upon adding 1 mM ATP, the Qdot was seen to move unidirectionally along the microtubule without detaching, which is indeed consistent with kinesin using ATP to move. The unidirectional movement of kinesin was composed of successive 8 nm steps. Taken together, FIONA using Qdots ensure a quantitative measurement for nano-scale tracking of proteins at the single molecular level.

3D single particle tracking using Qdots
The original FIONA only measured movement on a spatial plane, but it has since been expanded to three spatial dimensions. For this purpose, a 3D image under a microscope is obtained by scanning the objective lens along the focal axis with an actuator (Watanabe and Higuchi 2007, Wells *et al* 2008). This scanning, however, reduces the temporal resolution of the tracking. In order to solve this problem, 3D tracking methods without the objective scanning have been developed (Genovesio *et al* 2006, Holtzer *et al* 2007, Watanabe *et al* 2007, Ram *et al* 2008, 2012, Wells *et al* 2010, Jia *et al* 2014). Multifocal planes microscopy uses the difference of distinct optical pathways to estimate the Z-position by simultaneously obtaining the fluorescence intensities of several focal images (Toprak *et al* 2007, Watanabe *et al* 2007, Dalgarno *et al* 2010, Juette and Bewersdorf 2010, Ram *et al* 2012). Similarly, 3D tracking using a photon-limited double-helix response system with a spatial light modulator, which has two twisting lobes along the optical axis of the image, results in a single fluorescent probe appearing as two fluorescent spots from which the Z-position can be determined (Pavani *et al* 2009, Lew *et al* 2010).

One of the simplest 3D tracking methods intentionally generates astigmatism (Kao and Verkman 1994, Holtzer *et al* 2007, Izeddin *et al* 2012). Here, a pair of convex and concave cylindrical lenses is inserted into the optical pathway before the detection device (Watanabe *et al* 2013). In more detail, these lenses generate different optical path lengths along the X- and Y-axes, resulting in a measurable relationship between the Z-position of the particle and the ellipticity of the PSF. In order to calculate the ellipticity in addition to the 2D position, the approximation formula below is used, where σx and σy are the radial standard deviations of the Gaussian function along the X- and Y-axes, respectively. The ellipticity is defined as the ratio of the full width at half maximum (FWHM) of the 2D Gaussian in the X- and Y-axes due to the different focal lengths. Alteration of the distance between the convex and concave cylindrical lenses permits astigmatism for optimal tracking resolution. When the detection device receives 15,000 photons from a fluorescent probe, 3D tracking with precision of 2 nm in the X- and Y-axes and 5 nm along the Z-axis is reached. However, a reliable range is limited to a field view of between −800 and 800 nm. This drawback is common to several 3D tracking methods. A new 3D tracking method based on Airy beams, however, may overcome this. Here, a diffraction-free self-bending PSF is applied to a two-channeled detection system (Jia *et al* 2014) and the Z-position is translated to the distance difference of the two X-positions of the two channels. In particular, this method elongates the dynamic range of 3D tracking to 3 μm. Regardless of the 3D tracking method, the key is to extract Z information from the XY projection.

The 4D single particle using polarized Qdots
As significant as acquiring the third spatial dimension is, 3D single particle tracking ignores any rotational movement performed by the protein. In order to acquire the orientation, fluorescence anisotropy can be used, as the fluorescence emissions are of unequal intensities along the P and S polar axes (P- and S-polarization), which are defined by the polarizing beam-splitter, as described below (Werver 1953, Albrecht 1961, Harms *et al* 1999). In more detail, anisotropy is defined as $(I_P − I_S)/(I_P + I_S)$, where I_P and I_S are the intensities in P- and S-polarization, respectively (Harms *et al* 1999). Indeed, anisotropy measurements have successfully tracked the rotatory dynamics of single protein molecules *in vitro* (Sase *et al* 1997, Forkey *et al* 2003) and in cells (Mizuno *et al* 2011). The fluorescence anisotropy of a Qdot depends on the aspect ratio of its shape (Peng *et al* 2000, Hu *et al* 2001, Deka *et al* 2009). Taking advantage of this property, a highly polarized rod-shaped Qdot (Qrod) can be synthesized by elongating the CdS shell along one-axis of the CdSe core (Peng *et al* 2000, Hu *et al* 2001). The anisotropy alterations in Qrod fluorescence can be described as a sine function and the angular position by the arcsine function. The tracking precision of the orientation was about 1–2° when the photon number from a Qrod was 15,000. By utilizing this anisotropy technique, a fourth dimension, the angular (θ) component, can be added to the orthogonal three coordinate axes described by single particle tracking.

In the 4D tracking system, a polarizing beam splitter is set before the cylindrical lens pair in the 3D tracking optics to divide the fluorescent image into S- and P-polar channels (Watanabe *et al* 2013). For 3D tracking, the P- and S-polarized images are summed before the 3D position is calculated. However, a small gap is generated if the two channels are not completely overlapped, leading to an asymmetrical relationship between the respective FWHM values of the X- and Y-axes. The 3D position can be determined by fitting the merged PSF with a 2D Gaussian function, as mentioned above, and the orientation can be determined by the ratio of the intensities of the S- and P-polarized images. Thus, X, Y, Z and θ are obtained simultaneously with an acquired image. However, when the number of photons from a single Qrod is about 10 000 and the Z-position is near zero, the calculated precision for the X, Y, Z and θ-positions is at most 5, 7, 9 nm and 1°, respectively.

4D tracking is used to investigate the movement of a membrane protein conjugated with a Qrod via antibody affinity (Watanabe *et al* 2013). Isolated Qrods moving on the membrane are identified under a fluorescence microscope. The different intensities in the P- and S-polarized images indicate that the Qrod is inclined against the optical axis. One circular and one elliptical spot indicate that the two Qrods are at distinct Z-positions. In particular, one Qrod shows a half-moon like motion in the X- and Y-axes, accompanied by highly fluctuating movements along the Z-axis and fast rotational motion before endocytosis. However, this observation suggests that this protein's lateral diffusion is constrained by the membrane undercoat, but that it could rotate freely along the cell membrane. In the cytoplasm, a membrane protein seems to be moving along tracks such as microtubules, in three dimensions and slowly rotating helically.

Another 4D tracking method was developed to obtain X, Y, θ and var phi coordinates, the last of which provides information on the out-of-plane tilt angle (Ohmachi *et al* 2012). In this method, single Qrods are imaged as four crowded fluorescent spots by dividing the beam path using a beam splitter and two Wollaston prisms. Otherwise, the orientation of the individual fluorescent probe can be directly estimated using the dipole emission patterns of a defocused image (Bartko and Dickson 1999a, 1999b, Fourkas 2001, Böhmer and Enderlein 2003, Lieb *et al* 2004), an approach that was successfully applied to 4D tracking of a motor protein (Toprak *et al* 2006). The combination of the Wollaston prism method with defocusing could achieve comprehensive tracking of all rotatory and translational movements of a particular molecule in a living cell.

Qdots

Super-resolution microscopy describes the resolution of two objects closer than the diffraction limit of light (Schermelleh *et al* 2010, Galbraith and Galbraith 2011, Leung and Chou 2011). It can be classified into two main categories. The first is based on the photo-transition of a fluorescent probe between its radiative and non-radiative states in order to confine the fluorescence emission into a sub-diffraction-limit-sized volume. This approach is known as RESOLFT (REversible Saturable OpticaL Fluorescence Transitions) and was initially proposed and demonstrated by STED (STimulated Emission Depletion), which exploits the stimulated emission phenomenon of a fluorescent dye (Hell and Wichmann 1994, Klar and Hell 1999). RESOLFT can also be realized by other photoreactions, including those from a ground-state transition phenomenon (ground state depletion (GSD)) (Hell and Kroug 1995, Bretschneider *et al* 2007), the saturation of fluorescence excitation (SAX: SAturated eXcitation) (Fujita *et al* 2007) and reversibly photoswitchable fluorescent proteins (Hofmann *et al* 2005). RESOLFT can also be combined with structured illumination microscopy (SIM) (Heintzmann and Cremer 1999, Gustafsson 2000) in order to provide a wide field imaging capability with superresolution (Heintzmann 2003, Gustafsson 2005).

The second category is based on the separate detection of individual single fluorescent probes in the time or spectra domain, and can be further decomposed into different concepts. One, known as spectral precision distance microscopy (SPDM), localizes individual probes precisely over the many frames of sequentially obtained images (Bornfleth *et al* 1998, Lemmer *et al* 2008). Stochastic optical reconstruction microscopy (STORM) (Rust *et al* 2006) and fluorescence photoactivation localization microscopy (FPALM) (Betzig *et al* 2006) are both SPDM-based techniques that utilize repeated activation–deactivation cycles of photoswitchable fluorophores, such that the fluorescence spots on an obtained image are completely discrete.

Another method from the second category is blinking-based superresolution (BBS). In more detail, BBS relies on the randomness and non-Gaussian properties of blinking, which means stochastic processing can be performed to localize individual fluorescent probes. The first report of BBS used independent component analysis, which is a computational method that decomposes a multivariate signal into independent non-Gaussian signals (Lidke *et al* 2005). Other BBS-based techniques use temporal high-order cumulant (super-resolution optical fluctuation imaging (SOFI)) (Dertinger *et al* 2009), temporal high-order variance (variance imaging for superresolution (VISion)) (Watanabe *et al* 2010), spatial covariance (spatial covariance reconstructive (SCORE)) (Deng *et al* 2014) or Bayesian statistics (Cox *et al* 2011). A great advantage of SPDM and BBS is that they only need a relatively simple fluorescent microscope and no complicated optics.

Qdots are the most compatible with BBS owing to their strong blinking phenomenon. Supposing that there are two adjoining Qdots independently and randomly fluctuating, the moment that one Qdot emits and the other does not is a stochastic event (Dertinger *et al* 2009). To decrease the required number of images, a highly fluctuating Qdot has been developed in which the switching frequency between the on and off states has been increased greatly by optimizing the shell thickness to promote more interaction between the CdSe core and oxygen atoms in water. Although the quantum yield of this Qdot is less than that of standard Qdots, it still has sufficient intensity and stability when exposed to high power illumination and no long off state has been detected. Hence, we can easily obtain a super-resolved image by only labeling the target protein and calculating the fluctuation of the blinking-enhanced Qdots. In this case, the spatial resolution has improved from 267 to 154 nm using SOFI and only 100 images (Watanabe *et al* 2010).

However, conventional optical microscopy can quantitatively acquire 3D position and orientation information at the nano-scale from the shape of the PSF and the polarization characteristics of Qdots and Qrods. The amount of spatial information can be increased through analyzing the stochastic fluctuations of the fluorescence. Thus, the fluorescence of a probe attached to a molecule can provide information about the molecular phenomena or state. Increasing the intensity, stability and blinking of Qdots and its derivatives will improve the acquisition.

However, super-resolution microscopy and single particle tracking have made it possible to resolve and follow two objects closer than the diffraction limit of light. The result is quantitative information on the dynamics of biological phenomena at the nano-scale level. Even more details for the dynamics can be acquired with the above technologies by using Qdots and their derivatives as probes for labeling the molecules of interest. Moreover, the PSF and the polarization characteristics of the Qdots can be used to gain comprehensive information on both the position and orientation of the targeted molecule. As this information can be extracted from the stochastic properties of the fluorescence, increasing the intensity, stability and blinking of Qdots should provide even more quantitative details about the dynamics.

5.2 FRAP

After the genomic era, a major challenge in the field of cell biology is to understand the dynamics of proteins during signal transduction events regulating cellular responses and functions. In more detail, it is now possible to perform live-cell assays of protein diffusion utilizing the advances in protein labeling and fluorescence microscopy. Two classic techniques, fluorescence recovery after photo-bleaching (FRAP) and fluorescence correlation spectroscopy (FCS), are well-known and convenient tools for analyzing protein diffusion (Reits and Neefjes 2001, Medina and Schwille 2002). Even non-invasive fluorescent tagging with green fluorescent protein (GFP) or variants has increased the use of these techniques.

In FRAP, fluorescent proteins in a small region of interest in a cell expressing them are permanently photobleached using a high-powered laser beam and the subsequent movement of surrounding non-photobleached fluorescent proteins into the photobleached area is then recorded (Reits and Neefjes 2001). However, the rate of fluorescence recovery provides information about how quickly the fluorescent proteins can move into the bleached region. This 'mobility' is dependent on the rate of free diffusion and the active transport of the fluorescent protein. Mobility is also influenced by interactions with other proteins, which could even confine the movement of fluorescent molecules that would otherwise be freely mobile.

In FCS, the fluorescence intensity in the tiny confocal volume is measured with high repetition rates in order to detect fluctuations in the fluorescence intensity, including the contribution of Brownian motion (Medina and Schwille 2002). Through time correlation analysis of the fluorescence fluctuations, the diffusion coefficient, molecular concentration and molecular interactions can be estimated.

Although FRAP is adequate for measuring the mobile and immobile fractions of fluorescent molecules in a living cell, the measurability of diffusion rates is lower than for FCS. Moreover, the general photobleaching time required for FRAP (less than 10 ms) is sufficiently long that protein diffusion can start to occur during the photobleaching process itself, hence preventing an accurate calculation of the diffusion coefficient.

However, these problems can be overcome by using a photo-switchable fluorescent protein (PSFP) instead of a conventional fluorescent protein. The advantage of the usage of PSFPs is that they can change from dark to bright or undergo conversion of their absorption/emission wavelength when they are stimulated by light of a specific wavelength (Lukyanov et al 2005). Indeed, the rapid (less than 1 ms) photoswitching property of PSFPs encourages their use for protein mobility analyses. In particular, PSFPs have been used to investigate the diffusion time of a protein by using fluorescence decay measurements fitted to a single exponential curve (Theis-Febvre *et al* 2005) and to estimate the diffusion coefficient of a special protein from the alteration in the Gaussian profile of the fluorescence intensity (Ando *et al* 2002). Although these analyses are useful for relative comparison, they are not comparable to the values obtained by FCS, as the model equations to determine the diffusion coefficient are not based on the theoretical background. In more detail, the development of a method using rigorous analyses based on the FRAP theoretical model enables us to determine a wider range of diffusion coefficients, which cover the range of FCS to conventional FRAP (Matsuda 2008).

Photobleaching-based imaging
Although excitation irradiation of a sample is indispensable for fluorescence imaging, strong irradiation causes photobleaching, through which the chromophore becomes irreversibly non-fluorescent through a chemical structure alteration. In order to avoid photobleaching, excitation light power and time is reduced for general fluorescence imaging. However, for photobleaching-based imaging, bleaching is actively used. In more detail, once a region of fluorescence loss is obtained, the molecular movement is investigated by imaging the alterations in the non-fluorescent pattern.

In particular, FRAP visualizes the dynamics of fluorescent or fluorescent-labeled molecules. The method consists of photobleaching a fluorescent chromophore by irradiation of a limited region with strong excitation followed by imaging the fluorescence recovery caused by outflux of bleached molecules from that region and influx of fluorescent molecules from outside (Carrero *et al* 2003, Dundr and Misteli 2003). In general, confocal scanning-laser microscopy (CSLM), which can easily irradiate a specific region in the cell, is used for FRAP. In the past, switching the irradiation light intensity between imaging excitation and photobleaching required a significant time lag. However, recent CSLM techniques can simultaneously generate photobleaching and excitation light in order to ensure that it is possible to collect images immediately after photobleaching. By comparing the shapes of the fluorescence decay curves, qualitative evaluations can be performed to demonstrate that fast-moving molecules recover more quickly. It is even possible to perform quantitative evaluations to obtain the physiological constants associated with molecular dynamics by mathematical modeling of fluorescence recovery data (Carrero *et al* 2003, Dundr and Misteli 2003).

When the fluorescence-labeled molecules do not interact with other molecules or components in the cell, molecular dynamics are regulated by simple diffusion caused by Brownian motion. To determine the diffusion coefficients in simple diffusion, model equations based on Axelrod's model equation for 2D diffusion are generally used (Axelrod *et al* 1976). In this method, one must know the spatial distribution of photobleaching as a function of the distance from the bleach center, because the model equation for time recovery of fluorescence is expanded from the model equation for the initial bleaching profile. Thus, an image in which molecular movement is fixed (the sample is fixed in formaldehyde) is collected after photo-bleaching.

That image is analyzed by fitting of the model equation for the initial bleaching profile and parameters relating to the bleaching radius and bleach constant are obtained (Axelrod *et al* 1976). The diffusion coefficient is calculated by model fitting using the equation for fluorescence recovery, which contains these parameters (Axelrod *et al* 1976). However, this method assumes the ideal situation in which the concentration of the fluorescence molecule is homogenous, the initial bleaching profile is adapted to the known model and diffusion is isotropic. A recently developed method treats each cellular component in confocal images independently, such as the cytoplasm, nucleus, plasma membrane or nuclear membrane from a 3D cell model. In this model, a different position-dependent porosity is assigned to each component. Then the diffusion coefficient is calculated using the lattice Boltzmann method, which simulates molecular dynamics in the cell model as movements of virtual particles (Khün *et al* 2011). The development of analytical methods that can treat anisotropic diffusion will facilitate analyses based on the microenvironment in real cells. If molecular interactions are included in the system, fluorescence recovery becomes slower compared to simple diffusion because of the restricted motion of the labeled molecules.

When the counterpart of the interaction is immobile or has an extremely slow movement two situations may apply: (i) molecular diffusion is overwhelmingly faster than binding dissociation and the dynamics of the labeled molecule are regulated by binding dissociation (Sprague and McNally 2005); (ii) binding dissociation is much faster and the labeled molecules repeat binding and diffusion in the bleached area as they leave. Thus, different model equations specific to each case are required for precise analysis. There is indeed an equation for simple binding dissociation in the former case and an equation including diffusion for the latter case (Sprague and McNally 2005).

Diffusion analysis using time correlations such as FCS do not provide information about the immobile fraction in standard analysis. However, FRAP can extract information about the immobile fraction as the fraction that does not recover with time (Carrero *et al* 2003, Dundr and Misteli 2003). FRAP analysis is additionally applicable for diffusion rates $<1 \ \mu m^2 \, s^{-1}$, which are difficult to analyze even using FCS. However, FRAP cannot easily analyze diffusion rates faster than $10 \ \mu m^2 \, s^{-1}$ because of the diffusion during photobleaching (Matsuda 2008). However, this can be overcome by analysis using photoactivatable fluorescent proteins (PAFPs).

FLIP and iFRAP

Fluorescence loss in photobleaching (FLIP) and inverse FRAP (iFRAP) are other conventional photobleaching imaging techniques. For FLIP, instead of observing fluorescence recovery in the bleached area, the areas for measurement are set outside of the bleached area and measurement is performed with continuous photobleaching (Dundr and Misteli 2003). Generally, the bleached area (~1/4–1/2 of the whole cell area) is larger than that for FRAP. Differences in molecular dynamics can be observed as the difference in the starting time of the fluorescence reductions (for instance, faster molecules bleach earlier) (Dundr and Misteli 2003). There is no general method to determine the diffusion coefficient or dissociation constant for FLIP because fluorescence decay is related to the area of bleaching and the distance between the bleached and observed areas. However, it is useful for relative dynamic comparisons and is applicable for the fast diffusion, which is difficult to observe by FRAP as the initiation of the fluorescence decrease can be prolonged by setting the observation area farther from the bleaching area (Dundr and Misteli 2003). iFRAP is a method for which only a limited area remains unbleached and then the dynamics of the molecules in that area are traced (Dundr and Misteli 2003). This method is indeed effective for the analysis of molecules with small dissociation constants because molecular fluorescence is directly measured. However, it is also not suitable for the measurement of fast-diffusing molecules, as the photobleaching requires more time. In addition, there is the risk that strong irradiation over a larger area of the cell could be phototoxic and induce diffusion that may not appear under normal conditions (Dundr and Misteli 2003). As a similar image is obtained by the method based on PAFPs, which is even available for measurement of fast diffusion, iFRAP will be replaced by that novel method.

Photoactivation-based imaging
Live-cell imaging technology based on fluorescent proteins has been developed and it has become common to use PSFPs, in which the fluorescence properties can be altered by photostimulation. These fluorescent proteins make it possible to alter the fluorescence properties of molecules in a limited region of the living body by photostimulation at a specific time (Patterson *et al* 2010, Miyawaki 2011). Photostimulation of PAFPs can increase fluorescence intensity from the dark state, allowing for the collection of inverted images from photobleaching techniques such as FRAP. Only activated molecules appear on the non-fluorescent background, providing high-contrast imaging. Using this advantage, PAFPs are used to measure protein dynamics in the living body through region-specific highlighting and tracing (Miyawaki 2011).

PSFPs
After the development of the photoactivatable protein (PA-GFP) from *Aequorea victoria* GFP (Patterson and Lippincott-Schwartz 2002), many fluorescent proteins that can be altered by photostimulation have been developed by genetic modification or screening from new species and hence used for live-cell imaging (Patterson *et al* 2010, Miyawaki 2011). In the initial development, PSFPs posed problems such as the formation of oligomers or the need to use low temperatures for cell culturing, however, recent PSFPs have improved and are useful for imaging (Patterson *et al* 2010). PSFPs are classified into PAFPs, which can alter from non-fluorescent to fluorescent proteins, and photoconvertible fluorescent proteins, which alter fluorescence wavelength with photostimulation. Both can be categorized further as reversible or irreversible (Patterson *et al* 2010). Although a reversible photoconvertible fluorescent protein constructed by fusion of the yellow fluorescent protein EYFP and the reversible photoactivatable red fluorescent protein rsTagRFP have been reported (Subach *et al* 2010), no single fluorescent protein has been reported thus far. These photoswitchable proteins are used as genetically encoded fluorescence tags to

visualize dynamics at the cellular, organelle and protein levels by highlighting and tracing (Miyawaki 2011). Moreover, these proteins also find applications in super-resolution imaging (Patterson *et al* 2010, Miyawaki 2011).

FDAP

As with imaging using photobleaching, PSFPs are available for time-lapse imaging of molecular migration and for determination of diffusion coefficients. Because reverse images are taken, diffusion coefficients can be determined by PSFPs through using procedures similar to that of FRAP (Matsuda 2008). Contrary to FRAP, which analyzes fluorescence recovery after photobleaching, analysis by PSFPs measures the fluorescence decay after photostimulation caused by outflux of photoactivated molecules. Thus, the method is known as fluorescence decay after photoactivation (FDAP) (Matsuda 2008). The fluorescence decay curve of FDAP and the fluorescence recovery of FRAP have a symmetric relationship to the time axis. However, a model function for FRAP cannot be used directly. For normalization, it is necessary to know the fluorescence intensity where all PAFPs are photoactivated. Thus, the photostimulation proceeds until the fluorescence intensity becomes saturated and the measurement of fluorescence intensity is included in the method for FRAP (Matsuda 2008). The measurements and analysis are even based on FRAP, however, there is a difference in the detectable range for diffusion coefficients. Fluorescence recovery between photobleaching and first acquisition significantly affects analysis for fast-moving molecules over $10 \ \mu m^2 \ s^{-1}$ on FRAP. At the first image for fluorescence recovery, if the profile of the bleached molecule is significantly greater than that obtained by fixed cells, it breaks the basic assumption of the model equation for fluorescence recovery. Hence, the results of the analysis become inaccurate. On the other hand, the irradiation time for FDAP is much shorter than that for FRAP because photoswitching of PSFPs occurs with much less energy (approximately 10 times lower) compared to photobleaching. It is possible to analyze diffusion near $100 \ \mu m^2 \ s^{-1}$ in solution by FDAP, if stimulation is performed for 0.25 ms at a single spot and images are acquired with a high scanning speed of 4000 Hz (Matsuda 2008). In addition, it can be applied for measuring slow diffusion and binding dissociation (Plachta 2011).

FDAP does not require special equipment or software, hence, it is possible to measure a wide range of diffusion coefficients from FRAP's specialized slow diffusion to FCS's specialized fast diffusion with a general confocal microscope setup for cell observation. FDAP also contributes to research on protein dynamics as a convenient tool to obtain fluorescence decay curves that can easily be applied for relative comparison (Miyawaki 2011).

Imaging techniques using photobleaching and photoactivation of fluorescent proteins are a versatile technology for analyzing molecular dynamics. Data obtained from these methods comprise recovery or decay curves of fluorescence intensity that are easy to understand and compare. However, these curves include information on the simple diffusion of molecules and on molecular interactions. Thus, it is possible to observe interactions as incomplete recovery (or decay) due to the immobile fraction associated with intracellular structures or a significant inconsistency between the diffusion coefficient determined by curve fitting and the values estimated from molecular weight by using the Stokes–Einstein equation. However, such differences from simple diffusion can be regarded as a particular feature of the individual molecule that defines the molecular function in living cells. Although the conventional model equation was based on 2D diffusions, the real diffusion of intracellular molecules is definitely 3D and sometimes anisotropic. Thus, the development of a model equation considering these realities will allow deeper understanding of molecular dynamics in a living system.

Further reading

Levi V, Ruan Q and Gratton E 2005 3D particle tracking in a two-photon microscope: application to the study of molecular dynamics in cells *Biophys. J.* **88** 2919–28

References

Abbe E 1873 Contributions to the theory of the microscope and the microscopic perception *Arch. Mikr. Anat.* **9** 413–68

Adachi K, Yasuda R, Noji H, Itoh H, Harada Y, Yoshida M and Kinosita K Jr 2000 Stepping rotation of F1-ATPase visualized through angle-resolved single-fluorophore imaging *Proc. Natl Acad. Sci. USA* **97** 7243–37

Aguet F, Ville D V D and Unser M 2005 A maximum likelihood formalism for sub-resolution axial localization of fluorescent nanoparticles *Opt. Exp.* **13** 10503–22

Albrecht A 1961 Polarizations and assignments of transitions: the method of photoselection *J. Mol. Spectrosc.* **6** 84–108

Ando R, Hama H, Yamamoto-Hino M, Mizuno H and Miyawaki A 2002 An optical marker based on the UV-induced green-to-red photoconversion of a fluorescent protein *Proc. Natl Acad. Sci. USA* **99** 12651–6

Arhel N, Genovesio A, Kim K A, Miko S, Perret E, Olivo-Marin J C, Shorte S and Charneau P 2006 Quantitative four-dimensional tracking of cytoplasmic and nuclear HIV-1 complexes *Nat. Methods* **3** 817–24

Axelrod D, Koppel D E, Schlessinger J, Elson E and Webb W W 1976 Mobility measurement by analysis of fluorescence photo-bleaching recovery kinetics *Biophys. J.* **16** 1055–69

Barroso M M 2011 Quantum dots in cell biology *J. Histochem. Cytochem.* **59** 237–51

Bartko A P and Dickson R M 1999a Imaging three-dimensional single molecule orientations *J. Phys. Chem. B.* **103** 11237–41

Bartko A P and Dickson R M 1999b Three-dimensional orientations of polymer-bound single molecules *J. Phys. Chem. B.* **103** 3053–6

Betzig E, Patterson G H, Sougrat R, Lindwasser O W, Olenych S, Bonifacino J S, Davidson M W, Lippincott-Schwartz J and Hess H F 2006 Imaging intracellular fluorescent proteins at nanometer resolution *Science* **313** 1642–5

Böhmer M and Enderlein J 2003 Orientation imaging of single molecules by wide-field epifluorescence microscopy *J. Opt. Soc. Am. B.* **20** 554–9

Born M and Wolf E 1980 *Principles of Optics* (Oxford: Pergamon)

Bornfleth H, Satzler K, Elis R and Cremer C 1998 High-precision distance measurements and volume-conserving segmentation of objects near and below the resolution limit in three-dimensional confocal fluorescence microscopy *J. Microsc.* **189** 118–36

Bretschneider S, Eggeling S and Hell S W 2007 Breaking the diffraction barrier in fluorescence microscopy by optical shelving *Phys. Rev. Lett.* **98** (21) 218103

Bruchez M Jr, Moronne M, Gin P, Weiss S and Alivisatos A P 1998 Semiconductor nanocrystals as fluorescent biological labels *Science* **281** 2013–6

Bruchez M P 2011 Quantum dots find their stride in single molecule tracking *Curr. Opin. Chem. Biol.* **15** 775–80

Byers R J and Hitchman E R 2010 Quantum dots brighten biological imaging *Prog. Histochem. Cytochem.* **45** 201–37

Cao H, Chen B, Squier T C and Mayer M U 2006 CrAsH: a biarsenical multi-use affinity probe with low non-specific fluorescence *Chem. Commun.* **24** 2601–3

Carrero G, McDonald D, Crawford E, de Vries G and Hendzel M J 2003 Using FRAP and mathematical modeling to determine the *in vivo* kinetics of nuclear proteins *Methods* **29** 14–28

Chan W C W and Nie S 1998 Quantum dot bioconjugates for ultrasensitive nonisotopic detection *Science* **281** 2016–8

Chang Y P, Pinaud F, Antelman J and Weiss S 2008 Tracking bio-molecules in live cells using quantum dots *J. Biophotonics* **1** 287–98

Cheezum M K, Walker W F and Guilford W H 2001 Quantitative comparison of algorithms for tracking single fluorescent particles *Biophys. J.* **281** 2378–88

Chenouard N *et al* 2014 Objective comparison of particle tracking methods *Nat. Methods* **11** 281–9

Chien F C, Kuo C W and Chen P 2011 Localization imaging using blinking quantum dots *Analyst* **136** 1608–13

Chinnathambi S, Chen S, Ganesan S and Hanagata N 2014 Silicon quantum dots for biological applications *Adv. Healthc. Mater.* **3** 10–29

Choi H S and Frangioni J V 2010 Nanoparticles for biomedical imaging: fundamentals of clinical translation *Mol. Imaging* **9** 291–310 PMID: 21084027

Clausen M P and Lagerholm B C 2011 The probe rules in single particle tracking *Curr. Protein Pept. Sci.* **12** 699–713

Conner S D and Schmid S L 2003 Regulated portals of entry into the cell *Nature* **422** 37–44

Courty S, Luccardini C, Bellaiche Y, Cappello G and Dahan M 2006 Tracking individual kinesin motors in living cells using single quantum-dot imaging *Nano Lett.* **6** 1491–5

Cox S, Rosten E, Monypenny J, Jovanovic-Talisman T, Burnette D T, Lippincott-Schwartz J, Jones G E and Heintzmann R 2011 Bayesian localization microscopy reveals nanoscale podosome dynamics *Nat. Methods* **9** 195–200

Crane J and Verkman A 2008 Long-range nonanomalous diffusion of quantum dot-labeled aquaporin-1 water channels in the cell plasma membrane *Biophys. J.* **94** 702–13

Cui B, Wu C, Chen L, Ramirez A, Bearer E L, Li W, Mobley W C and Chu S 2007 One at a time, live tracking of NGF axonal transport using quantum dots *Proc. Natl Acad. Sci. USA* **104** 13666–71

Cupaioli F A, Zucca F A, Boraschi D and Zecca L 2014 Engineered nanoparticles. How brain friendly is this new guest? *Prog. Neurobiol.* **119–120** 20–38

Dabbousi B O, Rodriguez-Viej O J and Bawendi M G 1997 (CdSe)ZnS core-shell Qdots: synthesis and characterization of a size series of highly luminescent nanocrystallites *J. Phys. Chem. B.* **101** 9463–75

Dahan M, Levin S, Luccardini C, Rostaing P, Riveau B and Triller A 2003 Diffusion dynamics of glycine receptors revealed by single-quantum dot tracking *Science* **302** 442–5

Dalgarno P A, Dalgarno H I, Putoud A, Lambert R, Paterson L, Logan D C, Towers D P, Warburton R J and Greenaway A H 2010 Multiplane imaging and three dimensional nanoscale particle tracking in biological microscopy *Opt. Express* **18** 877–84

Deka S *et al* 2009 CdSe/CdS/ZnS double shell nanorods with high photoluminescence efficiency and their exploitation as biolabeling probes *J. Am. Chem. Soc.* **131** 2948–58

Deng Y, Sun M, Lin P H, Ma J and Shaevitz J W 2014 Spatial covariance reconstructive (SCORE) super-resolution fluorescence microscopy *PLoS One* **9** e94807

Dertinger T, Colyer R, Iyer G, Weiss S and Enderlein J 2009 Fast, background-free, 3D super-resolution optical fluctuation imaging (SOFI) *Proc. Natl Acad. Sci. USA* **106** 22287–92

Deschout H, Cella Zanacchi F, Mlodzianoski M, Diaspro A and Bewersdorf J 2014 Precisely and accurately localizing single emitters in fluorescence microscopy *Nat. Methods* **11** 253–66

Drummen G P 2012 Fluorescent probes and fluorescence (microscopy) techniques-illuminating biological and biomedical research *Molecules* **17** 14067–1490

Dubertret B, Skourides P, Norris D J, Noireaux V, Brivanlou A H and Libchaber A 2002 *In vivo* imaging of quantum dots encapsulated in phospholipid micelles *Science* **98** 1759–62

Dubois F, Mahler B, Dubertret B, Doris E and Mioskowski C 2007 A versatile strategy for quantum dot ligand exchange *J. Am. Chem. Soc.* **129** 482–3

Dundr M and Misteli T 2003 Measuring dynamics of nuclear proteins by photobleaching *Curr. Protoc. Cell Biol.* **13** Unit 13.5. 1–18 PMID: 18228420

Dupont A and Lamb D C 2011 Nanoscale three-dimensional single particle tracking *Nanoscale* **3** 4532–41

Egner A *et al* 2007 Fluorescence nanoscopy in whole cells by asynchronous localization of photoswitching emitters *Biophys. J.* **93** 3285–90

Ehrlich M, Boll W, Van Oijen A, Haniharan R, Chandran K, Nibert M L and Kirchhausen T 2004 Endocytosis by random initiation and stabilization of clathrin-coated pits *Cell* **118** 591–605

Ekimov A I and Onushchenko A A 1981 Quantum size effect in three-dimensional microscopic semiconductor crystals *JETP Lett.* **34** 345–9

Ellinger P 1940 Fluorescence microscopy in biology *Biol. Rev.* **15** 323–47

Erathodiyil N and Ying J Y 2011 Functionalization of inorganic nanoparticles for bioimaging applications *Acc. Chem. Res.* **44** 925–35

Folling J, Belov V, Kunetsky R, Medda R, Schonle A, Egner A, Eggeling C, Bossi M and Hell S W 2007 Photochromic rhodamines provide nanoscopy with optical sectioning *Angew. Chem.* **46** 6266–70

Fomenko V and Nesbitt D J 2008 Solution control of radiative and nonradiative lifetimes: a novel contribution to quantum dot blinking suppression *Nano Lett.* **8** 287–93

Forkey J N, Quinlan M E, Shaw M A, Corrie J E and Goldman Y E 2003 Three-dimensional structural dynamics of myosin V by single-molecule fluorescence polarization *Nature* **422** 399–404

Fourkas J T 2001 Rapid determination of the three-dimensional orientation of single molecules *Opt. Lett.* **26** 211–3

Fujita K, Kobayashi M, Kawano S, Yamanaka M and Kawata S 2007 High-resolution confocal microscopy by saturated excitation of fluorescence *Phys. Rev. Lett.* **99** 228105

Fujiwara T, Ritchie K, Murakoshi H, Jacobson K and Kusumi A 2002 Phospholipids undergo hop diffusion in compartmentalized cell membrane *J. Cell Biol.* **157** 1071–81

Galbraith C G and Galbraith J A 2011 Super-resolution microscopy at a glance *J. Cell Sci.* **124** 1607–11

Gao X, Yang L, Petros J A, Marshall F F, Simons J W and Nie S 2005 *In vivo* molecular and cellular imaging with quantum dots *Curr. Opin. Biotechnol.* **16** 63–72

Genovesio A, Liedl T, Emiliani V, Parak W J, Coppey-Moisan M and Olivo-Marin J C 2006 Multiple particle tracking in 3-D+t microscopy: method and application to the tracking of endocytosed quantum dots *IEEE Trans. Image Process* **15** 1062–70

Gerion D, Pinaud F, Williams S, Parak W, Zanchet D, Weiss S and Alivisatos A P 2001 Synthesis and properties of biocompatible water-soluble silica-coated CdSe/ZnS semiconductor quantum dots *J. Phys. Chem. B.* **105** 8861–71

Ghetie V and Ward E S 2000 Multiple roles for the major histocompatibility complex class I-related receptor FcRn *Annu. Rev. Immunol.* **18** 739–66

Guo W, Li J J, Wang Y A and Peng X 2003 Conjugation chemistry and bioapplications of semiconductor box nanocrystals prepared via dendrimer bridging *Chem. Mater.* **15** 3125–33

Gustafsson M G L 2000 Surpassing the lateral resolution limit by a factor of two using structured illumination microscopy *J. Microsc.* **198** 82–7

Gustafsson M G L 2005 Nonlinear structured-illumination microscopy: Wide-field fluorescence imaging with theoretically unlimited resolution *Proc. Natl Acad. Sci. USA* **102** 13081–6

Harms G S, Sonnleitner M, Schütz G J, Gruber H J and Schmidt T 1999 Single-molecule anisotropy imaging *Biophys. J.* **77** 2864–70

Han J J, Kiss C, Bradbury A R M and Werner J H 2012 Time-Resolved, Confocal Single Molecule Tracking of Individual Organic Dyes and Fluorescent Proteins in Three Dimensions *ACS Nano* **6** 8922–32

He K *et al* 2010 Intercellular transportation of quantum dots mediated by membrane nanotubes *ACS Nano* **4** 3015–22

Heintzmann R 2003 Saturated patterned excitation microscopy with two-dimensional excitation patterns *Micron* **34** 283–91

Heintzmann R and Cremer C 1999 Laterally modulated excitation microscopy: improvement of resolution by using a diffraction grating *Proceedings for the SPIE* vol 3568 *Optical Biopsies and Microscopic Techniques III, (Stockholm)* p 185

Hell S W and Kroug M 1995 Ground-state-depletion fluorescence microscopy: a concept for breaking the diffraction resolution limit *Appl. Phys. B.* **60** 495–7

Hell S W and Wichmann J 1994 Breaking the diffraction resolution limit by stimulated emission: stimulated-emission-depletion fluorescence microscopy *Opt. Lett.* **19** 780–2

Hess S T, Girirajan P K and Mason M D 2006 Ultra-high-resolution imaging by fluorescence photoactivation localization microscopy *Biophys. J.* **91** 4258–72

Hess S T, Gould T J, Gudheti M V, Maas S A, Mills K D and Zimmerberg J 2007 Dynamic clustered distribution of hemagglutinin resolved at 40 nm in living cell membranes discriminates between raft theories *Proc. Natl Acad. Sci. USA* **104** 17370–5

Hoffman M T, Sheung J and Selvin P R 2011 Fluorescence imaging with one nanometer accuracy: *in vitro* and *in vivo* studies of molecular motors *Methods Mol. Biol.* **778** 33–56

Hofmann M, Eggeling C, Jakobs S and Hell S W 2005 Breaking the diffraction barrier in fluorescence microscopy at low light intensities by using reversibly photoswitchable proteins *Proc. Natl Acad. Sci. USA* **102** 17565–9

Hohng S and Ha T 2004 Near-complete suppression of quantum dot blinking in ambient conditions *J. Am. Chem. Soc.* **126** 1324–5

Holtzer L, Meckel T and Schmidt T 2007 Nanometric three-dimensional tracking of individual quantum dots in cells *Appl. Phys. Lett.* **90** 053902

Hoyer P, Staudt T, Engelhardt J and Hell S W 2011 Quantum dot bluing and blinking enables fluorescence nanoscopy *Nano Lett.* **11** 245–50

Hu J, Li L S, Yang W, Manna L, Wang L W and Alivisatos A P 2001 Linearly polarized emission from colloidal semiconductor quantum rods *Science* **292** 2060–3

Hua W, Sheff D, Toomre D and Mellman I 2006 Vectorial insertion of apical and basolateral membrane proteins in polarized epithelial cells revealed by quantitative 3D live-cell imaging *J. Cell Biol.* **172** 1035–44

Huang B, Wang W, Bates M and Zhuang X 2008 Three-dimensional super-resolution imaging by stochastic optical reconstruction microscopy *Science* **319** 810–3

Huang Z L, Zhu H, Long F, Ma H, Qin L, Liu Y, Ding J, Zhang Z, Luo Q and Zeng S 2011 Localization-based super-resolution microscopy with an sCMOS camera *Opt. Express* **19** 19156–68

Izeddin I, El Beheiry M, Andilla J, Ciepielewski D, Darzacq X and Dahan M 2012 PSF shaping using adaptive optics for three-dimensional single-molecule super-resolution imaging and tracking *Opt. Express* **20** 4957–67

Jia S, Vaughan V C and Zhuang Z 2014 Isotropic three-dimensional super-resolution imaging with a self-bending point spread function *Nat. Photonics* **8** 302–6

Jiang W, Mardyani S, Fischer H and Chan W C W 2006 Design and characterization of lysine cross-linked mercapto-acid biocompatible quantum dots *Chem. Mater.* **18** 872–8

Jin T, Fujii F, Komai Y, Seki J, Seiyama A and Yoshioka Y 2008 Preparation and characterization of highly fluorescent, glutathione-coated near infrared quantum dots for *in vivo* fluorescence imaging *Int. J. Mol. Sci.* **9** 2044–61

Jones S A, Shim S H, He J and Zhuang X 2011 Fast, three-dimensional super-resolution imaging of live cells *Nat. Methods* **8** 499–508

Joo K I, Fang Y, Liu Y, Xiao L, Gu Z, Tai A, Lee C L, Tang Y and Wang P 2011 Enhanced real-time monitoring of adeno-associated virus trafficking by virus-quantum dot conjugates *ACS Nano* **5** 3523–35

Juette M F and Bewersdorf J 2010 Three-dimensional tracking of single fluorescent particles with submillisecond temporal resolution *Nano Lett.* **10** 4657–63

Juette M F, Gould T J, Lessard M D, Mlodzianoski M J, Nagpure B, Bennett B T, Hess S T and Bewersdorf J 2008 Three-dimensional sub-100 nm resolution fluorescence microscopy of thick samples *Nat. Methods* **5** 527–9

Juette M F, Rivera-Molina F E, Toomre D K and Bewersdorf J 2013 Adaptive optics enables three-dimensional single particle tracking at the sub-millisecond scale *Appl. Phys. Lett.* **102** 173702

Kairdolf B A, Smith A M, Stokes T H, Wang M D, Young A N and Nie S 2013 Semiconductor quantum dots for bioimaging and biodiagnostic applications *Annu. Rev. Anal. Chem.* **6** 143–62

Katayama Y, Burkacky O, Meyer M, Bräuchle C, Gratton E and Lamb D C 2009 Real-time nanomicroscopy via three-dimensional single-particle tracking *Chem. Phys. Chem.* **10** 2458–64

Kao H P and Verkman A S 1994 Tracking of single fluorescent particles in three dimensions: use of cylindrical optics to encode particle position *Biophys. J.* **67** 1291–300

Khün T, Ihalainen T O, Hyväluoma J, Dross N, Willman S F, Langowski J, Vihinen-Ranta M and Timonen J 2011 Protein diffusion in mammalian cell cytoplasm *PLoS One* **6** e22962

Kim S W, Kim S, Tracy J B, Jasanoff A and Bawendi M G 2005 Phosphine oxide polymer for water-soluble nanoparticles *J. Am. Chem. Soc.* **127** 4556–7

Klar T A and Hell S W 1999 Subdiffraction resolution in far-field fluorescence microscopy *Opt. Lett.* **24** 954–6

Kloepfer J A, Bradforth S E and Nadeau J L 2005 Photophysical properties of biologically compatible CdSe quantum dot structures *J. Phys. Chem. B.* **109** 9996–10003

Ko H C, Yuan C T and Tang J 2011 Probing and controlling fluorescence blinking of single semiconductor nanoparticles *Nano Rev.* **2** 5895

Kubitscheck U, Kückmann O, Kues T and Peters R 2000 Imaging and tracking of single GFP molecules in solution *Biophys. J.* **78** 2170–9

Kubitscheck U, Wedekind P, Zeidler O, Grote M and Peters R 1996 Single nuclear pores visualized by confocal microscopy and image processing *Biophys. J.* **70** 2067–77

Kuno M, Lee J K, Dabbousi B O, Mikulec F V and Bawendi M G 1997 The band edge luminescence of surface modified CdSe nanocrystallites: probing the luminescing state *J. Chem. Phys.* **106** 9869

Kural C, Kim H, Syed S, Goshima G, Gelfand V I and Selvin P R 2005 Kinesin and dynein move a peroxisome *in vivo*: a tug-of-war or coordinated movement? *Science* **308** 1469–72

Kusumi A, Ike H, Nakada C, Murase K and Fujiwara T 2005a Single-molecule tracking of membrane molecules: plasma membrane compartmentalization and dynamic assembly of raft-philic signaling molecules *Semin. Immunol.* **17** 3–21

Kusumi A, Nakada C, Ritchie K, Murase K, Suzuki K and Murakoshi H *et al* 2005b Paradigm shift of the plasma membrane concept from the two-dimensional continuum fluid to the partitioned fluid: high-speed single-molecule tracking of membrane molecules *Annu. Rev. Biophys. Biomol. Struct.* **34** 351–78

Kwik J, Boyle S, Fooksman D, Margolis L, Sheetz M P and Edidin M 2003 Membrane cholesterol, lateral mobility, and the phosphatidylinositol 4,5-bisphosphate-dependent organization of cell actin *Proc. Natl Acad. Sci. USA* **100** 13964–9

Lakadamyali M, Rust M J, Babcock H P and Zhuang X W 2003 Visualizing infection of individual influenza viruses *Proc. Natl Acad. Sci. USA* **100** 9280–5

Lasne D *et al* 2006 Single nanoparticle photothermal tracking (SNaPT) of 5-nm gold beads in live cells *Biophys. J.* **91** 4598–604

Lemmer P *et al* 2008 SPDM: light microscopy with single-molecule resolution at the nanoscale *Appl. Phys. B.* **93** 1–12

Leung B O and Chou K C 2011 Review of super-resolution fluorescence microscopy for biology *Appl. Spectrosc.* **65** 967–80

Levenberg K 1944 A method for the solution of certain non-linear problems in least squares *Q. Appl. Math.* **2** 164–8

Lew M D, Thompson M A, Badieirostami M and Moerner W E 2010 *In vivo* three-dimensional superresolution fluorescence tracking using a double-helix point spread function *Proc. Soc. Photo Opt. Instrum. Eng.* **7571** 75710Z

Li J, Wu D, Miao Z and Zhang Y 2010 Preparation of quantum dot bioconjugates and their applications in bio-imaging *Curr. Pharm. Biotechnol.* **11** 662–71

Liang Z Y, Xu N, Guan Y H, Xu M, He Q H, Han Q D, Zhang Y Y and Zhao X S 2007 The transport of alpha(1A)-adrenergic receptor with 33-nm step size in live cells *Biochem. Biophys. Res. Commun.* **353** 231–7

Lichtman J W and Conchello J A 2005 Fluorescence microscopy *Nat. Methods* **2** 910–9

Lidke K, Rieger B, Jovin T and Heintzmann R 2005 Superresolution by localization of quantum dots using blinking statistics *Opt. Express* **13** 7052–62

Lieb M A, Zavislan J M and Novotny L 2004 Single-molecule orientations determined by direct emission pattern imaging *J. Opt. Soc. Am. B.* **21** 1210–5

Liu H, Liu Y, Liu S, Pang D W and Xiao G 2011a Clathrin-mediated endocytosis in living host cells visualized through quantum dot labeling of infectious hematopoietic necrosis virus *J. Virol.* **85** 6252–62

Liu S L, Zhang Z L, Sun E Z, Peng J, Xie M, Tian Z Q, Lin Y and Pang D W 2011b Visualizing the endocytic and exocytic processes of wheat germ agglutinin by quantum dot-based single-particle tracking *Biomaterials* **32** 7616–24

Liu S L, Zhang Z L, Tian Z Q, Zhao H S, Liu H, Sun E Z, Xiao G F, Zhang W, Wang H Z and Pang D W 2012 Effectively and efficiently dissecting the infection of influenza virus by quantum-dot-based single-particle tracking *ACS Nano* **6** 141–50

Long F, Zeng S and Huang Z L 2012 Localization-based super-resolution microscopy with an sCMOS camera part II: experimental methodology for comparing sCMOS with EMCCD cameras *Opt. Express* **20** 17741–59

Loy G and Zelinsky A 2003 Fast radial symmetry for detecting points of interest IEEE *Trans. Pat. Rec. & Mach. Int.* **25** 959–73

Lukyanov K A, Chudakov D M, Lukyanov S and Verkhusha V V 2005 Photoactivatable fluorescent proteins *Nat. Rev. Mol. Cell Biol.* **6** 885–91

Ma H, Kawai H, Toda E, Zeng S and Huang Z L 2013 Localization-based super-resolution microscopy with an sCMOS camera part III: camera embedded data processing significantly reduces the challenges of massive data handling *Opt. Lett.* **38** 1769–71

Mandal A and Tamai N 2011 Suppressed blinking behavior of thioglycolic acid capped CdTe quantum dot by amine functionalization *Appl. Phys. Lett.* **99** 263111

Manley S, Gillette J M, Patterson G H, Shroff H, Hess H F, Betzig E and Lippincott-Schwartz J 2008 High-density mapping of single-molecule trajectories with photoactivated localization microscopy *Nat. Methods* **5** 155–7

Matsuda T, Miyawaki A and Nagai T 2008 Direct measurement of protein dynamics inside cells using a rationally designed photoconvertible protein *Nat. Methods* **5** 339–45

Maxfield F R and McGraw T E 2004 Endocytic recycling *Nat. Rev. Mol. Cell Biol.* **5** 121–32

Medina M A and Schwille P 2002 Fluorescence correlation spectroscopy for the detection and study of single molecules in *Biology Bioessays* **24** 758–64

Miyawaki A 2011 Proteins on the move: insights gained from fluorescent protein technologies *Nat. Rev. Mol. Cell Biol.* **12** 656–68

Miyawaki A 2013 Fluorescence imaging in the last two decades *Microscopy (Oxf.)* **62** 63–8

Mizuno H, Higashida C, Yuan Y, Ishizaki T, Narumiya S and Watanabe N 2011 Rotational movement of the formin mDia1 along the double helical strand of an actin filament *Science* **331** 80–3

Moerner W E 2007 New directions in single-molecule imaging and analysis *Proc. Natl Acad. Sci. USA* **104** 12596–602

Murase K, Fujiwara T, Umemura Y, Suzuki K, Iino R, Yamashita H, Saito M, Murakoshi H, Ritchie K and Kusumi A 2004 Ultrafine membrane compartments for molecular diffusion as revealed by single molecule techniques *Biophys. J.* **86** 4075–93

Nan X, Sims P A, Chen P and Xie X S 2005 Observation of individual microtubule motor steps in living cells with endocytosed quantum dots *J. Phys. Chem. B.* **109** 24220–4

Nann T 2005 Phase-transfer of CdSe@ZnS quantum dots using amphiphilic hyperbranched polyethylenimine *Chem. Commun.* **7** 1735–6

Nifosí R, Amat P and Tozzini V 2007 Variation of spectral, structural, and vibrational properties within the intrinsically fluorescent proteins family: a density functional study *J. Comput. Chem.* **28** 2366–77

Nirmal M, Dabbousi B O and Brus L E 1996 Fluorescence intermittency in single cadmium selenide nanocrystals *Nature* **383** 802–4

Nishikawa S, Arimoto I, Ikezaki K, Sugawa M, Ueno H and Komori T *et al* 2010 Switch between large hand-over-hand and small inchworm-like steps in myosin VI *Cell* **42** 879–88

Ober R J, Martinez C, Lai X, Zhou J and Ward E S 2004a Exocytosis of IgG as mediated by the receptor, FcRn: an analysis at the single molecule level *Proc. Natl Acad. Sci. USA* **101** 11076–81

Ober R J, Ram S and Ward E S 2004b Localization accuracy in single molecule microscopy *Biophys. J.* **86** 1185–200

Oheim M 2004 A deeper look into single-secretory vesicle dynamics *Biophys. J.* **87** 1403–5

Ohmachi M, Komori Y, Iwane A H, Fujii F, Jin T and Yanagida T 2012 Fluorescence microscopy for simultaneous observation of 3D orientation and movement and its application to quantum rod-tagged myosin V *Proc. Natl Acad. Sci. USA* **109** 5294–8

Palaniappan K, Hackney S A and Liu J 2004 Supramolecular control of complexation-induced fluorescence change of water-soluble, beta-cyclodextrin-modified CdS quantum dots *Chem. Commun.* **23** 2704–5

Palaniappan K, Xue C, Arumugam G, Hackney S A and Liu J 2006 Water-soluble, cyclodextrin-modified CdSe-CdS core-shell structured quantum dots *Chem. Mater.* **18** 1275–80

Park H, Toprak E and Selvin P R 2007 Single-molecule fluorescence to study molecular motors *Q. Rev. Biophys.* **40** 87–111

Parthasarathy R 2012 Rapid, accurate particle tracking by calculation of radial symmetry centers *Nat. Methods* **9** 724–6

Patterson G H and Lippincott-Schwartz J 2002 A photoactivatable GFP for selective photo-labeling of proteins and cells *Science* **297** 1873–7

Patterson G, Davidson M, Manley S and Lippincott-Schwartz J 2010 Superresolution imaging using single-molecule localization *Annu. Rev. Phys. Chem.* **61** 345–67

Patwardhan A 2003 Subpixel position measurement using 1D, 2D and 3D centroid algorithms with emphasis on applications in confocal microscopy *J. Microsc.* **186** 246–57

Pavani S R, Thompson M A, Biteen J S, Lord S J, Liu N and Twieg R J *et al* 2009 Three-dimensional, single-molecule fluorescence imaging beyond the diffraction limit by using a double-helix point spread function *Proc. Natl Acad. Sci. USA* **106** 2995–9

Pelkmans L and Zerial M 2005 Kinase-regulated quantal assemblies and kiss-and-run recycling of caveolae *Nature* **436** 128–33

Peng X *et al* 2000 Shape control of CdSe nanocrystals *Nature* **404** 59–61

Peter S, Harter K and Schleifenbaum F 2014 Fluorescence microscopy *Methods Mol. Biol.* **1062** 429–52

Petryayeva E, Algar W R and Medintz I L 2013 Quantum dots in bioanalysis: a review of applications across various platforms for fluorescence spectroscopy and imaging *Appl. Spectrosc.* **67** 215–52

Pierobon P and Cappello G 2012 Quantum dots to tail single bio-molecules inside living cells *Adv. Drug Deliv. Rev.* **64** 167–78

Pilla V, Munin E, Dantas N O, Silva A C A and Andrade A A 2012 Photothermal spectroscopic characterization in CdSe/ZnS and CdSe/CdS quantum dots: a review and new applications *Quantum Dots—A Variety of New Applications* ed A Al-Ahmadi (Rijeka, Croatia: InTech) pp 3–22

Pinaud F, King D, Moore H P and Weiss S 2004 Bioactivation and cell targeting of semiconductor CdSe/ZnS nanocrystals with phytochelatin-related peptides *J. Am. Chem. Soc.* **126** 6115–23

Plachta N, Bollenbach T, Pease S, Fraser S E and Pantazis P 2011 Oct4 kinetics predict cell lineage patterning in the early mammalian embryo *Nat. Cell Biol.* **13** 117–23

Planchon T A, Gao L, Milkie D E, Davidson M W, Galbraith J A, Galbraith C G and Betzig E 2011 Rapid three-dimensional isotropic imaging of living cells using Bessel beam plane illumination *Nat. Methods* **8** 417–23

Prabhat P, Gan Z, Chao J, Ram S, Vaccaro C, Gibbons S, Ober R J and Ward E S 2007 Elucidation of intracellular pathways leading to exocytosis of the Fc receptor, FcRn, using multifocal plane microscopy *Proc. Natl Acad. Sci. USA* **104** 5889–94

Prabhat P, Ram S, Ward E S and Ober R J 2004 Simultaneous imaging of different focal planes in fluorescence microscopy for the study of cellular dynamics in three dimensions *IEEE Trans. Nanobioscience* **3** 237–42

Quirin S., Pavani S. R. and Piestun R 2012 Optimal 3D single-molecule localization for superresolution microscopy with aberrations and engineered point spread functions *Proc. Natl Acad. Sci. U S A* **109** 675–9

Ram S, Kim D, Ober R J and Ward E S 2012 3D single molecule tracking with multifocal plane microscopy reveals rapid intercellular transferrin transport at epithelial cell barriers *Biophys. J.* **103** 1594–603

Ram S, Prabhat P, Chao J, Ward E S and Ober R J 2008 High accuracy 3D quantum dot tracking with multifocal plane microscopy for the study of fast intracellular dynamics in live cells *Biophys. J.* **95** 6025–43

Reits E A and Neefjes J J 2001 From fixed to FRAP: measuring protein mobility and activity in living cells *Nat. Cell Biol.* **3** E145–7

Ritchie K and Kusumi A 2003 Single-particle tracking image microscopy *Meth. Enzymol.* **360** 618–34

Rossetti R, Nakahara S and Brus L E 1980 Quantum size effects in the redox potentials, resonance Raman spectra, and electronic spectra of CdS crystallites in aqueous solution *J. Chem. Phys.* **79** 1086–8

Ruedas-Rama M J, Walters J D, Orte A and Hall E A 2012 Fluorescent nanoparticles for intracellular sensing: a review *Anal. Chim. Acta* **751** 1–23

Rust M J, Bate M and Zhuang X 2006a Sub-diffraction-limit imaging by stochastic optical reconstruction microscopy (STORM) *Nat. Methods* **3** 793–6

Rust M, Bates M and Zhuang X 2006b Sub-diffraction-limit imaging by stochastic optical reconstruction microscopy (STORM) *Nat. Methods* **3** 793–5

Ruthardt N, Lamb D C and Bräuchle C 2011 Single-particle tracking as a quantitative microscopy-based approach to unravel cell entry mechanisms of viruses and pharmaceutical nanoparticles *Mol. Ther.* **19** 1199–211

Rutter G A and Hill E V 2006 Insulin vesicle release: walk, kiss, pause … then run *Physiology (Bethesda)* **21** 189–96

Saadeh Y, Leung T, Vyas A, Chaturvedi L S, Perumal O and Vyas D 2014 Applications of nanomedicine in breast cancer detection, imaging, and therapy *J. Nanosci. Nanotechnol.* **14** 913–23

Sako Y, Minoghchi S and Yanagida T 2000 Single-molecule imaging of EGFR signaling on the surface of living cells *Nat. Cell Biol.* **2** 168–72

Sambongi Y, Iko Y, Tanabe M, Omote H, Iwamoto-Kihara A and Ueda I *et al* 1999 Mechanical rotation of the c subunit oligomer in ATP synthase (F0F1): direct observation *Science* **286** 1722–4

Sase I, Miyata H, Ishiwata S and Kinosita K Jr 1997 Axial rotation of sliding actin filaments revealed by single-fluorophore imaging *Proc. Natl Acad. Sci. USA* **94** 5646–50

Saxton M J 2008 Single-particle tracking: connecting the dots *Nat. Methods* **5** 671–2

Saxton M J 2009 Single particle tracking *Fundamental Concepts of Biophysics* ed T Jue (New York: Humana) pp 147–69

Saxton M J and Jacobson K 1997 Single-particle tracking: applications to membrane dynamics *Annu. Rev. Biophys. Biomol. Struct.* **26** 373–99

Schermelleh L, Heintzmann R and Leonhardt H 2010 A guide to super-resolution fluorescence microscopy *J. Cell Biol.* **190** 165–75

Schlegel G, Bohnenberger J, Potapova I and Mews A 2002 Fluorescence decay time of single semiconductor nanocrystals *Phys. Rev. Lett.* **88** 137401

Schütz G J, Axman M and Schindler H 2001 Imaging single molecules in three dimensions *Single Mol.* **2** 69–74

Shimomura O and Johnson F H 1692 Extraction, purification and properties of aequorin, a bioluminescent protein from the luminous hydromedusan. Aequorea *J. Cell. Comp. Physiol.* **59** 223–39

Shroff H, Galbraith C G, Galbraith J A and Betzig E 2008 Live-cell photoactivated localization microscopy of nanoscale adhesion dynamics *Nat. Methods* **5** 417–23

Small A and Stahlheber S 2014 Fluorophore localization algorithms for super-resolution microscopy *Nat. Methods* **11** 267–79

Speidel M, Jonas A and Florin E L 2003 Three-dimensional tracking of fluorescent nanoparticles with subnanometer precision by use of off-focus imaging *Opt. Lett.* **28** 69–71

Sprague B L and McNally J G 2005 FRAP analysis of binding: proper and fitting *Trends Cell Biol.* **15** 84–91

Subach F V, Zhang L, Gadella T W, Gurskaya N G, Lukyanov K A and Verkhusha V V 2010 Red fluorescent protein with reversibly photoswitchable absorbance for photochromic FRET *Chem. Biol.* **17** 745–55

Terai T and Nagano T 2013 Small-molecule fluorophores and fluorescent probes for bioimaging *Pflugers Arch.* **465** 347–59

Theis-Febvre N, Martel V, Laudet B, Souchier C, Grunwald D, Cochet C and Filhol O 2005 Highlighting protein kinase CK2 movement in living cells *Mol. Cell Biol.* **274** 15–22

Thomann D, Rines D R, Sorger P K and Danuser G 2002 Automatic fluorescent tag detection in 3D with super-resolution: application to the analysis of chromosome movement *J. Microsc.* **208** 49–64

Thompson R E, Larson D R and Webb W W 2002 Precise nanometer localization analysis for individual fluorescent probes *Biophys. J.* **82** 2775–83

Thompson M A, Lew M D, Badieirostami M and Moerner 2010 W E Localizing and tracking single nanoscale emitters in three dimensions with high spatio-temporal resolution using a double-helix point spread function *Nano Lett.* **10** 211–8

Tiwari D K, Tanaka S, Inouye Y, Yoshizawa K, Watanabe T M and Jin T 2009 Synthesis and characterization of anti-HER2 antibody conjugated CdSe/CdZnS quantum dots for fluorescence imaging of breast cancer cells *Sensors* **9** 9332–64

Tomczak N, Liu R and Vancso J G 2013 Polymer-coated quantum dots *Nanoscale* **5** 12018–32

Toprak E and Selvin P R 2007 New fluorescent tools for watching nanometer-scale conformational changes of single molecules *Annu. Rev. Biophys. Biomol. Struct.* **36** 349–69

Toprak E, Balci H, Blehm B H and Selvin P R 2007a Three-dimensional particle tracking via bifocal imaging *Nano Lett.* **7** 2043–5

Toprak E, Balci H, Blehm B H and Selvin P R 2007b Three-dimensional particle tracking via bifocal imaging *Nano Lett.* **7** 2043–5

Toprak E, Enderlein J, Syed S, McKinney S A, Petschek R G and Ha T *et al* 2006 Defocused orientation and position imaging (DOPI) of myosin V *Proc. Natl Acad. Sci. USA* **103** 6495–9

Tsien R Y 1998 The green fluorescent protein *Annu. Rev. Biochem.* **67** 509–44

van den Broek B, Ashcroft B, Oosterkamp T H and van Noort J 2013 Parallel nanometric 3D tracking of intracellular gold nanorods using multifocal two-photon microscopy *Nano Lett.* **13** 980–6

van Sark W G J H M, Frederix P L T, Bol A A, Gerritsen H C and Meijerink A 2002 Blueing, Bleaching, and Blinking of Single CdSe/ZnS Quantum Dots *Chem. Phys. Chem.* **3** 871–9

van Sark W G J H M, Frederix P L T M, den Heuvel D J V, Gerritsen H C J, Bol A A and van Lingen J N J *et al* 2001 Letter photooxidation and photobleaching of single CdSe/ZnS Quantum dots probed by room-temperature time-resolved spectroscopy *Phys. Chem. B.* **105** 8281–4

Wang E C and Wang A Z 2014 Nanoparticles and their applications in cell and molecular biology *Integr. Biol.* **6** 9–26

Wang Z G, Liu S L, Tian Z Q, Zhang Z L, Tang H W and Pang D W 2012 Myosin-Driven Intercellular Transportation of Wheat Germ Agglutinin Mediated by Membrane Nanotubes between Human Lung Cancer Cells *ACS Nano* **6** 10033–41

Watanabe T M and Higuchi H 2007 Stepwise movements in vesicle transport of HER2 by motor proteins in living cells *Biophys. J.* **92** 4109–20

Watanabe T M, Fujii F, Jin T, Umemoto E, Miyasaka M and Fujita H *et al* 2013 Four-dimensional spatial nanometry of single particles in living cells using polarized quantum rods *Biophys. J.* **105** 555–64

Watanabe T M, Fukui S, Jin T, Fujii F and Yanagida T 2010 Real-time nanoscopy by using blinking enhanced quantum dots *Biophys. J.* **99** L50–2

Watanabe T M, Sato T, Gonda K and Higuchi H 2007 Three-dimensional nanometry of vesicle transport in living cells using dual-focus imaging optics *Biochem. Biophys. Res. Commun.* **359** 1–7

Wells N P *et al* 2010 Time-resolved three-dimensional molecular tracking in live cells *Nano Lett.* **10** 4732–7

Wells N P, Lessard G A and Werner J H 2008 Confocal, three-dimensional tracking of individual quantum dots in high-background environments *Anal. Chem.* **80** 9830–4

Werver G 1953 Rotational Brownian motion and polarization of the fluorescence of solutions *Adv. Protein Chem.* **8** 415–59

Wombacher R and Cornish V W 2011 Chemical tags: applications in live cell fluorescence imaging *J. Biophotonics* **4** 391–402

Wysocki L M and Lavis L D 2011 Advances in the chemistry of small molecule fluorescent probes *Curr. Opin. Chem. Biol.* **15** 752–9

Yajima J, Mizutani K and Nishizaka T 2008 A torque component present in mitotic kinesin Eg5 revealed by three-dimensional tracking *Nat. Struct. Mol. Biol.* **15** 1119–21

Yildiz A and Selvin P R 2005 Fluorescence imaging with one nanometer accuracy: application to molecular motors *Acc. Chem. Res.* **38** 574–82

Yildiz A, Forkey J N, McKinney S A, Ha T, Goldman Y E and Selvin P R 2003 Myosin V walks hand-over-hand: single fluorophore imaging with 1.5 nm localization *Science* **300** 2061–5

Zenisek D, Steyer J A and Almers W 2000 Transport, capture and exocytosis of single synaptic vesicles at active zones *Nature* **406** 849–54

Zhang A, Dong C, Liu H and Ren J 2013 Blinking behavior of CdSe/CdS quantum dots controlled by alkylthiols as surface trap modifiers *J. Phys. Chem. C.* **117** 24592–600

Zhang Y and Clapp A 2011 Overview of stabilizing ligands for biocompatible quantum dot nanocrystals *Sensors* **11** 11036–55

IOP Publishing

Physics of Cancer

Claudia Tanja Mierke

Chapter 6

Intermediate filaments and nuclear deformability during matrix invasion

Summary

In this chapter the role of intermediate filaments and nuclear deformability is discussed, including how it may impact on the cell's matrix invasion through connective tissue. There is special emphasis on the role of the intermediate filaments that may be important for nuclear stiffness and hence cell motility. As intermediate filaments are integrated into a cytoskeletal network, the interaction between them and actin filaments in supporting cell motility is highlighted. Finally, it is suggested that cancer cells may overcome the steric restrictions of dense 3D extracellular matrix fiber networks by undergoing mitosis.

6.1 Intermediate filaments, cellular mechanical properties and cellular motility

Keratins (previously named cytokeratins) are intermediate filament-assembling proteins that provide cells' mechanical properties and fulfill a variety of functions in epithelial cells. Keratin genes encode most of the intermediate filaments in the human genome and build the two largest sequence homology groups of this large multigene family, the type I and II keratin groups. They are highly differentiation-specific in their expression patterns, which suggests different cellular functions. In particular, mutations in most of them are often associated with specific tissue-fragility disorders and hence antibodies to keratins are important markers of tissue differentiation as well as helpful tools in diagnostic pathology. The first comprehensive keratin nomenclature used 2D isoelectric focusing and SDS-PAGE to map the keratin profiles of normal human epithelia, tumors and cultured cells (Moll *et al* 1982). In more detail, they grouped the basic-to-neutral type II keratins as K1–K8 and the acidic type I keratins as K9–K19 (Moll *et al* 1982). The present naming of keratins is not systematic and hence a reorganized and durable scheme is overdue.

Genome analyses have shown that humans possess 54 functional keratin genes, 28 type I and 26 type II keratins, which form two clusters of 27 genes each on chromosomes 17q21.2 and 12q13.13, with the gene for the type I keratin K18 being located in the type II keratin gene domain (Hesse *et al* 2001, 2004, Rogers *et al* 2004, 2005). Knowledge of the extent of this mammalian gene family led to a revised nomenclature (Hesse *et al* 2004) based on an extended Moll system, K1–K8 and K9–K24 (Moll *et al* 1982, 1990 Chandler *et al* 1991, Zhang *et al* 2001, Sprecher *et al* 2002). Then, for both type I and II keratins, these four categories were arranged in the following numerical order:

1. human epithelial keratins;
2. human hair keratins;
3. non-human epithelial/hair keratins; and
4. human keratin pseudogenes.

For historical reasons and because of the extensive number of existing publications, the Moll designation for the epithelial keratins K1–K8 and K9–K24 has been retained (Moll *et al* 1982, 1990, Chandler *et al* 1991, Zhang *et al* 2001, Sprecher *et al* 2002).

Many keratins are found in epithelia, which are exposed to multiple forms of stress. These keratin intermediate filaments are abundant in epithelia and build cytoskeletal networks providing cell-type-specific functions, such as cellular adhesion, migration and metabolism. There is a perpetual keratin filament turnover cycle that supports these cellular functions. In particular, this multistep process regulates involvement of the cytoskeleton in motion, facilitating rapid protein biosynthesis, whereas the network is still intact during the remodeling process. However, the molecular mechanisms underlying the regulation of the keratin cycle are still elusive compared to actin and microtubule networks. What regulatory role does the keratin cycle play in epithelial tissue function?

The protection of tissue and organs against microenvironmental alterations is a major functional role of epithelial tissues. In more detail, the renewal and repair of epithelia involves continuous cycles of cellular proliferation, migration and differentiation. Thus, the epithelial cytoskeleton is constantly remodeled in order to optimize epithelial functions. Keratin intermediate filaments are the most diverse and abundant cytoskeletal components of epithelial cells. More than 50 isotypes of keratin intermediate filaments have been expressed in epithelia (Schweizer *et al* 2006, Moll *et al* 2008, Bragulla and Homberger 2009). The regulatory functions of keratins in organelle trafficking, motility, translation, signaling, immune response and cell survival have been revealed, demonstrating that keratin intermediate filaments have the plasticity and network architecture required to regulate the epithelial function precisely (Toivola *et al* 2005, Kim *et al* 2006, 2007, Long *et al* 2006, Kim and Coulombe 2007, Magin *et al* 2007, Vijayaraj *et al* 2009, Depianto *et al* 2010, Ku *et al* 2010). How can keratins provide rigidity and strength, while still remaining dynamic and flexible in their structure? The molecular mechanisms that regulate keratin assembly, disassembly and network architecture are not yet well understood. What are the properties of a biosynthesis-independent multistep assembly and disassembly cycle of keratins that provides rapid network remodeling without the disruption of the whole network?

Time-lapse imaging of the cultured monolayers of living cells expressing fluorescent keratins demonstrates that the keratin network is highly dynamic (Windoffer and Leube 1999, Yoon *et al* 2001, Windoffer *et al* 2004). These results suggest a perpetual cycle of keratin intermediate filament assembly and disassembly under standard 2D culture conditions (Kölsch *et al* 2010, Leube *et al* 2011). In particular, the cycle starts with the nucleation of keratin particles at the cell periphery, which is in most cases closely associated with lamellipodial focal adhesions. The next step is the elongation of newly assembled keratin particles during actin-dependent translocation to the peripheral 'old' keratin network. After the integration of precursor keratin particles into the existing network, keratin intermediate filaments are transported to the nucleus and they form bundles. In some cases, these novel bundles disassemble into soluble oligomers, which rapidly diffuse through the cytoplasm and are then available for another cycle of nucleation in the cell periphery near the cell membrane. In addition, other bundles mature into a stable network; they surround the nucleus and are anchored to desmosomes (cell–cell adhesions) or hemidesmosomes (cell–matrix adhesions). Taken together, cycling of the keratin bundles keeps the epithelial cytoskeleton in motion without losing structural integrity.

The assembly of the keratin intermediate filament network
In cultured epithelial cells, the assembly of keratin intermediate filaments (called nucleation) starts in the cell periphery in close neighborhood to focal adhesions (Windoffe *et al* 2006). The focal adhesions are known to connect actin bundles to the cell membrane and subsequently to the extracellular matrix microenvironment (Petit and Thiery 2000, Geiger *et al* 2001, Carragher and Frame 2004), thus inducing alterations in the microtubule network architecture (Krylyshkina *et al* 2003, Small and Kaverina 2003). Through the keratin intermediate filament nucleation at the cell periphery close to focal adhesions, the coordinated restructuring of the entire cytoskeleton is achieved. This spatial coordination of the restructuring is of special relevance for migrating cells. In line with this, a pronounced increase of keratin particle assembly has been reported in the lamellipodia of migrating cells (Wöll *et al* 2005, Kölsch *et al* 2010, Rolli *et al* 2010). An appropriate nucleus for the filament assembly has been identified *in vitro*: it should have an approximately 60 nm long unit length filament (ULF) containing 32 monomers (Herrmann *et al* 1999, Herrmann *et al* 2002). However, in living cells it is still not understood whether the particles that can assemble keratin intermediate filaments (named keratin intermediate filament precursors) (Windoffer *et al* 2004) are the same as ULFs, because the resolution of a standard light microscope is lower than a single ULF and hence it cannot distinguish between single ULFs.

As there is no keratin intermediate filament precursor polarity, both ends are equally suited to support the elongation by oligomer addition. *In vitro* observations of vimentin intermediate filaments indicate that single and multiple ULFs are added at either end without any differences (Kirmse *et al* 2007). In accordance with these observations, live cell imaging of keratins detects continuous particle elongation and fusion of larger particles (Windoffer *et al* 2004, 2006, Wöll *et al* 2005). As long as keratins retain their free ends, they still elongate. When particles approach the keratin intermediate filament network, they integrate via their ends and thus add another novel branch to the filament network (Windoffer *et al* 2004, 2006, Wöll *et al* 2005). However, in mutant keratins, which cause blistering skin diseases in humans, these elongated filaments cannot be assembled; instead, only short-lived spheroidal granules are built close to focal adhesions (Werner *et al* 2004, Windoffer *et al* 2006).

Keratin networks are heterogeneous, because they are assembled by two–ten different isotypes. However, the regulatory modes and mechanisms governing their organization and distribution are still elusive, and they seem to depend partly on cellular polarity. It has been suggested that the intracellular distribution of keratin isotypes is related to their primary sequence, but this has not been confirmed by experimental data. In particular, in the intestinal epithelia of mice the isotypes K20 and K8 are codistributed throughout the cell, whereas in the umbrella cells of the bladder K20 is located in the apical domain, indicating cell-type-specific regulatory mechanisms (Magin *et al* 2006). Another phenomenological feature of the network organization is the coexistence of individual filaments and bundles (interfilament assemblies). The bundling of intermediate filaments increases the mechanical stability and reduces the turnover (Flitney *et al* 2009, Lee and Coulombe 2009, Kim *et al* 2010a), two prerequisites for elastic and durable cytoskeletal scaffolding. However, this is not always observed, as in the absence of plectin, which acts as a cytoskeletal cross-linker, the bundling is enhanced, but seems to be dysfunctional due to reduced cellular tension (Osmanagic-Myers *et al* 2006). In cultured cells, the bundling is reflected by an increase to the keratin intermediate filament diameter toward the nucleus, which is evoked by lateral association of keratin intermediate filaments (Windoffer *et al* 2004, Lee and Coulombe 2009, Kölsch *et al* 2010). Several factors influence the bundling: (i) intermediate-filament-associated proteins (Krieg *et al* 1997, Xu *et al* 2000, Makino *et al* 2001, Listwan and Rothnagel 2004, Long *et al* 2006, Osmanagic-Myers *et al* 2006, Boczonadi *et al* 2007, Ishikawa *et al* 2010); (ii) intrinsic and isotype-specific properties of keratin intermediate filaments (Eichner *et al* 1986, Blessing *et al* 1993, Hofmann *et al* 2000), which bundle spontaneously *in vitro* (named self-organization) (Lee and Coulombe 2009, Kim *et al* 2010b); and (iii) phosphorylation, which is thought to coincide with bundling after mechanical and chemical stress application (Strnad *et al* 2001, Flitney *et al* 2009).

The disassembly of keratin intermediate filaments
As keratin intermediate filament assembly is highly favored over disassembly, there may be mechanisms to remove assembled filaments and especially dense filament bundles, which would otherwise interfere with cellular functions. Suitable mechanisms to regulate this balance are the degradation of keratin intermediate filament poly-peptides and the disassembly of keratin intermediate filaments into reusable subunits. One fact that supports the first mechanism is the ubiquitination of keratins and their subsequent proteasomal degradation (Ku and Omary 2000, Löffek *et al* 2010, Rogel *et al* 2010). This mechanism is elevated in stress and pathology as a consequence of increased network restructuring (Zatloukal *et al* 2007, Jaitovich *et al* 2008, Na *et al* 2010). This process has been used to reduce aggregates typical of keratinopathies by the

addition of chemical chaperones and chaperone-associated ubiquitin ligases into tissue (Lee *et al* 2008, Chamcheu *et al* 2010, Löffek *et al* 2010). However, the second mechanism seems to be the most prominent mode in rapidly proliferating cultured cells, as time-lapse fluorescence recordings have demonstrated that keratin intermediate filament assembly occurs independently of and in the absence of protein biosynthesis (Windoffer *et al* 2004, Kölsch *et al* 2010). The detection of single inward-moving keratin intermediate filament bundles further highlights that they dissolve over time without the appearance of distinct fragments, which leads to the suggestion that the released subunits are non-filamentous. The release of soluble subunits seems to be similar to the lateral subunit exchange, which has been observed for intermediate filaments at the equilibrium state (Eriksson *et al* 2009). Although how the disassembly is regulated remains elusive, the involvement of phosphorylation seems to be most likely, because inhibition of p38 MAPK or PKCζ activities leads to enhanced network stability and, unexpectedly, to increased kinase activity, which results in increased keratin intermediate filament network turnover (Wöll *et al* 2007, Sivaramakrishnan *et al* 2009). Moreover, the non-filamentous keratin pool is elevated during mitosis and in different stress conditions, such as elevated network remodeling, which occur simultaneously with increased keratin phosphorylation (Chou *et al* 1993, Liao and Omary 1996, Omary *et al* 1998, Strnad *et al* 2002, Ridge *et al* 2005). Interestingly, sumoylation has also been implicated in keratin network dynamics (Snider *et al* 2011).

What are the properties of the soluble keratin fraction?
Heterotypic, non-filamentous keratins build up the biochemically defined soluble pool of tetramers and/or small oligomeric keratin assemblies (Soellner *et al* 1985, Chou *et al* 1993, Bachant and Klymkowsky 1996). In order to inhibit immediate assembly after biosynthesis or directly after filament disassembly, the soluble, non-assembled state needs to be stabilized. This stabilization may be performed by protein modification, association with chaperones such as Hsp70 and Hsc70, interaction with intermediate filament-associated proteins (IFAPs) or binding to 14-3-3 proteins (Liao and Omary 1996, Wiche 1998, Planko *et al* 2007, Mashukova *et al* 2009). 14-3-3 proteins preferrentially bind to phosphorylated target proteins and are then able to alter their conformation (Kjarland *et al* 2006, Díaz-Moreno *et al* 2009). Thus, Ser phosphorylation of keratin subunits along the head domain occurs immediately after biosynthesis or disassembly in order to inhibit the assembly at non-permissive sites in the cytoplasm. In particular, hyperphosphorylation by Cdk1, Plk1, Rho-kinase and Aurora B is important for the local breakdown of several intermediate filament classes during mitosis and is hence essential for the efficient segregation of intermediate filament networks into the two daughter cells (Izawa and Inagaki 2006). Due to the small size of the disassembled subunits and their solubility in the aqueous cytoplasm, it is suggested that they can be distributed rapidly within the cytoplasmic space by diffusion. Indeed, a fast diffusible pool has been detected by fluorescence recovery after photobleaching (Kölsch *et al* 2010). The nucleation continues even in the presence of substances that disrupt actin filaments and microtubules. This suggests that diffusive delivery of keratins to peripheral nucleation sites is sufficient (Wöll *et al* 2005, Kölsch *et al* 2009).

How is the keratin cycle regulated regarding space and time?
A continuous transport of filamentous keratins to the nucleus is necessary for keratin cycling. The keratin isotype, cell-type-specific properties and other factors such as filament-associated proteins seem to regulate this transport process. In particular, growing keratin particles move along actin stress fibers with a velocity of approximately 300 nm min^{-1} (Wöll *et al* 2005, Kölsch *et al* 2009). This movement is supposed to be directly coupled to lamellar actin treadmilling through plectin-facilitated linkage (Litjens *et al* 2003, Rezniczek *et al* 2004). In addition, keratin particles are also transported along microtubules (Yoon *et al* 2001, Liovic *et al* 2003, Wöll *et al* 2005, Windoffer *et al* 2006).

The molecular mechanism of the subsequent inward-directed movement of the keratin network (Windoffer and Leube 1999, Yoon *et al* 2001, Kölsch *et al* 2010) is not yet understood. The intrinsic elasticity of the filaments (Kreplak *et al* 2008), in combination with their nuclear anchorage, may be responsible through the interaction between plectin and the cytoplasmic nuclear membrane protein nesprin-3 (Wilhelmsen *et al* 2005). As an alternative, actin filaments and/or microtubules are also supposed to be involved in this movement. In more detail, it has been observed that energy depletion inhibits keratin intermediate filament motility, which suggests an energy-requiring and motor-protein-driven active movement process (Hollenbeck *et al* 1989, Strnad *et al* 2001, Yoon *et al* 2001).

Focal adhesion-dependent nucleation is *in vitro* another major cycle determinant. It is known that the cytoskeletal cross-linker plectin (particularly isoform 1f) is in close proximity to focal adhesions and is able to bind to keratins (Nikolic *et al* 1996, Steinböck *et al* 2000, Litjens *et al* 2003; Rezniczek *et al* 2003). Other candidate proteins that may regulate keratin dynamics are integrins, vinculin, metavinculin, talin and zyxin. All these focal adhesion proteins can bind to intermediate filaments (Kreis *et al* 2005, Ivaska *et al* 2007, Kostan *et al* 2009, Sun *et al* 2008a, 2008b, 2010). Besides these structural components, focal adhesion-dependent signal transduction seems to be involved in keratin nucleation. Among additional factors, PKCs and MAPK have been discussed (Omary *et al* 1992, Ridge *et al* 2005, Osmanagic-Myers *et al* 2006, Akita *et al* 2007, Wöll *et al* 2007, Bordeleau *et al* 2008, 2010, Sivaramakrishnan *et al* 2009). However, the spatiotemporal interaction of these components with the keratin assembly at focal adhesions is not well understood.

Cell shape alterations have been seen frequently in motile and dividing cells and are additionally suggested to enhance cycling, whereas stably anchored and non-moving cells in mature tissues are not supposed to cycle extensively. In order to also account for the gradual transition between the moving and non-moving cells, regulators need to be there to provide graded and locally restricted responses. In line with this, cycling of wild-type and mutant keratins is slowed down by p38 MAPK inhibitory drugs, which are supposed to affect keratin phosphorylation (Wöll *et al* 2007). Indeed, this seems to be a possible mechanism for the regulation of keratin cycling through stress-induced signaling. In addition, the compartmentalization of kinases and phosphatases, such as focal adhesion kinase and protein kinase C, can further help to regulate the localized alteration of the network configuration precisely. Members of the plakin family, which link cytoskeletal networks and members of the plakophilin family, which are localized to adhesion sites, are both good candidates to regulate these organizational functions, as they possess keratin binding sites and can impact on kinase/phosphatase activity (Osmanagic-Myers *et al* 2006, Bass-Zubek *et al* 2009, Kostan *et al* 2009, Bordeleau *et al* 2010). Moreover, the shear stress increased PKCζ-dependent phosphorylation of K18-S33 leads to an increased exchange rate of the keratin intermediate filament network (Sivaramakrishnan *et al* 2009). In particular, the relevance of signaling-dependent keratin phosphorylation for dynamic network organization is most prominent in the case of a disease: autoantibodies from the skin blistering disease (*Pemphigus vulgaris*) have been demonstrated to facilitate p38 MAPK-dependent keratin retraction (Berkowitz *et al* 2005). Moreover, toxic liver injury induced by the antifungal drug griseofulvin mediates increased keratin phosphorylation, increases soluble keratins and leads to enhanced aggregate formation (Ku *et al* 1996, Stumptner *et al* 2001, Toivola *et al* 2004, Fortier *et al* 2010), which can be inhibited *in vitro* by p38 MAPK inhibitors (Nan *et al* 2006).

How can the keratin intermediate filaments exist from the turnover cycle?
The plasticity of cells needs to be balanced by the stabilizing properties of the keratin network promoting mechanical strength for resting cells. In cultured interphase cells, the desmosome- and hemidesmosome-anchored filaments as well as the perinuclear cage-like structures are stabilized under this condition and no longer dynamic. Moreover, desmosome-anchored filaments are more tensile compared to the rest of the keratin intermediate filament network and are able to withstand disruption by the tyrosine phosphatase inhibitor vanadate (Strnad *et al* 2002). In a stable tissue, cycling is less important, because mechanical functions are more important.

Anchorage-dependent mechanosensing seems to affect keratin intermediate filament stability by altered interaction with regulatory proteins such as desmoplakin, BPAG1, plectin, periplakin and epiplakin facilitating keratin intermediate filament dynamics (Bornslaeger *et al* 1996, Wan *et al* 2004, Osmanagic-Myers *et al* 2006, Boczonadi *et al* 2007, Spazierer *et al* 2008, Ishikawa *et al* 2010). For the perinuclear network, attachment to the nuclear envelope through the plectin–nesprin-3a connection (Wilhelmsen *et al* 2005) provides filament stabilization. In the case of junction-associated keratin intermediate filaments, special keratins such as K80 (Langbein *et al* 2010) and keratin-binding proteins such as desmoplakin, plakophilin, plectin and BPAG1 seem to be involved in the keratin intermediate filament stabilization (Guo *et al* 1995, Eger *et al* 1997, Holthöfer *et al* 2007, Kostan *et al* 2009, Green *et al* 2010). All these results demonstrate that anchorage protects keratin intermediate filaments against disassembly, whereas the filament turnover has not been analyzed in detail at the single bundle or filament level. In more detail, chemical and/or biophysical keratin modification seems to be the result of the coupling of keratin intermediate filaments to the mechanotransducive systems and hence may support keratine intermediate filament stabilization, similar to the phosphorylation of the tail domain of the intermediate filament glial fibrillary acidic protein (GFAP), which impairs the turnover (Takemura *et al* 2002). Finally, this property can be seen as maturation, which has been observed for the other cytoskeletal filaments, such as the association with proteins and the detyrosination of tubulin (Bulinski and Gundersen 1991, Arce *et al* 2008, Konishi and Setou 2009, Ikegami and Setou 2010).

What are the advantages of cytoskeletal cycling?
The cycling of keratins is more efficient than degradation and *de novo* biosynthesis. The cycling process has also been recognized for other cytoskeletal components, such as for the actin system, which is characterized by filament treadmilling and retrograde flow in migrating cells (Small and Resch 2005, Schaus *et al* 2007, Michalski and Carlsson 2010). Another example of the cycling process is the dynamic instability of microtubules, for example the switching between growing and shrinking (Mitchison and Kirschner 1984, Gardner *et al* 2008). Time-dependent, cyclic alterations of the amount of cellular components increase the diversification of functional states and enhance the probability for the cell to adapt to microenvironmental alterations through immediate responses, ensuring survival and function (Wolf *et al* 2005, Vogel and Sheetz 2009).

The cycling of keratins is a mechanism for probing and sensing the cell periphery for the occurrence of new cell adhesions. The presence/absence of keratins alters the stability of desmosomes and hemidesmosomes (Long *et al* 2006). In the case of focal adhesions, the cycle is accelerated by enhanced nucleation and thus promotes network growth toward the leading edge. It has been suggested that in migrating cells, this cycle participates in the complex assembly and disassembly mechanisms needed to move the cell body (Proux-Gillardeaux *et al* 2005). Indeed, upon filament attachment to hemidesmosomes and desmosomes, the cycle decreases and hence favors mechanical stability. This mechanism may additionally support *in vivo* distribution patterns, such as localization in the terminal web, a dense filamentous network below the apical surface of the polarized epithelial cells of the gut (Oriolo *et al* 2007). Collectively, keratin cycling may be regarded as a continuous sensing of the immediate extracellular surroundings until new physical adhesion contacts with other cells and/or the extracellular matrix can be established and stabilized through desmosomes and hemidesmosomes. Without *de novo* protein biosynthesis, the cell has a variety of ways to respond to microenvironmental alterations within a small time frame. Thus, it is dynamical remodeling of keratin intermediate filament networks that guarantees the adaptability and subsequently the function of moving cells.

Another function of keratin cycling is to maintain an intact network during epithelial differentiation, which promotes gradual polypeptide exchange without the disruption of filaments. Different admixtures of keratins are found in basal versus suprabasal epidermal keratinocytes and thus, basal type keratins can be detected in cells that possess no corresponding mRNAs (Lersch and Fuchs 1988, Reichelt *et al* 2001).

How does the keratin cycling process alter the epithelial functions?
Keratin intermediate filaments have been seen as a rather static component of the cytoskeleton, providing the mechanical strength of the epithelia. Moreover, this property is essential for resting cells, providing the whole epithelium with mechanical strength. Disruption of the keratin intermediate filament network integrity and,

consequently, of epithelial rigidity induces skin-blistering disease (Coulombe *et al* 2009). Indeed, keratin intermediate filaments fulfill an important role in dynamic processes such as wound healing and cancer metastasis (Paladini *et al* 1996, Mazzalupo *et al* 2003, Knösel *et al* 2006, Ptitsyn *et al* 2008, Karantza 2011). The keratin cycle can run at different length scales, ranging from diffusible filament precursors to macromolecular network components, involving the entire cytoplasm and hence it seems to be a key promoting factor.

Importantly, the keratin cycling regulates epithelial motility, migration and vesicle trafficking, which is known to be important for the epithelial stress response. The regulation of cycling is connected to keratin modification, such as phosphorylation, and is hence a target of signaling pathways. A hypothesis is that the predisposition of transgenic mice carrying phosphorylation-deficient K8 and K18 mutants leads to liver disease (Ku and Omary 2006) and, conversely, the cytoprotective effects of glycosylated keratins (Ku *et al* 2010), are associated with differences in keratin dynamics. The extent to which keratins impact on protein biosynthesis through 14-3-3 proteins or glucose transporters is determined through the keratin dynamics (Kim *et al* 2006, Vijayaraj *et al* 2009). The activity of 14-3-3 proteins depends on targeted protein phosphorylation and their nucleo-cytoplasmic distribution (Mackintosh 2004). The expression of K17 (and possibly additional type I keratins) recruits 14-3-3 to the cytoplasm, where it activates the mammalian target of the rapamycin (mTOR) pathway (Kim *et al* 2006). However, the presence of keratins provides the correct localization and function for the glucose transporter GLUT, whereas in the absence of keratins increased levels of glucose can be taken up and an AMPK-facilitated down-regulation of mTORC1 occurs (Vijayaraj *et al* 2009). Taken together, the degree of translational stimulation through elongation factor 4A (eIF4A)–plakophilin-1 seems to be guided through the interaction of keratins with plakophilin-1 (Wolf *et al* 2010). Do distinct keratin assembly forms, such as filamentous or non-filamentous keratin, possess different specific regulatory functions? In more detail, is it true that filamentous subdomains fulfill specific regulatory roles, while non-filamentous keratins function solely in a context-specific manner?

Another prediction is that the accumulation of several cycle intermediates leads to functional consequences for the behavior of the epithelial cell. In line with this, cells lacking keratins or producing mutant keratins, which are characterized by reduced keratin intermediate filaments and increased soluble keratins, are able to migrate even faster in 2D scratch assays (Morley *et al* 2003, Long *et al* 2006, Seltmann *et al* 2013).

Does the cycling process exist in other intermediate filament systems?
Other intermediate filament networks have been shown to utilize similar subunit exchange mechanisms to those found for keratins (Tsuruta and Jones 2003, Mignot *et al* 2007, Burgstaller *et al* 2010). There is evidence for intense cross-talk between vimentin intermediate filaments and focal adhesions (Seifert *et al* 1992, Gonzalez *et al* 2001, Tsuruta and Jones 2003, Spurny *et al* 2008, Bhattacharya *et al* 2009, Burgstaller *et al* 2010) and in addition it has been demonstrated that plectin 1f is involved in the recruitment of growing vimentin particles to focal adhesions

(Burgstaller *et al* 2010). Particle elongation occurs through end-to-end annealing for vimentin and neurofilament intermediate filament proteins (Colakoğlu and Brown 2009). Moreover, the cytoplasmic transport of filament precursor particles has been reported for many intermediate filament types (Brown 2003, Helfand *et al* 2004, Wöll *et al* 2005, Barry *et al* 2007, Kölsch *et al* 2009). However, differences have been observed: although actin filaments play an important role for keratin intermediate filament particles, microtubules seem to be more important for other intermediate filaments, such as vimentin, peripherin and neurofilaments (Roy *et al* 2000, Wang *et al* 2000, Yoon *et al* 2001, Helfand *et al* 2003). In addition, bundling has been found for neuronal type IV intermediate filaments. In more detail, highly phosphorylated intermediate filaments are tightly packed and located at the center of the axoplasm, whereas less phosphorylated and hence more dynamic intermediate filaments are found to be localized in the peripheral axoplasm (Sihag *et al* 2007, Kushkuley *et al* 2009).

The cycling of the intermediate filament cytoskeleton is also important for many cell types, including neurons for growth cone probing (Chan *et al* 2003), mesenchymal cells and astrocytes for motility (Eckes *et al* 1998, Lepekhin *et al* 2001, Nieminen *et al* 2006, Pan *et al* 2008) and endothelial cells for adaptation to fluid shear stress (Helmke *et al* 2001). In particular, intermediate filament cycling seems to be essential for the regeneration occurring upon neuronal axon injury or in diverse stress situations, such as in reactive gliosis (Pekny and Lane 2007).

The strongest evidence for the importance of intermediate filament cycling comes from observations in *C. elegans*, where the inhibition of sumoylation leads to a reduction in the intermediate filament network turnover associated with enhanced intermediate filament bundling and aggregation (Kaminsky *et al* 2009). However, it has also been demonstrated that intermediate filament phosphorylation is linked to network organization, and these are in turn both coupled to hemidesmosome-facilitated mechanotransduction (Zhang *et al* 2011).

Finally, several observations support the view that the plasticity of the epithelial keratin intermediate filament network not only relies on alterations in biosynthesis and degradation, but is solely provided by the cycles of assembly and disassembly. Although this concept is supported by microscopic observations in cultured living cells, the underlying molecular mechanisms still need to be confirmed *in vivo* and correlated with observations on *in vitro* keratin intermediate filament assembly (Herrmann *et al* 2002) as well as alternative concepts of intermediate filament network dynamics (Ngai *et al* 1990, Miller *et al* 1991, Chang *et al* 2006). Some questions remain unanswered. What is the driving force of the cycle? How do keratin-associated proteins such as plectin, 14-3-3, Akt-1, Hsp70 and others affect the cycle? How do microenvironmental factors such as mechanical force, cytokines and microbes modulate the keratin cycle and what effects does the cycle have on the cell? How can keratin isotypes affect the keratin intermediate filament cycling? What impact does keratin modification have on the cycling process? Which cellular processes are associated with the keratin cycling and how is this achieved? Is the cycling necessary for the stress protection facilitated by keratins? Is the cycling concept universal and hence observed in other intermediate filaments?

6.2 The role of intermediate and actin filaments interactions

Cellular mechanical activity generated by the connection between the extracellular matrix and the actin cytoskeleton seems to be essential for the regulation of cell adhesion, spreading and migration during normal and cancerous development. In particular, keratins are the intermediate filament proteins of epithelial cells, expressed as pairs in a lineage/differentiation manner. The hallmarks of all relatively simple epithelia are intermediate filaments that contain only keratins 8/18 (K8/K18). In addition, in these epithelia the intermediate filaments K8/K18 play a key regulatory role in adhesion and migration, through facilitating integrin interactions with the extracellular matrix, actin adaptor proteins and signaling molecules at focal adhesions. Using a K8 knockdown in rat H4 hepatoma cells and their K8/K18-containing wild-type counterparts seeded on fibronectin-coated substrata of different rigidities, it has been shown that the K8/K18 intermediate filament-lacking cells are unable to spread and hence display an altered actin fiber organization when seeded on a low-rigidity substrate. In addition, a concomitant reduction of local cellular stiffness, especially at focal adhesions generated by fibronectin-coated microbeads attached to the dorsal cell surface, has been detected (Bordeleau *et al* 2012). Furthermore, this K8/K18 intermediate filament modulation of cellular stiffness and actin fiber organization is mediated by RhoA-ROCK signaling (Bordeleau *et al* 2012). Taken together, these results reveal that K8/K18 intermediate filaments contribute to the cellular stiffness–extracellular matrix rigidity interaction through alteration of their Rho-dependent actin organization and cytoskeletal dynamics.

It has been shown that the ability of cells to sense and adapt to mechanical properties from the extracellular matrix is crucial for many biological processes, such as the involvement of mechanical force in supporting embryonic development (Butcher *et al* 2009). In particular, embryonic stem cells progressively stiffen when they undergo differentiation and hence adapt their individual stiffness to the rigidity of the underlying extracellular matrix (Pajerowski *et al* 2007). Similarly, there is evidence for the involvement of increased extracellular matrix rigidity in promoting the occurrence and development of primary tumors, and subsequently in supporting cancer metastasis by enhancing the migration of certain escaping cancer cells (Lopez *et al* 2008). In more detail, aggressive and invasive cancer cells in suspension (where they are independent of interaction with the extracellular matrix) are more compliant than less aggressive cancer cells, which in turn are more compliant than healthy cells (Guck *et al* 2005). However, when these tumorigenic cancer cells are adhered on a rigid extracellular matrix substratum, they display increased contractility (Paszek *et al* 2005). Alterations in the matrix rigidity can cause great differences in cell behavior and thus highlight how important changes in the mechanical properties of the microenvironment are for the cells to adapt and counterbalance the extracellular matrix constraints. Indeed, an extracellular matrix-produced stress is perceived and integrated intracellularly through the involvement of integrin receptors, which act as mechanotransducers by inducing signaling cascades in order to facilitate cellular responses such as cell migration and contractility through modulation of the actin cytoskeleton at focal adhesions (Butcher *et al* 2009, Matthews *et al* 2006).

Cell contractility and the associated internal stiffness can be measured by tracking the force-induced displacement of fibronectin-coated beads that are bound to integrin receptors and subsequently to focal adhesions generated at the dorsal cell surface (Bausch *et al* 1999, Matthews *et al* 2004). These measurements at the cellular level have shown that a de-polymerization of the actin cytoskeleton reduces cell stiffness, which demonstrates that this cytoskeletal network can respond to the mechanical force applied at focal adhesions (Matthews *et al* 2004, Bordeleau *et al* 2011). There is a molecular-based balance between internal stiffness and extracellular force exerted at focal adhesions, which is maintained by alteration of the fibrillar actin contractility (Paszek *et al* 2005, Matthews *et al* 2006, McBeath *et al* 2004) In particular, this is mediated by activation of Rho and the effector ROCK, which is in turn a regulator of the myosin light chain (Asparuhova *et al* 2009, Amano *et al* 2010). This Rho-dependent contractility suggests prominent actin cytoskeleton involvement in the interaction between cell stiffness and extracellular matrix rigidity.

Keratins belong to the largest family of cytoskeletal proteins and are grouped into type I (K9–28) and type II (K1–K8 and K71–K80) subfamilies (Amano *et al* 2010). Keratin intermediate filaments are heteropolymers that include at least one type I and one type II keratin. Due to cellular differentiation and special cell linage, these keratins are coordinately expressed as specific pairs. Intermediate filaments from all simple epithelial cells contain K8/K18 and most of them also express two–three other keratins (Pekny and Lane 2007, Omary and Ku 1997). In more detail, K8 and K18 are the ancestral genes for the multiple specialized type II and type I keratin classes, respectively, and are the first cytoplasmic intermediate filament genes expressed in the embryo when stem cells differentiate into different cell lineages (Oshima *et al* 1996, Coulombe and Wong 2004). In cancer disease, there is indeed some evidence that the persistence of K8/K18 intermediate filaments is a hallmark of invasive squamous cell carcinoma, where such perturbed K8/K18 expression contributes to cell invasiveness based on actin-dependent motility (Alam *et al* 2011a). In addition, point mutations in K8 and K18 genes cause intermediate filament disorganization and consequently a predisposition to liver cirrhosis (Omary *et al* 2009), while in turn cirrhosis enhances hepatic tissue stiffness. This extracellular matrix-linked mechanical alteration is mostly associated with the occurrence of the hepatocellular carcinoma (Jung *et al* 2011, Kuo *et al* 2010). Keratin intermediate filaments assemble an elastic and flexible cytoskeletal network, which is responsible for the ability of epithelial cells to withstand mechanical stress (Coulombe and Wong 2004). Moreover, it can be hypothesized that K8/K18 intermediate filaments need to be included as the factor responsible for providing the interaction between actin-mediated cell stiffness and extracellular matrix rigidity in simple epithelial cells.

This hypothesis has been addressed by using monolayer cultures of K8-knock-down H4-II-E-C3 (shK8b) rat hepatoma cells and their K8/K18-containing wild-type counterparts (H4ev) (Bordeleau *et al* 2010). The intermediate filaments of hepatocytes are made up solely of K8/K18 (Omary and Ku 1997, Oshima 2002), which means that a loss of K8 leads to the degradation of K18, leading to the epithelial cells lacking K8/K18 intermediate filaments (Galarneau *et al* 2007). H4ev cells and shK8b cells have been cultured on fibronectin-coated substrates of diverse rigidity (Wang and Pelham 1998). After their adhesion, a laser-tweezers-driven force was exerted on a fibronectin-coated microbead attached to an integrin (Matthews *et al* 2004, Bordeleau *et al* 2008). Then the cell stiffness at focal adhesions was determined by measuring the bead displacement (Matthews *et al* 2004). In addition, the involvement of the Rho-ROCK was investigated in shK8b cells and compared to H4ev cells regarding cell stiffness and fibrillar actin organization. The results revealed an intervention of K8/K18 intermediate filaments in the cell stiffness–extracellular matrix rigidity interaction through alteration of the Rho-dependent actin organization and dynamics.

Intermediate filaments consist of a heterogenic group of evolutionarily conserved proteins and are specified in a tissue-, cell type- and context-dependent manner. Intermediate filaments feature in multiple cellular processes, such as the maintenance of cell and tissue integrity and the response and adaptation to various external stresses. Inherited mutations in intermediate filament coding sequences can lead to a broad array of clinical disorders. As a consequence, the expression, assembly and organization of intermediate filaments are precisely regulated (Chung *et al* 2014). Indeed, cell migration is a cell-based phenomenon in which intermediate filaments act as effectors and regulators. In the following the focus is on vimentin and keratin and how intermediate filaments provide the cell's mechanical properties, cytoarchitecture assembly and adhesion, and how they regulate the pathways, which impact cellular motility.

Ten-nanometer-wide intermediate filaments were first described in muscle (Ishikawa *et al* 1968). They are composed of the most diverse and heterogeneous group of proteins among intracellular cytoskeletal fibers. There are approximately 70 genes that encode for intermediate filament-forming proteins in the human genome. Among them are 54 genes encoding for keratin proteins that occur in epithelia (Schweizer *et al* 2006, Pan *et al* 2013). Intermediate filaments can be subdivided into six major subtypes based on their gene substructure or sequence homology regarding their signature within the central rod domain. All intermediate filaments proteins self-assemble into approximately 10 nm wide filaments by forming obligatory or facultative heteropolymers, along with a defining tripartite domain structure with a central α-helical rod domain with long-range, coiled-coil forming hepta repeats, flanked by variable end domains located at their N- and C-termini. Intermediate filament proteins display pronounced heterogeneity, because their molecular mass ranges from 40 kDa (type I keratin 19) to 240 kDa (type IV nestin). However, focusing on individual keratins, their primary structure is evolutionarily well conserved. Intermediate filament systems are present across nearly all multi-cellular eukaryotes (Erber *et al* 1998). The evidence suggests that they appeared as nuclear proteins similar to lamins in lower eukaryotes such as *Dictyostelium* (Batsios *et al* 2012). The presence of the intermediate filament-like crescentin in *Caulobacter crescenti* (Ausmees *et al* 2003) raises the question of whether intermediate filaments might have existed earlier, in prokaryotes.

Another important signature feature of the intermediate filament superfamily of genes and proteins is the tissue-type-, differentiation-program- and context-dependent nature of their regulation. Accordingly, the list of functions fulfilled by intermediate filaments in their natural biological setting is growing longer as intermediate filaments and their associated proteins have been linked to cell motility (Pan *et al* 2013, Burke and Stewart 2013, Toivola *et al* 2010). Due to their status as abundant fibrous elements inside cells, intermediate filaments regulate cellular migration on the mechanical and cytoarchitectural levels. Intermediate filaments also impact on migration as regulators, because they can interact with and regulate various cellular effectors, such as signaling molecules (Pan *et al* 2013).

There are intermediate filament proteins such as vimentin that consistently stimulate cell migration and invasion independently of the setting, whereas others, such as various keratins, exert a more variable, fine-tuned and complex impact on these processes. Beyond the type of intermediate filament protein, additional determinants such as the expression level, associated proteins, intracellular organization and covalent modifications such as phosphorylation, act together to regulate cell migration. In addition, the cellular and biological features are crucially important. The nature and impact of various intermediate filaments during migration under normal and disease settings reflect their pervasive integration in a microenvironment-dependent manner within the broader background of the cell.

Which basic attributes of intermediate filaments are relevant for their properties and function *in vivo*? Similar to F-actin and microtubules, intermediate filaments depend on an array of partner proteins for their assembly, organization, function and regulation. Moreover, plakin family proteins are 'cytoskeletal organizers' that link intermediate filaments, microtubules and actin at several strategic locations within the cell's interior (Leung *et al* 2001). Beyond their signature plakin domain, plakin family members are large and possess a modular substructure that helps them act as key organizers of the cytoskeleton (Leung *et al* 2002). Plakin proteins facilitate intermediate filament attachment to the cytoplasmic so-called plaque domain present in cell–cell desmosome adhesions and cell–matrix hemidesmosome adhesions to the other two main cytoskeletal protein filaments, such as F-actin and microtubules, as well as to the surface of the nucleus (Leung *et al* 2001, Leung *et al* 2002, Desai *et al* 2009).

Intermediate filament proteins are modified by post-translational modifications such as phosphorylation, O-glycosylation, ubiquitination, sumoylation and acetylation (Izawa and Inagaki 2006, Hyder *et al* 2008, Zencheck *et al* 2012). In particular, these modifications are site-specific within the intermediate filament protein backbone, reversible and regulate nearly all aspects of their assembly, organization, and biochemical and mechanical properties, as well as their function (Pan *et al* 2013, Omary *et al* 2004, Rogel *et al* 2010). Associated proteins and post-translational modifications of intermediate filaments help to define the polymerization status and their intracellular organization. Actively migrating, polarized cells possess a specific intermediate filament system reorganized around the nucleus or at their rear, trailing end (Paladini *et al* 1996, Helfand *et al* 2011, Weber *et al* 2012) and when mutant intermediate filament proteins are expressed (Morley *et al* 2003).

What is the interplay between intermediate filaments, cell adhesion and other cytoskeletal components?

Desmosomes are composed of transmembrane cadherins, armadillo proteins such as plakoglobin and plakophilins, and also plakin proteins such as desmoplakin, which anchors desmosomal plaques to intermediate filaments on the intracellular site (Desai *et al* 2009). Thus, desmosomes maintain tissue integrity even under mechanical stress conditions (Desai *et al* 2009) that start at an early stage during mouse embryogenesis (Gallicano *et al* 1998). In particular, pro-migratory stimuli such as epidermal growth factor (EGF) support the assembly and functional state of desmosomes, hemi-desmosomes and the whole network architecture of the intermediate filaments (Keski-Oja *et al* 1981, Baribault *et al* 1989, Chung *et al* 2012, Felkl *et al* 2012). However, the induction of cell migration is typically linked to weaker desmosome-facilitated cell–cell adhesion (Kitajima 2013). Moreover, increased turnover of desmosomes and their decreased colocalization with keratins have been revealed in migrating oral squamous cell carcinoma cells (Roberts *et al* 2011).

Moreover, cell migration is a function of dynamic interactions between extracellular matrix components and the cell's cortex. The transmembrane, adhesion-facilitating entity in hemidesmosomes is the α6β4 integrin heterodimer, which binds to extracellular laminin (Giancotti *et al* 1999). Intracellularly, the integrin connection to intermediate filaments is provided by plakin proteins such as the bullous pemphigoid antigens 1 and 2 (BPAG1 and BPAG2), as well as plectin (Leung *et al* 2002). In the complete absence of keratins, the hemidesmosome components are still present, but they are scattered in skin keratinocytes. Surprisingly, these cells lacking keratins adhere faster to the extracellular matrix microenvironment and show increased migration compared to wild-type cells (Seltmann *et al* 2013). Re-expression of the K5–K14 keratin pair alone, which is typically expressed in progenitor basal keratinocytes, in such keratin-free skin keratinocytes reverses this phenotype, even when the K5-K14 expression is at a sub-physiological level. By contrast, keratinocytes with no expression for BPAG1 show a normal density of hemidesmosomes at the cell–matrix interface, but lack a cytoplasmic plaque as well as attachment to keratin intermediate filaments and display a delayed wound healing response (Guo *et al* 1995). The knockdown of actinin-4, which can bind to actin, leads to a loss of directionality during migration of keratinocytes. In more detail, this behavior is correlated with a mislocalization of α6β4 integrin as well as BPAG1e and defects in cellular polarity and lamellipodial dynamics (Hamill *et al* 2013). It has been suggested that the p90 ribosomal protein S6 kinase (RSK) is involved in hemidesmosome remodeling (Frijns *et al* 2010, Faure *et al* 2012), and also in the regulation of wound-inducible K17 (Pan *et al* 2011). What is the influence of K17 in complete keratin-null and/or actinin-4 knockdown keratinocytes? It has also been shown that the knock-down of K8 in cultured hepatoma cells decreases cell migration in a 2D scratch-wound assay (Bordeleau *et al* 2010), reduces cell spreading, alters Rho-dependent actin fiber organization and reduces local stiffness at focal adhesions, suggesting that an interaction between K8/K18 intermediate filaments and Rho-facilitated actin dynamics occurs through plectin, RACK1 and Src (Bordeleau *et al* 2012).

Linker of nucleoskeleton and cytoskeleton (LINC) is a protein complex located at the nuclear membrane and it participates in linking the nuclear lamina to cytoskeletal proteins on the cytoplasmic side (Mellad *et al* 2011). In particular, Nesprin-3 is a component of LINC and can associate with plectin (Wilhelmsen *et al* 2005). However, the disruption of LINC via expression of mutated nesprin reduces the transmission of intracellular forces, alters the organization of F-actin as well as vimentin intermediate filaments and causes reduced migration and polarization in mouse embryonic fibroblasts (Lombradi *et al* 2011). Similarly, depletion of Nesprin-3 in human aortic endothelial cells impairs the organization of vimentin intermediate filaments and reduces cell migration (Morgan *et al* 2011). In line with this, depletion of other major intermediate filaments, such as nestin, vimentin and glial fibrillary acidic protein in astrocytes, changes the localization and rotation of the nucleus during astrocyte migration (Dupin *et al* 2011) and subsequently reduces their migration (Lepekhin *et al* 2001). Cell migration has been shown to depend on dynamic alterations of the position and shape of the nucleus (Friedl *et al* 2011) and additionally it has been demonstrated that intermediate filaments contribute to nuclear architecture in skin keratinocytes (Pan *et al* 2013) and migrating cells (Friedl *et al* 2011).

How is the keratinocyte migration regulated?
Although there is an interaction between intermediate filaments and their associated proteins in supporting cell motility, it obvious that there converging migration phenotypes are still exhibited by many genetic null mutants in mouse skin keratinocytes. Genetic loss of epiplakin, a plakin family member, in mice results in increased skin keratinocyte migration (Goto *et al* 2006), associated with the loss of keratin intermediate filament bundling in post-wounding (Ishikawa *et al* 2010). In line with this, increased migration occurs in mouse keratinocytes that are genetically null for plectin (Osmanagic-Myers *et al* 2006), plakoglobin (Yin *et al* 2005), plakophilin (Yin *et al* 2005) and keratin 6 (K6a/K6b) (Yin *et al* 2005, Rotty and Coulombe 2012). Moreover, the loss of K6a/K6b, plectin or plakoglobin leads to Src family kinase activation and changes to the F-actin reorganization (Goto *et al* 2006, Osmanagic-Myers *et al* 2006, Todorovic *et al* 2010). A plectin deficiency phenotype demonstrated that the intermediate filament is supposed to indirectly regulate the organization and stability of microtubules through an interaction with the plectin1c isoform. In addition, this then has an associated impact on focal adhesion dynamics and the directional migration of keratinocytes (Valencia *et al* 2013).

Finally, the Src kinase is known to regulate leading edge protrusion through Rac and Cdc42 signaling and thus facilitate focal adhesion dynamics and building of invadopodia. In addition, Src is able to induce epithelial-to-mesenchymal transitions (EMT) directly.

Intermediate filaments, the epithelial-to-mesenchymal transition, tumor growth, invasiveness and metastasis
Invasion of the local microenvironment (such as the connective tissue stroma) by cancer cells is a critical initial step in determining in cancer metastasis. In addition, epithelial cancer cell invasion is mediated by the EMT. It is called this because epithelial cells lose their polarity and other characteristics, such as E-cadherin and keratin expression, and can adopt a fibroblast-like morphology, which includes vimentin expression and leads to aggressive migratory properties. One of the key differences between the epithelial and mesenchymal phenotypes is based on the tight

cell–cell and cell–matrix contacts provided by epithelial cells compared to the weak contacts of mesenchymal cells (Thiery *et al* 2009, Nakamura and Tokura 2011). The process of EMT seems to be an important mechanism to account for the increased motility and invasiveness of epithelial-derived cancer cells (Mendez *et al* 2010, Yilmaz and Christofori 2009, Kalluri and Weinberg 2009).

Because epithelial cells can undergo EMT, their intermediate filament system switches from being keratin-dominated to vimentin-dominated, which is another characteristic feature of mesenchymal cells. Indeed, many cancer cell lines display both a keratin-based and a vimentin-based intermediate filament network and these possess distinct intracellular organization and regulatory mechanisms (Gilles *et al* 1999). Cell-culture-based (Chu *et al* 1993, 1996) and xenograft-assays-based *in vivo* studies (Hendrix *et al* 1997) revealed that vimentin has a pro-migratory potential, which is facilitated by the interplay between vimentin and keratins 8/18 and the regulatory role of focal adhesions.

Src has been shown to interact directly with keratin intermediate filaments in a K6-dependent fashion through a novel, non-phosphotyrosine-dependent contact involving Src's SH2 domain using K6a/K6b null keratinocytes (Rotty and Coulombe 2012). Also, Src's localization in the membrane, where it is transiently inactive, is reduced in K6a/K6b null keratinocytes (Rotty and Coulombe 2012). Whether such findings also occur in epiplakin, plakophilin, plakoglobin and/or plectin null keratinocytes is still elusive (Bordeleau *et al* 2012). However, the keratin-containing multi-protein partnership, its mechanisms and the effectors through which it is so precisely regulated during keratinocyte migration, are not yet fully understood. The apparent paradox between the wound-inducible character of K6 and its negative influence for pure cell migration has also been found for several other cytoskeletal proteins. How is the optimal speed and mode of cellular migration achieved under certain conditions? It seems that this has to be determined for each case separately (Weber *et al* 2012).

Keratin 6 and K16 are often up-regulated in various types of carcinomas, suggesting that they may serve as useful diagnostic markers (Karantza 2011). K6's impact on keratinocyte migration (and possibly also its interaction with Src) may help to explain clinical correlations (Depianto *et al* 2010). For example, the lack of K6 expression correlates with an aggressive behavior for endometrial carcinomas (Stefansson *et al* 2006), whereas decreased K6 expression is coupled with reoccurrence of K8/K18 expression and consequently correlates with the acquisition of malignancy in mouse skin subjected to chemical carcinogenesis (Larcher *et al* 1992). These findings suggest that the functional significance of inducing or modifying K6 (and possibly K16, plakoglobin, plectin and others) may be a natural strategy to withstand dedifferentiation- and malignancy-promoting signaling and cellular processes, such as EMT (Rotty and Coulombe 2012). However, the link between keratins and cancer has not yet been clearly revealed and hence cannot be regarded as simple (Karantza 2011). Higher levels of K16 lead to poorer survival rates among breast cancer patients with metastatic reoccurrence (Joosse *et al* 2012), while higher levels of keratins 5, 6 and 17 have been associated with a poor prognosis in breast cancer (van de Rijn *et al* 2002, Abd El-Rehim *et al* 2004, de Silva Rudland *et al* 2011).

Whether a keratin or a conglomerate of intermediate filament proteins and binding partners promotes or inhibits cell migration and displays cancer cell properties is determined by the overall microenvironemtal background within the cell, such as associated proteins, post-translational modifications and biological setting.

How do intermediate filaments affect cellular mechanics and migration behavior?
Cells develop a polarized cytoarchitecture upon the initiation of cell migration, as their front and rear contain different molecular components and possess different functional properties (Lammermann 2009, Friedl and Wolf 2010). As a cell senses microenvironmental properties, signaling events, actin polymerization and myosin motor function can be spatially regulated in order to generate membrane protrusions at the leading edge and retractive forces at the trailing edge. Mechanical signals facilitate polarized cell protrusions and directional migration and hence there is evidence that intermediate filaments regulate cell migration from the cellular mechanics point of view (Seltmann *et al* 2013, Gilles *et al* 1999, Eckes *et al* 1998).

In addition, mechanotransduction processes involving intermediate filaments are important for the attachment of epithelial cells to their microenvironment, such as the extracellular matrix. A mechanotransduction pathway has been revealed in *C. elegans* that includes hemidesmosome-like elements regulating intermediate filaments (Zhang *et al* 2011). In particular, it has been observed that muscle contraction mechanically alters the epidermis and activates p21-activated kinase (PAK). Then PAK phosphorylates intermediate filament proteins, which induces hemidesmosome biogenesis. Thus, hemidesmosomes act as mechanosensors, which, when under tension, trigger intracellular signaling pathways that promote epithelial morphogenesis. Cell–cell junctions are involved in local traction force generation, providing long-range gradients of intra- and inter-cellular tension during collective cell migration (Trepat and Fredberg 2011). When investigating *Xenopus gastrulation*, the application of a punctual mechanical force to single *Xenopus mesendoderm* cells through magnetic tweezers and cadherin-coated beads connected to the cells promoted polarized protrusions at the opposite end of the force (and the cell) and subsequently induced persistent directional cell migration (Weber *et al* 2012). It has been suggested that this localized tension increases induces plakoglobin-dependent redistribution of the keratin intermediate filaments at the cell's rear. Indeed, these events based on keratin and plakoglobin are needed for force-induced, polarized cell protrusions and normal mesendoderm polarity, as well as organization *in vivo*.

The role of vimentin as a promotor of cell migration during cancer progression
Vimentin is a type III intermediate filament protein that is expressed significantly throughout embryogenesis, whereas it is restricted to mesenchymal cell types in the adult setting, such as fibroblasts, bone marrow-derived blood cells and endothelial cells (Hansson *et al* 1984, Dellagi *et al* 1985). Vimentin can reoccur in adults, as it is drastically up-regulated after injury to diverse tissues, such as musclse, the central nervous system and various connective tissues, and during EMT. Vimentin exerts pleiotropic and context-dependent roles in cells (Satelli and Li 2011) and, in particular, has a marked impact on cell migration in several physiologically normal

settings (Ivaska *et al* 2007). For instance, vimentin is essential for lymphocyte adhesion to the endothelial cell lining of blood and lymphoid vessels and their subsequent transmigration through the endothelium into the connective tissue (Nieminen *et al* 2006), for fibroblast or individual breast cancer cell motility (Mendez *et al* 2010) and for the *in vitro* wound healing and closure processes of alveolar epithelial cells (Rogel *et al* 2011).

Vimentin has been detected at unusually high levels in many types of epithelial cancers (Satelli and Li 2011). Vimentin expression is necessary for the invasive phenotypes of prostate cancer cells (Singh *et al* 2003, Wei *et al* 2008), soft tissue sarcoma cells and breast cancer cells, in *in vitro* assays (Zhu *et al* 2011). In line with this, inhibition of vimentin expression in a squamous carcinoma cell model not only decreases motility (McInroy and Maatta 2007), it also promotes a more pronounced epithelial phenotype, as supported by the up-regulation of K13, K14 and K15 (Paccione *et al* 2008) and alteration of the cellular shape (Mendez *et al* 2010). Conversely, vimentin overexpression has been shown to increase prostate cancer cell invasion (Zhao *et al* 2008) and invadopodia elongation (Schoumacher *et al* 2010).

Numerous studies support the idea that vimentin and the process of cellular migration mutually regulate one another. The tumor suppressor adenomatous polyposis coli (APC), which is frequently mutated or lost in colorectal cancer, directly binds to and thus regulates vimentin organization (Sakamoto *et al* 2013). In addition, in migrating astrocytes APC is required for vimentin intermediate filament alignment with the microtubule network (Sakamoto *et al* 2013). A C-terminal APC truncation mutant can bind to and disorganize vimentin, whereas the keratin intermediate filament organization remains unaltered when expressed in human SW480 colon cancer cells. It has been suggested that the loss of APC in cancer cells that have undergone EMT alters vimentin intermediate filament organization and impacts on motility and invasiveness. In cultured breast epithelial cells, over-expression of oncogenic H-Ras-V12G or the transcription factor Slug, which both favor cell migration and EMT, induces vimentin expression. In turn, vimentin expression is necessary for H-Ras-V12G- and Slug-induced migration and expression of the receptor tyrosine kinase Axl, whereas epithelial markers such as K6 are suppressed (Vuoriluoto *et al* 2011). Overexpression of Axl is able to rescue the decreased migration phenotype of a breast cancer cell line expressing vimentin siRNA, suggesting that vimentin acts through interaction with Axl.

An RNAi screen has identified regulators of vimentin expression such as the mitochondrial enzyme methylenetetrahydrofolate dehydrogenase 2 (MTHFD2), which is required for vimentin expression and network organization (Lehtinen *et al* 2013). Similarly to vimentin, the siRNA-mediated knock-down of MTHFD eliminates the migration of breast cancer cells and their invasion into the extrac-ellular matrix, suggesting their interdependence in cell motility. Moreover, vimentin expression is also regulated by miRNAs. In more detail, overexpression of mir-138, which is down-regulated in several primary tumors, leads to decreased vimentin expression and reduced cell migration and invasion in renal cell carcinoma cell lines (Yamasaki *et al* 2012). Similarly, mir-30a represses vimentin expression and thus consequently cell migration and invasion in breast cancer cell lines (Cheng *et al*

2012). As some cancer cells show decreased expression of mir-138 and mir-30a (Yamasaki *et al* 2012, Cheng *et al* 2012), these findings may provide evidence regarding how vimentin expression is up-regulated in EMT and cancer.

The impact of keratins 8/18 on epithelial cell migration
Type II keratin 8 has also been shown to be involved in cell migration and cancer metastasis (Bordeleau *et al* 2010, Raul *et al* 2004). Similarly to vimentin, K8 is expressed broadly during development, but is then restricted to simple epithelial lineages, such as in the liver, gut, kidney and lungs, in adults (Moll *et al* 1982). Moreover, K8 expression is induced or enhanced in special cancer types, such as breast, lung and pancreatic cancers, and specific tumor-derived cell lines (Moll *et al* 1982). Unlike vimentin, K8's impact on cancer cell migration and invasion is inhibitory. Thus, the balance between vimentin and K8/K18 expression is supposed to be a key determinant of the migratory properties and invasiveness of various types of cancer cells *in vitro* and *in vivo*.

In line with this, it has been reported that treatment of pancreatic cancer Panc-1 cells with sphingosylphosphorylcholine (SPC), which is a bioactive lipid, leads to keratin phosphorylation, promotes a dramatic reorganization of the keratins' intermediate filaments to the perinuclear region, reduces cellular elasticity and induces pronounced cell migration (Beil *et al* 2003). These observations are supported by other studies. One study showed that SPC treatment facilitates the expression of transglutaminase-2 expression in Panc-1 cells, which precedes JNK kinase activation and phosphorylation of K8 at Ser 431 (Park *et al* 2011). Another study reported that SPC can activate ERK kinase upstream of keratin intermediate filament reorganization, and provides phosphorylation of K8 and K18 at Ser 431 and Ser 52, respectively, in pancreatic and gastric cancer cells (Busch *et al* 2012). However, do these events contribute to the so-called mechanical softening of the cytoplasm in SPC-treated Panc-1 cells (Beil *et al* 2003), which contributes to their enhanced motile behavior?

The expression and/or site-specific phosphorylation of K8 (and of its partner K18) alters the migratory behavior and invasiveness of various types of cancer cells. An inhibitory influence for K8 on cell migration was observed by studies in which pancreatic cancer cells (Busch *et al* 2012) and a weakly invasive subclone of MDA-MB-468 breast cancer cells were subjected to K8 knockdown (Iyer *et al* 2013), a highly invasive subclone of MDA-MB-435 breast cancer cells was transfected to overexpress K8 (Iyer *et al* 2013) and KLE endometrial cancer cells and also HepG2 hepatocellular cancer cells expressed no K8/K18 due to silencing (Fortier *et al* 2013). By contrast, other studies reported that the loss of K8 phosphorylation at either Ser 73 or Ser 431 led to elevated migration and enhanced metastatic potential for oral squamous cell carcinoma cells (Alam *et al* 2011b) and colorectal cancer cells (Mizuuchi *et al* 2009). In addition, K8-dependent stimulation of cell migration led to reduced collective migration of hepatoma cells after K8 silencing (Bordeleau *et al* 2010). Taken together, the silencing of the desmosomal plaque protein plakophilin 3 enhances the migration and metastasis of human colon carcinoma cells (Kundu *et al* 2008), while another study speculated that this result is caused by increased levels of K8 protein and phosphatase PRL-3, which evokes K8 de-phosphorylation (Khapare *et al* 2012).

Keratin-dependent activation of Akt signaling seems to be important during tumorigenesis. Indeed, lactotransferrin has anti-tumor activity and is even down-regulated in cancer (Bezault *et al* 1994). In particular, interaction with lactotransferrin blocks the binding of K18 to 14-3-3σ and hence suppresses K18-induced Akt activation and its impact on cancer cell proliferation and invasion (Deng *et al* 2013). However, others have shown that K17 interacts with 14-3-3σ and regulates the Akt-mTOR signaling pathway (Kim *et al* 2006), whereas vimentin interacts with and becomes activated by Akt to promote cancer cell invasion (Zhu *et al* 2011). Although not related to migration so far, O-linked N-acetylglucosamine modification of K18 induces Akt activity and consequently protects against liver injury (Ku *et al* 2010).

In addition to vimentin and keratin, the increased expression of nestin, which is a class IV intermediate filament protein and a marker of stem/progenitor cells (Lendahl *et al* 1990), has been found in multiple types of tumors (Florenes *et al* 1994, Parry *et al* 2008). Moreover, nestin regulates cellular motility and metastatic properties, whereas it does not affect the growth of prostate (Kleeberger *et al* 2007) and pancreatic cancer cells (Matsuda *et al* 2011). Based on its connection to stem/progenitor cells, it may be possible that nestin expression can be useful to identify cancer stem cells (Matsuda *et al* 2012). However, this is still a hypothesis that needs to be tested in future studies.

Taken together, the complex role of intermediate filaments during cell migration has been detected. In particular, intermediate filaments affect cell migration because (i) they are intrinsic determinants of cellular micromechanical properties and (ii) they contribute to the regulation of certain migratory pathways. Finally, the impact of intermediate filaments during cell migration in normal and disease settings reflects their full integration into the cytoskeletal framework of migratory and invasive cells.

The role of filamin A and vimentin in cell migration
Cell adhesion and spreading are regulated by diverse and complex interactions between the cytoskeleton and extracellular matrix proteins. Thus it has been suggested that the interaction of the intermediate filament protein vimentin with the actin cross-linking protein filamin A may regulate the spreading in HEK-293 and 3T3 cells (Kim *et al* 2010c). In more detail, filamin A and vimentin-expressing cells spread on collagen and exhibit numerous cell extensions containing filamin A and vimentin. However, when the cells are treated with small interfering RNA (siRNA) in order to knock-down filamin A or vimentin, they spread badly. Both of these filamin A and vimentin knock-down cell populations have exhibited a less than 50% decrease in cell adhesion, cell surface expression and β1 integrin activation. In more detail, the knock-down of filamin A decreases vimentin phosphorylation and blocks recruitment of vimentin to cell extensions, whereas the knock-down of filamin A and/or vimentin diminishes the exertion of cell protrusions. Decreased vimentin phosphorylation, impaired cell spreading and reduced β1 integrin surface expression and decreased activation occur in cells that have been cultured with the protein kinase C inhibitor bisindolylmaleimide. In addition, cell spreading has been shown to be reduced by siRNA-facilitated knock-down of protein kinase C-ε. Immunoprecipitation of cell lysates and pull-down assays with purified proteins has revealed a direct binding

interaction between filamin A and vimentin. Filamin A is additionally associated with the protein kinase C-ε, which has been observed to increase in cell extensions. Taken together, these data propose that filamin A associates with vimentin and protein kinase C-ε, thereby inducing vimentin phosphorylation, which is crucial for β1 integrin activation and cell spreading on collagen substrates.

The adhesion and spreading of fibroblasts is critical for development, tissue remodeling and wound healing. Cell spreading is characterized by the extension of cell membrane protrusions, including filopodia and lamellipodia (Albrecht-Buehler and Lancaster 1976, Guillou et al 2008, Johnston et al 2008, Ladwein and Rottner 2008). The exertion of protrusions during cell spreading on extracellular matrix proteins such as collagen and fibronectin is regulated by complex cell–matrix signaling pathways involving β1 integrins, small GTPases and actin-binding proteins, which facilitate acto-myosin remodeling (Li et al 2005, Price et al 1998, Zimmerman et al 2004). Whereas the actin cytoskeleton is crucially involved in cell adhesion and spreading (Serrels et al 2007, Vicente-Manzanares et al 2009), the intermediate filament cytoskeleton and, in particular, vimentin seems to be additionally involved in cell adhesion and spreading (Eckes et al 1998, Fortin et al 2010, Ivaska et al 2002, 2007, Tsuruta and Jones 2003). As actin and vimentin cytoskeletons colocalize at the leading edge of spreading cells (Correia et al 1999), the two systems may also be functionally correlated during cell adhesion and spreading. However, the mechanisms through which the two cytoskeletal components act together in order to promote cell spreading are not yet well understood. Numerous proteins can bind to both actin and intermediate filaments and thus may functionally coordinate the two different cytoskeletal systems. Candidate proteins for the interaction are plectin (Seifert et al 1992, Svitkina et al 1996), BPAG1 (Yang et al 1996), fimbrin (Correia et al 1999) and filamin A (Brown and Binder 1992). Filamin A is an actin cross-linking protein inducing gelation, membrane stability, adhesion and motility (Brotschi et al 1978, Feng and Walsh 2004, Glogauer et al 1998, Popowicz et al 2006, Stossel et al 2001). In addition, filamin A is appropriately located in order to regulate cell–matrix inter-actions, as it binds to actin and regulates the properties of β1 integrins (Loo et al 1998, Popowicz et al 2006, Stossel et al 2001). Moreover, filamin A is concentrated in focal adhesions (Campbell 2008, Critchley 2000, Glogauer et al 1998) and hence regulates cell adhesion through the control of the cell surface expression (Meyer et al 1998) and activation (Kim et al 2008) of β1 integrins.

Vimentin has several functions in common with filamin A, such as the regulation of cell adhesion and motility (Eckes et al 1998, McInroy and Maata 2007, Nieminen et al 2006, Tsuruta and Jones 2003, Whipple et al 2008), the clustering of β1 integrin in focal adhesions (Kreis et al 2005) and the mechanical stabilization of cells (Eckes et al 1998). In more detail, the expression of filamin A and vimentin is elevated during shear stress in osteoblasts (Jackson et al 2008), indicating that they protect against external mechanical stimulation. As with filamin A, many functions of vimentin are facilitated upon its phosphorylation (Chou et al 2007, Eriksson et al 2004, Izawa and Inagaki 2006, Sihag et al 2007). In particular, the phosphorylation of vimentin by PKC is necessary for the recycling of β1 integrins to the cell membrane and, consequently, for cell migration (Fortin et al 2010, Ivaska et al 2002). In addition, filamin A can bind PKC (Feng and Wash 2004, Glogauer *et al* 1998, Tigges *et al* 2003), which then facilitates the phosphorylation of mediates vimentin (Ando *et al* 1989, Ivaska *et al* 2005, Ogawara *et al* 1995, Yasui *et al* 2001).

Taken together, it has been demonstrated that filamin A and vimentin coregulate the activation and cell surface expression of β1 integrin in order to provide cell spreading in fibroblasts. Indeed, PKC-facilitated phosphorylation of vimentin is involved in β1 integrin activation as well as cell spreading and consequently this phosphorylation is dependent on filamin A, which can bind directly to vimentin.

Does filamin A interact with vimentin during cell spreading?

Filamin A is an important determinant for cell spreading (Kim *et al* 2008). In addition to revealing novel filamin A binding proteins involved in filamin A-mediated cell spreading, an isotope-coded affinity tag analysis was performed on filamin A immunoprecipitates. Twenty-one peptides have been identified with >99% certainty and at least six different proteins were predicted to be differentially expressed in suspended and spreading cells. Among these six preotins was vimentin. Starting from the previously identified roles of filamin A and vimentin in cell adhesion and migration (Bellanger *et al* 2000, Eckes *et al* 1998, Nieminen *et al* 2006, Tsuruta and Jones 2003, Vadlamudi *et al* 2002), subsequent investigations have focused on the roles of filamin A and vimentin in the onset of cell spreading.

The coregulation of filamin A and vimentin in cell spreading

The extrusion of filopodia and/or lamellipodia by cells is an important marker of early cell spreading (Li *et al* 2005, Price *et al* 1998, Zimmerman *et al* 2004). It has been shown whether filamin A and/or vimentin are present in the earliest build extensions of spreading cells. Indeed, immunofluorescence revealed that in control HEK cells spreading on collagen, filamin A and vimentin colocalized within these extensions. As true filopodial and lamellipodial extensions consist of cortactin, vinculin and paxillin in addition to actin filaments (Artym *et al* 2006, Nobes and Hall 1995, Perrin *et al* 2006, Webb *et al* 2007), the presence of these proteins in cell extensions could be confirmed using immunofluorescence. To investigate the individual roles of filamin A and vimentin in cell spreading, RNA interference was performed to diminish filamin A and vimentin expression in HEK cells. To evaluate the combined roles of filamin A and vimentin in cell spreading, siRNA was used for vimentin in cells stably transfected with a filamin A shRNA, thereby creating a double knock-down cell line. As the initial extension of filopodia and lamellipodia normally occurs within 30 min of postplating (Arthur and Burridge 2001, Defilippi *et al* 1999, Ren *et al* 1999), time points between 15 and 120 min were assessed for differences in cell spreading between control and filamin/vimentin-deficient cells. Unexpectedly, knock-down of either filamin A or vimentin expression leads to a pronounced reduction in the number of cell extensions. Concurrent knock-down of filamin A and vimentin did not reveal further reductions in the number of cell extensions compared with the two individual knock-downs of filamin A or vimentin, suggesting that the role of vimentin in mediating cell spreading is not additive to the role of filamin A. Similar results have been obtained for the reductions in cell extension numbers upon silencing of filamin A and/or vimentin in mouse 3T3 fibroblasts. In more detail, filamin A expression was then rescued by introducing a filamin cDNA into filamin-knock-down cells. The vimentin expression was restored in vimentin knock-down cells after four days following transient and hence reversible siRNA knock-down. In addition, the formation of cell extensions by filamin- and vimentin-deficient HEK-293 cells was rescued after the restoration of filamin or vimentin expression.

What is the difference between filamin A and vimentin during actin assembly in spreading cells?
The single and combined effects of filamin A and vimentin have been analyzed on the generation of actin filament free barbed ends in spreading cells. Filamin A can regulate the GTPase activity of Rac and Cdc42 (Bellanger *et al* 2000, Ohta *et al* 1999, Vadlamudi *et al* 2002), which then control cell spreading through determining the generation of free barbed ends, which are required for actin filament assembly (Sun *et al* 2007, Winokur and Hartwig 1995). Indeed, the incorporation of rhodamine-labeled actin monomers into growing actin filaments was pronouncedly decreased in filamin A-deficient cells. No difference was found in the number of free barbed ends in vimentin knock-down cells compared to their wild-type counterparts. However, incorporation of actin monomers into actin filaments was hardly detectable after concurrent knock-down of filamin A and vimentin. Taken together, these data indicate that vimentin contributes to the generation of free barbed ends in the absence of filamin A, whereas vimentin is not able to regulate actin assembly independently.

As filamin A plays a role in binding and regulating the activity of Rac and Cdc42 (Bellanger *et al* 2000, Stossel *et al* 2001, Vadlamudi *et al* 2002), it has been investigated whether vimentin supports the association between filamin A and Rac and/or Cdc42. Thus, filamin A was immunoprecipitated from spreading control and vimentin-deficient cells and then the immunoprecipitates were immunoblotted for Rac and Cdc42. As expected, the amount of filamin A-bound Rac and Cdc42 did not differ significantly between the control and vimentin-deficient cells. In particular, the levels of active, GTP-bound Rac and Cdc42 were similar in the control and vimentin-deficient cells. Finally, the data suggest that vimentin may regulate cell spreading independently of the filamin A-small GTPase-actin assembly pathway.

Do filamin A and vimentin facilitate $\beta 1$ integrin expression and activation?
As cell spreading is $\beta 1$ integrin-dependent (Frame and Norman 2008, Jovic *et al* 2007, Price *et al* 1998) and filamin A-null cells express fewer cell surface $\beta 1$ integrins (Meyer *et al* 1998), it has been suggested that filamin A and vimentin coregulate cell surface $\beta 1$ integrin expression. To address the individual and collective effects of filamin A and vimentin knock-down on $\beta 1$ integrin expression and activation, HEK cells were spread on collagen for 15 min to allow initial adhesion (Zimmerman *et al* 2004). Then, cells were suspended and incubated with the 4B4 antibody in order to detect cell surface $\beta 1$ integrins. In addition, active, ligand-bound $\beta 1$ integrins were labeled with the 12G10 antibody. Flow cytometric analysis revealed that cell surface $\beta 1$ integrin was reduced to

50% upon knock-down of filamin A and/or vimentin. This phenomenon was rescued after reexpression of filamin or vimentin, which is still consistent with the cell spreading results. In particular, 12G10 staining intensity was reduced four-fold after filamin A and/or vimentin knock-down. Taken together, these data showed that filamin and vimentin are important determinants of β1 integrin expression, while they are especially important for regulating β1 integrin activation.

Which proteins are associated with nascent cell adhesions? To investigate this collagen-coated magnetic beads were incubated with cells and isolated and bead-associated proteins were immunoblotted (Glogauer *et al* 1998, Glogauer *et al* 1998, Zhao *et al* 2007). There was indeed more bead-associated vinculin and paxillin in control cells compared to filamin A- or vimentin-deficient cells, suggesting that filamin A and vimentin are both necessary for the generation of stable adhesions. Moreover, these data suggest that filamin A and vimentin regulate β1 integrin activation, cell adhesion and the early steps of cell spreading, possibly using the same mechanism.

The role of vimentin phosphorylation for β1 integrin expression, activity and cell spreading

As expected, the phosphorylation of vimentin facilitated by PKC-ε is a major regulatory step for β1 integrin trafficking and cell migration (Fortin *et al* 2010, Ivaska *et al* 2002). In particular, the role of PKC-mediated phosphorylation of vimentin in early cell spreading was analyzed. Cells were treated for 90 min with BIM (1 μM), which inhibits PKC-ε (Ivaska *et al* 2002, Ivaska *et al* 2002) and then plated on collagen-coated planar substrates. In order to verify the specificity of BIM as a PKC inhibitor, all experiments were additionally performed with cells pretreated with calphostin C (100 nM). In BIM- and calphostin-C-treated cells undergoing early spreading, vimentin was decreasingly phosphorylated at serines-6, -38, and -50, which are PKC phosphorylation sites (Ando *et al* 1989, Ivaska *et al* 2005, Ogawara *et al* 1995, Yasui *et al* 2001). Indeed, BIM-treated and calphostin-C-treated cells both saw a significant reduction in the number of cell extensions compared to untreated controls. In more detail, the BIM- and calphostin-C-induced effects on cell spreading were comparable to siRNA knock-down of vimentin in magnitude. In order to validate the use of BIM as an inhibitor of PKC-ε, the phenotype of BIM-treated and calphostin-C-treated cells was compared to the phenotype of PKC-ε-deficient cells obtained by siRNA knock-down. In line with this, PKC-ε-knock-down cells displayed reduced numbers of cell extensions compared to those detected in vimentin-knock-down, BIM-treated and calphostin C-treated cells. Although vimentin is not a direct substrate for PKC-ε (Ivaska *et al* 2002), its phosphorylation is PKC-ε dependent. Thus, these findings suggest that the phosphorylation of vimentin regulated by PKC-ε may be a key regulatory step for initial cell adhesion and spreading.

PKC-ε-facilitated vimentin phosphorylation is a critical for cell surface β1 integrin expression (Ivaska *et al* 2002). In flow cytometric analysis, non-permeabilized cells performing initial spreading displayed less active and total cell surface β1 integrin after knock-down of filamin A and/or vimentin, BIM treatment, calphostin C treatment or PKC-ε knock-down. Thus, these data indicate that the phosphorylation of vimentin by PKC seems to be crucial for cell surface β1 integrin expression and activation.

Does filamin A bind directly to vimentin in spreading cells under in vitro conditions?

As filamin A can bind PKC (Feng and Wash 2004, Glogauer *et al* 1998, Tigges *et al* 2003), it has been hypothesized that filamin A facilitates vimentin phosphorylation by interacting with it. In order to investigate the filamin A–vimentin interaction, filamin A was immunoprecipitated from the lysates of spreading HEK-293 cells and subsequently blotted for vimentin. Consistently, a small increase in vimentin–filamin A association in response to spreading has been seen, and a small increase of filamin A-associated vimentin in spreading versus suspended cells has been found by immunoprecipitation. These data suggest that filamin A–vimentin interactions are mostly constitutive and depend only moderately on cell adhesion.

Dot-blot analysis revealed that purified filamin A and vimentin seem to interact directly *in vitro*. In addition, the question of whether there is direct interaction was explored using purified GST-tagged vimentin bound to glutathione beads, with the beads being incubated with purified FLAG-tagged filamin A. A direct filamin A–vimentin interaction was detected by immunoblotting the bead-associated proteins for filamin A. Conversely, purified vimentin associated with purified FLAG-filamin A bound to beads was detected by immunoblotting for vimentin. As filamin A has been shown to bind to vimentin directly, it has been suggested that filamin A regulates the phosphorylation of vimentin by PKC.

The role of filamin A in vimentin phosphorylation and reorganization

Indeed, the phosphorylation by PKC is needed for the assembly and disassembly of vimentin filaments and distribution in the cytoplasm (Eriksson *et al* 2004, Ivaska *et al* 2005, Sihag *et al* 2007), and for processes that affect cell adhesion and motility through possible regulation of the recycling and trafficking of $\beta 1$ integrins to the cell membrane (Fortin *et al* 2010, Ivaska *et al* 2002). The spatial colocalization of vimentin and PKC-ε has been determined through immunofluorescence and it has been confirmed that vimentin and $\beta 1$ integrins colocalize in spreading control cells. However, this colocalization can be disrupted after knock-down of filamin A or treatment with bisindolylmaleimide (BIM), which is a cell permeable and reversible inhibitor of PKC, to block PKC. Moreover, the cytoplasmic distribution of vimentin was pronouncedly altered among filamin-A-expressing, filamin-A-deficient and BIM-treated cells. In particular, control cells displayed a well-defined vimentin filament network throughout the cytoplasm, in which vimentin was located in cell extensions. In contrast, the location of vimentin in filamin A-deficient or BIM-treated cells appeared as a single focal accumulation in the cytoplasm. These data support the hypothesis that filamin A expression and PKC activity are both required for the reorganization and redistribution of vimentin.

What is the role of filamin A in PKC-driven phosphorylation of vimentin in respect of the association between PKC and filamin A? Coimmunoprecipitation revealed that filamin A associates with phosphorylated vimentin as well as with phosphorylated PKC-ε, which leads to it being catalytically active through its phosphorylation at serine-729 (Cenni *et al* 2002). Indeed, using coimmunofluorescence, the spatial colocalization of filamin A and PKC-ε can be found after 30 min of cell spreading on collagen-coated planar substrates.

The effect of filamin A knock-down on PKC-mediated phosphorylation of vimentin was analyzed using phospho-specific antibodies on serine-6, serine-33, serine-38 and serine-50, all of which are known to be phosphorylated by PKC (Ando *et al* 1989, Ivaska *et al* 2005, Ogawara *et al* 1995, Yasui *et al* 2001). Knock-down of filamin A eliminated vimentin phosphorylation at serines-6, -38 and -50 in spreading cells, whereas there was no detectable alteration of phosphorylation at serine-33. In more detail, filamin A-deficient cells revealed lower levels of the phosphorylated, catalytically active form of PKC-ε. These findings emphasize that filamin A is crucial for PKC-mediated phosphorylation of vimentin. However, phosphorylation of vimentin at serines-6, -38 and -50 can be restored after pretreatment of filamin-knock-down cells with the PKC activator, bryostatin. In addition, the impaired assembly of cell extensions and reduced cell surface β1 integrin expression found in filamin-deficient cells is reversed after bryostatin treatment.

The major finding is that in spreading fibroblasts, filamin A can bind to vimentin and hence act as a scaffold for PKC-provided phosphorylation of vimentin. Moreover, these functions regulate the activation of β1 integrins and their trafficking to the cell surface. The direct interaction between filamin and vimentin leads to a novel mechanism by which the actin and intermediate filament cytoskeletons coregulate cell adhesion and spreading. Although actin and intermediate filament cytoskeletons have distinct roles in regulating cell adhesion, spreading and motility (Ivaska *et al* 2007, Serrels *et al* 2007, Vicente-Manzanares *et al* 2009), the functions of these two cytoskeletal systems may be interdependent and coregulated (Arocena 2006, Chang and Goldmann 2004, Green *et al* 1987, Kasas *et al* 2005). Proteins that bind both actin and intermediate filaments such as plectin (Seifert *et al* 1992, Svitkina *et al* 1996), BPAG1 (Yang *et al* 1996) and fimbrin (Correia *et al* 1999) can combine the functions of these different cytoskeletal systems. Taken together, these findings provide a new mechanism in which the actin binding protein filamin A serves as a linker protein between vimentin and actin filaments.

Filamin A colocalizes with vimentin (Brown and Binder 1992) and filamin A and vimentin can bind directly *in vitro*. In particular, the two proteins have been reported to colocalize at the extensions of spreading cells, which is indeed consistent with the colocalization of filamin A and vimentin in pseudopodia of cancer cells (Jia *et al* 2005). As there is direct interaction between these proteins and their roles in regulating cell motility (Eckes *et al* 1998, 36, McInroy and Maata 2007, Stossel *et al* 2001), it seems to be likely that filamin A and vimentin also coregulate early events in cell spreading.

Filamin A and vimentin coregulate cell spreading
Cell spreading and migration require the extrusion of cell extensions such as filopodia and lamellipodia (Guillou *et al* 2008, Johnston *et al* 2008, Ladwein and Rottner 2008). Indeed, the formation of cell extensions can be evaluated by recording the numbers, and they can even be validated as authentic filopodia/lamellipodia through

immunostaining for vinculin, paxillin and cortactin. The assembly of these actin-rich cell extensions is facilitated by β1 integrins, small GTPases and actin-binding proteins (Li *et al* 2005, Zimmerman *et al* 2004) such as filamin A regulate β1 integrin activity and small GTPases by binding to the actin filaments (Bellanger *et al* 2000, Stossel *et al* 2001, Vadlamudi *et al* 2002). Knock-down of filamin A and/or vimentin inhibits the extrusion of cell extensions, which is consistent with hindered migration in vimentin-deficient cells (Eckes *et al* 1998, McInroy and Maata 2007). In addition, knock-down of filamin A and vimentin blocked cell spreading comparable to cells, in which either filamin A or vimentin was knocked down, suggesting that filamin A and vimentin coregulate early processes in cell spreading.

Actin-free barbed ends have been reported to decrease in number after filamin A knock-down, whereas a knock-down of vimentin has no effect. Surprisingly, combined knock-down of filamin A and vimentin further decreases free barbed end formation, suggesting that vimentin may also support actin assembly in the absence of filamin A, partially superseding the role of filamin A. In particular, vimentin associates with PAK, which is a protein that facilitates actin free barbed end formation via the LIM kinase–cofilin pathway (Chan *et al* 2002, Goto *et al* 2002). Indeed, mutation of vimentin's PAK-binding site eliminated PAK activation (Li *et al* 2005), indicating that vimentin is involved in PAK activation and actin free barbed end formation. However, vimentin does not modulate free barbed end formation independently of filamin A and in addition knock-down of vimentin does not interfere with binding between filamin A and Rac or Cdc42, the small GTPases regulating filopodia and lamellipodia formation, respectively (Nobes and Hall 1995). These findings show that filamin A–vimentin interactions do not affect actin assembly directly, while they seem to regulate β1 integrin function as an upstream target protein in the cell-spreading process.

The coregulation of β1 integrin expression and activation by filamin A and vimentin
In the early stages of cell spreading and migration, cell surface β1 integrins are endocytosed and still need to be transported back to the cell membrane (Dunphy *et al* 2006, Jovic *et al* 2007). As the location, availability and activation of cell surface integrins are crucial regulatory factors in cell adhesion and spreading (Guillou *et al* 2008), it has been supposed that filamin A and vimentin regulate β1 integrin activation in spreading. Moreover, β1 integrin activation assays have shown that initial β1 integrin–ligand binding is dependent on the presence of both filamin A and vimentin. In line with cell spreading, ligand-bound β1 integrins decrease equivalently after knock-down of filamin A, vimentin, or both proteins, indicating that they coregulate β1 integrin activation. In particular, focal adhesion proteins have revealed that filamin A and vimentin are both required for the targeting and localization of vinculin and paxillin to focal adhesions. However, other studies have shown the importance of vimentin for cell adhesion (Eckes *et al* 1998, Nieminen *et al* 2006, Tsuruta and Jones 2003), thus filamin A has been identified as a novel regulator of cell adhesion. Whereas some reports suggest that cell surface expression of β1 integrins is regulated by filamin A (Meyer *et al* 1998) or vimentin (Nieminen *et al* 2006,

Rizki *et al* 2007), other results show that filamin A and vimentin coregulate β1 integrin trafficking, β1 integrin activation, cell adhesion and spreading.

The role of vimentin phosphorylation by PKC in delivering β1 integrin activity and cell spreading

PKC-ε-facilitated vimentin phosphorylation, β1 integrin recycling and directed cell migration seem to be linked processes (Ivaska *et al* 2005), while the regulatory role of vimentin phosphorylation during the onset of cell spreading is not yet well understood. In particular, it has been reported that vimentin knock-down (as well as inhibition of PKC) impairs the formation of cell extensions and reduces the β1 integrin–ligand binding and expression on the cell surface. Moreover, cell extensions in the early stages of spreading have been found to be enriched with filamin A, vimentin and PKC-ε. Thus, filamin A, vimentin and phospho-vimentin seem to be involved in the same pathway for regulating cell spreading. As β1 integrins associate with filamin A (Kiema *et al* 2006, Loo *et al* 1998) and with vimentin (Ivaska *et al* 2005, Kreis *et al* 2005), while filamin A additionally associates with PKC (Glogauer *et al* 1998, Tigges *et al* 2003), it has been proposed that filamin A regulates the activity of β1 integrins and cell spreading through facilitating the phosphorylation of vimentin, which is performed by PKC.

The role of filamin A in PKC-facilitated vimentin phosphorylation

There is a close spatial relationship between filamin A and PKC, which is in line with the finding that filamin A is phosphorylated by PKC (Glogauer *et al* 1998, Tigges *et al* 2003). In more detail, the catalytically active, phosphorylated form of PKC-ε can be coimmunoprecipitated with filamin A, which is important since the PKC-ε isoform is essential for integrin recycling (Ivaska *et al* 2005). Moreover, knockdown of filamin A impairs vimentin phosphorylation at serines-6, -38 and -50, which are known PKC targets for phosphorylation (Ando *et al* 1989, Ivaska *et al* 2005, Ogawara *et al* 1995). While these finding do not indicate a direct interaction between filamin A and PKC-ε, they strongly suggest a role for filamin A in promoting vimentin phosphorylation by PKC. Moreover, knock-down of filamin A prevents the translocation of vimentin to the cell periphery. The same result is observed after inhibition of PKC by the PKC inhibitor, BIM. These results support the model in which phosphorylation by PKC-ε is necessary for vimentin reorganization and for β1 integrin trafficking to the cell membrane (Ivaska *et al* 2005). In addition, filamin A regulates this crucial vimentin phosphorylation step. Taken together, filamin A, vimentin and PKC regulate β1 integrin activity, cell adhesion and the exertion of early cell extensions caused by the binding of filamin A to vimentin.

6.3 The role of nuclear intermediate filaments in cell invasion

Mutations in nuclear lamins or other proteins of the nuclear envelope are the cause of a group of phenotypically diverse genetic disorders called laminopathies. These laminopathies have diverse symptoms, ranging from muscular dystrophy through neuropathy to premature aging syndromes. The precise disease mechanisms are not yet well understood. However, there has been substantial progress in our understanding of the biological nature of the nuclear structure and thus also of laminopathies. In particular, the dysfunction of the nuclear envelope is based on altered nuclear activity, impaired structural dynamics and signal transduction. As a result of these observations, small molecules are used as effective therapeutic substances.

Since the discovery of the constituents of the nuclear lamina in 1978 (Gerace *et al* 1978), nuclear lamins have been discussed in respect of and proposed for diverse roles in almost everything that takes place in the nucleus. These early studies were based on biochemistry and cell biology, with the goal of revealing the basic principles that regulate nuclear organization. The nuclear envelope became a medical research field in the mid-1990s, as mutations in emerin were found in patients with Emery–Dreifuss muscular dystrophy (EDMD) (Bione *et al* 1994). The LMNA gene that encodes all A-type nuclear lamins has also been linked to EDMD (Bonne *et al* 2000). Therefore, extensive connections between nuclear structure and human diseases have been identified. With approximately 15 diseases, including dystrophic and progeroid syndromes, being associated with LMNA mutations and mutations in the genes encoding associated nuclear envelope proteins, the following questions have been raised. Why do alterations in nuclear envelope proteins lead directly to a disease? What are the precise mechanisms underlying the pathology of the disease? Do A-type lamins play a role in aging? Although many discoveries have been made in this field, many aspects are still not well known.

Nuclear lamins
The nuclear envelope consists of two membranes: (i) the outer nuclear membrane, which is continuously connected to the endoplasmic reticulum, and (ii) the inner nuclear membrane, which is associated with the nuclear lamina. The nuclear pore complexes are holes in the nuclear envelope that facilitate transport between the cytoplasm and nucleus. The nuclear lamina consists of nuclear lamins, which have been identified as lamins A, B and C (Gerace *et al* 1978). These proteins belong to the only class of intermediate filament proteins in the nucleus and build the associated filamentous structures that underlie the nuclear envelope and interact with neighboring proteins (Gerace and Huber 2012). Lamins A and C are classed as A-type lamins and are encoded by the LMNA gene through alternative splicing. In addition, three different lamin B family members (B-type lamins) are encoded by only two genes (lamin B1 by LMNB1 and lamins B2 and B3 by LMNB2).

In more detail, A- and B-type lamins have totally different properties, possibly due to their different isoelectric points, which enable B-type lamins to be associated with the nuclear envelope during mitosis, whereas A-type lamins are soluble. Moreover, the expression patterns are different; B-type lamins are expressed in most or all cell types and A-type lamins are expressed only during cell differentiation in many developmental lineages (Röber *et al* 1989). At the cellular level, it has been proposed that both A-type and B-type lamins have structural roles in the nucleus, and also roles in a range of other activities, such as coordination of transcription and replication. While the specific functions of A-type lamins are still elusive, a number of discoveries point to key interactions between lamins and cell proliferation, differentiation and stress response pathways. In more detail, both A- and B-type lamins undergo post-translational processing based on a C-terminal CaaX motif that leads to a series of modifications (Weber *et al* 1989). Only lamin C avoids this, as it lacks the C terminus due to alternative splicing of the LMNA transcript. As a first step, the cysteine residue is farnesylated. Next, proteolytic processing evokes the cleavage after the cysteine residue, followed by a carboxymethylation of the new C-terminal residue. Many membrane-associated proteins such as Ras undergo this processing event. In more detail, in the case of lamin A, isoprenylation is a transient event, as a second proteolytic event facilitated by the zinc metalloproteinase Zmpste24 leads to excision of another 15 amino acids. Thus, mature lamin A lacks the modified cysteine. This process is clearly important for pathologic states, because laminopathies are linked to altered processing of lamin A and loss-of-function mutations in ZMPSTE24.

The purpose of farnesylation of lamin A is still unclear. It has been thought that the transient farnesylation event was needed, through association of the hydrophobic farnesyl group with the nuclear envelope, to provide initial recruitment of lamin A to the nuclear periphery (Hennekes and Nigg 1994). However, after assembly into filaments, farnesylation of lamin A may no longer be essential. Consistent with this, the nucleus has been reported to be the site of both lamin A carboxymethylation and proteolytic cleavage by ZMPSTE24. However, several studies using mice and/or cells have been engineered to express mutant forms of lamin A, indicating that farnesylation is not required for recruitment (Davies *et al* 2011). When only a non-farnesylated version of lamin A is expressed, normal localization of the lamin A variant to the nuclear periphery has been detected (Davies *et al* 2010, Lee *et al* 2010), although mice generated in this way will develop cardiomyopathy. In addition, mice expressing only lamin C (not farnesylated) or a mature (preprocessed) lamin A are unexpectedly normal and have the correct localization of the respective protein to the nuclear periphery. Although these studies do not preclude a more subtle role for lamin A processing in filament assembly or envelope association, they raise questions about the importance of these events in the mouse.

Since the identification of diseases caused by mutations of genes encoding nuclear lamina proteins, studies have focused on understanding the molecular mechanisms causing these specific phenotypes. In particular, it seem to be crucial to understand how the nuclear lamina interacts with structural proteins such as chromatin, transcription factors and other signaling partners, which will reveal the mechanistic links to disease. The overall picture of how lamins regulate all of these pathways and how this regulation leads to disease is becoming clearer, but it is not yet sharp. Knowing the mechanisms by which mutations in lamins cause these rare diseases will help to reveal common conditions and help to provide a good laminopathies model for muscle diseases and cardiomyopathy. As mutations in the nuclear lamina lead to rapid aging-like diseases, defining the precise role of the nuclear lamina in controlling human aging will be of great interest.

6.4 The role of cell division in cellular motility

As discussed above, there is an interaction between intermediate filaments and actin filaments and they work together to provide cellular functions such as cell motility.

An interaction between the three major components of the cytoskeleton was proposed a long time ago, and it is becoming likely that there is such an interaction between intermediate filaments, actin filaments and microtubules. The latter cytoskeletal components play an essential role in effecting cell division by assembling the mitotic spindle. However, microtubules can interfere with the acto-myosin cytoskeleton within the cells in order to regulate cell motility.

Further reading

Akhmanova A and Steinmetz M O 2008 Tracking the ends: a dynamic protein network controls the fate of microtubule tips *Nat. Rev. Mol. Cell Biol.* **9** 309–22

Campellone K G and Welch M D 2010 A nucleator arms race: cellular control of actin assembly *Nat. Rev. Mol. Cell Biol.* **11** 237–51

Chan M W, Arora P D and McCulloch C A 2007 Cyclosporin inhibition of collagen remodeling is mediated by gelsolin *Am. J. Physiol. Cell Physiol.* **293** C1049–58

Chen S C, Kennedy B K and Lampe P D 2013 Phosphorylation of connexin43 on S279/282 may contribute to laminopathy-associated conduction defects *Exp. Cell Res.* **319** 888–96

Chhabra E S and Higgs H N 2007 The many faces of actin: matching assembly factors with cellular structures *Nat. Cell Biol.* **9** 1110–21

Chong S A, Lee W, Arora P D, Laschinger C, Young E W, Simmons C A, Manolson M, Sodek J and McCulloch C A 2007 Methylglyoxal inhibits the binding step of collagen phagocytosis *J. Biol. Chem.* **282** 8510–20

Franke W W, Schiller D L, Hatzfeld M and Winter S 1983 Protein complexes of intermediate-sized filaments: melting of cytokeratin complexes in urea reveals different polypeptide separation characteristics *Proc. Natl Acad. Sci. USA* **80** 7113–7 PMID: 6196784

Gerace M D 2012 Huber Nuclear lamina at the crossroads of the cytoplasm and nucleus *J. Struct. Biol.* **177** 24–31

Glogauer M, Arora P, Yao G, Sokholov I, Ferrier J and McCulloch C A 1997 Calcium ions and tyrosine phosphorylation interact coordinately with actin to regulate cytoprotective responses to stretching *J. Cell Sci.* **110** 11–21 PMID: 9010780

Herrmann H, Bär H, Kreplak L, Strelkov S V and Aebi U 2007 Intermediate filaments: from cell architecture to nanomechanics *Nat. Rev. Mol. Cell Biol.* **8** 562–73

Johansen L D *et al* 2008 IKAP localizes to membrane ruffles with filamin A and regulates actin cytoskeleton organization and cell migration *J. Cell Sci.* **121** 854–64

Knowles G C, McKeown M, Sodek J and McCulloch C A 1991 Mechanism of collagen phagocytosis by human gingival fibroblasts: importance of collagen structure in cell recognition and internalization *J. Cell Sci.* **98** 551–8 PMID: 1650378

Lammermann T and Sixt M 2009 Mechanical modes of 'amoeboid' cell migration *Curr. Opin. Cell Biol.* **21** 636–44

Lee W, Sodek J and McCulloch C A 1996 Role of integrins in regulation of collagen phagocytosis by human fibroblasts *J. Cell Physiol.* **168** 695–704

Li Q F, Spinelli A M, Wang R, Anfinogenova Y, Singer H A and Tang D D 2006 Critical role of vimentin phosphorylation at Ser-56 by p21-activated kinase in vimentin cytoskeleton signaling *J. Biol. Chem.* **281** 34716–24

Meng X, Yuan Y, Maestas A and Shen Z 2004 Recovery from DNA damage-induced G2 arrest requires actin-binding protein filamin-A/actin-binding protein 280 *J. Biol. Chem.* **279** 6098–105

Muchir A, Bonne G, van der Kooi A J, van Meegen M, Baas F, Bolhuis P A, de Visser M and Schwartz K 2000 Identification of mutations in the gene encoding lamins A/C in autosomal dominant limb girdle muscular dystrophy with atrioventricular conduction disturbances (LGMD1B) *Hum. Mol. Genet.* **9** 1453–9

Muchir A, Shan J, Bonne G, Lehnart S E and Worman H J 2009 Inhibition of extracellular signal-regulated kinase signaling to prevent cardiomyopathy caused by mutation in the gene encoding A-type lamins *Hum. Mol. Genet.* **18** 241–7

Muchir A, Wu W and Worman H J 2009 Reduced expression of A-type lamins and emerin activates extracellular signal-regulated kinase in cultured cells *Biochim. Biophys. Acta.* **1792** 75–81

Nakamura F, Osborn T M, Hartemink C A, Hartwig J H and Stossel T P 2007 Structural basis of filamin A functions *J. Cell Biol.* **179** 1011–25

Skalski M and Coppolino M G 2005 SNARE-mediated trafficking of alpha5beta1 integrin is required for spreading in CHO cells *Biochem. Biophys. Res. Commun.* **335** 1199–210

Sun N, Huiatt T W, Paulin D, Li Z and Robson R M 2010 Synemin interacts with the LIM domain protein zyxin and is essential for cell adhesion and migration *Exp. Cell Res.* **316** 491–505

Wade R H 2009 On and around microtubules: an overview *Mol. Biotechnol.* **43** 177–91

Wong P and Coulombe P A 2003 Loss of keratin 6 (K6) proteins reveals a function for intermediate filaments during wound repair *J. Cell Biol.* **163** 327–37

Yang S H, Qiao X, Fong L G and Young S G 2008 Treatment with a farnesyltransferase inhibitor improves survival in mice with a Hutchinson–Gilford progeria syndrome mutation *Biochim. Biophys. Acta.* **1781** 36–9

References

Abd El-Rehim D M, Pinder S E, Paish C E, Bell J, Blamey R W, Robertson J F, Nicholson R I and Ellis I O 2004 Expression of luminal and basal cytokeratins in human breast carcinoma *J. Pathol.* **203** 661–71

Akita Y, Kawasaki H, Imajoh-Ohmi S, Fukuda H, Ohno S, Hirano H, Ono Y and Yonekawa H 2007 Protein kinase C epsilon phosphorylates keratin 8 at Ser8 and Ser23 in GH4C1 cells stimulated by thyrotropin-releasing hormone *FEBS J.* **274** 3270–85

Alam H *et al* 2011a Loss of keratin 8 phosphorylation leads to increased tumor progression and correlates with clinico-pathological parameters of OSCC patients *PLoS ONE* **6** e27767

Alam H, Kundu S T, Dalal S N and Vaidya M M 2011b Loss of keratins 8 and 18 leads to alterations in alpha6beta4-integrin-mediated signaling and decreased neoplastic progression in an oral-tumour-derived cell line *J. Cell Sci.* **124** 2096–106

Albrecht-Buehler G and Lancaster R M 1976 A quantitative description of the extension and retraction of surface protrusions in spreading 3T3 mouse fibroblasts *J. Cell Biol.* **71** 370–82

Amano M, Nakayama M and Kaibuchi K 2010 Rho-kinase/ROCK: A key regulator of the cytoskeleton and cell polarity *Cytoskeleton (Hoboken)* **67** 545–54

Ando S, Tanabe K, Gonda Y, Sato C and Inagaki M 1989 Domain- and sequence-specific phosphorylation of vimentin induces disassembly of the filament structure *Biochemistry* **28** 2974–9

Arce C A, Casale C H and Barra H S 2008 Submembraneous microtubule cytoskeleton: regulation of ATPases by interaction with acetylated tubulin *FEBS J.* **275** 4664–74

Arocena M 2006 Effect of acrylamide on the cytoskeleton and apoptosis of bovine lens epithelial cells *Cell Biol. Int.* **30** 1007–12

Arthur W T and Burridge K 2001 RhoA inactivation by p190RhoGAP regulates cell spreading and migration by promoting membrane protrusion and polarity *Mol. Biol. Cell* **12** 2711–20

Artym V V, Zhang Y, Seillier-Moiseiwitsch F, Yamada K M and Mueller S C 2006 Dynamic interactions of cortactin and membrane type 1 matrix metalloproteinase at invadopodia: defining the stages of invadopodia formation and function *Cancer Res.* **66** 3034–43

Asparuhova M B, Gelman L and Chiquet M 2009 Role of the actin cytoskeleton in tuning cellular responses to external mechanical stress *Scand. J. Med. Sci. Sports* **19** 490–9

Ausmees N, Kuhn J R and Jacobs-Wagner C 2003 The bacterial cytoskeleton: an intermediate filament-like function in cell shape *Cell* **115** 705–13

Bachant J B and Klymkowsky M W 1996 A nontetrameric species is the major soluble form of keratin in Xenopus oocytes and rabbit reticulocyte lysates *J. Cell Biol.* **132** 153–65

Baribault H, Blouin R, Bourgon L and Marceau N 1989 Epidermal growth factor-induced selective phosphorylation of cultured rat hepatocyte 55-kD cytokeratin before filament reorganization and DNA synthesis *J. Cell Biol.* **109** 1665–76

Barry D M, Millecamps S, Julien J P and Garcia M L 2007 New movements in neurofilament transport, turnover and disease *Exp. Cell Res.* **313** 2110–20

Bass-Zubek A E, Godsel L M, Delmar M and Green K J 2009 Plakophilins: multifunctional scaffolds for adhesion and signaling *Curr. Opin. Cell Biol.* **21** 708–16

Batsios P, Peter T, Baumann O, Stick R, Meyer I and Graf R 2012 A lamin in lower eukaryotes? *Nucleus* **3** 237–43

Bausch A R, Moller W and Sackmann E 1999 Measurement of local viscoelasticity and forces in living cells by magnetic tweezers *Biophys. J.* **76** 573–9

Beil M *et al* 2003 Sphingosylphosphorylcholine regulates keratin network architecture and visco-elastic properties of human cancer cells *Nat. Cell Biol.* **5** 803–11

Bellanger J M, Astier C, Sardet C, Ohta Y, Stossel T P and Debant A 2000 The Rac1- and RhoG-specific GEF domain of Trio targets filamin to remodel cytoskeletal actin *Nat. Cell Biol.* **2** 888–92

Berkowitz P, Hu P, Liu Z, Diaz L A, Enghild J J, Chua M P and Rubenstein D S 2005 Desmosome signaling. Inhibition of p38 MAPK prevents pemphigus vulgaris IgG-induced cytoskeleton reorganization *J. Biol. Chem.* **280** 23778–84

Bezault J, Bhimani R, Wiprovnick J and Furmanski P 1994 Human lactoferrin inhibits growth of solid tumors and development of experimental metastases in mice *Cancer Res.* **54** 2310–2 PMID: 8162571

Bhattacharya R, Gonzalez A M, Debiase P J, Trejo H E, Goldman R D, Flitney F W and Jones J C 2009 Recruitment of vimentin to the cell surface by beta3 integrin and plectin mediates adhesion strength *J. Cell Sci.* **122** 1390–400

Bione S, Maestrini E, Rivella S, Mancini M, Regis S, Romeo G and Toniolo D 1994 Identification of a novel X-linked gene responsible for Emery–Dreifuss muscular dystrophy *Nat. Genet.* **8** 323–7

Blessing M, Rüther U and Franke W W 1993 Ectopic synthesis of epidermal cytokeratins in pancreatic islet cells of transgenic mice interferes with cytoskeletal order and insulin production *J. Cell Biol.* **120** 743–55

Boczonadi V, McInroy L and Määttä A 2007 Cytolinker cross-talk: periplakin N-terminus interacts with plectin to regulate keratin organisation and epithelial migration *Exp. Cell Res.* **313** 3579–91

Bonne G *et al* 2000 Clinical and molecular genetic spectrum of autosomal dominant Emery-Dreifuss muscular dystrophy due to mutations of the lamin A/C gene *Ann Neurol.* **48** 170–80

Bordeleau F, Bessard J, Marceau N and Sheng Y 2011 Measuring integrated cellular mechanical stress response at focal adhesions by optical tweezers *J. Biomed. Opt.* **16** 095005

Bordeleau F, Bessard J, Sheng Y and Marceau N 2008 Keratin contribution to cellular mechanical stress response at focal adhesions as assayed by laser tweezers *Biochem. Cell Biol.* **86** 352–9

Bordeleau F, Galarneau L, Gilbert S, Loranger A and Marceau N 2010 Keratin 8/18 modulation of protein kinase C-mediated integrin-dependent adhesion and migration of liver epithelial cells *Mol. Biol. Cell* **21** 1698–713

Bordeleau F, Myrand Lapierre M E, Sheng Y and Marceau N 2012 Keratin 8/18 regulation of cell stiffness-extracellular matrix interplay through modulation of Rho-mediated actin cytoskeleton dynamics *PLoS ONE* **7e38780**

Bornslaeger E A, Corcoran C M, Stappenbeck T S and Green K J 1996 Breaking the connection: displacement of the desmosomal plaque protein desmoplakin from cell–cell interfaces disrupts anchorage of intermediate filament bundles and alters intercellular junction assembly *J. Cell Biol.* **134** 985–1001

Bragulla H H and Homberger D G 2009 Structure and functions of keratin proteins in simple, stratified, keratinized and cornified epithelia *J. Anat.* **214** 516–59

Brotschi E A, Hartwig J H and Stossel T P 1978 The gelation of actin by actin-binding protein *J. Biol. Chem.* **253** 8988–93 PMID: 721823

Brown A 2003 Axonal transport of membranous and nonmembranous cargoes: a unified perspective *J. Cell Biol.* **160** 817–21

Brown K D and Binder L I 1992 Identification of the intermediate filament-associated protein gyronemin as filamin. Implications for a novel mechanism of cytoskeletal interaction *J. Cell Sci.* **102** 19–30

Bulinski J C and Gundersen G G 1991 Stabilization of post-translational modification of microtubules during cellular morphogenesis *Bioessays* **13** 285–93

Burgstaller G, Gregor M, Winter L and Wiche G 2010 Keeping the vimentin network under control: cell–matrix adhesion-associated plectin 1f affects cell shape and polarity of fibroblasts *Mol. Biol. Cell* **21** 3362–75

Burke B and Stewart C L 2013 The nuclear lamins: flexibility in function *Nat. Rev. Mol. Cell Biol.* **14** 13–24

Busch T, Armacki M, Eiseler T, Joodi G, Temme C, Jansen J, von Wichert G, Omary M B, Spatz J and Seufferlein T 2012 Keratin 8 phosphorylation regulates keratin reorganization and migration of epithelial tumor cells *J. Cell Sci.* **125** 2148–59

Butcher D T, Alliston T and Weaver V M 2009 A tense situation: forcing tumor progression *Nat. Rev. Cancer* **9** 108–22

Campbell I D 2008 Studies of focal adhesion assembly *Biochem. Soc. Trans.* **36** 263–6

Carragher N O and Frame M C 2004 Focal adhesion and actin dynamics: a place where kinases and proteases meet to promote invasion *Trends Cell Biol.* **14** 241–9

Cenni V, Doppler H, Sonnenburg E D, Maraldi N, Newton A C and Toker A 2002 Regulation of novel protein kinase C epsilon by phosphorylation *Biochem. J.* **363** 537–45

Chamcheu J C, Virtanen M, Navsaria H, Bowden P E, Vahlquist A and Törmä H 2010 Epidermolysis bullosa simplex due to KRT5 mutations: mutation-related differences in cellular fragility and the protective effects of trimethylamine N-oxide in cultured primary keratinocytes *Br. J. Dermatol.* **162** 980–9

Chan W K, Yabe J T, Pimenta A F, Ortiz D and Shea T B 2003 Growth cones contain a dynamic population of neurofilament subunits *Cell Motil. Cytoskeleton* **54** 195–207

Chan W, Kozma R, Yasui Y, Inagaki M, Leung T, Manser E and Lim L 2002 Vimentin intermediate filament reorganization by Cdc42: involvement of PAK and p70 S6 kinase *Eur. J. Cell Biol.* **81** 692–701

Chandler J S, Calnek D and Quaroni A 1991 Identification and characterization of rat intestinal keratins. Molecular cloning of cDNAs encoding cytokeratins 8, 19 and a new 49 kDa type I cytokeratin (cytokeratin 21) expressed by differentiated intestinal epithelial cells *J. Biol. Chem.* **266** 11932–8 PMID: 1711044

Chang L and Goldman R D 2004 Intermediate filaments mediate cytoskeletal crosstalk *Nat. Rev. Mol. Cell Biol.* **5** 601–13

Chang L, Shav-Tal Y, Trcek T, Singer R H and Goldman R D 2006 Assembling an intermediate filament network by dynamic cotranslation *J. Cell Biol.* **172** 747–58

Cheng C W, Wang H W, Chang C W, Chu H W, Chen C Y, Yu J C, Chao J I, Liu H F, Ding S L and Shen C Y 2012 MicroRNA-30a inhibits cell migration and invasion by down-regulating vimentin expression and is a potential prognostic marker in breast cancer *Breast Cancer Res. Treat* **134** 1081–93

Chou C F, Riopel C L, Rott L S and Omary M B 1993 A significant soluble keratin fraction in 'simple' epithelial cells. Lack of an apparent phosphorylation and glycosylation role in keratin solubility *J. Cell Sci.* **105** 433–44 PMID: 7691841

Chou Y H, Flitney F W, Chang L, Mendez M, Grin B and Goldman R D 2007 The motility and dynamic properties of intermediate filaments and their constituent proteins *Exp. Cell Res.* **313** 2236–43

Chu Y W, Runyan R B, Oshima R G and Hendrix M J 1993 Expression of complete keratin filaments in mouse L cells augments cell migration and invasion *Proc. Natl Acad. Sci. USA* **90** 4261–5

Chu Y W, Seftor E A, Romer L H and Hendrix M J 1996 Experimental coexpression of vimentin and keratin intermediate filaments in human melanoma cells augments motility *Am. J. Pathol.* **148** 63–9 PMID: 8546227

Chung B-M, Rotty J D and Coulombe P A 2014 Networking galore: intermediate filaments and cell migration *Curr. Opin. Cell Biol.* **25** 600–12

Chung B M, Murray C I, Van Eyk J E and Coulombe P A 2012 Identification of novel interaction between annexin A2 and keratin 17: evidence for reciprocal regulation *J. Biol. Chem.* **287** 7573–81

Colakoğlu G and Brown A A 2009 Intermediate filaments exchange subunits along their length and elongate by end-to-end annealing *J. Cell Biol.* **185** 769–77

Correia I, Chu D, Chou Y H, Goldman R D and Matsudaira P 1999 Integrating the actin and vimentin cytoskeletons. Adhesion-dependent formation of fimbrin-vimentin complexes in macrophages *J. Cell Biol.* **146** 831–42

Coulombe P A and Wong P 2004 Cytoplasmic intermediate filaments revealed as dynamic and multipurpose scaffolds *Nat. Cell Biol.* **6** 699–706

Coulombe P A, Kerns M L and Fuchs E 2009 Epidermolysis bullosa simplex: a paradigm for disorders of tissue fragility *J. Clin. Invest.* **119** 1784–93

Critchley D R 2000 Focal adhesions–the cytoskeletal connection *Curr. Opin. Cell Biol.* **12** 133–9

Davies B S, Barnes R H 2nd, Tu Y, Ren S, Andres D A, Spielmann H P, Lammerding J, Wang Y, Young S G and Fong L G 2010 An accumulation of non-farnesylated prelamin A causes cardiomyopathy but not progeria *Hum. Mol. Genet.* **19** 2682–694

Davies B S, Coffinier C, Yang S H, Barnes R H 2nd, Jung H J, Young S G and Fong L G 2011 Investigating the purpose of prelamin A processing *Nucleus* **2** 4–9

de Silva Rudland S, Platt-Higgins A, Winstanley J H, Jones N J, Barraclough R, West C, Carroll J and Rudland P S 2011 Statistical association of basal cell keratins with metastasis-inducing proteins in a prognostically unfavorable group of sporadic breast cancers *Am. J. Pathol.* **179** 1061–72

Defilippi P, Olivo C, Venturino M, Dolce L, Silengo L and Tarone G 1999 Actin cytoskeleton organization in response to integrin-mediated adhesion *Micro. Res. Tech.* **47** 67–78

Dellagi K, Tabilio A, Portier M M, Vainchenker W, Castaigne S, Guichard J, Breton-Gorius J and Brouet J C 1985 Expression of vimentin intermediate filament cytoskeleton in acute nonlymphoblastic leukemias *Blood* **65** 1444–52 PMID: 3888314

Deng M *et al* 2013 Lactotransferrin acts as a tumor suppressor in nasopharyngeal carcinoma by repressing AKT through multiple mechanisms *Oncogene* **32** 4273–83

Depianto D, Kerns M L, Dlugosz A A and Coulombe P A 2010 Keratin 17 promotes epithelial proliferation and tumor growth by polarizing the immune response in skin *Nat. Genet.* **42** 910–4

Desai B V, Harmon R M and Green K J 2009 Desmosomes at a glance *J. Cell Sci.* **122** 4401–7

Díaz-Moreno I, Hollingworth D, Frenkiel T A, Kelly G, Martin S, Howell S, García-Mayoral M, Gherzi R, Briata P and Ramos A 2009 Phosphorylation-mediated unfolding of a KH domain regulates KSRP localization via 14-3-3 binding *Nat. Struct. Mol. Biol.* **16** 238–46

Dunphy J L, Moravec R, Ly K, Lasell T K, Melancon P and Casanova J E 2006 The Arf6 GEF GEP100/BRAG2 regulates cell adhesion by controlling endocytosis of beta1 integrins *Curr. Biol.* **16** 315–20

Dupin I, Sakamoto Y and Etienne-Manneville S 2011 Cytoplasmic intermediate filaments mediate actin-driven positioning of the nucleus *J. Cell Sci.* **124** 865–72

Eckes B *et al* 1998 Impaired mechanical stability, migration and contractile capacity in vimentin-deficient fibroblasts *J. Cell Sci.* **111** 1897–907 PMID: 9625752

Eger A, Stockinger A, Wiche G and Foisner R 1997 Polarization-dependent association of plectin with desmoplakin and the lateral submembrane skeleton in MDCK cells *J. Cell Sci.* **110** 1307–16

Eichner R, Sun T T and Aebi U 1986 The role of keratin subfamilies and keratin pairs in the formation of human epidermal intermediate filaments *J. Cell Biol.* **102** 1767–77

Erber A, Riemer D, Bovenschulte M and Weber K 1998 Molecular phylogeny of metazoan intermediate filament proteins *J. Mol. Evol.* **47** 751–62

Eriksson J E, Dechat T, Grin B, Helfand B, Mendez M, Pallari H M and Goldman R D 2009 Introducing intermediate filaments: from discovery to disease *J. Clin. Invest.* **119** 1763–71

Eriksson J E, He T, Trejo-Skalli A V, Harmala-Brasken A S, Hellman J, Chou Y H and Goldman R D 2004 Specific *in vivo* phosphorylation sites determine the assembly dynamics of vimentin intermediate filaments *J. Cell Sci.* **117** 919–32

Faure E, Garrouste F, Parat F, Monferran S, Leloup L, Pommier G, Kovacic H and Lehmann M 2012 P2Y2 receptor inhibits EGF-induced MAPK pathway to stabilize keratinocyte hemi-desmosomes *J. Cell Sci.* **125** 4264–77

Felkl M, Tomas K, Smid M, Mattes J, Windoffer R and Leube R E 2012 Monitoring the cytoskeletal EGF response in live gastric carcinoma cells *PLoS ONE* **7e45280**

Feng Y and Walsh C A 2004 The many faces of filamin: a versatile molecular scaffold for cell motility and signalling *Nat. Cell Biol.* **6** 1034–8

Flitney E W, Kuczmarski E R, Adam S A and Goldman R D 2009 Insights into the mechanical properties of epithelial cells: the effects of shear stress on the assembly and remodeling of keratin intermediate filaments *FASEB J.* **23** 2110–9

Florenes V A, Holm R, Myklebost O, Lendahl U and Fodstad O 1994 Expression of the neuroectodermal intermediate filament nestin in human melanomas *Cancer Res.* **54** 354–6 PMID: 8275467

Fortier A M, Asselin E and Cadrin M 2013 Keratin 8 and 18 Loss in epithelial cancer cells increases collective cell migration and cisplatin sensitivity through Claudin1 up-regulation *J. Biol. Chem.* **288** 11555–71

Fortier A M, Riopel K, Désaulniers M and Cadrin M 2010 Novel insights into changes in biochemical properties of keratins 8 and 18 in griseofulvin-induced toxic liver injury *Exp. Mol. Pathol.* **89** 117–25

Fortin S, Le Mercier M, Camby I, Spiegl-Kreinecker S, Berger W, Lefranc F and Kiss R 2010 Galectin-1 is implicated in the protein kinase C epsilon/vimentin-controlled trafficking of integrin-beta1 in glioblastoma cells *Brain Pathol.* **20**(1) 39–49

Frame M and Norman J 2008 A tal(in) of cell spreading *Nat. Cell Biol.* **10** 1017–9

Friedl P and Wolf K 2010 Plasticity of cell migration: a multiscale tuning model *J. Cell Biol.* **188** 11–9

Friedl P, Wolf K and Lammerding J 2011 Nuclear mechanics during cell migration *Curr. Opin. Cell Biol.* **23** 55–64

Frijns E, Sachs N, Kreft M, Wilhelmsen K and Sonnenberg A 2010 EGF-induced MAPK signaling inhibits hemidesmosome formation through phosphorylation of the integrin beta4 *J. Biol. Chem.* **285** 37650–62

Galarneau L, Loranger A, Gilbert S and Marceau N 2007 Keratins modulate hepatic cell adhesion, size and G1/S transition *Exp. Cell Res.* **313** 179–94

Gallicano G I, Kouklis P, Bauer C, Yin M, Vasioukhin V, Degenstein L and Fuchs E 1998 Desmoplakin is required early in development for assembly of desmosomes and cytoskeletal linkage *J. Cell Biol.* **143** 2009–22

Gardner M K, Hunt A J, Goodson H V and Odde D J 2008 Microtubule assembly dynamics: new insights at the nanoscale *Curr. Opin. Cell Biol.* **20** 64–70

Geiger B, Bershadsky A, Pankov R and Yamada K M 2001 Transmembrane crosstalk between the extracellular matrix-cytoskeleton crosstalk *Nat. Rev. Mol. Cell Biol.* **2** 793–805

Gerace L and Huber M D 2012 Nuclear lamina at the crossroads of the cytoplasm and nucleus *J. Struct. Biol.* **177** 24–31

Gerace L, Blum A and Blobel G 1978 Immunocytochemical localization of the major polypeptides of the nuclear pore complex-lamina fraction Interphase and mitotic distribution *J. Cell Biol.* **79** 546–66

Giancotti F G and Ruoslahti E 1999 Integrin signaling *Science* **285** 1028–32

Gilles C, Polette M, Zahm J M, Tournier J M, Volders L, Foidart J M and Birembaut P 1999 Vimentin contributes to human mammary epithelial cell migration *J. Cell Sci.* **112**(Pt 24) 4615–25 PMID: 10574710

Glogauer M, Arora P, Chou D, Janmey P A, Downey G P and McCulloch C A 1998 The role of actin-binding protein 280 in integrin-dependent mechanoprotection *J. Biol. Chem.* **273** 1689–98

Gonzalez A M, Otey C, Edlund M and Jones J C 2001 Interactions of a hemidesmosome component and actinin family members *J. Cell Sci.* **114** 4197–206 PMID: 11739652

Goto H, Tanabe K, Manser E, Lim L, Yasui Y and Inagaki M 2002 Phosphorylation and reorganization of vimentin by p21-activated kinase (PAK) *Genes. Cells* **7** 91–7

Goto M, Sumiyoshi H, Sakai T, Fassler R, Ohashi S, Adachi E, Yoshioka H and Fujiwara S 2006 Elimination of epiplakin by gene targeting results in acceleration of keratinocyte migration in mice *Mol. Cell Biol.* **26** 548–58

Green K J, Geiger B, Jones J C, Talian J C and Goldman R D 1987 The relationship between intermediate filaments and microfilaments before and during the formation of desmosomes and adherens-type junctions in mouse epidermal keratinocytes *J. Cell Biol.* **104** 1389–402

Green K J, Getsios S, Troyanovsky S and Godsel L M 2010 Intercellular junc- tion assembly, dynamics, and homeostasis *Cold Spring Harb. Perspect Biol.* **2** a000125

Guck J *et al* 2005 Optical deformability as an inherent cell marker for testing malignant transformation and metastatic competence *Biophys. J.* **88** 3689–98

Guillou H, Depraz-Depland A, Planus E, Vianay B, Chaussy J, Grichine A, Albiges-Rizo C and Block M R 2008 Lamellipodia nucleation by filopodia depends on integrin occupancy and downstream Rac1 signaling *Exp. Cell Res.* **314** 478–88

Guo L, Degenstein L, Dowling J, Yu Q C, Wollmann R, Perman B and Fuchs E 1995 Gene targeting of BPAG1: abnormalities in mechanical strength and cell migration in stratified epithelia and neurologic degeneration *Cell* **81** 233–43

Hamill K J, Hopkinson S B, Skalli O and Jones J C 2013 Actinin-4 in keratinocytes regulates motility via an effect on lamellipodia stability and matrix adhesions *FASEB J* **27** 546–56

Hansson G K, Starkebaum G A, Benditt E P and Schwartz S M 1984 Fc-mediated binding of IgG to vimentin-type intermediate filaments in vascular endothelial cells *Proc. Natl Acad. Sci. USA* **81** 3103–7

Helfand B T, Chang L and Goldman R D 2004 Intermediate filaments are dynamic and motile elements of cellular architecture *J. Cell Sci.* **117** 133–41

Helfand B T, L. Chang L and R.D. Goldman R D 2003 The dynamic and motile properties of intermediate filaments *Annu. Rev. Cell Dev. Biol.* **19** 445–67

Helfand B T *et al* 2011 Vimentin organization modulates the formation of lamellipodia *Mol. Biol. Cell* **22** 1274–89

Helmke B P, Thakker D B, Goldman R D and Davies P F 2001 Spatiotemporal analysis of flow-induced intermediate filament displacement in living endothelial cells *Biophys. J.* **80** 184–94

Hendrix M J, Seftor E A, Seftor R E and Trevor K T 1997 Experimental co-expression of vimentin and keratin intermediate filaments in human breast cancer cells results in phenotypic interconversion and increased invasive behavior *Am. J. Pathol.* **150** 483–95 PMID: 9033265

Hennekes H and Nigg E A 1994 The role of isoprenylation in membrane attachment of nuclear lamins. A single point mutation prevents proteolytic cleavage of the lamin A precursor and confers membrane binding properties *J. Cell Sci.* **107** 1019–29 PMID: 8056827

Herrmann H, Häner M, Brettel M, Ku N O and Aebi U 1999 Characterization of distinct early assembly units of different intermediate filament proteins *J. Mol. Biol.* **286** 1403–20

Herrmann H, Wedig T, Porter R M, Lane E B and Aebi U 2002 Characterization of early assembly intermediates of recombinant human keratins *J. Struct. Biol.* **137** 82–96

Hesse M, Magin T M and Weber K 2001 Genes for intermediate filament proteins and the draft sequence of the human genome: novel keratin genes and a surprisingly high number of pseudogenes related to keratins 8 and 18 *J. Cell Sci.* **114** 2569–75 PMID: 11683385

Hesse M, Zimek A, Weber K and Magin T M 2004 Comprehensive analysis of keratin gene clusters in humans and rodents *Eur. J. Cell Biol.* **83** 19–26

Hofmann I, Mertens C, Brettel M, Nimmrich V, Schnölzer M and Herrmann H 2000 Interaction of plakophilins with desmoplakin and intermediate filament proteins: an *in vitro* analysis *J. Cell Sci.* **113** 2471–83

Hollenbeck P J, Bershadsky A D, Pletjushkina O T, Tint I S and Vasiliev J M 1989 Intermediate filament collapse is an ATP-dependent and actin-dependent process *J. Cell Sci.* **92** 621–31

Holthöfer B, Windoffer R, Troyanovsky S and Leube R E 2007 Structure and function of desmosomes *Int. Rev. Cytol.* **264** 65–163

Hyder C L, Pallari H M, Kochin V and Eriksson J E 2008 Providing cellular signposts–post-translational modifications of intermediate filaments *FEBS Lett.* **582** 2140–8

Ikegami K and Setou M 2010 Unique post-translational modifications in special- ized microtubule architecture *Cell Struct. Funct.* **35** 15–22

Ishikawa H, Bischoff R and Holtzer H 1968 Mitosis and intermediate-sized filaments in developing skeletal muscle *J. Cell Biol.* **38** 538–55

Ishikawa K *et al* 2010 Epiplakin accelerates the lateral organization of keratin filaments during wound healing *J. Dermatol. Sci.* **60** 95–104

Ivaska J, Pallari H M, Nevo J and Eriksson J E 2007 Novel functions of vimentin in cell adhesion, migration, and signaling *Exp. Cell Res.* **313** 2050–62

Ivaska J, Vuoriluoto K, Huovinen T, Izawa I, Inagaki M and Parker P J 2005 PKCepsilon-mediated phosphorylation of vimentin controls integrin recycling and motility *EMBO J.* **24** 3834–45

Ivaska J, Whelan R D, Watson R and Parker P J 2002 PKC epsilon controls the traffic of beta1 integrins in motile cells *EMBO J.* **21** 3608–19

Iyer S V, Dange P P, Alam H, Sawant S S, Ingle A D, Borges A M, Shirsat N V, Dalal S N and Vaidya M M 2013 Understanding the role of keratins 8 and 18 in neoplastic potential of breast cancer derived cell lines *PLoS ONE* **8** e53532

Izawa I and Inagaki M 2006 Regulatory mechanisms and functions of intermediate filaments: a study using site- and phosphorylation state-specific antibodies *Cancer Sci.* **97** 167–74

Jackson W M, Jaasma M J, Tang R Y and Keaveny T M 2008 Mechanical loading by fluid shear is sufficient to alter the cytoskeletal composition of osteoblastic cells *Am. J. Physiol. Cell Physiol.* **295** C1007–15

Jaitovich A, Mehta S, Na N, Ciechanover A, Goldman R D and Ridge K M 2008 Ubiquitin-proteasome-mediated degradation of keratin intermediate filaments in mechanically stimu-lated A549 cells *J. Biol. Chem.* **283** 25348–55

Jia Z, Barbier L, Stuart H, Amraei M, Pelech S, Dennis J W, Metalnikov P, O'Donnell P and Nabi I R 2005 Tumor cell pseudopodial protrusions. Localized signaling domains coordi-nating cytoskeleton remodeling, cell adhesion, glycolysis, RNA translocation, and protein translation *J. Biol. Chem.* **280** 30564–73

Johnston S A, Bramble J P, Yeung C L, Mendes P M and Machesky L M 2008 Arp2/3 complex activity in filopodia of spreading cells *BMC Cell Biol.* **9** 65

Joosse S A, Hannemann J, Spotter J, Bauche A, Andreas A, Muller V and Pantel K 2012 Changes in keratin expression during metastatic progression of breast cancer: impact on the detection of circulating tumor cells *Clin. Cancer Res.* **18** 993–1003

Jovic M, Naslavsky N, Rapaport D, Horowitz M and Caplan S 2007 EHD1 regulates beta1 integrin endosomal transport: effects on focal adhesions, cell spreading and migration *J. Cell Sci.* **120** 802–14

Jung K S, Kim S U, Ahn S H, Park Y N, Kim do Y, Park J Y, Chon C Y, Choi E H and Han K H 2011 Risk assessment of hepatitis B virus-related hepatocellular carcinoma development using liver stiffness measurement (FibroScan) *Hepatology* **53** 885–94

Kalluri R and Weinberg R A 2009 The basics of epithelial–mesenchymal transition *J. Clin. Invest.* **119** 1420–8

Kaminsky R, Denison C, Bening-Abu-Shach U, Chisholm A D, Gygi S P and Broday L 2009 SUMO regulates the assembly and function of a cytoplasmic intermediate filament protein in C. elegans *Dev. Cell* **17** 724–35

Karantza V 2011 Keratins in health and cancer: more than mere epithelial cell markers *Oncogene* **30** 127–38

Kasas S *et al* 2005 Superficial and deep changes of cellular mechanical properties following cytoskeleton disassembly *Cell Motil Cytoskel* **62** 124–32

Keski-Oja J, Lehto V P and Virtanen I 1981 Keratin filaments of mouse epithelial cells are rapidly affected by epidermal growth factor *J. Cell Biol.* **90** 537–41

Khapare N *et al* 2012 Plakophilin3 loss leads to an increase in PRL3 levels promoting K8 dephosphorylation, which is required for transformation and metastasis *PLoS ONE* **7** e38561

Kiema T, Lad Y, Jiang P, Oxley C L, Baldassarre M, Wegener K L, Campbell I D, Ylanne J and Calderwood D A 2006 The molecular basis of filamin binding to integrins and competition with talin *Mol. Cell* **21** 337–47

Kim J S, Lee C H and Coulombe P A 2010a Modeling the self-organization property of keratin intermediate filaments *Biophys. J.* **99** 2748–56

Kim H, Nakamura F, Lee W, Shifrin Y, Arora P and McCulloch C A 2010b Filamin A is required for vimentin-mediated cell adhesion and spreading *Am. J. Phys. – Cell Phys.* **298**(2) C221–36

Kim H, Nakamura F, Lee W, Hong C, Pérez-Sala D and McCulloch C A 2010c Regulation of cell adhesion to collagen via beta1 integrins is dependent on interactions of filamin A with vimentin and protein kinase C epsilon *Exp. Cell Res.* **316** 1829–44

Kim H, Sengupta A, Glogauer M and McCulloch C A 2008 Filamin A regulates cell spreading and survival via beta1 integrins *Exp. Cell Res.* **314** 834–46

Kim S and Coulombe P A 2007 Intermediate filament scaffolds fulfill mechanical, organizational and signaling functions in the cytoplasm *Genes Dev.* **21** 1581–97

Kim S, Kellner J, Lee C H and Coulombe PA 2007 Interaction between the keratin cytoskeleton and eEF1Bgamma affects protein synthesis in epithelial cells *Nat. Struct. Mol. Biol.* **14** 982–3

Kim S, Wong P and Coulombe P A 2006 A keratin cytoskeletal protein regulates protein synthesis and epithelial cell growth *Nature* **441** 362–5

Kirmse R, Portet S, Mücke N, Aebi U, Herrmann H and Langowski J 2007 A quantitative kinetic model for the *in vitro* assembly of intermediate filaments from tetrameric vimentin *J. Biol. Chem.* **282** 18563–72

Kitajima Y 2013 New insights into desmosome regulation and pemphigus blistering as a desmosome-remodeling disease Kaohsiung *J. Med. Sci.* **29** 1–13 PMID: 23257250

Kjarland E, Keen T J and Kleppe R 2006 Does isoform diversity explain functional differences in the 14-3-3 protein family? *Curr. Pharm. Biotechnol.* **7** 217–23

Kleeberger W, Bova G S, Nielsen M E, Herawi M, Chuang A Y, Epstein J I and Berman D M 2007 Roles for the stem cell associated intermediate filament Nestin in prostate cancer migration and metastasis *Cancer Res.* **67** 9199–206

Knösel T, Emde V, Schlüns K, Schlag P M, Dietel M and Petersen I 2006 Cytokeratin profiles identify diagnostic signatures in colorectal cancer using multiplex analysis of tissue microarrays *Cell Oncol.* **28** 167–75 PMID: 16988472

Kölsch A, Windoffer R and Leube R E 2009 Actin-dependent dynamics of keratin filament precursors *Cell Motil. Cytoskeleton* **66** 976–85

Kölsch A, Windoffer R, Würflinger T, Aach T and Leube R E 2010 The keratin-filament cycle of assembly and disassembly *J. Cell Sci.* **123** 2266–72

Konishi Y and Setou M 2009 Tubulin tyrosination navigates the kinesin-1 motor domain to axons *Nat. Neurosci.* **12** 559–67

Kostan J, Gregor M, Walko G and Wiche G 2009 Plectin isoform-dependent regulation of keratin-integrin alpha6beta4 anchorage via Ca2 + /calmodulin *J. Biol. Chem.* **284** 18525–36

Kreis S, Schonfeld H J, Melchior C, Steiner B and Kieffer N 2005 The inter- mediate filament protein vimentin binds specifically to a recombinant integrin alpha2/beta1 cytoplasmic tail complex and co-localizes with native alpha2/beta1 in endothelial cell focal adhesions *Exp. Cell Res.* **305** 110–21

Kreplak L, Herrmann H and Aebi U 2008 Tensile properties of single desmin intermediate filaments *Biophys. J.* **94** 2790–9

Krieg P, Schuppler M, Koesters R, Mincheva A, Lichter P and Marks F 1997 Repetin (Rptn), a new member of the 'fused gene' subgroup within the S100 gene family encoding a murine epidermal differentiation protein *Genomics* **43** 339–48

Krylyshkina O, Anderson K I, Kaverina I, Upmann I, Manstein D J, Small J V and Toomre D K 2003 Nanometer targeting of microtubules to focal adhesions *J. Cell Biol.* **161** 853–9

Ku N O and Omary M B 2000 Keratins turn over by ubiquitination in a phosphorylation-modulated fashion *J. Cell Biol.* **149** 547–52

Ku N O and Omary M B 2006 A disease- and phosphorylation-related non- mechanical function for keratin 8 *J. Cell Biol.* **174** 115–25

Ku N O, Michie S A, Soetikno R M, Resurreccion E Z, Broome R L, Oshima R G and Omary M B 1996 Susceptibility to hepatotoxicity in transgenic mice that express a dominant-negative human keratin 18 mutant *J. Clin. Invest.* **98** 1034–46

Ku N O, Toivola D M, Strnad P and Omary M B 2010 Cytoskeletal keratin glycosylation protects epithelial tissue from injury *Nat. Cell Biol.* **12** 876–85

Kundu S T *et al* 2008 Plakophilin3 down-regulation leads to a decrease in cell adhesion and promotes metastasis *Int. J. Cancer* **123** 2303–14

Kuo Y H, Lu S N, Hung C H, Kee K M, Chen CH, Hu T H, Lee C M, Changchien C S and Wang J H 2010 Liver stiffness measurement in the risk assessment of hepatocellular carcinoma for patients with chronic hepatitis *Hepatol Int.* **4** 700–6

Kushkuley J, Chan W K, Lee S, Eyer J, Leterrier J F, Letournel F and Shea T B 2009 Neurofilament cross-bridging competes with kinesin-dependent association of neurofilaments with microtubules *J. Cell Sci.* **122** 3579–86

Ladwein M and Rottner K 2008 On the Rho'd: the regulation of membrane protrusions by Rho-GTPases *FEBS Lett.* **582** 2066–74

Langbein L, Eckhart L, Rogers M A, Praetzel-Wunder S and Schweizer J 2010 Against the rules: human keratin K80: two functional alternative splice variants, K80 and K80.1, with special cellular localization in a wide range of epithelia *J. Biol. Chem.* **285** 36909–21

Larcher F, Bauluz C, Diaz-Guerra M, Quintanilla M, Conti C J, Ballestin C and Jorcano J L 1992 Aberrant expression of the simple epithelial type II keratin 8 by mouse skin carcinomas but not papillomas *Mol. Carcinog.* **6** 112–21

Lee C H and Coulombe P A 2009 Self-organization of keratin intermediate filaments into cross-linked networks *J. Cell Biol.* **186** 409–21

Lee D, Santos D, Al-Rawi H, McNeill A M and Rugg E L 2008 The chemical chaperone trimethylamine N-oxide ameliorates the effects of mutant keratins in cultured cells *Br. J. Dermatol.* **159** 252–5

Lee R, Chang S Y, Trinh H, Tu Y, White A C, Davies B S, Bergo M O, Fong L G, Lowry W E and Young S G 2010 Genetic studies on the functional relevance of the protein prenyl-transferases in skin keratinocytes *Hum. Mol. Genet.* **19** 1603–17

Lehtinen L, Ketola K, Makela R, Mpindi J P, Viitala M, Kallioniemi O and Iljin K 2013 High-throughput RNAi screening for novel modulators of vimentin expression identifies MTHFD2 as a regulator of breast cancer cell migration and invasion *Oncotarget* **4** 48–63 PMID: 23295955

Lendahl U, Zimmerman L B and McKay R D 1990 CNS stem cells express a new class of intermediate filament protein *Cell* **60** 585–95

Lepekhin E A, Eliasson C, Berthold C H, Berezin V, Bock E and Pekny M 2001 Intermediate filaments regulate astrocyte motility *J. Neurochem.* **79** 617–25

Lersch R and Fuchs E 1988 Sequence and expression of a type II keratin, K5, in human epidermal cells *Mol. Cell Biol.* **8** 486–93 PMID: 2447486

Leube R E, Moch M, Kölsch A and Windoffer R 2011 "Panta rhei": Perpetual cycling of the keratin cytoskeleton *BioArchitecture* **1** 39–44

Leung C L, Green K J and Liem R K 2002 Plakins: a family of versatile cytolinker proteins *Trends. Cell Biol.* **12** 37–45

Leung C L, Liem R K, Parry D A and Green K J 2001 The plakin family *J. Cell Sci* **114** 3409–10

Li S, Guan J L and Chien S 2005 Biochemistry and biomechanics of cell motility *Annu. Rev. Biomed. Eng.* **7** 105–50

Liao J and Omary M B 1996 14-3-3 proteins associate with phosphorylated simple epithelial keratins during cell cycle progression and act as a solubility cofactor *J. Cell Biol.* **133** 345–57

Liovic M, Mogensen M M, Prescott A R and Lane E B 2003 Observation of keratin particles showing fast bidirectional movement colocalized with microtubules *J. Cell Sci.* **116** 1417–27

Listwan P and Rothnagel J A 2004 Keratin bundling proteins *Methods. Cell Biol.* **78** 817–27

Litjens S H, Koster J, Kuikman I, van Wilpe S, de Pereda J M and Sonnenberg A 2003 Specificity of binding of the plectin actin-binding domain to beta4 integrin *Mol. Biol. Cell* **14** 4039–50

Löffek S, Wöll S, Höhfeld J, Leube R E, Has C, Bruckner-Tuderman L and Magin T M 2010 The ubiquitin ligase CHIP/STUB1 targets mutant keratins for degradation *Hum. Mutat.* **31** 466–76

Lombardi M L, Jaalouk D E, Shanahan C M, Burke B, Roux K J and Lammerding J 2011 The interaction between nesprins and sun proteins at the nuclear envelope is critical for force transmission between the nucleus and cytoskeleton *J. Biol. Chem.* **286** 26743–53

Long H A, Boczonadi V, McInroy L, Goldber M and Määttä A 2006 Periplakin-dependent re-organisation of keratin cytoskeleton and loss of collective migration in keratin-8-down-regulated epithelial sheets *J. Cell Sci.* **119** 5147–59

Loo D T, Kanner S B and Aruffo A 1998 Filamin binds to the cytoplasmic domain of the beta1-integrin. Identification of amino acids responsible for this interaction *J. Biol. Chem.* **273** 23304–12

Lopez J I, Mouw J K and Weaver V M 2008 Biomechanical regulation of cell orientation and fate *Oncogene* **27** 6981–93

Mackintosh C 2004 Dynamic interactions between 14-3-3 proteins and phosphoproteins regulate diverse cellular processes *Biochem. J.* **381** 329–42

Magin T M, Reichelt J and Chen J 2006 The role of keratins in epithelial homeostasis *Skin Barrier* ed P M Elias and K R Feingold (New York: Taylor and Francis) pp 141–70

Magin T M, Vijayaraj P and Leube R E 2007 Structural and regulatory functions of keratins *Exp. Cell Res.* **313** 2021–32

Makino T, Takaishi M, Morohashi M and Huh N H 2001 Hornerin, a novel profilaggrin-like protein and differentiation-specific marker isolated from mouse skin *J. Biol. Chem.* **276** 47445–52

Mashukova A, Oriolo A S, Wald F A, Casanova M L, Kröger C, Magin T M, Omary M B and Salas P J 2009 Rescue of atypical protein kinase C in epithelia by the cytoskeleton and Hsp70 family chaperones *J. Cell Sci.* **122** 2491–503

Matsuda Y, Kure S and Ishiwata T 2012 Nestin and other putative cancer stem cell markers in pancreatic cancer *Med. Mol. Morphol.* **45** 59–65

Matsuda Y, Naito Z, Kawahara K, Nakazawa N, Korc M and Ishiwata T 2011 Nestin is a novel target for suppressing pancreatic cancer cell migration, invasion and metastasis *Cancer Biol. Ther.* **11** 512–23

Matthews B D, Overby D R, Alenghat F J, Karavitis J, Numaguchi Y, Allen P G and Ingber D E 2004 Mechanical properties of individual focal adhesions probed with a magnetic micro-needle *Biochem. Biophys. Res. Commun.* **313** 758–64

Matthews B D, Overby D R, Mannix R and Ingber D E 2006 Cellular adaptation to mechanical stress: role of integrins, Rho, cytoskeletal tension and mechanosensitive ion channels *J. Cell Sci.* **119** 508–18

Mazzalupo S, Wong P, Martin P and Coulombe P A 2003 Role for keratins 6 and 17 during wound closure in embryonic mouse skin *Dev. Dyn.* **226** 356–65

McBeath R, Pirone D M, Nelson C M, Bhadriraju K and Chen C S 2004 Cell shape, cytoskeletal tension, and RhoA regulate stem cell lineage commitment *Dev. Cell* **6** 483–95

McInroy L and Maatta A 2007 Down-regulation of vimentin expression inhibits carcinoma cell migration and adhesion *Biochem. Biophys. Res. Commun.* **360** 109–14

Mellad J A, Warren D T and Shanahan C M 2011 Nesprins LINC the nucleus and cytoskeleton *Curr. Opin. Cell Biol.* **23** 47–54

Mendez M G, Kojima S and Goldman R D 2010 Vimentin induces changes in cell shape, motility, and adhesion during the epithelial to mesenchymal transition *FASEB J.* **24** 1838–51

Meyer S C, Sanan D A and Fox J E 1998 Role of actin-binding protein in insertion of adhesion receptors into the membrane *J. Biol. Chem.* **273** 3013–20

Michalski P J and Carlsson A E 2010 The effects of filament aging and annealing on a model lamellipodium undergoing disassembly by severing *Phys. Biol.* **7** 026004

Mignot C *et al* 2007 Dynamics of mutated GFAP aggregates revealed by real-time imaging of an astrocyte model of Alexander disease *Exp. Cell Res.* **313** 2766–79

Miller R K, Vikstrom K and Goldman R D 1991 Keratin incorporation into intermediate filament networks is a rapid process *J. Cell Biol.* **113** 843–55

Mitchison T and Kirschner M 1984 Dynamic instability of microtubule growth *Nature* **312** 237–42

Mizuuchi E, Semba S, Kodama Y and Yokozaki H 2009 Down-modulation of keratin 8 phosphorylation levels by PRL-3 contributes to colorectal carcinoma progression *Int. J. Cancer* **124** 1802–10

Moll R, Divo M and Langbein L 2008 The human keratins: biology and pathology *Histochem. Cell Biol.* **129** 705–33

Moll R, Franke W W, Schiller D L, Geiger B and Krepler R 1982 The catalog of human cytokeratins: patterns of expression in normal epithelia, tumors and cultured cells *Cell* **31** 11–24

Moll R, Schiller D L and Franke W W 1990 Identification of protein IT of the intestinal cytoskeleton as a novel type I cytokeratin with unusual properties and expression patterns *J. Cell Biol.* **111** 567–80

Morgan J T, Pfeiffer E R, Thirkill T L, Kumar P, Peng G, Fridolfsson H N, Douglas G C, Starr D A and Barakat A I 2011 Nesprin-3 regulates endothelial cell morphology, perinuclear cytoskeletal architecture, and flow-induced polarization *Mol. Biol. Cell* **22** 4324–34

Morley S M *et al* 2003 Generation and characterization of epidermolysis bullosa simplex cell lines: scratch assays show faster migration with disruptive keratin mutations *Br. J. Dermatol.* **149** 46–58

Na N, Chandel N S, Litvan J and Ridge K M 2010 Mitochondrial reactive oxygen species are required for hypoxia-induced degradation of keratin intermediate filaments *FASEB J.* **24** 799–809

Nakamura M and Tokura Y 2011 Epithelial–mesenchymal transition in the skin *J. Dermatol. Sci.* **61** 7–13

Nan L, Dedes J, French B A, Bardag-Gorce F, Li J, Wu Y and French S W 2006 Mallory body (cytokeratin aggresomes) formation is prevented *in vitro* by p38 inhibitor *Exp. Mol. Pathol.* **80** 228–40

Ngai J, Coleman T R and Lazarides E 1990 Localization of newly synthesized vimentin subunits reveals a novel mechanism of intermediate filament assembly *Cell* **60** 415–27

Nieminen M, Henttinen T, Merinen M, Marttila-Ichihara F, Eriksson J E and Jalkanen S 2006 Vimentin function in lymphocyte adhesion and transcellular migration *Nat. Cell Biol.* **8** 156–62

Nikolic B, Mac Nulty E, Mir B and Wiche G 1996 Basic amino acid residue cluster within nuclear targeting sequence motif is essential for cytoplasmic plectin–vimentin network junctions *J. Cell Biol.* **134** 1455–67

Nobes C D and Hall A 1995 Rho, rac, and cdc42 GTPases regulate the assembly of multi-molecular focal complexes associated with actin stress fibers, lamellipodia, and filopodia *Cell* **81** 53–62

Ogawara M *et al* 1995 Differential targeting of protein kinase C and CaM kinase II signalings to vimentin *J. Cell Biol.* **131** 1055–66

Ohta Y, Suzuki N, Nakamura S, Hartwig J H and Stossel T P 1999 The small GTPase RalA targets filamin to induce filopodia *Proc. Natl Acad. Sci. USA* **96** 2122–8

Omary M B and Ku N O 1997 Intermediate filament proteins of the liver: emerging disease association and functions *Hepatology* **25** 1043–8

Omary M B, Baxter G T, Chou C F, Riopel C L, Lin W Y and Strulovici B 1992 PKC epsilon-related kinase associates with and phosphorylates cytokeratin 8 and 18 *J. Cell Biol.* **117** 583–93

Omary M B, Coulombe P A and McLean W H 2004 Intermediate filament proteins and their associated diseases *N. Engl. J. Med.* **351** 2087–100

Omary M B, Ku N O, Liao J and Price D 1998 Keratin modifications and solubility properties in epithelial cells and *in vitro Subcell. Biochem.* **31** 105–40

Omary M B, Ku N O, Strnad P and Hanada S 2009 Toward unraveling the complexity of simple epithelial keratins in human disease *J. Clin. Invest.* **119** 1794–805

Oriolo A S, Wald F A, Ramsauer V P and Salas P J 2007 Intermediate filaments: a role in epithelial polarity *Exp. Cell Res.* **313** 2255–64

Oshima R G 2002 Apoptosis and keratin intermediate filaments *Cell Death Differ.* **9** 486–92

Oshima R G, Baribault H and Caulin C 1996 Oncogenic regulation and function of keratins 8 and 18 *Cancer Metastasis Rev.* **15** 445–71

Osmanagic-Myers S, Gregor M, Walko G, Burgstaller G, Reipert S and Wiche G 2006 Plectin-controlled keratin cytoarchitecture affects MAP kinases involved in cellular stress response and migration *J. Cell Biol.* **174** 557–68

Paccione R J, Miyazaki H, Patel V, Waseem A, Gutkind J S, Zehner Z E and Yeudall W A 2008 Keratin down-regulation in vimentin-positive cancer cells is reversible by vimentin RNA interference, which inhibits growth and motility *Mol. Cancer Ther.* **7** 2894–903

Pajerowski J D, Dahl K N, Zhong F L, Sammak P J and Discher D E 2007 Physical plasticity of the nucleus in stem cell differentiation *Proc. Natl Acad. Sci. USA* **104** 15619–24

Paladini R D, Takahashi K, Bravo N S and Coulombe P A 1996 Onset of re-epithelialization after skin injury correlates with a reorganization of keratin filaments in wound edge keratinocytes: defining a potential role for keratin 16 *J. Cell Biol.* **132** 381–97

Pan X, Hobbs R P and Coulombe P A 2013 The expanding significance of keratin intermediate filaments in normal and diseased epithelia *Curr. Opin. Cell Biol.* **25** 47–56

Pan X, Kane L A, Van Eyk J E and Coulombe P A 2011 Type I keratin 17 protein is phosphorylated on serine 44 by p90 ribosomal protein S6 kinase 1 (RSK1) in a growth- and stress-dependent fashion *J. Biol. Chem.* **286** 42403–13

Pan Y, Jing R, Pitre A, Williams B J and Skalli O 2008 Intermediate filament protein synemin contributes to the migratory properties of astrocytoma cells by influencing the dynamics of the actin cytoskeleton *FASEB J.* **22** 3196–206

Park M K, Lee H J, Shin J, Noh M, Kim S Y and Lee C H 2011 Novel participation of transglutaminase-2 through c-Jun N-terminal kinase activation in sphingosylphosphorylcholine-induced keratin reorganization of PANC-1 cells *Biochim. Biophys. Acta.* **1811** 1021–9

Parry S, Savage K, Marchio C and Reis-Filho J S 2008 Nestin is expressed in basal-like and triple negative breast cancers *J. Clin. Pathol.* **61** 1045–50

Paszek M J *et al* 2005 Tensional homeostasis and the malignant phenotype *Cancer Cell* **8** 241–54

Pekny M and Lane E B 2007 Intermediate filaments and stress *Exp. Cell Res.* **313** 2244–54

Perrin B J, Amann K J and Huttenlocher A 2006 Proteolysis of cortactin by calpain regulates membrane protrusion during cell migration *Mol. Biol. Cell* **17** 239–50

Petit V and Thiery J P 2000 Focal adhesions: structure and dynamics *Biol. Cell* **92** 477–94

Planko L, Böhse K, Höhfeld J, Betz R C, Hanneken S, Eigelshoven S, Kruse R, Nöthen M M and Magin T M 2007 Identification of a keratin-associated protein with a putative role in vesicle transport *Eur. J. Cell Biol.* **86** 827–39

Popowicz G M, Schleicher M, Noegel A A and Holak T A 2006 Filamins: promis- cuous organizers of the cytoskeleton *Trends Biochem. Sci.* **31** *411–9*

Price L S, Leng J, Schwartz M A and Bokoch G M 1998 Activation of Rac and Cdc42 by integrins mediates cell spreading *Mol. Biol. Cell* **9** 1863–71

Proux-Gillardeaux V, Gavard J, Irinopoulou T, Mège R M and Galli T 2005 Tetanus neurotoxin-mediated cleavage of cellubrevin impairs epithelial cell migration and integrin-dependent cell adhesion *Proc. Natl Acad. Sci. USA* **102** 6362–7

Ptitsyn A A, Weil M M and Thamm D H 2008 Systems biology approach to identification of biomarkers for metastatic progression in cancer *BMC Bioinformatics* **9** S8

Raul U, Sawant S, Dange P, Kalraiya R, Ingle A and Vaidya M 2004 Implications of cytokeratin 8/18 filament formation in stratified epithelial cells: induction of transformed phenotype *Int. J. Cancer* **111** 662–8

Reichelt J, Büssow H, Grund C and Magin T M 2001 Formation of a normal epidermis supported by increased stability of keratins 5 and 14 in keratin 10 null mice *Mol. Biol. Cell* **12** 1557–68

Ren X D, Kiosses W B and Schwartz M A 1999 Regulation of the small GTP- binding protein Rho by cell adhesion and the cytoskeleton *EMBO J.* **18** 578–85

Rezniczek G A, Abrahamsberg C, Fuchs P, Spazierer D and Wiche G 2003 Plectin 5'-transcript diversity: short alternative sequences determine stability of gene products, initiation of translation and subcellular localization of isoforms *Hum. Mol. Genet.* **12** 3181–94

Rezniczek G A, Janda L and Wiche G 2004 Plectin *Methods Cell Biol.* **78** 721–55

Ridge K M, Linz L, Flitney F W, Kuczmarski E R, Chou Y H, Omary M B, Sznajder J I and Goldman R D 2005 Keratin 8 phosphorylation by protein kinase C delta regulates shear stress-mediated disassembly of keratin intermediate filaments in alveolar epithelial cells *J. Biol. Chem.* **280** 30400–5

Rizki A, Mott J D and Bissell M J 2007 Polo-like kinase 1 is involved in invasion through extracellular matrix *Cancer Res.* **67** 11106–10

Röber R A, Weber K and Osborn M 1989 Differential timing of nuclear lamin A/C expression in the various organs of the mouse embryo and the young animal: a developmental study *Development* **105** 365–78 PMID: 2680424

Roberts B J, Pashaj A, Johnson K R and Wahl J K 2011 Desmosome dynamics in migrating epithelial cells requires the actin cytoskeleton *Exp. Cell Res.* **317** 2814–22

Rogel M R, Jaitovich A and Ridge K M 2010 The role of the ubiquitin proteasome pathway in keratin intermediate filament protein degradation *Proc. Am. Thorac. Soc.* **7** 71–6

Rogel M R, Soni P N, Troken J R, Sitikov A, Trejo H E and Ridge K M 2011 Vimentin is sufficient and required for wound repair and remodeling in alveolar epithelial cells *FASEB J.* **25** 3873–83

Rogers M A, Edler L, Winter H, Langbein L, Beckman I and Schweizer J 2005 Characterization of new members of the human type II keratin gene family and a general evaluation of the keratin gene domain on chromo- some 12q13.13 *J. Invest. Dermatol.* **124** 536–44

Rogers M A, Winter H, Langbein L, Bleiler R and Schweizer J 2004 The human type I keratin gene family: Characterization of new hair follicle specific members and evaluation of the chromosome 17q21.2 gene domain *Differentiation* **72** 527–40

Rolli C G, Seufferlein T, Kemkemer R and Spatz J P 2010 Impact of tumor cell cytoskeleton organization on invasiveness and migration: a microchannel-based approach *PLoS ONE* **5** e8726

Rotty J D and Coulombe P A 2012 A wound-induced keratin inhibits Src activity during keratinocyte migration and tissue repair *J. Cell Biol.* **197** 381–9

Roy S, Coffee P, Smith G, Liem R K, Brady S T and Black M M 2000 Neurofilaments are transported rapidly but intermittently in axons: implications for slow axonal transport *J. Neurosci.* **20** 6849–61

Sakamoto Y, Boeda B and Etienne-Manneville S 2013 APC binds intermediate filaments and is required for their reorganization during cell migration *J. Cell Biol.* **200** 249–58

Satelli A and Li S 2011 Vimentin in cancer and its potential as a molecular target for cancer therapy *Cell Mol. Life Sci.* **68** 3033–46

Schaus T E, Taylor E W and Borisy G G 2007 Self-organization of actin filament orientation in the dendritic-nucleation/array-treadmilling model *Proc. Natl Acad. Sci. USA* **104** 7086–91

Schoumacher M, Goldman R D, Louvard D and Vignjevic D M 2010 Actin, microtubules, and vimentin intermediate filaments cooperate for elongation of invadopodia *J. Cell Biol.* **189** 541–56

Schweizer J *et al* 2006 New consensus nomenclature for mammalian keratins *J. Cell Biol.* **174** 169–74

Seifert G J, Lawson D and Wiche G 1992 Immunolocalization of the intermediate filament-associated protein plectin at focal contacts and actin stress fibers *Eur. J. Cell Biol.* **59** 138–47

Seltmann K, Roth W, Kröger C, Loschke F, Lederer M, Hüttelmaier S and Magin T M 2013 Keratins mediate localization of hemidesmosomes and repress cell motility *J. Invest. Dermatol.* **133**(1) 181–90

Serrels B, Serrels A, Brunton V G, Holt M, McLean G W, Gray C H, Jones G E and Frame M C 2007 Focal adhesion kinase controls actin assembly via a FERM-mediated interaction with the Arp2/3 complex *Nat. Cell Biol.* **9** 1046–56

Sihag R K, Inagaki M, Yamaguchi T, Shea T B and Pant H C 2007 Role of phosphorylation on the structural dynamics and function of types III and IV intermediate filaments *Exp. Cell Res.* **313** 2098–109

Singh S, Sadacharan S, Su S, Belldegrun A, Persad S and Singh G 2003 Overexpression of vimentin: role in the invasive phenotype in an androgen-independent model of prostate cancer *Cancer Res.* **63** 2306–11

Sivaramakrishnan S, Schneider J L, Sitikov A, Goldman R D and Ridge K M 2009 Shear stress induced reorganization of the keratin intermediate filament network requires phosphorylation by protein kinase C zeta *Mol. Biol. Cell* **20** 2755–65

Small J V and Kaverina I 2003 Microtubules meet substrate adhesions to arrange cell polarity *Curr. Opin. Cell Biol.* **15** 40–7

Small J V and Resch G P 2005 The comings and goings of actin: coupling protrusion and retraction in cell motility *Curr. Opin. Cell Biol.* **17** 517–23

Snider N T, Weerasinghe S V, Iñiguez-Lluhí J A, Herrmann H and Omary M B 2011 Keratin hypersumoylation alters filament dynamics and is a marker for human liver disease and keratin mutation *J. Biol. Chem.* **286** 2273–84

Soellner P, Quinlan R A and Franke W W 1985 Identification of a distinct soluble subunit of an intermediate filament protein: tetrameric vimentin from living cells *Proc. Natl Acad. Sci. USA* **82** 7929–33

Spazierer D, Raberger J, Gross K, Fuchs P and Wiche G 2008 Stress-induced recruitment of epiplakin to keratin networks increases their resistance to hyperphosphorylation-induced disruption *J. Cell Sci.* **121** 825–33

Sprecher E, Itin P, Whittock N V, McGrath A, Meyer R, DiGiovanna J J, Bale S J, Uitto J and Richard G 2002 Refined mapping of Naegeli- Franceschetti-Jadasohn syndrome to a 6 cM interval on chromosome 17q11.2-q21 and investigation of candidate genes *J. Invest. Dermatol.* **119** 692–8

Spurny R, Gregor M, Castañón M J and Wiche G 2008 Plectin deficiency affects precursor formation and dynamics of vimentin networks *Exp. Cell Res.* **314** 3570–80

Stefansson I M, Salvesen H B and Akslen L A 2006 Loss of p63 and cytokeratin 5/6 expression is associated with more aggressive tumors in endometrial carcinoma patients *Int. J. Cancer* **118** 1227–33

Steinböck F A, Nikolic B, Coulombe P A, Fuchs E, Traub P and Wiche G 2000 Dose-dependent linkage, assembly inhibition and disassembly of vimentin and cytokeratin 5/14 filaments through plectin's intermediate filament-binding domain *J. Cell Sci.* **113** 483–91 PMID: 10639335

Stossel T P, Condeelis J, Cooley L, Hartwig J H, Noegel A, Schleicher M and Shapiro S S 2001 Filamins as integrators of cell mechanics and signalling *Nat. Rev. Mol. Cell Biol.* **2** 138–45

Strnad P, Windoffer R and Leube R E 2001 *In vivo* detection of cytoker- atin filament network breakdown in cells treated with the phosphatase inhibitor okadaic acid *Cell Tissue Res.* **306** 277–93

Strnad P, Windoffer R and Leube R E 2002 Induction of rapid and reversible cytokeratin filament network remodeling by inhibition of tyrosine phosphatases *J. Cell Sci.* **115** 4133–48

Stumptner C, Fuchsbichler A, Lehner M, Zatloukal K and Denk H 2001 Sequence of events in the assembly of Mallory body components in mouse liver: clues to the pathogenesis and significance of Mallory body formation *J. Hepatol.* **34** 665–75

Sun C X, Magalhaes M A and Glogauer M 2007 Rac1 and Rac2 differentially regulate actin free barbed end formation downstream of the fMLP receptor *J. Cell Biol.* **179** 239–45

Sun L P, Wang L, Wang H, Zhang Y H and Pu J L 2010 Connexin 43 remodeling induced by LMNA gene mutation Glu82Lys in familial dilated cardiomyopathy with atrial ventricular block *Chin. Med. J. (Engl.)* **123** 1058–62

Sun N, Critchley D R, Paulin D, Li Z and Robson R M 2008a Human alpha-synemin interacts directly with vinculin and metavinculin *Biochem. J.* **409** 657–67

Sun N, Critchley D R, Paulin D, Li Z and Robson R M 2008b Identification of a repeated domain within mammalian alpha-synemin that interacts directly with talin *Exp. Cell Res.* **314** 1839–49

Svitkina T M, Verkhovsky A B and Borisy G G 1996 Plectin sidearms mediate interaction of intermediate filaments with microtubules and other components of the cytoskeleton *J. Cell Biol.* **135** 991–1007

Takemura M, Gomi H, Colucci-Guyon E and Itohara S S 2002 Protective role of phosphorylation in turnover of glial fibrillary acidic protein in mice *J. Neurosci.* **22** 6972–9

Thiery J P, Acloque H, Huang R Y and Nieto M A 2009 Epithelial–mesenchymal transitions in development and disease *Cell* **139** 871–90

Tigges U, Koch B, Wissing J, Jockusch B M and Ziegler W H 2003 The F-actin cross-linking and focal adhesion protein filamin A is a ligand and *in vivo* substrate for protein kinase C alpha *J. Biol. Chem.* **278** 23561–9

Todorovic V, Desai B V, Patterson M J, Amargo E V, Dubash A D, Yin T, Jones J C and Green K J 2010 Plakoglobin regulates cell motility through Rho- and fibronectin-dependent Src signaling *J. Cell Sci.* **123** 3576–86

Toivola D M, Ku N O, Resurreccion E Z, Nelson D R, Wright T L and Omary M B 2004 Keratin 8 and 18 hyperphosphorylation is a marker of progression of human liver disease *Hepatology* **40** 459–66

Toivola D M, Strnad P, Habtezion A and Omary M B 2010 Intermediate filaments take the heat as stress proteins *Trends. Cell Biol.* **20** 79–91

Toivola D M, Tao G Z, Habtezion A J, Liao J and Omary M B 2005 Cellular integrity plus: organelle-related and protein-targeting functions of intermediate filaments *Trends Cell Biol.* **15** 608–17

Trepat X and Fredberg J J 2011 Plithotaxis and emergent dynamics in collective cellular migration *Trends Cell Biol.* **21** 638–46

Tsuruta D and Jones J C 2003 The vimentin cytoskeleton regulates focal contact size and adhesion of endothelial cells subjected to shear stress *J. Cell Sci.* **116** 4977–84

Vadlamudi R K, Li F, Adam L, Nguyen D, Ohta Y, Stossel T P and Kumar R 2002 Filamin is essential in actin cytoskeletal assembly mediated by p21- activated kinase 1 *Nat. Cell Biol.* **4** 681–90

Valencia R G, Walko G, Janda L, Novacek J, Mihailovska E, Reipert S, Andra-Marobela K and Wiche G 2013 Intermediate filament-associated cytolinker plectin 1c destabilizes microtubules in keratinocytes *Mol. Biol. Cell* **24** 768–84

van de Rijn M *et al* 2002 Expression of cytokeratins 17 and 5 identifies a group of breast carcinomas with poor clinical outcome *Am. J. Pathol.* **161** 1991–6

Vicente-Manzanares M, Choi C K and Horwitz A R 2009 Integrins in cell migration—the actin connection *J. Cell Sci.* **122** 199–206

Vijayaraj P, Kröger C, Reuter U, Windoffer R, Leube R E and Magin T M 2009 Keratins regulate protein biosynthesis through localization of GLUT1 and -3 upstream of AMP kinase and Raptor *J. Cell Biol.* **187** 175–84

Vogel V and Sheetz M P 2009 Cell fate regulation by coupling mechanical cycles to biochemical signaling pathways *Curr. Opin. Cell Biol.* **21** 38–46

Vuoriluoto K, Haugen H, Kiviluoto S, Mpindi J P, Nevo J, Gjerdrum C, Tiron C, Lorens J B and Ivaska J 2011 Vimentin regulates EMT induction by Slug and oncogenic H-Ras and migration by governing Axl expression in breast cancer *Oncogene* **30** 1436–48

Wan H *et al* 2004 Striate palmoplantar keratoderma arising from desmoplakin and desmoglein 1 mutations is associated with contrasting perturbations of desmosomes and the keratin filament network *Br. J. Dermatol.* **150** 878–91

Wang L, Ho C L, Sun D, Liem R K and Brown A 2000 Rapid movement of axonal neurofilaments interrupted by prolonged pauses *Nat. Cell Biol.* **2** 137–41

Wang Y L and Pelham R J Jr 1998 Preparation of a flexible, porous polyacrylamide substrate for mechanical studies of cultured cells *Methods Enzymol.* **298** 489–96

Webb B A, Jia L, Eves R and Mak A S 2007 Dissecting the functional domain requirements of cortactin in invadopodia formation *Eur. J. Cell Biol.* **86** 189–206

Weber G F, Bjerke M A and DeSimone D W 2012 A mechanoresponsive cadherin-keratin complex directs polarized protrusive behavior and collective cell migration *Dev. Cell* **22** 104–15

Weber K, Plessmann U and Traub P 1989 Maturation of nuclear lamin A involves a specific carboxy-terminal trimming, which removes the polyisoprenylation site from the precursor; implications for the structure of the nuclear lamina *FEBS Lett.* **257** 411–4

Wei J *et al* 2008 Overexpression of vimentin contributes to prostate cancer invasion and metastasis via Src regulation *Anticancer Res.* **28** 327–34 PMID: 18383865

Werner N S, Windoffer R, Strnad P, Grund C, Leube R E and Magin T M 2004 Epidermolysis bullosa simplex-type mutations alter the dynamics of the keratin cytoskeleton and reveal a contribution of actin to the transport of keratin subunits *Mol. Biol. Cell* **15** 990–1002

Whipple R A, Balzer E M, Cho E H, Matrone M A, Yoon J R and Martin S S 2008 Vimentin filaments support extension of tubulin-based microtentacles in detached breast tumor cells *Cancer Res.* **68** 5678–88

Wiche G 1998 Role of plectin in cytoskeleton organization and dynamics *J. Cell Sci.* **111** 2477–86

Wilhelmsen K, Litjens S H, Kuikman I, Tshimbalanga N, Janssen H, van den Bout I, Raymond K and Sonnenberg A 2005 Nesprin-3, a novel outer nuclear membrane protein, associates with the cytoskeletal linker protein plectin *J. Cell Biol.* **171** 799–810

Windoffer R and Leube R E 1999 Detection of cytokeratin dynamics by time-lapse fluorescence microscopy in living cells *J. Cell Sci.* **112** 4521–34

Windoffer R, Kölsch A, Wöll S and Leube R E 2006 Focal adhesions are hotspots for keratin filament precursor formation *J. Cell Biol.* **173** 341–8

Windoffer R, Wöll S, Strnad P and Leube R E 2004 Identification of novel principles of keratin filament network turnover in living cells *Mol. Biol. Cell* **15** 2436–48

Winokur R and Hartwig J H 1995 Mechanism of shape change in chilled human platelets *Blood* **85** 1796–804

Wolf A, Krause-Gruszczynska M, Birkenmeier O, Ostareck-Lederer A, Hüttelmaier S and Hatzfeld M 2010 Plakophilin 1 stimulates translation by promoting eIF4A1 activity *J. Cell Biol.* **188** 463–71

Wolf D M, Vazirani V V and Arkin A P 2005 Diversity in times of adversity: probabilistic strategies in microbial survival games *J. Theor. Biol.* **234** 227–53

Wöll S R, Windoffer R and Leube R E 2007 p38 MAPK-dependent shaping of the keratin cytoskeleton in cultured cells *J. Cell Biol.* **177** 795–807

Wöll S, Windoffer R and Leube R E 2005 Dissection of keratin dynamics: different contributions of the actin and microtubule systems *Eur. J. Cell Biol.* **84** 311–28

Xu Z, Wang M R, Xu X, Cai Y, Han Y L, Wu K M, Wang J, Chen B S, Wang X Q and Wu M 2000 Novel human esophagus-specific gene c1orf10: cDNA cloning, gene structure, and frequent loss of expression in esophageal cancer *Genomics* **69** 322–30

Yamasaki T, Seki N, Yamada Y, Yoshino H, Hidaka H, Chiyomaru T, Nohata N, Kinoshita T, Nakagawa M and Enokida H 2012 Tumor suppressive microRNA138 contributes to cell migration and invasion through its targeting of vimentin in renal cell carcinoma *Int. J. Oncol.* **41** 805–17

Yang Y, Dowling J, Yu Q C, Kouklis P, Cleveland D W and Fuchs E 1996 An essential cytoskeletal linker protein connecting actin microfilaments to intermediate filaments *Cell* **86** 655–65

Yasui Y, Goto H, Matsui S, Manser E, Lim L, Nagata K and Inagaki M 2001 Protein kinases required for segregation of vimentin filaments in mitotic process *Oncogene* **20** 2868–76

Yilmaz M and Christofori G 2009 EMT, the cytoskeleton and cancer cell invasion *Cancer. Metastasis. Rev.* **28** 15–33

Yin T, Getsios S, Caldelari R, Kowalczyk A P, Muller E J, Jones J C and Green K J 2005 Plakoglobin suppresses keratinocyte motility through both cell–cell adhesion-dependent and -independent mechanisms *Proc. Natl Acad. Sci. USA* **102** 5420–5

Yoon K H, Yoon M, Moir R D, Khuon S, Flitney F W and Goldman R D 2001 Insights into the dynamic properties of keratin intermediate filaments in living epithelial cells *J. Cell Biol.* **153** 503–16

Zatloukal K, French S W, Stumptner C, Strnad P, Harada M, Toivola D M, Cadrin M and Omary M B 2007 From Mallory to Mallory–Denk bodies: what, how and why? *Exp. Cell Res.* **313** 2033–49

Zencheck W D, Xiao H and Weiss L M 2012 Lysine post-translational modifications and the cytoskeleton *Essays Biochem.* **52** 135–45

Zhang H, Landmann F, Zahreddine H, Rodriguez D, Koch M and Labouesse M 2011 A tension-induced mechanotransduction pathway promotes epithelial morphogenesis *Nature* **471** 99–103

Zhang J S, Wang L, Huang H M, Nelson M and D.I. Smith D I 2001 Keratin 23 (K23), a novel acidic keratin, is highly induced by histone deacetylase inhibitors during differentiation of pancreatic cancer cells *Genes. Chromosomes Cancer* **30** 123–35

Zhao X H, Laschinger C, Arora P, Szaszi K, Kapus A and McCulloch C A 2007 Force activates smooth muscle alpha-actin promoter activity through the Rho signaling pathway *J. Cell Sci.* **120** 1801–9

Zhao Y, Yan Q, Long X, Chen X and Wang Y 2008 Vimentin affects the mobility and invasiveness of prostate cancer cells *Cell Biochem. Funct.* **26** 571–7

Zhu Q S *et al* 2011 Vimentin is a novel AKT1 target mediating motility and invasion *Oncogene* **30** 457–70

Zimerman B, Volberg T and Geiger B 2004 Early molecular events in the assembly of the focal adhesion-stress fiber complex during fibroblast spreading *Cell Motil Cytoskel* **58** 143–59

IOP Publishing

Physics of Cancer

Claudia Tanja Mierke

Chapter 7

Cell surface tension, the mobility of cell surface receptors and their location in specific regions

Summary

In this chapter the impact of cell surface tension on the selection of an aggressive and invasive cancer cell subpopulation that is able to disseminate from the primary tumor in the microenvironment and follow the steps of the metastatic cascade for malignant tumor progression is discussed. In particular, the classical differential adhesion hypothesis is presented and its applicability to experimental results is proved. In addition, the mobility of cell surface receptors and their structural organization in special compartments (regions) within the cell surface membrane and the impact of this on the motility of cancer cells is addressed. The impact that location of cell surface receptors within the membrane has on cellular mechanical properties is highlighted. There is a particular focus on surface tension, the lateral mobility of cell surface receptors and special membrane regions called lipid rafts.

7.1 Surface tension

During animal morphogenesis, large-scale cellular movements occur, which lead to the rearrangement, mutual spreading and compartmentalization of cell populations in specific structural arrangements. The morphogenetic cell rearrangements, including cell sorting and mutual tissue spreading, have been compared with the behavior of other matter, such as immiscible liquids, which are very similar. Based on this strong similarity, it has been suggested that whole tissues act as liquids and are characterized by a specific surface tension, which is a collective, macroscopic property of groups of mobile, cohering (connected) cells. How can tissues generate surface tension? Several different theories have been formulated to explain how mesoscopic cell properties such as cell–cell adhesion and the contractility of cell interfaces may the regulate surface tension of tissues. It has been proposed that cell–cell adhesion and cellular contractility contribute to tissue surface tension, but there is not yet a model that describes and predicts the dependence of tissue surface tension on these two mesoscopic parameters. However, there is an approach that shows that the ratio of adhesion to cortical tension regulates the amount of tissue surface tension. This approach seems to be based on a minimal model that relies on the regulatory feedback between mechanical energy and geometry and predicts the shapes of aggregate surface cells. In addition, this model has been experimentally verified. Moreover, this model indicates that there is a crossover from adhesion-driven to cortical-tension-driven behavior as a function of the ratio between these two influencing factors.

As mentioned above, it is well established that many tissues act similarly to liquids over long timescales. Cell tracking *in vivo* and *in vitro* has revealed large-scale flows, exchange of nearest neighbors in a cellular aggregate and rounding-up and fusion of cell aggregates (Schoetz 2008). Surface tension is a macroscopic rheological property and thus can be determined by a tissue surface tensiometer (TST) (Schoetz 2008, Foty *et al* 1994, 1996, Forgacs *et al* 1998, Schoetz *et al* 2008, Mgharbel *et al* 2009, Norotte *et al* 2008, Davis *et al* 1997) or the biophysical micropipette aspiration technique (Guevorkian *et al* 2010). In particular, the surface tension may be the driving factor for tissue self-organization in embryogenesis (Davis *et al* 1997, Armstrong 1989, Holtfreter 1944, Steinberg 1996) and cancer (Foty *et al* 1998, Foty and Steinberg 1997). In more detail, the special manner of cell sorting within tissues and tissue spreading can be indeed explained in terms of tissue surface tensions, which alter between different cell types (Schoetz 2008, Foty *et al* 1996, Forgacs *et al* 1998, Schoetz *et al* 2008, Davis *et al* 1997, Duguay *et al* 2003, Borghi and Nelson 2009).

The role of tissue surface tension has to be revealed, as tissue surface tension is supposed to act as a driving force for many biological processes. Knowing its cellular origins would help to manipulate tissue organization using special drug treatments. Over the last three decades, there have been only two opposing theories describing the mesoscopic origin of tissue surface tension. One theory, the differential adhesion hypothesis (DAH), predicts that, as with ordinary fluids, tissue surface tension is proportional to the intensity of the adhesive energy between the constituent cells. The constituent cells are treated as simple point objects. The DAH explains the motility of cells during tissue morphogenesis using thermodynamic principles. The theory treats tissues as liquids consisting of migrating cells with different amounts of surface tension. In order to minimize their interfacial free energy, the cells assemble spontaneously into aggregates, leading to tissue reorganization. During tissue morphogenesis, the migrating cells behave very similarly to a mixture of different liquids according to the DAH theory. The theory was originally primarily developed to understand cell sorting behavior in the embryos of vertebrates, but it can also be applied to other morphogenic phenomena, for example those occurring during the malignant progression of cancer, such as metastasis, epithelial-to-mesenchymal transition and wound-healing processes. The DAH relies on quantitative differences in the strength of cell surface adhesion.

The DAH has proven successful in numerous studies with different cell lines (Foty *et al* 1994, 1996, Forgacs *et al* 1998, Schoetz *et al* 2008, Duguay *et al* 2003), malignant tissues (Foty *et al* 1998, Foty and Steinberg 1997) and embryonic tissues (Schoetz 2008, Schoetz *et al* 2008, Davis *et al* 1997, Borghi and Nelson 2009) and hence is widely accepted in the field of biophysical research (Steinberg 1996, Lecuit and Lenne 2007). In addition, it has been experimentally verified that there is a linear relationship between the expression of adhesion molecules on the cell surface and the levels of tissue surface tension (Foty and Steinberg 2005).

However, experimental data obtained with AFM (Krieg *et al* 2008) and TST (Manning *et al* 2010) postulate a dependence of the surface tension on activity of the actomyosin cytoskeleton in the cell, supporting an alternative theory to DAH in which cortical tension in individual cells is supposed to be the important factor for providing the surface tension of tissues. This second hypothesis on the origin of tissue surface tension is called the differential interfacial tension hypothesis (DITH), and it was formulated by Harris (Harris 1976), Brodland (Broadland 2003) and Graner (Graner 1993). In particular, the DITH relates tissue surface tension to the tension along individual cellular interfaces. The DITH theories are appealing because they recognize, in contrast to the DAH theory, that individual cells cannot be treated as point objects. Thus, a cell's mechanical energy alterations due to its shape and cortical tension obviously play a role in this energy balance.

It has been reported that interfacial tensions arise from a balance of adhesion, cortical tension and cortical elasticity (Lecuit and Lenne 2007, Krieg *et al* 2008, Farhadifar *et al* 2007, Paluch and Heisenberg 2009). However, the exact nature of this interplay is still elusive. A model has been developed that specifies a relationship between surface tension and the ratio of adhesion to cortical tension; this needs to be experimentally verified. With this model, it can be explained why the 'simple' DAH is so successful. In addition, this model predicts regimes where the DAH breaks down and shows that alterations in tissue surface tension are accompanied by alterations in the shapes of surface cells as a function of the ratio between adhesion and cortical tension. One of two main implications is that when $\gamma/\beta < 2$, surface cells do not stretch out and the DAH is essentially correct. Indeed, surface cells form fewer focal adhesions compared to interior cells and this effect is the primary contribution to the surface tension (as in fluids). Although cells alter their shapes, the surface tension varies nearly linearly with the effective adhesion. Thus, adhesion in the DAH must correspond to the net energetic contribution of contacting surfaces, which depends on the free energy of cadherin binding complexes and local alterations to the cortical tension caused by these bonds. Taken together, the balance between the cortical tension based on the actomyosin cytoskeleton and cell adhesion seems to be crucial for determining surface cell morphologies.

The second implication is that the analogy with fluids is no longer applicable when surface cells stretch to form additional nearest-neighbor contacts. In contrast to fluids, if the surface cells (outermost cells) stretch to have the same net contact perimeter compared to cells in the bulk, there is no adhesive contribution to the surface tension and the DAH must fail. However, this observation does not depend on a theoretical model. In particular, it has been shown that in high adhesion zebrafish aggregates, surface cells possess a precisely defined, reproducible cross-sectional area that is significantly larger than that in bulk cells. In more detail, this areal fraction is consistent with the assumption that surface cells provide the same surface area of contact when they are located in the bulk. Thus, the surface tension of these aggregates seems not to be strongly dependent on the cell adhesion.

In particular, the observation that stretched surface cells exist leads to the consideration of cadherin diffusion and its impact on surface tension. In another model, the adhesive energy density is considered to be constant along contacting interfaces and thus in stretched cells cadherin molecules need to diffuse to the much larger contact interface to maintain the same cadherin density. However, in compact surface cells cadherins are not recruited additionally to the contact interface by inducing their migration. This may be a good first approximation because there are no excess cadherins to bind with on the surface of the bulk cells and hence compact cells can obtain no additional energy from such a migration. Indeed, heterogeneities in cadherin density could lead to unique minimum energy cell shapes. However, these interactions have not yet been investigated and hence are still elusive.

The theoretical model

A minimal mechanical model has been developed that is based on two experimental systems: (i) zebra-fish embryonic tissues and (ii) a P-cadherin-transfected L-cell line (LP2). A confocal section of a zebra-fish aggregate shows that cells in the bulk are roughly polyhedral with sharp corners, an aspect ratio of unity and without obvious polarization. The rate of cell divisions in zebra-fish aggregates is low and cells within a single tissue type have approximately the same size. The model hence enforces a constant volume for individual cells of $V = 1$ and all cell volumes are normalized by the average volume of a single cell (Manning *et al* 2010).

As the goal is to understand the collective behavior of cell populations, one can focus on the coarse-grained mechanical properties of individual cells, such as cortical tension and adhesion. As in other cell models (Simons 1997, Bouchaud 1990), the energy is associated with cell–cell contacts, where the energy is proportional to the surface area of contact between cells: $W_{ad} = (\Gamma/2)\, PC$, where PC is the surface area (perimeter in 2D) in contact with other cells and Γ is the free energy per unit surface area for cadherin (adhesive) bonds.

Moreover, the response of single cells to low-frequency pressures and forces may be characterized by a cortical tension (Simons 1997, Slattery 1995): $W_{cort} = \beta PT$, where PT is the total surface area of a cell and β is the aggregate surface tension. Indeed, feedbacks between the adhesion molecule and cytoskeletal dynamics are abundant, which indicates that the cortical tension along contacting interfaces (βC) may be different than that along non-contacting interfaces (βNC). As the cortical-tension energy and adhesive energy both scale linearly with the surface area, we can adjust feedbacks in a simple manner. In more detail, the term $\beta = \beta NC$ is the cortical tension of a cell in the absence of any cell–cell adhesions, which is the quantity that is measured in single-cell pipette aspiration experiments (Slattery 1995). The effective adhesion γ represents the total energetic contribution of contacting surfaces. In particular, this is defined as the difference between the free energy of the adhesive bonds per unit area (Γ) and local alterations of the cortical tension near an interface

$2(\beta C - \beta NC)$. Moreover, the coarse-grained mechanical energy for each cell in an aggregate is given by equation (7.1):

$$W_{\text{cell}} = (\beta - \gamma/2)P_C + \beta P_{\text{NC}} | V = 1, \qquad (7.1)$$

where $P_{NC} = PT - PC$ is the surface area of the non-contacting interface. Furthermore, $\beta - \gamma/2$ is half the interfacial tension of cell–cell adhesion and β is the interfacial tension of cell–culture medium interfaces. It has been found that cells in the interior of aggregates exhibit polyhedral shapes with sharp corners. This means that when $\gamma/\beta < 2$, the cortical elasticity only contributes less to the energy on the relevant long timescales over which the surface tension is a meaningful quantity (Manning *et al* 2010). Thus equation (7.1) omits elastic terms and is valid when $\gamma/\beta < 2$.

Previous methods for approximating tissue surface tension assumed that individual cells do not alter their individual shapes. If the shapes are unaltered, the surface tension is the difference between the interfacial tensions of cell–cell and cell–culture medium interfaces (Simson *et al* 1998). However, confocal images of the equatorial plane of spherical zebra-fish aggregates indicate that the surface cell shape depends on the tissue surface tension. Moreover, there seems to be an interplay between the mechanical energy and geometry, which cannot be ignored. Thus, the strategy will be to find cellular shapes and configurations that locally minimize the mechanical energy and then calculate the surface tension of those configurations.

When an aggregate of cells is compressed, the surface area of the aggregate increases and the cells, which were formerly located in the bulk, are then exposed to the surface. Thus, in an exact analogy with fluids, the response of the tissue to alterations in surface area is the difference in energy ΔW between a cell in the bulk and a cell on the surface multiplied by the number of cells per unit area (or length in 2D) at the surface. The number of cells per unit area is equal to unity divided by the projected area A_{proj} (or projected length in 2D) of a single cell onto the surface of the aggregate (equation (7.2)):

$$\sigma = \left(W_{\text{surf}} - W_{\text{bulk}} \right)/A_{\text{proj}}, \qquad (7.2)$$

where σ is the surface tension.

This quantity is difficult to calculate for large aggregates, as the total energy depends on the precise geometry of each individual cell, which is in force balance with other cells under the constant volume constraint. However, this can be solved for an ordered 2D system.

Active processes enable the cells to exchange neighbors and explore many possible configurations, indicating that they are not confined to the global minimum energy configuration. By contrast, the collection of cells explores a large number of local minima and these configurations are disordered, such as the typical configurations of a fluid or a jammed granular material. Numerical simulations have revealed that the surface tension for 2D disordered aggregates is related in a simple manner to ordered structures.

Two minimal structures generated by this procedure have been found (Feder *et al* 1996). For small values of γ/β, surface cells are rounded, as they minimize their total perimeter at the expense of cell–cell contacts, whereas for large values of γ/β the cells are flat because they maximize neighbor cell–cell adhesions.

Because the surface tension is a change in energy divided by a projected length, both of these quantities are calculated for individual cells in each numerical simulation. Indeed, the projected length, which corresponds to the macroscopic perimeter of the entire aggregate, is the only quantity altered by the disorder in the limit of small γ/β. The total area cannot change because the number of cells with a fixed area is constant. The macroscopic shape of a 2D hexagonal ordered crystal has hexagonal symmetry, whereas the disordered structure is spherically symmetric, suggesting that the macroscopic perimeter should change as the ratio of the perimeter of a hexagon to a circle of the same area: σ disorder $= 1.05\sigma$ order.

Indeed, there is excellent agreement at small values of γ/β and systematic deviations at large values of γ/β can be explained easily. The geometry serves as a major constraint on the macroscopic surface tension when $\gamma = 2\beta$. In more detail, as γ approaches 2β, the tension along cell–cell interfaces approaches zero and the force balance requires that even the cell-culture medium interface is flat. Then the macroscopic surface tension is identical to the cortical tension at $\gamma = 2\beta$

LP2 and zebra-fish cell-shape alterations

This prediction of surface cell-shape alterations can be investigated experimentally in LP2 cells by applying actin-depolymerizing drugs such as cytochalasin D and latrunculin A to the cell aggregates and determining their surface tension using a TST and imaging the aggregates with SEM. The control aggregate has a relatively high surface tension of 3.16 erg cm^{-2} (work per area). In particular, the cells on the surface of this aggregate are so flat that they cannot be distinguished from each other and the model predicts that the ratio of adhesion to cortical tension is high (Manning *et al* 2010). As expected, cell aggregates treated with actin-depolymerizing drugs showed reduced cortical tension and cell–cell adhesion as the actin anchor of cadherin bonds is weakened. Thus, the macroscopic surface tension is pronouncedly lower. In addition, it is important that the effect of actin-depolymerizing drugs on tissue surface tension is reversible and that rounded surface cells are viable cells.

Confocal images of zebra-fish surface cells indicate that this shape alteration is more substantial than going from round to flat. Although this model suggests that structures with nstretch $= 1$ are the minimum energy states, aggregates with high surface tension often contain surface cells that stretch over multiple bulk cells. In more detail, these stretched surface cells, however, do not express epithelial markers on their cell surface, they rather express the same tissue-type-specific markers as bulk cells (Jovin and Vaz 1989). In addition, they are also indistinguishable from bulk cells in their behavior, as stretched surface cells become intermixed with bulk cells during aggregate fusion (Jovin and Vaz 1989) and the surface cells continue to diffuse in and out of the surface layer. Due to these observations, the model will be improved in order to account for these stretched cells and thus investigate the phenomenon quantitatively in control experiments (Manning *et al* 2010).

Modeling of stretched cell shapes

When the adhesion is larger than the cortical tension, the mechanical model equation (7.1) predicts that cells will begin to spread out because the line tension is negative, $dW/dP = -\gamma/2 + \beta < 0$, thus cells will continue to increase their perimeter without limit. However, this perimeter increase cannot continue until infinity, as other mechanical forces may cause the cells to abolish their expansion (Manning *et al* 2010).

What additional restoring force explains the experimental findings? Adhesion molecule regulation generates a plausible restoring force. In more detail, the amount of energy a single cell can obtain by increasing its perimeter seems to be restricted by the number of cadherins on its cell surface, which is in turn regulated. Although the exact form of regulation is still elusive, its energetic contribution can be expanded as a Taylor series in the difference between the actual adhesive energy and the adhesive energy of a cell with the preferred (optimal) number of cadherins. Keeping to second-order terms, this adds a term $\alpha P^2{}_C$ to equation (7.1). Another possible restoring force is the elasticity, which is generated by a denser cortical network underneath the cell–cell interfaces. In line with this, it would also add a term proportional to $P^2{}_C$ (Sheets *et al* 1997, Edidin 1997). The analysis does not depend on the mechanistic origin of these forces, but requires that the first higher-order term is of the form $P^2{}_C$ (equation (7.3)):

$$W_{\text{cell}} = (\beta - \gamma/2)P_C + \alpha P_C^2 + \beta P_{NC} \mid V = 1 \qquad (7.3)$$

The lowest energy states corresponding to this energy functional are complicated, because the $P^2{}_C$ introduces a length scale that is not necessarily the same as the length scale introduced by the incompressibility constraint. For instance, in 2D the smallest perimeter structure is hexagonal, whereas the preferred perimeter introduced by the second-order term does not need to be a hexagon. However, in agreement with experimental observations, a simple assumption is that both length scales are the same and therefore α increases linearly from zero with γ/β for $\gamma/\beta > 2$ (Manning *et al* 2010). Then the surface cells stretch to reach the same contact area as cells in the bulk and thus the 'covering ratio' for ordered 2D packings can be calculated. Each bulk cell contacts a surface cell over a length $6L$ and hence a surface cell covers approximately three bulk cells. The calculation for the 3D microenvironment is similar and the projected area of surface cells is 3.7 times greater than for bulk cells.

Zebra-fish tissue surface cells

The covering ratio for the stretched surface cells has been determined in disordered 3D zebra-fish ectoderm aggregates. After analysis of hand-segmented confocal slices in order to determine an estimate of the projected areas of each cell, the first result is that stretched surface cells possess a preferred size that is reproducible from aggregate to aggregate and there is a pronounced difference between surface cells and bulk cells. Indeed, the projected areas for the surface cells are much larger than for the bulk cells; for instance, the ratio between the projected areas is 3.7 ± 0.4, which is in line with the theoretical prediction of 3.7 (Manning *et al* 2010). However, this stretching effect is not due to an increase in the volume of surface cells, as using $V_{cell} \sim A_{projh}$ (where h is the distance between the cell's top and bottom), it has been found that bulk cells span on average eight–nine z-direction slices and surface cells three slices, and A_{proj} for surface cells is about three times larger than for bulk cells.

What are the theoretical predictions for the surface tension in this case? In particular, a specific value for α is used to adapt the contact length for bulk and surface cells to equal values and calculate the surface tension of ordered 2D aggregates for a wide range of values of γ/β and nstretch. As expected, the experimentally observed surface tension and surface cell shape will correspond to the minimum solution at each value of γ/β. In more detail, for $\gamma/\beta < 2$ the compact surface cell shapes are optimal, whereas for $\gamma/\beta > 2$ the surface cells stretch across three interior bulk cells. The surface tension exhibits a crossover at $\gamma/\beta \sim 2$ from adhesion-dominated behavior (according to DAH) to a dependence on cortical tension and other mechanical effects.

Can the stretched surface cell states be explained by a minimal model? In a regime where the adhesion is stronger than the cortical tension, stretched surface cells have been shown to be the minimum energy structures as long as there is still a restoring force regulating the areas of cells in contact. Although there seems to be at least two plausible mechanisms for such a restoring force, such as (i) adhesion molecule regulation and (ii) the elasticity of the cortical network, these assumptions have not been directly confirmed experimentally. In addition, when calculating the surface tension, a particular value for the magnitude of this restoring force (α) is chosen, which is based on the assumption that the contact area preferred by the restoring force is the same area as that for bulk cells with sharp corners. If this assumption is relaxed, the surface tension would still exhibit a crossover (γ/β approximately 2), however, the exact nature of the crossover would be altered. In a future experiment, both the existence and the magnitude of the restoring force can be investigated by using laser ablation in order to destroy individual cell–cell interfaces in low- and high-adhesion aggregates and analyzing the structural relaxation. It has been proposed that the anisotropy of the network response and the magnitude of the structural relaxation can be used to extract the relative magnitudes of cortical elasticity compared to interfacial tension (Farhadifar et al 2007).

For a model that includes only adhesion and cortical tension (see equation (7.1)), stretched surface cells are not the minimal energy structures. Despite this, as γ/β approaches two the energies of stretched states become closer to that for the unstretched states and active processes would allow surface cells to explore these so-called metastable configurations. However, these metastable stretched states are supposed to have a wide range of projected areas, which are weighted toward smaller area ratios, as these possess lower energy. The fact that the zebra-fish surface cells have a specific, reproducible area fraction, which is significantly greater than unity, indicates that these structures are not metastable, whereas they have a preferred contact area with other neighboring cells as predicted by the model (see equation (7.3)).

In order to fully interpret the available experimental data with this model, the magnitude of the adhesive tension should be compared to that of the cortical tension in individual cells. However, the net effect of adhesive contacts on interfacial tension (which is denoted γ) depends on both the free energy of adhesive molecule bonds and the alterations to the cortical tension along the contacting interface. Thus it is difficult to determine how alterations to the expression levels or the activity of cadherins, actin or myosin affect γ. Surface tension has been reported to increase linearly with the numbers of surface cadherins (Foty and Steinberg 2005), which is in agreement with the DAH and this novel model, if γ/β increases linearly with the number of cadherins. However, because the interaction between actin and cadherin-facilitated adhesion is a highly dynamic process that is controlled by α-catenin (Drees *et al* 2005), it seems possible to argue that elevated cadherin expression significantly increases cortical tension and γ/β remains unknown. In line with this, when the surface tension is decreased by cytoskeletal drugs both cortical tension and adhesive energy are reduced, because cadherin bonds are stabilized by the cortical network (Imamura *et al* 1999, McClay *et al*, Chu *et al* 2004). At least one attempt to dissect the connection of cortical tension and adhesion of individual cells has been reported, using AFM measurements (Krieg *et al* 2008). Although in principle AFM is a suitable technique for those measurements, thermal drift remains a technical challenge that still prevents long timescale measurements. Indeed, the results are on timescales of seconds and thus are too short to be relevant for the interpretation of tissue surface tension, which unfortunately becomes valid only for long timescales on the order of tens of minutes (Krieg *et al* 2008). At these short timescales, the actin cytoskeleton cannot remodel (Adams *et al* 1996, Angres *et al* 1996) and thus the AFM probes the cytoskeletal elasticity and not exclusively the cortical tension. In addition, cadherin bonds strengthen pronouncedly over time after initiation, thus the adhesion dynamics over long timescales are substantially different from those on short timescales (Borghi and Nelson 2009, Imamura *et al* 1999, McClay *et al* 1981). Taken together, the latter seem to be major obstacles for measuring cortical tension.

In order to solve this problem, a set of experiments able to evaluate the cortical-tension β and the effective adhesion γ is presented. One possible approach for measuring the cortical tension is the micropipette aspiration technique (Evans and Yeung 1989, Chu *et al* 2004). However, in the near future it may be possible to reduce AFM drift in order to be able to perform single-cell AFM experiments on the relevant long timescales, allowing one to measure the repulsive force generated by the pure cortical tension. The determination of the effective adhesion γ is more difficult. A semi-quantitative approach for investigating the effect of γ/β would be the use of cell lines that are engineered to express a controlled number of fluorescently labeled adhesion molecules, actin, myosin and actin-associated proteins as performed previously for Madin–Darby canine kidney cells (Yamada and Nelson 2007). Tissue surface tension and surface cell geometries can then be calculated as a function of the ratio of the densities of these molecules and used to investigate the predicted crossover in surface cell shapes and energy contributions at $\gamma/\beta = 2$ in this novel model. Another point is that laser ablation experiments can be used to estimate the interfacial tension along cell–cell interfaces (Farhadifar *et al* 2007). Moreover, it would be interesting to adapt the shape–energy functional given by equation (7.1) to a Monte Carlo or cellular Potts model (Graner and Glazier 1992, Mombach *et al* 1995) approach with activated dynamics and compare simulated cell sorting based on this interaction potential with the experimentally available data. However, these approaches will need to be confirmed by future studies and for the present model it is suggested that surface cell shapes can be used to estimate the ratio between adhesion and cortical tension in an experimental aggregate.

Disordered cellular structures have been detected in many problems in physics and biology. The fact that an analytic expression for their surface energy can be calculated is unexpected, but very useful. For instance, equation (7.1) with $\gamma = \beta$ describes a dry foam and thus this method can be used to calculate the surface energies of finite 2D foams as a function of the cluster size. An extension of this work also predicts how cells in the bulk alter their shape from spheres when there is no adhesion, to polyhedra with sharp corners when the adhesion is high.

Thus, a minimal model has been developed that relates tissue surface tension to the mechanical properties of the individual cells, such as cortical tension, cell–cell adhesion and incompressibility. In addition, this model predicts how a crossover from the DAH to significant cortical-tension dependence may occur, which is an important consideration when designing and developing novel drugs to modulate the mechanical behavior of cells. Both the DAH and the DITH have been developed in order to explain cell-sorting experiments *in vitro*. Integrating both surface-tension-based hypotheses into a single framework, the model then predicts not only that cells sort out according to the surface tension of their aggregates, but also that this surface tension exhibits a crossover from a regime where intercellular adhesion is dominant to another regime where cortical tension is dominant.

Osmotic stress evokes one of the most fundamental challenges to living cells. In more detail, the largely inextensible plasma membrane of eukaryotic cells can easily rupture under in-plane tension, searching for sophisticated strategies to respond immediately to osmotic stress. How do epithelial cells react and adapt mechanically to the exposure to hypotonic and hypertonic solutions in the context of a confluent monolayer? In order to answer this question, site-specific indentation experiments in conjunction with tether pulling on individual cells have been performed with an AFM to reveal the spatio-temporal alterations in membrane tension and surface area. It has been observed that cells compensate for an increase in lateral tension due to hypoosmotic stress by sacrificing excess of membrane area, which is stored in protrusions and invaginations such as microvilli and caveolae. At mild hypotonic conditions lateral tension increases and is hence partly compensated for by surface regulation, for example, the cell sacrifices some of its membrane reservoirs. A loss of membrane–actin anchors occurs upon exposure to stronger hypotonic solutions, giving rise to a drop in lateral tension. Tension release recovers on longer timescales by an increasing endocytosis, which efficiently removes excess membrane from the apical side in order to restore the initial pre-stress. Hypertonic solutions lead to the shrinkage of cells and a collapse of the apical membrane onto the cortex. In addition, exposure to distilled water leads to stiffening of cells due to removal of excess surface area and tension increase, which is caused by elevated osmotic pressure across the cell membrane.

The plasma membrane of cells is a highly dynamic and strongly regulated 2D liquid crystal. Many cellular processes, such as endo- and exocytosis (Dai *et al* 1997, Apodaca 2002), cell migration (Sheetz and Dai 1996), cell spreading (Gauthier *et al* 2011, Raucher and Sheetz 2000) and mitosis (Raucher and Sheetz 1999), are regulated by an intrinsic feature of the cell membrane and in particular by the membrane tension. The cell membrane tension encompasses the in-plane tension of the lipid bilayer and the membrane–cytoskeleton adhesion, which is actively regulated by the contractile actomyosin cortex (Dai and Sheetz 1999). The interplay between the cell membrane and its actin cortex enables cells to withstand mechanical alterations from the microenvironment. Due to the fact that cell membranes are thin and fragile structures, sophisticated feedback mechanisms are necessary based on tension homeostasis to maintain an intact cellular shell. For instance, osmotic stress is a physiologically relevant mechanical stimulus, as animal cells have to withstand substantial fluctuations in the osmolarity of external fluids, which leads to a considerable pressure difference between the cytosol and the microenvironment. Osmotic pressure forces the cell to adapt quickly in order to avoid damage to the largely inextensible cell membrane. When cells are exposed to a hypoosmotic solution, they increase their volume due to the influx of water and hence they extend their projected surface area. Animal cells have been reported to increase their surface area by a factor of three and ten times their volume, depending on the particular cell type (Groulx *et al* 2006). As membranes cannot bear large strains (3–4%), they require regulatory processes to maintain the overall cell membrane tension below the lysis tension. In a more precise manner, tension-driven surface area regulation is necessary to accommodate alterations in tension. In more detail, high tension can be buffered through an excess of membrane area, whereas a decrease in membrane area is triggered when the tension lowers (Morris and Homann 2001). In order to provide sufficient membrane area, the cells store excess membrane in reservoirs such as microvilli and caveolae, which by virtue of unfolding can buffer membrane tension. Indeed, it has been shown that caveolae can serve as membrane reservoirs to compensate for an increase in membrane tension, which has been induced by swelling (Kozera *et al* 2009).

However, a comprehensive picture of membrane homeostasis is still missing, as it is difficult to measure surface area and tension at the same time in order to analyze the regulative relationship between the two parameters. To solve this, a unique combination of site-specific indentation followed by tether pulling has been combined to acquire the tension and excess area of the apical side of epithelial cells, which can be spatio-temporally resolved. Indeed, polar epithelial cells such as intestine- or kidney-derived epithelial cells are suitable for investigating the impact of micoenvironmental physicochemical stimuli on alterations in membrane tension, as they frequently face alterations in chemical potential. In particular, Madin–Darby canine kidney cells II (MDCK II) are grown to confluence and challenged by different osmotic solutions to determine the mechanical response of the cell membrane with respect to tension-buffering membrane reservoirs. Thus, indentation experiments, which are then analyzed with an extended liquid droplet model for adherent cells, have been performed with an AFM to simultaneously assess local alterations in the membrane tension, the time and the location to monitor the excess membrane area as a function of osmotic pressure. The latter task is performed by measuring the apparent area modulus of the cell membrane mirroring the amount of stored excess area. In conjunction with membrane-tether-pulling experiments, which are an independent mechanical approach, it has been possible to show how epithelial cells adjust their surface area in response to tension alterations. In addition, morphological alterations due to cell swelling and shrinking have been observed, showing distinct alteration in tension buffering membrane reservoirs such as microvilli. Moreover, it has been found that MDCK II cells use their microvilli to generate excess membrane that is readily consumed to accommodate increasing tension. The lysis of the cell membrane is prevented through loosening membrane–actin connections and sacrificing membrane reservoirs. However, exposure of the cells to distilled water exhausts all existing reservoirs and consequently leads to cell lysis and death. The long-term observation of tension and surface area reveals that membrane tension largely recovers at the expense of membrane area taken up via an increased endocytosis rate.

7.2 The mobility of surface receptors

It has been revealed from fluorescence recovery after photobleaching experiments that the mobility of most cell surface receptors is much smaller than expected for a free diffusion of proteins within a fluid lipid bilayer. In more detail, single-particle tracking experiments have been performed to uncover the complexity of the local constraints to free diffusion. Indeed, evidence has been obtained for several different processes, such as domain-limited diffusion, temporary confinement and anomalous diffusion. The type of motion of a given cell surface receptor will pronouncedly influence the rate of any functional process, which requires movement in the plane of the membrane. For instance, anomalous diffusion greatly reduces the distance travelled by a receptor on a timescale of minutes.

The lateral movement of several membrane proteins is essential to their appropriate function. This may include movement toward a specific site on the cell membrane as in receptor-facilitated endocytosis or the formation of transient as well as long-lived interactions between different cell surface receptors. Restrictions on receptor movement, confining functionally related proteins to close proximity, are necessary for their function and regulation (Mierke *et al* 2001). An understanding of the mobility of membrane proteins is essential for revealing the mechanisms and especially the kinetics of several membrane-associated functions.

The lateral mobility of a wide variety of membrane proteins has been determined by the method of fluorescence recovery after photobleaching (FRAP) (Jovin 1989, Peters 1991). Typically, a small area (1–2 μm diameter) of fluorescent-labeled receptors is photobleached using a focused laser beam. The fluorescence in the bleached area is recorded and recovers in a certain time interval due to diffusion of unbleached molecules into this area. The recovery of the fluorescence is normally incomplete when measurements are performed on the cell membrane of living cells. These experiments are usually interpreted using two components. One fraction is immobile on the timescale of the experiment and another is mobile, and this is characterized by a diffusion coefficient smaller than that for the unconfined diffusion.

It has become possible to record the movements of individual cell surface receptors using the technique of single-particle tracking (SPT) (Cherry 1992, Sheetz et al 1995, Saxton and Jacobson 1997). SPT involves the attachment of a small particle (typically 11–40 nm in diameter) to the targeted protein. In more detail, two types of particles have been used: fluorescent particles, which are imaged by low-light-level fluorescence microscopy; and gold particles, which can be imaged using differential interference contrast microscopy. The movement of proteins in the cell membrane of living cells can be observed by tracking the individual particle positions through a sequence of images. When the particles are well separated compared with the resolution of the optical microscope, the positions of the particles can be determined with high precision, hence the spatial resolution of SPT is 10–20 nm (compared with ~1 μm in a FRAP experiment).

SPT measurements have been performed with a number of receptors on different cell types. In particular, these experiments have revealed a considerable complexity in the motion of individual molecules. The methods for analyzing SPT data mostly depend on comparing movements over different timescales (Anderson et al 1992, Saxton 1993, 1994, 1995, 1996, Kusumi et al 1993, Simson et al 1995): for random diffusion, $\langle r^2 \rangle / t$ is independent of time, where $\langle r^2 \rangle$ is the mean square displacement measured over a time interval t. In particular, an increase in $\langle r^2 \rangle / t$ with time is indicative of directed motion, whereas a decrease corresponds to some form of constrained diffusion. Individual tracks have to be carefully analyzed. Using Monte Carlo simulations it has been shown that random movements produce apparently non-random behavior with high probability (Saxton 1993). Various statistical tests, however, have revealed that non-random movements of membrane proteins occur commonly. These statistical analysis can be applied to individual tracks when there is a sufficiently large number of data points (Qian et al 1991), which ensures classification into receptor sub-populations exhibiting random diffusion, constrained diffusion, directed motion or immobility. As an alternative method, different populations on the same cell may be inferred from analysis of the experimental probability distribution of particle displacements (Anderson et al 1992, Wilson et al 1996, Schutz et al 1997).

Constrained diffusion: domain-limited diffusion
The simplest conceptual model for constrained diffusion seems to be the domain model. According to this model, barriers to free diffusion split the cell membrane into domains. In more detail, receptors undergo random diffusion within a certain domain, with a diffusion coefficient similar to that of free diffusion in a lipid bilayer. In particular, long-range diffusion occurs much more slowly and depends on the rate at which receptors can hop or switch between special domains. The first evidence for the existence of these domains was obtained from a FRAP experiment in which the 'immobile fraction' was revealed to increase with increasing size of the bleached spot (Yechiel and Edidin 1987). In line with this, SPT experiments with gold-labeled transferrin receptors on normal rat kidney epithelial cells (NRK cells) revealed

trajectories, which visually give a strong impression of the receptors hopping between different domains of a few hundred nanometers' diameter (Sako and Kusumi 1994). Barriers that can build the walls of domains have been detected by experiments in which gold-labeled receptors are dragged across the cell surface by laser tweezers (Edidin *et al* 1991, Sako and Kusumi 1995, Sako *et al* 1998). In addition, SPT and laser tweezer experiments with E-cadherin indicate that these molecules may be either corralled by the cytoskeleton or tethered to it (Sako *et al* 1998).

Constrained diffusion: temporary confinement
The trajectories of both the neural cell adhesion molecule (NCAM) and the T-lymphocyte differentiation marker Thy-1 have been analyzed using a temporary confinement model (Simson *et al* 1995, 1998). In this model, molecules perform free random diffusion, which is interspersed with periods of confinement within regions of about 300 nm diameter. These regions seem to be similar to the domains proposed by Sako and Kusumi (Sako and Kusumi 1994), but they are also explainable using a variety of other mechanisms (Simson *et al* 1998). Thus the temporary confinement domains may consist of clusters of integral membrane proteins, between which the mobile protein is entangled. In the case of the GPI-anchored receptor Thy-1, it has been suggested (Sheetz *et al* 1997) that transient confinement zones may be glycolipid-rich regions, and these may correspond to the detergent-insoluble membrane fractions observed in biochemical experiments (Simons and Ikonenn 1997). In addition, the evidence for lipid microdomains in cell membranes has recently been presented in another study (Edidin 1997).

Constrained diffusion: anomalous diffusion
Alternatively, the constrained diffusion may be interpreted using an anomalous diffusion model. Studies of transport in disordered systems have revealed many examples where anomalous diffusion occurs (Bouchaud and Georges 1990). In particular, anomalous diffusion in cell membranes may result from obstacles and traps, such as binding sites with a broad distribution of binding energies and escape times (Saxton 1996).

SPT measurements with fluorescent LDL bound to LDL receptors or via Fab to IgE receptors have been interpreted and confirmed using an anomalous diffusion model (Ghosh 1991, Slattery 1995, Feder *et al* 1996). Additional data for LDL receptors are indeed consistent with anomalous diffusion (Anderson *et al* 1992). However, these experiments are difficult to perform as LDL can bind non-specifically and this binding has to be eliminated (Goldstein and Brown 1977). These findings suggest that specifically bound LDL may bind weakly to other cellular components, such as the extracellular matrix, which would make it complicated to interpret the SPT data.

Detailed studies of the mobility of MHC class I molecules have been performed on HeLa cells (Smith *et al* 1998). In more detail, SPT studies were performed using R-phycoerythrin coupled to Fab derived from a monoclonal antibody to MHC class I (Smith *et al* 1998). However, this probe is the smallest (11 × 8 nm) used for SPT experiments on cells and hence the non-specific binding is negligible, so that the risk of a perturbing effect of the particle is minimized. Another advantage of this probe is that it is monovalent, as it is purified as a 1:1 complex of Fab and R-phycoerythrin. This circumvents complications caused by crosslinking, which may occur with multivalent probes.

A disadvantage of R-phycoerythrin is that it photobleaches rather readily, thus limiting the number of images that can be obtained in an SPT experiment (Sako *et al* 1998). In experiments with HeLa cells, the time interval between images has varied from 4 to 60 s, so data were obtained for time intervals from 4 s to 20 min. The displacements r over a given time interval t were fitted to the probability distribution (equation (7.4)):

$$P(r)dr = [r/2Dt]\left[\exp\left(-r^2/4Dt\right)\right]dr. \qquad (7.4)$$

In more detail, equation (7.4) assumes random diffusion and yields the diffusion coefficient D. Determination of D over a range of times provides a prediction for different types of motion (Saxton 1996). For normal diffusion, D is independent of the time, whereas for anomalous diffusion D (strictly $\langle r^2\rangle/4t$) decreases over all times. For instance, the domain-hopping model predicts anomalous diffusion whilst molecules are confined within a domain, followed by a crossover to normal diffusion as long-range motion becomes limited by the rate of hopping between domains. However, if a population of molecules performs directed motion, the histogram develops a second peak at longer times (Wilson *et al* 1996).

The SPT experiments with MHC class I on HeLa cells revealed strong evidence for anomalous diffusion. In particular, plots of log D versus log t for data obtained from two experiments on different timescales show that the negative slope of these plots demonstrates that diffusion is anomalous over all times covered by the experiments. However, as is usual for single-cell experiments, there is indeed some cell-to-cell variability in the parameters, while the negative slope was consistently observed. Fitting the data to equation (7.4) (Feder *et al* 1996)

$$D = D_o t^{\alpha-1}$$

gave a value of α of about 0.5 ($\alpha = 1$ for normal diffusion).

However, the domain-hopping and anomalous diffusion models are not necessarily incompatible. The results obtained for MHC class I on HeLa cells can be explained by domains when the distribution of escape times from the domains is sufficiently broad. In addition, a range of models involving obstacles and binding sites can also predict these results (Saxton 1996). It has been proposed several times that diffusion confinements in membranes have to arise from binding to cytoskeletal components, immobile transmembrane proteins or the extracellular matrix (Sheetz 1995). These constraints may be caused by binding or obstruction. However, it is plausible that there is considerable variability amongst receptors and cell types in the hindrance factors, which constrain mobility, and it is unlikely that a single model would be universally applicable.

The relationship between FRAP and SPT

The results of SPT experiments raise questions about the interpretation of FRAP experiments. It has been proposed that anomalous diffusion might occur in cell membranes as a consequence of long-time tails in the jump rate of diffusing molecules (Nagle 1992). Moreover, the effect of long tail kinetics has been analyzed on FRAP measurements and indeed this showed that the diffusion coefficient as well as the immobile fraction determined by conventional means depends on the length and timescale of the individual experiment. In addition, FRAP data for IgE receptors have been analyzed on rat basophilic leukemia cells using both the conventional model of random diffusion with an immobile fraction and a model in which all receptors simply undergo anomalous diffusion (Feder *et al* 1996). They found that the two models fitted the experimental data equally well. In addition, they also performed simulations hypothesizing that FRAP experiments are not suitable to distinguish between the two models.

A further issue is whether there is quantitative agreement between FRAP and SPT measurements. There is some evidence that there may be an agreement, but, due to a number of problems, it is difficult to perform a valid comparison. The simplest experimental system consists of lipid diffusion in models for lipid membranes. Indeed, a fair agreement has been revealed between values of D measured by SPT and FRAP for lipids labeled with a single fluorophore (Schmidt *et al* 1996), but another study observed a two-to-four times lower D for SPT of lipids labeled with gold particles (Lee *et al* 1991). The lower D seems to be related to the multivalency of the particles.

An advantage of fluorescence SPT is that comparison with FRAP is feasible under exactly identical experimental conditions. In addition, experiments with fluorescent LDL particles attached via IgE to IgE receptors are performed (Feder *et al* 1996). In SPT experiments, 27% of receptors have been classified as immobile. In particular, the mobile receptors mainly exhibited anomalous diffusion with mean values of $\alpha = 0.64$ and D measured over 1 s of $0.96 \times 10^{-10} \, \mathrm{cm^2 \, s^{-1}}$. By comparison, FRAP experiments using the same probe lead to $\alpha = 0.15$ and $D(1 \, \mathrm{s}) = 1.4 \times 10^{-10} \, \mathrm{cm^2 \, s^{-1}}$, when the data are fitted to the anomalous diffusion model. A complication in these experiments was that the IgE receptors seemed to be much more mobile when FRAP experiments were performed with fluorescein-labeled IgE serving as the probe. Moreover, this suggests that the attachment of the LDL particle pronouncedly perturbs the receptor mobility. However, it will be important to determine whether this problem is restricted to LDL or occurs even with other particles used for SPT.

The new insights obtained regarding the movement of cell surface receptors have profound implications for membrane function. In particular, a variety of functional processes require receptors to form associations or move to specific sites, such as coated pits. Theoretical analyses of such processes have generally assumed normal diffusion with diffusion coefficients derived from the mobile fraction observed in FRAP experiments. When, however, receptors undergo anomalous diffusion, then processes occur over longer distances and timescales may be dramatically slowed down. It has been worked out that the receptors are often coupled through

G proteins in order to overcome slow diffusion in the membrane (Peters 1988). The existence of anomalous diffusion supports this case strongly. Moreover, it has been proposed that altered mobility of integrins can impair cell adhesion by reducing the ability of the receptors to build clusters (Yauch *et al* 1997). On the other hand, restrictions on diffusion, by whatever mechanism, seem to be advantageous. Indeed, receptors, which are delivered to the cell membrane through vesicle transport may not move far from the fusion site. This could provide an explanation as to why functionally related receptors remain in close proximity to each other.

Additional experiments need to be performed to investigate whether SPT and FRAP data can be incorporated within the same theoretical model. SPT experiments indicate that the random diffusion plus the immobile fraction model is inappropriate for evaluating FRAP experiments. However, doubts about possible effects of the particle in SPT experiments have yet to be fully resolved. One advantage will be to dispense with particles by tracking fluorescent-labeled antibodies. In line with this, it has become possible to image and track single fluorophores, though only so far in a model system (Schmidt *et al* 1996). In addition, IgG bound to MHC class I has been imaged on HeLa cells. In more detail, the IgG was labeled at a ratio of 10 fluorophores per IgG without loss of specific binding, indicating that tracking experiments for cells with fluorescent-labeled antibodies is indeed feasible.

7.3 Specific membrane regions as a location for surface receptors

Nearly one third of the human genome encodes proteins that are membrane-associated, such as transmembrane, lipid-anchored or peripheral membrane proteins. The protein diffusion (described by the diffusion constant D) not only provides the distribution of proteins across the membrane surface, it also determines their ability to interact. In particular, the forward reaction kinetics of two generic membrane proteins is ascertained by the sum of their individual diffusion constants. Thus, the mobility of proteins within the cell membrane determines their interaction capabilities, for example in signal transduction events. The interaction efficiency is further determined by the local protein concentration, which determines the collision probability. Thus, two slowly diffusing proteins interact inefficiently unless their local protein concentration is very high. To draw a complete picture of membrane proteins in a complex cellular process such as receptor signaling, it is not sufficient to know the binding affinities of ligands and kinases, but only by quantifying protein dynamics and concentrations can the interactions and functions be fully revealed.

The kinetic properties of a membrane protein and the types of motion it can undergo are critically influenced by the lateral organization of the cell membrane. The fluid mosaic model of Singer and Nicolson proposed that the cell membrane is laterally homogenous and hence membrane-associated proteins can diffuse freely using Brownian motion in the plane of the lipid bilayer (Singer and Nicolson 1972) and hence they are randomly distributed on the cell surface. In this random diffusion model, the diffusion constant D is only dependent on the hydrodynamic radius of the protein, which is assumed to be cylindrical, and the viscosity of the membrane can be described by the Saffman–Delbruck model (Saffman 1975). In practical terms, Brownian motion means that the distance a molecule diffuses, measured as the mean square displacement (MSD) in time t in a 2D membrane, relates linearly to the diffusion coefficient D: $MSD = 4D\,t$ (Singer and Nicolson 1972).

However, cell membranes such as the plasma membrane are complex lipid–protein composites in which protein diffusion becomes anomalous (Weiss *et al* 2003), with diffusion coefficients an order of magnitude lower than those observed in artificial membrane systems (Bacia *et al* 2004). This retardation is possibly due to protein–protein interactions (Douglass and Vale 2005), active transport, cytoskeleton interactions and confinement (Kusumi *et al* 2005), as well as membrane lipid microdomains (Simons and Ikonen 1997). In particular, lipid rafts are defined as small, heterogeneous, highly dynamic, sterol- and sphingolipid-enriched domains (Pike and Rafts 2006) that are hypothesized to facilitate protein–protein interactions through their ability to selectively accumulate proteins and control the diffusion via their internal viscosity and/or boundary properties.

In heterogeneous cell membranes, the diffusion coefficient D can be divided into local and global values, as diffusion over a short distance and time period such as within a raft is clearly different from that over longer distances and timescales such as when covering many rafts. Thus, by simply changing the observation area and recording time intervals, one can obtain different values. Indeed, this phenomenon has been observed and hence it has been concluded that there seem to be protein-rich membrane domains of ~1 μm in size in the plasma membrane of endothelial cells (Yechiel and Edidin 1987). Hence by quantifying the local and global diffusion behavior of proteins and lipids, it is possible to build more detailed physical models of cell membranes and understand the underlying principles of their organization. Indeed, there seems to be a link between membrane organization and protein diffusion in different scenarios that are neither comprehensive nor mutually exclusive.

Accuracy in measuring protein dynamics in living cells has been a major restriction. However, fluorescence microscopy is the only suitable approach to characterize protein dynamics in living cells and tremendous progress has been made in hardware, automated acquisition and analysis algorithms. In principle, there are three different techniques that can be used for the measurement and analysis of molecular diffusion in cell membranes. These are FRAP, SPT and fluorescence correlation spectroscopy (FCS). The goal is to explain how these have deepened our knowledge of membrane rafts and the global organization of the cell membrane.

Brownian protein diffusion and membrane viscosity
FRAP has been used to measure diffusion coefficients in living cells. In more depth, an area of the sample is first quickly photobleached by intense laser exposure. As mobile fluorophores from outside the bleached region diffuse into the dark area, the fluorescence intensity is observed to recover. Indeed, the speed of this recovery is related to the fluorophore mobility and thus the diffusion coefficient can be determined. The fraction of the fluorescence that is recovered after long timescales is the so-called mobile fraction (Mf) and this gives information on the percentage of

molecules that can freely diffuse. By using the volume of the laser focal spot for bleaching, one derives more localized diffusion parameters, distinct from the global measurements, in which the laser spot is used to bleach a larger area such as a disk or rectangle. In more detail, the size and geometry of the bleached area has to be taken into account when interpreting and comparing FRAP results.

FRAP studies were employed to analyze the hypothesized effects of lipid rafts on diffusion (Kenworthy *et al* 2004). If rafts are small immobile domains of high viscosity, raft-favoring proteins with a stable raft association would have low mobile fractions and diffusion coefficients and hence different raft markers would have similar diffusion properties. Proteins with a high affinity for membrane rafts, including the dual-palmitoylated transmembrane protein linker for activation of T cells (LAT) and farnesylated/palmitoylated H-Ras, and proteins that target raft constituents, such as cholera toxin subunit B (CTxB), which binds and clusters ganglioside GM1 (Brown 2006), were determined, however, to have pronouncedly diverse diffusion coefficients. It was thus concluded that the type of membrane anchor, and not purely the raft association phenomenon, determines the diffusion coefficient of proteins in the cell membrane (Kenworthy *et al* 2004).

Interestingly, in the apical membranes of Madin–Darby canine kidney (MDCK) epithelial cells at room temperature, only the raft-associated proteins LAT and glycosylphosphatidylinositol (GPI)-anchored proteins showed complete fluorescence recovery in bleached regions, whereas 30% of non-raft proteins were immobile (Meder *et al* 2006). However, the latter formerly immobile proteins became mobile at 37°C. This temperature dependence was explained as the formation of a discontinuous liquid-ordered phase at lower temperatures, which segregates and confines membrane proteins. Lipid raft proteins can diffuse freely in either phase, whereas non-raft proteins are excluded from ordered-phase domains, which hence act as obstacles for their own free diffusion. Here, FRAP was sensitive enough to detect differences between proteins that could dynamically participate in raft domains, but only because raft coverage was close to the percolation threshold. However, there is currently no direct evidence that lipid phase separation does indeed occur in native *in vivo* cell membranes.

Due to spatial resolution restrictions, FRAP studies cannot detect the impact of small raft domains on protein diffusion, while they are able to confirm the general relationship between membrane viscosity and protein dynamics. As expected, diffusion rates for both raft and non-raft markers slow down with decreasing temperature, as the membranes become less fluid (Kenworthy *et al* 2004). Cholesterol has been reported to reduce the diffusion rate of Ras proteins due to increased membrane viscosity, but so does cholesterol depletion using methyl-ß-cyclodextrin (mßCD) (Goodwin *et al* 2005). It is known that the removal of cholesterol, particularly with mßCD, not only alters membrane viscosity, but it can also hinder diffusion inside the membrane (Shvartsman *et al* 2006), possibly due to induction of gel-like regions (Nishimura *et al* 2006) and reorganization of the actin cytoskeleton (Kwik *et al* 2003).

The spatial limitation has been overcome by simultaneously recording FRAP and anisotropy recovery after photobleaching (ARAP) of GPI-anchored proteins

(Goswami *et al* 2008). Alterations in fluorescence anisotropy are an indication of the fluorescence resonance energy transfer (FRET) between identical fluorophores (homoFRET). While monomers diffuse freely, nanoclusters of only a few GPI-anchored proteins are essentially immobile and thus are non-randomly distributed across the membrane. Together with the concentration-independent and temperature- as well as cholesterol-dependent nature of the conversion between monomers and clusters, this study indicates that the formation of clusters is actively regulated by cortical actin and myosin contractility (Goswami *et al* 2008). However, the extent to which protein multimerization correlates with lipid domains can only be resolved when both processes are recorded simultaneously but independently from each other.

The ability of FRAP to detect the global behavior of macromolecular complexes was demonstrated for the epidermal growth factor receptor (EGFR) (Lajoie *et al* 2007). By systematically measuring the diffusion rate and the mobile fraction, it was discovered that the EGFR association with a galectin lattice overrides the immobilization of the receptor in caveolae and hence the tumor suppressor function of caveolin-1. Subsequently, the regulation of EGFR diffusion rates determines its ligand responsiveness, which seems to be of high importance. EGFR signaling was suppressed when the receptor was sequestered into immobile caveolae, whereas EGFR signaling was enhanced, relative to unconfined diffusion, when the receptor's diffusion rate was retarded by the galectin lattice. Taken together, the data thus indicate that a receptor's mobility, as regulated by the membrane properties, has a strong influence on its receptor–ligand interactions and function.

The major restriction of FRAP studies is that sub-populations that diffuse differently cannot easily be distinguished, as FRAP is essentially an ensemble or bulk measurement. In an elegant theoretical paper, it was shown that the existence of high-viscosity islands such as rafts affects the long-range mobility of both raft- and non-raft-partitioning proteins (Nicolau *et al* 2006). Moreover, the theoretical model predicted that all proteins, with or without raft affinity, diffuse pronouncedly more slowly when rafts are present in the cell membrane. It is interesting to note that protein diffusion in cell membranes is several times slower than in artificial bilayer constructs composed solely of a few lipid species. For instance, the lipid dye diI-C18 has diffusion coefficients of $\sim 1.4 \ \mu m^2 s^{-1}$ in the cell membranes of human embryonic kidney (HEK) cells (Bacia *et al* 2004), whereas the same dye has a diffusion coefficient of $\sim 6.3 \ \mu m^2 s^{-1}$ in dioleoylphosphatidylcholine (DOPC)-containing artificial bilayers (Kahya *et al* 2003). In addition to raft domains, the diffusion within the membrane of cells seems to be confined by the underlying membrane cytoskeleton and the extracellular matrix, which suggest a more complex diffusion behavior, with at least two different sub-populations. However, these systems must be tackled using microscopy techniques with single-molecule sensitivity.

Anomalous sub-diffusion: single molecule techniques
SPT can be used to characterize the anomalous sub-diffusion and the relative diffusion rates in different regions of the cell membrane. In SPT (Schmidt *et al* 1996), individual molecules are labeled, imaged and tracked, and the trajectories are

analyzed (Saxton and Jacobson 1997, Sergé et al 2008). When only a few, precisely separated molecules are fluorescent, their location can be pinpointed with great accuracy by simply calculating the center of the observed Gaussian point-spread function (Ober et al 2004). The localization in consecutive image frames is then connected to form trajectories and additionally the diffusion properties can be determined from the mean-square displacement (MSD).

However, the first publication that showed that raft proteins are anchored into ~30 nm domains that diffuse as an entity across the cell surface employed tracking of membrane proteins with latex beads and a laser trap (Pralle et al 2000). Due to the fact that the diffusion coefficient of membrane constituents is related to their hydrodynamic radius, large tags such as antibody-coated gold beads may pronouncedly alter diffusion characteristics, particularly if protein cross-linking cannot be omitted. Most studies have been tailored to track only single proteins (Suzuki et al 2005) or have exploited the improved S/N ratio of total internal reflection fluorescence (TIRF) microscopy to image fluorescent fusion constructs or small molecule dyes directly.

The main disadvantages of SPT are that there is an inherent trade-off between spatial and temporal resolution and that statistical analysis is somehow limited unless a large number of molecules are tracked in each experiment. Thus, to map the entire cell surface, localization and connecting molecules into trajectories have to be achieved for a high density of molecules. To date, there has been no standardized tracking algorithm that can simultaneously account for high particle density, heterogeneous particle motion, particle interactions (merging and splitting) and temporary disappearance, such as from blinking. Progress has been made in addressing this (Sergé et al 2008, Jaqaman et al 2008), but the interpretation of trajectories will continue to rely on simulations (Wieser et al 2008) until robust algorithms have been established.

The advantages of SPT are that it enables molecular diffusion coefficients to be calculated and different diffusion modes of the same molecule to be identified. Monte Carlo simulations have identified how modes of movement differ from Brownian motion when molecules are actively transported or confined in immobile as well as in diffusing domains (Wieser et al 2008).

Many examples that demonstrate the power of SPT have addressed the influence of post-translational lipid modifications and raft affinity (Douglass and Vale 2005), protein–protein interactions (Suzuki et al 2007) and the actin cytoskeleton (Andrews et al 2008) on protein diffusion. Similar to FRAP experiments, SPT was used to assess the contribution of lipid modifications to protein dynamics on the surface of T cells. Palmitoylation of LAT or dual acylation of the Src-family kinase Lck did not alter their diffusion pronouncedly, but the trapping of these signaling proteins in cholesterol-independent clusters of the co-receptor CD2 did slow them by ~ twofold (Douglass and Vale 2005). Such differential modes of movement can only by detected by dual-color SPT, and not with FRAP. In the case of the T cell plasma membrane, protein–protein interactions have to be factored into the interpretation of modes of motion, which seems to be the case for other cell types where stimulation of surface receptors results in multi-molecular protein complexes. Dual-color tracking of the GPI-anchored receptor CD59 in the outer leaflet and signaling proteins anchored to the inner leaflet also revealed how ligand clustering of the receptor induces temporary immobilizations (named STALLs (stimulation-induced temporary arrest of lateral diffusion) that serve as short-lived platforms for signal transduction activities (Suzuki *et al* 2007).

Another recent example of SPT revealing the mechanism of receptor dynamics is the tracking of individual Fcε RI receptors using non-bleachable quantum dots. Different modes of motion (such as immobile, free, directed and confined) were all detected for the same receptor. Indeed, the confined motion was attributed to the presence of the actin cytoskeleton, which was found to dynamically restrict location of the receptor into micron-sized domains.

Transient confinement zones (TCZs), detected by SPT, define areas where an observed molecule stays much longer than expected from the average diffusion coefficient and are thought to bear some resemblance to lipid rafts. TCZs are typically 200–300 nm in diameter (Simson *et al* 1998), preferentially trap GPI-anchored proteins and glycosphingolipids (Sheets *et al* 1997) and are cholesterol-dependent (Dietrich *et al* 2002). Whether the viscosity differential inside and outside the raft is a sufficient mechanical diffusion barrier is still under investigation (Dietrich *et al* 2001), particularly since TCZs appear to be temperature-independent (Dietrich *et al* 2002). Fluorescent fusion constructs of only the membrane anchor regions of H-Ras (raft), K-Ras (non-raft) and Lck (raft) have been tracked. These three constructs showed similar diffusion coefficients, assuming Brownian motion, however two populations of diffusing molecules were observed (Lommerse *et al* 2006). The major population displayed similar diffusion times to small molecule membrane dyes ($0.6 - 1.6$ $\mu m^2 s^{-1}$). However, a second population (16% for Lck and 27% for K-Ras) was confined to domains of roughly 200 nm. In this instance, cholesterol depletion did not affect confinement. For H-Ras, the mobile fraction was 73%, with a diffusion coefficient of 0.53 $\mu m^2 s^{-1}$ (SPT) or 0.48 $\mu m^2 s^{-1}$(FRAP) (Lommerse *et al* 2004). Interestingly, the 200 nm confinement of the slow-diffusing fraction was only observed for active GTP-bound and 'raft-associated' Ras, but not the inactive GDP-bound form (observed by both mutagenesis and insulin activation of H-Ras), indicating that confinement is not purely controlled by lipid–lipid interactions (Lommerse *et al* 2006, Prior *et al* 2001). A similar confinement was shown for a G-protein coupled odorant receptor with ~50% of the receptor being confined to larger domains (300–550 nm), ~30% to small domains (180–200 nm) and the remaining 20% being either immobile or freely diffusing (Jacquier *et al* 2006). For a different G-protein-coupled receptor, the μ-opioid one, it was found that 90% of the receptors diffuse within a domain that diffuses itself by a motion, which has been called 'walking confined diffusion' (Daumas *et al* 2003). Taken together, these studies suggest that confinement areas are unique to each of the analyzed receptors, in turn suggesting that specific receptor interactions define these zones rather than generic 'raft domains'.

As with the observation area size in FRAP, the sampling frequency in SPT experiments can have an impact on the values of the diffusion coefficients and data interpretation. By developing SPT with ultra-high temporal resolution (25 μs instead of the typical video rate of 30 ms), it was possible to detect a different type of diffusion, 'hop diffusion', in which the entire membrane is compartmentalized into 30–250 nm compartments (Kusumi *et al* 2005, Fujiwara *et al* 2002, Murase *et al* 2004). In particular, proteins and lipids are slowed down when they hop from one compartment to another, whereas within each compartment the diffusion is not pronouncedly lower compared to that observed for free diffusion. Thus, these membrane compartments are distinctly different to the TCZs formed by lipid microdomains (Kusumi *et al* 2004) and are considered to be the result of an underlying membrane cytoskeleton to which the transmembrane proteins are anchored, creating a 'picket fence' within the membrane (Morone *et al* 2006). However, it should be kept in mind that in order to distinguish between Brownian motion and hop diffusion a time resolution of tens of microseconds is required, without which hop diffusion is simply interpreted as free diffusion with a slower average diffusion coefficient.

Multi-parameter imaging: combining diffusion with concentration
A key finding for raft characteristics through theoretical calculation is that global diffusion retardation caused by high-viscosity islands increases the rate of protein–protein interactions (Nicolau *et al* 2006). Moreover, the existence of raft domains with high viscosity can elevate the local concentration of proteins with a low average density but high raft-affinity due to their specific membrane anchors. The maximal collision rate, which is a measure of signaling efficiency, is achieved when 10% of the membrane is covered by small (approximately 6 nm diameter), mobile domains and $D_{\text{raft}}/D_{\text{non-raft}} = 0.5$. Indeed, values of $D_{\text{raft}}/D_{\text{non-raft}} = 0.3–0.5$ have been reported previously in cells (Pralle *et al* 2000, Dietrich *et al* 2002). Raft domains of that nature also increase the collision rate of proteins modestly, even if they have no affinity for the raft domain. It has been demonstrated that this may have important biological consequences for different types of membrane domains, such as caveolae (large and immobile) and GPI-protein-enriched domains (small and mobile). However, it has been predicted that experiments require a time resolution of 5–10 ms in order to detect these small raft domains by SPT, which is at the technical limit of most set-ups.

An alternative single molecule technique is FCS, which was first demonstrated in 1974 as a method for analyzing molecular kinetics (Elson and Magde 1974). In its most basic form, the femto-liter illumination volume of a laser scanning confocal microscope can be held stationary. The intensity fluctuations as a low concentration of fluorescent molecules diffuse in and out of this spot are then recorded using high-speed detectors (Schwille *et al* 1999). From this time series of intensity measurements, an auto-correlation analysis allows molecular diffusion rates to be obtained with high statistical accuracy and over a wide range of timescales. In more detail, modern detectors achieve a range from microseconds up to seconds (Digman *et al* 2005). In particular, this method uses low laser powers and low probe concentrations, making it applicable even to living cell measurements, with minimized perturbation to the cell membrane. Several related techniques have been developed recently (and continue to be developed), such as two-color cross-correlation spectroscopy (FCCS) (Bacia *et al* 2006), image correlation spectroscopy (ICS) (Hebert *et al* 2005) and raster-scanned FCS (raster ICS or RICS) (Digman *et al* 2005). Indeed, these novel methods have the potential to map the spatial distribution of diffusion coefficients at selected membrane regions, as demonstrated for the mobility of paxillin at focal adhesions (Digman *et al* 2008).

An advantage of FCS is that it not only provides extract diffusion coefficients and identifies anomalous diffusion, it can also measure the number of molecules in the observed spot. For instance, fluorescence fluctuation analysis has been used to determine the degree of clustering of the epithelial growth factor (EGF) receptor (Saffarian *et al* 2007) and the stoichiometry of protein complexes (Chen and Muller 2007). Thus, FCS has further developed into a multi-parameter imaging mode that has been used to describe the dynamics and oligomerization of GPI-anchored receptors (Malengo *et al* 2008), adhesion receptors and tetraspanins (Barreiro *et al* 2008).

In order to reveal the relative contributions of membrane lipid microdomains and the actin cytoskeleton, variable spot-size FCS and a defined 'FCS diffusion law' have been used, extracting structural information from below the diffraction limit of optical microscopy (Lenne *et al* 2006). Microdomains and cytoskeletal confinement impart different deviations to the relationship between the diameter of the observation spot and the transit time that a molecule requires to diffuse through that spot (Wawrezinieck *et al* 2005). On a plot of diffusion time against area, dynamic partitioning results in a positive intercept with the y-axis, whereas a meshwork produces a negative intercept (unhindered diffusion passes through the origin) (Lenne *et al* 2006). Indeed, this was shown theoretically and confirmed by observation of Thy-1 (raft associated means positive intercept) and transferrin receptors (transmembrane protein hindered by the cytoskeleton means negative intercept). In particular, this approach has been used to demonstrate that the protein kinase Akt dynamically partitions into microdomains in cells via its pleckstrin homology (PH) domain (Lasserre *et al* 2008).

FCS has been combined with total internal reflection (TIR-FCS) excitation (Starr and Thompson 2001, Thompson and Steele 2007). Such a set-up enables measurement of the local probe concentration and the local translational mobility within the membranes, as well as kinetic rate constants, which describe the association and dissociation of ligands with membrane receptors. The improved accuracy of TIR-FCS over FCS has been demonstrated by imaging a farnesylated variant of green fluorescent protein (GFP) (Ohsugi *et al* 2006). While SPT under TIR illumination cannot easily track cytosolic proteins due to their 3D motion, TIR-FCS was able to quantify the proportion of the GFP construct bound to the membrane. Further technical improvements to FCS measurements specific to membranes have been accomplished (Ries and Schwille 2008).

An exciting prospect is the integration of time-resolved single photon counting (TCSPC) techniques with unrestricted data acquisition to store spatial, temporal, spectral and intensity information (Koberling *et al* 2008). In particular, this enables the correlation of dependences between various fluorescence parameters.

For instance, FCS autocorrelation curves can be constructed for fluorescent events of a certain fluorescence lifetime. This not only helps to remove background noise and autofluorescence, thus providing more accurate diffusion measurements, it also enables a single experiment to give information about the kinetics of, for instance, monomers and dimers, local concentrations of proteins and their different states, their relative stoichiometry and affinities. However, can this information be combined with mathematical models to reveal a more detailed understanding of how proteins function within the cell membrane?

Taken together, the technical advances in fluorescence microscopy hardware and analysis have led to the quantification of the dynamics of membrane proteins in living cells. FRAP, FCS and SPT can go beyond traditional fluorescence imaging and have turned microscopes into molecular measurement devices that can operate on many spatial and temporal length scales. Synergistically, all these approaches have developed a picture where the diffusion characteristics of membrane proteins display complex and subtle behavior. As the influence of the membrane association type on protein diffusion has been suggested, diffusion restrictions imparted by the membrane cytoskeleton and anchored-protein fences, protein oligomerization, clustering and protein–protein interactions, and dynamic partitioning into small, heterogeneous and mobile lipid microdomains have been measured with remarkable accuracy, given the resolution limits of optical microscopy. The arrival of super-resolution microscopy approaches, such as stimulated emission depletion (STED), photoactivated localization microscopy (PALM) and stochastic optical reconstruction microscopy (STORM) have further improved the accuracy of dynamic measurements, as demonstrated by PALM being combined with SPT (Manley *et al* 2008) and STED with FCS (Eggeling *et al* 2009). The reduced observation volume of STED detected differences between the FCS autocorrelation curve of phosphoethanolamine and that of sphingomyelin, whose diffusion is transiently trapped in cholesterol-enriched complexes (Eggeling *et al* 2009).

However, the interdependence of membrane organization and protein diffusion is not yet well understood. Moreover, current models do not adequately take into account the active transport systems, such as directed flow, membrane budding and recycling. In order to investigate these parameters multi-dimensional microscopy techniques have to be performed, as they have the ability to extract more precise information about molecular diffusion rates and other membrane parameters. To date, there has only been a limited understanding of how local variations in membrane fluidity affect protein concentration, oligomerization and diffusion. However, these questions can be addressed by integrating FCS or SPT with the imaging of the membrane lipid order using microenvironmentally sensitive probes such as Laurdan (Gaus *et al* 2006) or di-4-ANEPPDHQ (Jin *et al* 2006). Taken together, quantitative multi-parameter microscopy can further reveal how global cellular manipulations and treatments act on the localized and molecular organization of membrane domains and compartments, which in turn influence protein dynamics, interactions and functions.

Further reading

Lommerse P H M, Snaar-Jagalska B E, Spaink H P and Schmidt T 2005 Single-molecule diffusion measurements of H-Ras at the plasma membrane of live cells reveal microdomain localization upon activation *J. Cell Sci.* **118** 1799–809

Saxton M J and Jacobson K 1997 *Annu. Rev. Biomol. Struct.* **26** 373–99

Schmidt T, Schutz G J, Baumgartner W, Gruber H J and Schindler H 1996 Imaging of single molecule diffusion *Proc. Natl Acad. Sci.* **93** 2926–9

Serge A, Bertaux N, Rigneault H and Marguet D 1997 Dynamic multiple-target tracing to probe spatiotemporal cartography of cell membranes *Nat. Meth.* **5** 687–94

Sheets E D, Simson R and Jacobson K 1995 *Curr. Opin. Cell Biol.* **7** 707–14

Suzuki K G N, Fujiwara T K, Edidin M and Kusumi A 2007 Dynamic recruitment of phospholipase Cγ at transiently immobilized GPI-anchored receptor clusters induces IP3-Ca2 + signaling: single-molecule tracking study 2 *J. Cell Biol.* **177** 731–42

Yamada S, Pokutta S, Drees F, Weis W I and Nelson W J 2005 Deconstructing the cadherin–catenin–actin complex *Cell* **123** 889–901

References

Adams C L, Nelson J W and Smith S J 1996 Quantitative analysis of cadherin–catenin–actin reorganization during development of cell–cell adhesion *J. Cell Biol.* **135** 1899–911

Anderson C M, Georgiou G N, Morrison I E G, Stevenson G V and Cherry R J 1992 *J. Cell Sci.* **101** 415–25

Andrews N L, Lidke K A, Pfeiffer J R, Burns A R, Wilson B S, Oliver J M and Lidke D S 2008 Actin restricts Fcε RI diffusion and facilitates antigen-induced receptor immobilization *Nat. Cell Biol.* **10** 955–63

Angres B, Barth A and Nelson W J 1996 Mechanism for transition from initial to stable cell–cell adhesion: kinetic analysis of E-cadherin-mediated adhesion using a quantitative adhesion assay *J. Cell Biol.* **134**(2) 549–57

Apodaca G 2002 Modulation of membrane traffic by mechanical stimuli *Am. J. Physiol. Renal. Physiol.* **282** 179–90

Armstrong P B 1989 Cell sorting out: The self-assembly of tissues *in vitro* CRC Cr Rev. Biochem. Mol. Biol. **24** 119–49

Bacia K, Kim S A and Schwille P 2006 Fluorescence cross-correlation spectroscopy in living cells *Nat. Meth.* **3** 83–9

Bacia K, Scherfeld D, Kahya N and Schwille P 2004 Fluorescence correlation spectroscopy relates rafts in model and native membranes *Biophys. J.* **87** 1034–43

Barreiro O, Zamai M, Yanez-Mo M, Tejera E, Lopez-Romero P, Monk P N, Gratton E, Caiolfa V R and Sanchez-Madrid F 2008 Endothelial adhesion receptors are recruited to adherent leukocytes by inclusion in preformed tetraspanin nanoplatforms *J. Cell Biol.* **183** 527–42

Borghi N and Nelson J W 2009 Intercellular adhesion in morphogenesis: molecular and biophysical considerations *Curr. Top. Dev. Biol.* **89** 1–32 PMID: 19737640

Bouchaud J-P and Georges A 1990 *Phys. Rep.* **195** 127–293

Brodland G W 2003 New information from aggregate compression tests and its implications for theories of cell sorting *Biorheology* **40** 273–7 PMID: 12454416

Brown D A 2006 Lipid rafts, detergent-resistant membranes and raft targeting signals *Physiology* **21** 430–9

Chen Y and Muller J D 2007 Determining the stoichiometry of protein heterocomplexes in living cells with fluorescence fluctuation spectroscopy *Proc. Natl Acad. Sci.* **104** 3147–52

Cherry R J 1992 *Trends Cell Biol.* **2** 242–4

Chu Y-S, Thomas W A, Eder O, Pincet F, Perez E, Thiery J P and Dufour S 2004 Force measurements in E-cadherin-mediated cell doublets reveal rapid adhesion strengthened by actin cytoskeleton remodeling through Rac and Cdc42 *J. Cell Biol.* **167** 1183–94

Dai J and Sheetz M P 1999 Membrane tether formation from blebbing cells *Biophys. J.* **77** 3363–70

Dai J, Ting-Beall H P and Sheetz M P 1997 The secretion-coupled endocytosis correlates with membrane tension changes in RBL 2H3 cells *J. Gen. Physiol.* **110** 1–10

Daumas F, Destainville N, Millot C, Lopez A, Dean D and Salome L 2003 Confined diffusion without fences of a G-protein-coupled receptor as revealed by single particle tracking *Biophys. J.* **84** 356–66

Davis G S, Phillips H M and Steinberg M S 1997 Germ-layer surface tensions and 'tissue affinities' in *Rana pipiens gastrulae*: quantitative measurements *Dev. Biol.* **192** 630–44

Dietrich C, Bagatolli L A, Volovyk Z N, Thompson N L, Levi M, Jacobson K and Gratton E 2001 Lipid rafts reconstituted in model membranes *Biophys. J.* **80** 1417–28

Dietrich C, Yang B, Fujiwara T, Kusumi A and Jacobson K 2002 Relationship of lipid rafts to transient confinement zones detected by single particle tracking *Biophys. J.* **82** 274–84

Digman M A, Brown C M, Horwitz A R, Mantulin W W and Gratton E 2008 Paxillin dynamics measured during adhesion assembly and disassembly by correlation spectroscopy *Biophys. J.* **94** 2819–31

Digman M A, Brown C M, Sengupta P, Wiseman P W, Horwitz A R and Gratton E 2005 Measuring fast dynamics in solutions and cells with a laser scanning microscope *Biophys. J.* **89** 1317–27

Douglass A D and Vale R D 2005 Single-molecule microscopy reveals plasma membrane microdomains created by protein–protein networks that exclude or trap signaling molecules in T cells *Cell* **121** 937–50

Drees F, Pokutta S, Yamada S, Nelson W J and Weis W I 2005 α-Catenin is a molecular switch that binds E-cadherin-β-catenin and regulates actin-filament assembly *Cell* **123** 903–15

Duguay D, Foty R A and Steinberg M S 2003 Cadherin-mediated cell adhesion and tissue segregation: qualitative and quantitative determinants *Dev. Biol.* **253** 309–23

Edidin M 1997 *Curr. Opin. Cell Biol.* **7** 528–32

Edidin M, Kuo S C and Sheetz M P 1991 *Science* **254** 1379–82

Eggeling C *et al* 2009 Direct observation of the nanoscale dynamics of membrane lipids in a living cell *Nature* **457** 1159–63

Elson E L and Magde D 1974 Fluorescence correlation spectroscopy: Conceptual basis and theory *Biopolymers* **13** 1–27

Evans E and Yeung A 1989 Apparent viscosity and cortical tension of blood granulocytes determined by micropipet aspiration *Biophys. J.* **56** 151–60

Farhadifar R, Röper J C, Aigouy B, Eaton S and Jülicher F 2007 The influence of cell mechanics, cell–cellinteractions, and proliferation on epithelial packing *Curr. Biol.* **17** 2095–104

Feder T J, Brust-Mascher I, Slattery J P, Baird B and Webb W W 1996 *Biophys. J.* **70** 2767–73

Forgacs G, Foty R A, Shafrir Y and Steinberg M S 1998 Viscoelastic properties of living embryonic tissues: a quantitative study *Biophys. J.* **74** 2227–34

Foty R A and Steinberg M S 2005 The differential adhesion hypothesis: a direct evaluation *Dev. Biol.* **278** 255–63

Foty R A, Corbett S A, Schwarzbauer J E and Steinberg M S 1998 Effects of dexamethasone on cadherin-mediated cohesion of human fibrosarcoma HT-1080 cells *Cancer Res.* **58** 3586–9

Foty R A, Forgacs G, Pfleger C M and Steinberg M S 1994 Liquid properties of embryonic tissues: measurement of interfacial tensions *Phys. Rev. Lett.* **72** 2298–301

Foty R A, Pfleger C M, Forgacs G and Steinberg M S 1996 Surface tensions of embryonic tissues predict their mutual envelopment behavior *Development* **122** 1611–20 PMID: 8625847

Foty R A and Steinberg M S 1997 Measurement of tumor cell cohesion and suppression of invasion by E- or P-cadherin *Cancer Res.* **57** 5033–6 PMID: 9371498

Fujiwara T, Ritchie K, Murakoshi H, Jacobson K and Kusumi A 2002 Phospholipids undergo hop diffusion in compartmentalized cell membrane *J. Cell Biol.* **157** 1071–82

Gaus K, Zech T and Harder T 2006 Visualizing membrane microdomains by Laurdan 2-photon microscopy *Mol. Membr. Biol.* **23** 41–8

Gauthier N C, Fardin M A, Roca-Cusachs P and Sheetz M P 2011 Temporary increase in plasma membrane tension coordinates the activation of exocytosis and contraction during cell spreading *Proc. Natl Acad. Sci. USA* **108** 14467–72

Ghosh R N 1991 *PhD. Thesis* (Ithaca, NY: Cornell University)

Goldstein J L and Brown M S 1977 *Annu. Rev. Biochem.* **46** 897–930

Goodwin J S, Drake K R, Remmert C L and Kenworthy A K 2005 Ras diffusion is sensitive to plasma membrane viscosity *Biophys. J.* **89** 1398–410

Goswami D, Gowrishankar K, Bilgrami S, Ghosh S, Raghupathy R, Chadda R, Vishwakarma R, Rao M and Mayor S 2008 Nanoclusters of GPI- anchored proteins are formed by cortical actin-driven activity *Cell* **135** 1085–97

Graner F 1993 Can surface adhesion drive cell-rearrangement? Part I: biological cell-sorting *J. Theor. Biol.* **164** 455–76

Graner F and Glazier J A 1992 Simulation of biological cell sorting using a two-dimensional extended Potts model *Phys. Rev. Lett.* **69** 2013–6

Groulx N, Boudreault F, Orlov S and Grygorczyk R 2006 Membrane reserves and hypotonic cell swelling *J. Membr. Biol.* **214** 43–56

Guevorkian K, Colbert M-J, Durth M, Dufour S and Brochard-Wyart F 2010 Aspiration of biological viscoelastic drops *Phys. Rev. Lett.* **104** 218101

Harris A K 1976 Is cell sorting caused by differences in the work of intercellular adhesion? A critique of the Steinberg hypothesis *J. Theor. Biol.* **61** 267–85

Hebert B, Costantino S and Wiseman P W 2005 Spatiotemporal image correlation spectroscopy (STICS) theory, verification, and application to protein velocity mapping in living CHO cells *Biophys. J.* **88** 3601–14

Holtfreter J 1944 A study of the mechanics of gastrulation *J. Exp. Zool.* **95** 171–212

Imamura Y, Itoh M, Maeno Y, Tsukita S and Nagafuchi A 1999 Functional domains of alpha-catenin required for the strong state of cadherin-based cell adhesion *J. Cell Biol.* **144** 1311–22

Jacquier V, Prummer M, Segura J-M, Pick H and Vogel H 2006 Visualizing odorant receptor trafficking in living cells down to the single-molecule level *Proc. Natl Acad. Sci.* **103** 14325–30

Jaqaman K, Loerke D, Mettlen M, Kuwata H, Grinstein S, Schmid S L and Danuser G 2008 Robust single-particle tracking in live-cell time-lapse sequences *Nat. Meth.* **5** 695–702

Jin L, Millard A C, Wuskell J P, Dong X, Wu D, Clark H A and Loew L M 2006 Characterization and application of a new optical probe for membrane lipid domains *Biophys. J.* **90** 2563–75

Jovin T and Vaz W L C 1989 Rotational and translational diffusion in membranes measured by fluorescence and phosphorescence methods *Methods Enzymol.* **172** 471–513 PMID: 2747540

Kahya N, Scherfeld D, Bacia K, Poolman B and Schwille P 2003 Probing lipid mobility of raft-exhibiting model membranes by fluorescence correlation spectroscopy *J. Biol. Chem.* **278** 28109–15

Kenworthy A K, Nichols B J, Remmert C L, Hendrix G M, Kumar M, Zimmerberg J and Lippincott-Schwartz J 2004 Dynamics of putative raft-associated proteins at the cell surface *J. Cell Biol* **165** 735–46

Koberling F, Kramer B, Tannert S, Ruttinger S, Ortmann U, Patting M, Wahl M, Ewers B, Kapusta P and Erdmann R 2008 Recent advances in time-correlated single-photon counting *Proc. SPIE* **6862** 09

Kozera L, White S and Calaghan S 2009 Caveolae act as membrane reserves which limit mechanosensitive I(Cl, swell) channel activation during swelling in the rat ventricular myocyte *PLoS One* **4** e8312

Krieg M, Arboleda-Estudillo Y, Puech P H, Käfer J, Graner F, Müller D J and Heisenberg C P 2008 Tensile forces govern germ-layer organization in zebra fish *Nat. Cell Biol.* **10** 429–36

Kusumi A, Koyama-Honda I and Suzuki K 2004 Molecular dynamics and interactions for creation of stimulation-induced stabilized rafts from small unstable steady-state rafts *Traffic* **5** 213–30

Kusumi A, Nakada C, Ritchie K, Murase K, Suzuki K, Murakoshi H, Kasai R S, Kondo J and Fujiwara T 2005 Paradigm shift of the plasma membrane concept from the two-dimensional continuum fluid to the partitioned fluid: high-speed single-molecule tracking of membrane molecules *Ann. Rev. Biophys. Biomol. Struct.* **34** 351–78

Kusumi A, Sako Y and Yamamoto M 1993 *Biophys. J.* **65** 2021–40

Kwik J, Boyle S, Fooksman D, Margolis L, Sheetz M P and Edidin M 2003 Membrane cholesterol, lateral mobility, and the phosphatidylinositol 4,5-bisphosphate-dependent organization of cell actin *Proc. Natl Acad. Sci.* **100** 13964–9

Lajoie P, Partridge E A, Guay G, Goetz J G, Pawling J, Lagana A, Joshi B, Dennis J W and Nabi I R 2007 Plasma membrane domain organization regulates EGFR signaling in tumor cells *J. Cell Biol.* **179** 341–56

Lasserre R *et al* 2008 Raft nanodomains contribute to Akt/PKB plasma membrane recruitment and activation *Nat. Chem. Biol.* **4** 538–47

Lecuit T and Lenne P F 2007 Cell surface mechanics and the control of cell shape, tissue patterns and morphogenesis *Nat. Rev. Mol. Cell Biol.* **8** 633–44

Lee G M, Ishihara A and Jacobson K A 1991 *Proc. Natl Acad. Sci. USA* **88** 6274–8

Lenne P -F, Wawrezinieck L, Conchonaud F, Wurtz O, Boned A, Guo X- J, Rigneault H, He H-T and Marguet D 2006 Dynamic molecular confinement in the plasma membrane by microdomains and the cytoskeleton meshwork *EMBO J.* **25** 3245–56

Lommerse P H M, Blab G A, Cognet L, Harms G S, Snaar-Jagalska B E, Spaink H P and Schmidt T 2004 Single-molecule imaging of the H-Ras membrane-anchor reveals domains in the cytoplasmic leaflet of the cell membrane *Biophys. J.* **86** 609–16

Lommerse P H M, Vastenhoud K, Pirinen N J, Magee A I, Spaink H P and Schmidt T 2006 Single-molecule diffusion reveals similar mobility for the Lck, H-Ras and K-Ras membrane anchors *Biophys. J.* **91** 1090–7

Malengo G, Andolfo A, Sidenius N, Gratton E, Zamai M and Caiolfa V R 2008 Fluorescence correlation spectroscopy and photon counting histogram on membrane proteins: functional dynamics of the glycosylphosphatidylinositol-anchored urokinase plasminogen activa-tor receptor *J. Biomed. Opt.* **13** 031215–4

Manley S, Gillette J M, Patterson G H, Shroff H, Hess H F, Betzig E and Lippincott-Schwartz J 2008 High-density mapping of single-molecule trajectories with photoactivated localization microscopy *Nat. Meth.* **5** 155–7

Manning M L, Foty R A, Steinberg M S and Schoetz E-M 2010 Coaction of intercellular adhesion and cortical tension specifies tissue surface tension. *Proc. Natl Acad. Sci. USA* **107** 12517–22

McClay D R, Wessel G M and Marchase R B 1981 Intercellular recognition: quantitation of initial binding events *Proc. Natl Acad. Sci. USA* **78** 4975–9

Meder D, Moreno M J, Verkade P, Vaz W L C and Simons K 2006 Phase coexistence and connectivity in the apical membrane of polarized epithelial cells *Proc. Natl Acad. Sci.* **103** 329–34

Mierke C T, Bretz N and Altevogt P 2011 Contractile forces contribute to increased glycosyl-phosphatidylinositol-anchored receptor CD24-facilitated cancer cell invasion *J. Biol. Chem.* **286** 34858–71

Mgharbel A, Delanoë, Ayari H and Rieu J-P 2009 Measuring accurately liquid and tissue surface tension with a compression plate tensiometer *HFSP J.* **3** 213–21

Mombach J C M, Glazier J A, Raphael R C and Zajac M 1995 Quantitative comparison between differential adhesion models and cell sorting in the presence and absence of fluctuations *Phys. Rev. Lett.* **75** 2244–7

Morone N, Fujiwara T, Murase K, Kasai R S, Ike H, Yuasa S, Usukura J and Kusumi A 2006 Three-dimensional reconstruction of the membrane skeleton at the plasma membrane interface by electron tomography *J. Cell Biol.* **174** 851–62

Morris C E and Homann U 2001 Cell surface area regulation and membrane tension *J. Membr. Biol.* **179** 79–102 PMID: 11220366

Murase K, Fujiwara T, Umemura Y, Suzuki K, Lino R, Yamashita H, Saito M, Murakoshi H, Ritchie K and Kusumi A 2004 Ultrafine membrane compartments for molecular diffusion as revealed by single molecule techniques *Biophys. J.* **86** 4075–93

Nagle J F 1992 *Biophys. J.* **63** 366–70

Nicolau D V Jr., Burrage K, Parton R G and Hancock J F 2006 Identifying optimal lipid raft characteristics required to promote nanoscale protein–protein interactions on the plasma membrane *Mol. Cell Biol.* **26** 313–23

Nishimura S Y, Vrljic M, Klein L O, McConnell H M and Moerner W E 2006 Cholesterol depletion induces solid-like regions in the plasma membrane *Biophys. J.* **90** 927–38

Norotte C, Marga F, Neagu A, Kosztin I and Forgacs G 2008 Experimental evaluation of apparent tissue surface tension based on the exact solution of the Laplace equation *Europhys. Lett.* **81** 46003.1–6

Ober R J, Ram S and Ward E S 2004 Localization accuracy in single-molecule microscopy *Biophys. J.* **86** 1185–200

Ohsugi Y, Saito K, Tamura M and Kinjo M 2006 Lateral mobility of membrane-binding proteins in living cells measured by total internal reflection fluorescence correlation spectroscopy *Biophys. J.* **91** 3456–64

Paluch E and Heisenberg C-P 2009 Biology and physics of cell shape changes in development *Curr. Biol.* **19** R790–9

Peters R 1988 *FEBS Lett.* **234** 1–7

Peters R 1991 *New Techniques of Optical Microscopy and Microspectroscopy* ed R J Cherry (Basingstoke: Macmillan) 199–228

Pike L J 2006 Rafts defined: a report on the Keystone symposium on lipid rafts and cell function *J. Lipid. Res.* **47** 1597–8

Pralle A, Keller P, Florin E-L, Simons K and Horber J K H 2000 Sphingolipid–cholesterol rafts diffuse as small entities in the plasma membrane of mammalian cells *J. Cell Biol* **148** 997–1008

Prior I A, Harding A, Yan J, Sluimer J, Parton R G and Hancock J F 2001 GTP-dependent segregation of H-ras from lipid rafts is required for biological activity *Nat. Cell Biol.* **3** 368–75

Qian H, Sheetz M P and Elson E 1991 *Biophys. J.* **60** 910–21

Raucher D and Sheetz M P 1999 Membrane expansion increases endocytosis rate during mitosis *J. Cell Biol.* **144** 497–506

Raucher D and Sheetz M P 2000 Cell spreading and lamellipodial extension rate is regulated by membrane tension *J. Cell Biol.* **148** 127–36

Ries J and Schwille P 2008 New concepts for fluorescence correlation spectroscopy on membranes *Phys. Chem. Chem. Phys.* **10** 3487–97

Saffarian S, Li Y, Elson E L and Pike L J 2007 Oligomerization of the EGF receptor investigated by live cell fluorescence intensity distribution analysis *Biophys. J.* **93** 1021–31

Saffman P G and Delbruck M 1975 Brownian motion in biological membranes *Proc. Natl Acad. Sci.* **72** 3111–3

Sako Y, Nagafuchi A, Tsukita S, Takeich M and Kusumi A 1998 *J. Cell Biol.* **140** 1227–40

Sako Y and Kusumi A 1994 *J. Cell Biol.* **125** 1251–64

Sako Y and Kusumi A 1995 *J. Cell Biol.* **129** 1559–74

Saxton M 1995 *Biophys. J.* **69** 389–98

Saxton M J 1993 *Biophys. J.* **64** 1766–80

Saxton M J 1994 *Biophys. J.* **66** 394–401

Saxton M J 1996 *Biophys. J.* **70** 1250–62

Saxton M J and Jacobson K 1997 Single-particle tracking: applications to membrane dynamics *Ann. Rev. Biophys. Biomol. Struct.* **26** 373–99

Schmidt T, Schutz G J, Baumgartner W, Gruber H J and Schindler H 1996 *Proc. Natl Acad. Sci. USA* **93** 2926–9

Schoetz E M, Burdine R D, Jülicher F, Steinberg M S, Heisenberg C P and Foty R A 2008 Quantitative differences in tissue surface tension influence zebra fish germlayer positioning *HFSP J.* **2** 1–56 PMID: 19404452

Schoetz E-M 2008 *Dynamics and Mechanics of Zebra Fish Embryonic Tissues—a Study of the Physical Properties of Zebra Fish Germlayer Cells and Tissues and Cell Dynamics During Early Embryogenesis* (Saarbruecken: Mueller)

Schutz G J, Schindler H and Schmidt T 1997 *Biophys. J.* **73** 1073–80

Schwille P, Korlach J and Webb W W 1999 Fluorescence correlation spectroscopy with single-molecule sensitivity on cell and model membranes *Cytometry* **36** 176–82

Sergé A, Bertaux N, Rigneault H and Marguet D 2008 Dynamic multiple-target tracing to probe spatiotemporal cartography of cell membranes *Nat. Methods* **5** 687–94

Sheetz M P 1995 Cellular plasma membrane domains *Mol. Membr. Biol.* **12** 89–91

Sheets E D, Lee G M, Simson R and Jacobson K 1997 *Biochemistry* **36** 12449–58

Sheetz M P and Dai J 1996 Modulation of membrane dynamics and cell motility by membrane tension *Trends Cell Biol.* **6** 85–9

Shvartsman D E, Gutman O, Tietz A and Henis Y I 2006 Cyclodextrins but not compactin inhibit the lateral diffusion of membrane proteins independent of cholesterol *Traffic* 917–26

Simons K and Ikonen E 1997 Functional rafts in cell membranes *Nature* **387** 569–72

Simson R, Yang B, Moore S E, Doherty P, Walsh F S and Jacobson K A 1998 *Biophys. J.* **74** 297–308

Simson R, Sheets E D and Jacobson K 1995 *Biophys. J.* **69** 989–93

Simson R, Yang B, Moore S, Doherty P, Walsh F and Jacobson K 1998 Structural mosaicism on the submicron scale in the plasma membrane *Biophys. J.* **74** 297–308

Singer S J and Nicolson G L 1972 The fluid mosaic model of the structure of cell membranes *Science* **175** 720–31

Slattery J P 1995 *PhD Thesis* (Ithaca, NY: Cornell University)

Smith P R, Wilson K M, Morrison I E G, Cherry R J and Fernandez N 1998 *MHC Biochemistry and Genetics* ed N Fernandez and G Butcher (Oxford: Oxford University Press) pp 133–51

Starr T E and Thompson N L 2001 Total internal reflection with fluorescence correlation spectroscopy: combined surface reaction and solution diffusion *Biophys. J.* **80** 1575–84

Steinberg M S 1996 Adhesion in development: an historical overview *Dev. Biol.* **180** 377–88

Suzuki K, Ritchie K, Kajikawa E, Fujiwara T and Kusumi A 2005 Rapid hop diffusion of a G-protein-coupled receptor in the plasma membrane as revealed by single-molecule techniques *Biophys. J.* **88** 3659–80

Suzuki K G N, Fujiwara T K, Sanematsu F, Iino R, Edidin M and Kusumi A 2007 GPI-anchored receptor clusters transiently recruit Lyn and Gαfor temporary cluster immobilization and Lyn activation: single-molecule tracking study 1 *J. Cell Biol* **177** 717–30

Thompson N L and Steele B L 2007 Total internal reflection with fluorescence correlation spectroscopy *Nat. Protoc.* **2** 878–90

Wawrezinieck L, Rigneault H, Marguet D and Lenne P-F 2005 Fluorescence correlation spectroscopy diffusion laws to probe the submicron cell membrane organization *Biophys. J.* **89** 4029–42

Weiss M, Hashimoto H and Nilsson T 2003 Anomalous protein diffusion in living cells as seen by fluorescence correlation spectroscopy *Biophys. J.* **84** 4043–52

Wieser S, Axmann M and Schutz G J 2008 Versatile analysis of single- molecule tracking data by comprehensive testing against Monte Carlo simulations *Biophys. J.* **95** 5988–6001

Wilson K W, Morrison I E G, Smith P R, Fernandez N and Cherry R J 1996 *J. Cell Sci.* **109** 2101–9 PMID: 8856506

Yamada S and Nelson W J 2007 Localized zones of Rho and Rac activities drive initiation and expansion of epithelial cell–cell adhesion *J. Cell Biol.* **178** 517–27

Yauch R L, Felsenfeld D P, Kraeft S K, Chen L B, Sheetz M P and Hemler M E 1997 Mutational evidence for control of cell adhesion through integrin diffusion/clustering independent of ligand binding *J. Exp. Med.* **186** 1347–55

Yechiel E and Edidin M 1987 Micrometer-scale domains in fibroblast plasma membranes *J. Cell Biol.* **105** 755–60

IOP Publishing

Physics of Cancer

Claudia Tanja Mierke

Chapter 8

The mechanical and structural properties of the microenvironment

Summary

The mechanical properties of the microenvironment (for example, the connective tissue consisting of extracellular matrix proteins) play an important role in providing the conditions for cellular motility and invasiveness. In this chapter the question of how the physical limits of the extracellular matrix regulate cellular invasion is discussed. In particular, how the structural composition, mechanical properties and steric hindrance impact on the malignant progression of cancer is considered. How each of the parameters of the extracellular matrix facilitates cellular invasiveness is highlighted. The focus here is on the proteins building up the extracellular matrix, while the effect of the embedded cells is discussed in chapters 11 and 12.

What effects do the mechanical and structural properties of the local tissue microenvironment have on the ability of caner cells to migrate?

The general description of cellular motility through 3D connective tissue is based on a physicochemical balance between cellular deformability and physical as well as mechanical tissue constraints. Migration rates are determined by the capacity of the cells to degrade the extracellular matrix through proteolytic enzymes, such as membrane-bound or secreted MMPs, and mechanocoupling between the integrin transmembrane receptors and the actomyosin cytoskeleton. How these parameters cooperate when the space is confined is not yet well understood. Using MMP-degradable collagen lattices or non-degradable substrates of varying porosity, the limits of cell migration can be quantitatively identified by the physical arrest of the cells within the tissue. In particular, the MMP-independent migration decreases linearly with decreasing pore size and with the deformation of the nucleus, which can be deformed at 10% of the nuclear cross-section until the cell is caught within the matrix (cancer cells 7 μm^2, T-lymphocytes 4 μm^2 and neutrophile granulocytes 2 μm^2)

(Wolf *et al* 2013). This suggests that the residual migration under space restriction depends upon MMP-dependent degradation of the extracellular matrix through enlarging the matrix's pore diameters and enhancing integrin- and actomyosin-dependent force generation, which together push the nucleus. The main restrictions for interstitial cell migration are scaffold porosity, deformability of the nucleus, pericellular collagen degradation and mechanocoupling between the cell's cytoskeleton and the extracellular matrix. In principle, the nucleus can be deformed by much more than 10%, as has been reported for endothelial cells cultured on micropatterned substrates (Versaevel *et al* 2012).

Cell migration along and through the 3D extracellular matrix is a fundamental process involved in the formation and regeneration of tissues, immune cell trafficking and diseases, such as cancer invasion and metastasis. Interstitial migration is a cyclic process consisting of multiple consecutive steps: (i) actin polymerization-dependent pseudopod protrusion at the cell's leading edge; (ii) integrin-facilitated adhesion to the extracellular matrix; (iii) contact-dependent extracellular matrix cleavage by the cell's membrane-bound proteases; (iv) actomyosin-facilitated contraction of the cell's body increasing longitudinal tension and cellular polarity; and (5) the retraction of the cell's rear followed by the translocation of the cell's body (Ridley *et al* 2003, Friedl and Wolf 2009, Friedl and Alexander 2011). This type of migration is constitutively active in mesenchymal cells such as fibroblasts and solid cancer cells (Wolf *et al* 2007, Sanz-Moreno *et al* 2008, Sabeh *et al* 2009, Grinnell and Petroll 2010). In particular, these mesenchymal cells show prominent protrusions and possess a spindle-shaped morphology, adhere strongly to the extracellular matrix and remodel the tissue proteolytically. In contrast to the mesenchymal movement of cells, the interstitial movement of leukocytes can be described by an ellipsoid cell shape and a rapidly deforming cellular morphology with short protrusions, weak adhesion strength and no proteolytic degradation of the extracellular matrix (Wolf *et al* 2003b, Sabeh *et al* 2009). Finally, each migration step is supposed to be adaptive due to cell-intrinsic and extracellular chemical or mechanical signals, such as regulators of adhesion, cytoskeletal dynamics, proteolysis, deformability of the cells and the matrix's extracellular geometry and material properties (Berton *et al* 2009, Lautenschläger *et al* 2009, Friedl and Wolf 2010, Friedl *et al* 2011, Tong *et al* 2012).

The interstitial invasion of mesenchymal cells such as fibroblasts and cancer cells into highly concentrated collagen-based extracellular matrices is controlled by MMPs. They are particularly membrane-bound, with, for example, MT1-MMP (formerly MMP-14) serving as the most important enzyme for degrading intact fibrillar collagen (Sabeh *et al* 2004, Wolf *et al* 2007 Rowe and Weiss 2009). Active MT1-MMP is concentrated at the membrane sites at which the migrating cell is in close contact with the extracellular matrix confinements, such as the collagen fibrils, in order to cleave the fibrils acting as steric hindrances for cancer cell migration. However, inhibition of MT1-MMP eliminates collagen cleavage and remodeling of the extracellular matrix (Sabeh *et al* 2004, Wolf *et al* 2007). Thus, non-proteolytic migration is performed by amoeboid cellular deformation (Wolf *et al* 2003a) or ceases entirely (Sabeh *et al* 2004), depending on the type of collagen scaffold used as a migration substrate (Packard *et al* 2009; Sodek *et al* 2008, Sabeh *et al* 2009).

Collagen scaffolds reconstituted from different collagen animal and tissue sources may vary in their physicochemical properties, such as porosity and stiffness (Zaman *et al* 2006, Sabeh *et al* 2009, Wolf *et al* 2009, Yang and Kaufman 2009, Miron-Mendoza *et al* 2010, Yang *et al* 2010). To date, there have been numerous studies that have investigated the effect of the collagen matrices on cellular motility, but there is no overall integrative concept that explains the differences reported for the particular types of collagen matrices. Such an integrative concept seems to be necessary to understand how the properties of the extracellular matrix allow or restrict the migration due to the MMP activity or the matrix's stiffness.

What are the rate-limiting substrate conditions that regulate the migration of different cell types in 3D extracellular matrices?
In order to answer this question live-cell microscopy can be performed to monitor migration rates and the deformation of the cell's body and nucleus in 3D extracellular matrices. These matrices vary from low to high collagen density. Thus, the subtotal and absolute migration limits are mapped in order to address the important molecular regulators of migration efficacy in a constrained environment (Wolf *et al* 2013). Using multi-parameter analyses, the ratio between extracellular matrix density and cellular deformability has been identified as key parameters that are influenced by MMP activity, actomyosin-based contractility and integrin-facilitated mechanocoupling all of which are modulators of invasion efficacy as they regulate cellular migration in dense tissue microenvironments.

In vitro reconstitution of 3D collagen matrices
The structure of 3D collagen matrices depends on the type of collagen (such as types I to XIX) and the animal source (such as a rat or bovine) from which the collagen is isolated. In most studies, the 3D hydrogels were reconstituted from either telopeptide-intact covalently cross-linked collagen obtained from rat-tail tendons using only acid extraction, or telopeptide- and cross-link-reduced bovine dermal collagen isolated using acid and pepsin treatment (Wolf *et al* 2003a, Sabeh *et al* 2004, Sodek *et al* 2007, Packard *et al* 2009). In order to investigate the effect of telopeptide-intact covalently cross-linked collagen and telopeptide- and cross-link-reduced collagen, equal collagen concentrations (such as $1.7\,mg\,ml^{-1}$) were compared in respect of collagen fibril assembly speed, fibril architecture, matrix porosity and matrix stiffness. For controlled preparation and imaging conditions, the collagen matrices were anchored using custom glass chambers. The time needed to polymerize the collagen matrices was monitored and recorded using confocal backscatter microscopy at $37\,°C$. The rat-tail collagen (half-maximum polymerization after 30 s) assembled 16-fold more quickly than the bovine dermal collagen (half-maximum polymerization after 8 min). The different speeds of collagen fibril assembly correlate with the different telopeptide contents within the collagen preparations (Helseth and Veis 1981, Sabeh *et al* 2009). The fibrillar matrix architecture was observed by confocal reflection microscopy and it was additionally analyzed using collagen type-I immunofluorescence. There was only a negligible detection error from backscatter-negative fibrils in vertical orientation, below 3% of

signal-containing pixels. This result is in contrast to another report in which twice as many fibers were detected with collagen I specific fluorescence compared to confocal reflection microscopy (Jawerth *et al* 2010). Thus, the method for analyzing the collagen fiber within a 3D collagen matrix has to be carefully chosen. Rat-tendon-derived collagen assembles thin fibrils with a diameter of 20 nm and a narrow pore size range of 2–5 μm^2 (1–2 μm pore diameters), whereas bovine-dermis-derived collagen matrices form fibrils with a diameter of 60 nm and wider pore cross-sections, ranging from 6–30 μm^2 (2-6 μm pore diameters) (Wolf *et al* 2013). In order to control the fibril density–dependent alterations in collagen matrix stiffness, AFM was used with a 10 μm bead as a cantilever probe to approximate the size of a cell. The surface of the collagen matrix was probed by the cantilever with the connected 10 μm bead and then the bead penetration and the force were determined. The AFM measurements revealed a two-fold lower elastic modulus (28 Pa) for a bovine dermal collagen matrix compared to a rat-tail collagen matrix (both at 1.7 mg ml^{-1} (51 Pa)) (Wolf *et al* 2013, Stein *et al* 2008, Yang and Kaufman 2009). Although reconstituted at equivalent 1.7 mg ml^{-1} collagen type I concentrations, the collagen matrices differ substantially in fibril diameter as well as interfibrillar space and consequently in network stiffness.

8.1 Pore size

The pore size of 3D extracellular matrices is important for providing a scaffold in which cancer cells are able to invade. If the pore size is too small, cancer cells are not even able to migrate into the matrix when they have adhered to the surface of the 3D extracellular matrix (Mierke *et al* 2011, Mierke 2013). If the pore size is too small, another option for cells is to degrade and thus restructure the collagen matrix in a manner that allows them to move deeper in. This will be addressed in the following.

The role of pore size during proteolytic and non-proteolytic migration of HT1080 cells

In order to investigate differences in MMP-independent cell migration rates between the collagen preparations, MT1-MMP expressing HT1080 (HT/MT1) cells were used as a model for collagenolytic invasion, which can be inhibited and thus turned to collagenolysis-independent migration after addition of a broad-spectrum MMP inhibitor such as GM6001, or by MT1-MMP silencing (Wolf *et al* 2007, Sabeh *et al* 2009). As reported, the HT/MT1 cells have a higher migration speed of 0.7 μm min^{-1} in bovine dermal collagen matrices compared to a migration speed of 0.3 μm min^{-1} in rat-tail collagen matrices, with a collagen concentration of 1.7 mg ml^{-1} collagen type I for both. These results lead to the suggestion that MMPs support the migration in collagen scaffolds of different pore sizes and the migration speed depends on the pore size rather than the type of collagen. Conversely, the broad-spectrum MMP inhibitor GM6001 inhibited cell motility in rat-tail collagen matrices, whereas no inhibition of cellular motility in bovine dermal collagen matrices was observed. In particular, individual cell migration was quantified by cell tracking, cell emigration from multicellular spheroids and vertical invasion after seeding the cells on top of the 3D collagen matrices. Moreover, similar results have been obtained using a transient knockdown of MT1-MMP expression, confirming that MT1-MMP is indeed the invasion-promoting collagenase (Sabeh *et al* 2009). How can this difference in the inhibition of cell motility be explained?

In order to analyze whether the difference of MMP-independent migration is caused by alterations in the fibril density and/or matrix porosity, different collagen concentrations have been used. Increased concentration of bovine dermal collagen (from 1.7 to 15 mg ml^{-1}) led to significantly smaller pore cross-sections (median 5 μm^2), higher stiffness and a dose-dependent decreased migration. Finally, the migration of HT/MT1 cells in the presence of GM6001 was totally inhibited. This graded response was subdivided into subtotal and absolute migration limits of 90 and 99% speed delay, respectively. As expected, decreased migration in bovine dermis collagen was linearly correlated with the pore size for both MMP-dependent and MMP-independent migration, with similar slopes for the linear functions. The difference shown was a higher offset with MMP-dependent migration. By contrast, when the concentration of rat-tail collagen was reduced to a minimum of 0.3 mg ml^{-1}, causing decreased collagen scaffold stiffness but providing larger pore cross-sections of 20–30 μm^2, a totally restored MMP-independent migration was observed (the speed distribution was similar to that of the control cells). The migration correlated linearly with the pore size, but with a significantly steeper slope than for the migration of cells in which the MMP activity was inhibited.

What is the effect of varying collagen concentrations on cellular motility and are there side-effects? Similar experiments have been performed, but without varying the collagen concentrations in order to minimize the indirect effects unrelated to porosity, such as altered density of the adhesive ligands (Zaman *et al* 2006) or uncontrolled collagen cross-link density in bovine dermal collagen, and the porosity of rat-tail collagen matrices was altered by reducing the polymerization temperature (Raub *et al* 2007) while keeping the collagen concentration constant (1.7 mg ml^{-1}). At 9 °C, the fibrillogenesis was delayed in time and hence, the pore diameters and related cross-sections increased (median diameter 8 μm and median cross-section 30 μm^2) and the fractal box count decreased, while the diameter of the fibrils (80 nm) and the mechanical stiffness increased. As the pore size increased, MMP-independent migration in rat-tail collagen was rescued with single-cell movement with peak speeds close to the proteolytic migration rates. This behavior was seen for cell emigration from multicellular spheroids and vertical invasion cultures. As reported previously, the speed of MMP-independent migration was linearly dependent on the pore size, with a significantly steeper slope than for protease-dependent migration. In the method where the pore size and the type of migration assay were altered, the cell speed was in all cases a linearly correlated with the pore size, independently of collagen preparation. This ability to perform MMP-dependent motility when cells are sterically hindered by small pores that otherwise impede or arrest migration leads to maintenance of the migration.

The role of the deformability of the cell's nucleus during MMP-independent migration in dense extracellular matrices

What types of subcellular compartments regulate migration in dense 3D micro-environments? In order to answer this question, the morphokinetic alterations of the cell's body, leading edge and nucleus were analyzed during cell migration as a function of collagen density. HT/MT1 cells embedded in bovine dermal collagen in the presence of the MMP-inhibitor GM6001 moved the leading edge and cell bodies at equal velocities. However, these cells stopped their migration in rat-tail collagen matrices and the cells had a spherical central body with dynamic, dendrite-like extensions that were able to deform the collagen at the front pole without promoting cell movement. In addition, it was observed that pseudopods could break from cell bodies and then move with a snake-like morphology (Wolf *et al* 2013). The immobile fraction of the cell's body still consisted of cytoplasm and nucleus, forming occasional small protrusions pointing toward the extension of the cell's leading edge. In non-moving cells in the matrices, the nuclear prolapses measured $1-3\,\mu m$ in diameter ($1-7\,\mu m^2$ cross-section), whereas the nuclei of cells during MMP-independent migration adopted deformations that were $3-7\,\mu m$ in diameter ($7-40\,\mu m^2$ cross-section). By contrast, these morphological nuclear structures were distinct from the ellipsoid nuclear shapes, with a diameter of $8-11\,\mu m$ and a cross-section of $50-100\,\mu m^2$ being found in MMP-dependent migrating HT1080 cells able to generate proteolytic tracks with a diameter close to that of the cell's diameter. Independently of the collagen preparation or matrix scaffold porosity, the nuclear diameters of MT1-MMP expressing control cells possess average cross-sections of $40-90\,\mu m^2$. However, the addition of GM6001 leads to a decrease in nuclear diameters until the cell sticks in the collagen matrix without any movement. Then the nucleus displays a single small nuclear prolapse or a non-deformed spherical shape, suggesting that the non-moving nuclei cycle between prolapse and rounding. Finally, hourglass-shaped deformations of the cell's nucleus supported the MMP-independent migration of HT/MT1 cells through 3D collagen matrices with pore diameters above $7\,\mu m^2$. By contrast, smaller pore sizes break down nuclear deformability and lead to a physical arrest, while persistent leading-edge kinetics and force generation and transmission are still there. Taken together, the MMP-independent migration into dense 3D extracellular matrices is regulated through the deformation of the cell's nucleus. However, the role of the contractile forces in providing cancer cell motility is still controversial. There are reports that have found a connection between cellular motility and the generation of contractile forces (Mierke *et al* 2011).

The impact of actomyosin contractility and integrin-mediated mechanocoupling of cells migrating in a dense matrix confinement

Mechanocoupling seems to have an active role in confined space, as cytoskeletal activity is observed regulating the migration of cells. Thus, it was investigated whether cell motility is altered by (1) the integrin-facilitated leading-edge traction on a substrate (using a β1 integrin-perturbing mAb 4B4) (Wolf *et al* 2003a) and (2) actomyosin-dependent contractility (using a ROCK inhibitor, such as Y-27632, which inhibits myosin light chain (MLC) phosphorylation and contraction of the cell rear) (Ren *et al* 2004, Lämmermann *et al* 2008). However, these two approaches gradually reduced HT/MT1 cell-facilitated contraction of collagen matrices. When using 3D collagen matrices with $20\,\mu m^2$ pore area, which pose only a moderate physical challenge to nuclear deformability, both mAb 4B4 (inhibiting the function of the $\beta 1$ integrin subunit) and Y27632 reduced migration rates in a dose-dependent manner. In particular, the effect was even more pronounced in the presence of GM6001-treated cells compared to buffer-treated control cells. Mechanically perturbed force generation from mAb 4B4 caused alterations in the cell elongation. When the mAb 4B4 was used at a concentration that inhibited the contraction of the collagen matrix by approximately 50% ($1\,\mu g\,ml^{-1}$), the MMP-independent migration was almost completely inhibited due to impaired capacity to generate sufficient adhesion and traction force to move the nucleus through the cell's cytoplasm. In line with these observations, reducing space constraints by enlarging pore diameters to approximately $55\,\mu m^2$ almost completely rescued the migration of cells. When the migration speed was decreased in a confined 3D collagen matrix ($1.7\,mg\,ml^{-1}$ bovine collagen) in the presence of GM6001, increased nuclear deformation was observed, which could be reversed when matrix porosity was increased or MMPs were not inhibited by GM6001. When cell–matrix adhesion receptors such as integrins are active, the importance of actomyosin-facilitated cell contraction in pushing the nucleus through the 3D extracellular matrix was mirrored in the time-delayed rear retraction and increased cell length in the presence of Y27632. In addition, a concentration of Y27632 ($2\,\mu M$) led to a half-maximum collagen contraction, which effected a partial cell migration arrest and at the same time caused a strong deformation of the nucleus in the presence of GM6001. When the cross-sections of the pores were increased, these two effects could be reversed. Taken together, both integrin-facilitated traction force and actomyosin contractility are needed to squeeze the nucleus forward when a dense extracellular matrix is transmigrated in concert with MMP-facilitated pore generation.

Are there different kinetics and rate-limits in mononuclear and polymorphonuclear cells?

To investigate how different nuclear shape types regulate migration in confined matrices, the nuclear mechanics of different cell types with mononuclear or polymorphonuclear organization were analyzed. The intermediate filament lamin A/C is a central nuclear protein that is required for nuclear membrane organization and stability (Goldberg *et al* 2008), and in addition it is expressed in cancer cells such as HT/MT1, HT/wt and MDA-MB-231/MT1 (MDA/MT1) breast cancer cells, whereas it cannot be detected in human CD4+ T-lymphocytes or polymorphonuclear neutrophils (PMNs). Independently of cell type, migration of mononuclear cells through low- to intermediate-density 3D collagen matrices is not affected by GM6001. Migration is still possible by deformation of the nucleus with cross-section distributions matching the available pore-size range of the 3D extracellular matrix. In HT/wt cells and T-lymphocytes migrating through dense 3D extracellular matrices, hourglass nuclear shapes predominate, whereas MDA/MT1 cells show a broader spectrum, from hourglass to cigar-like shapes. When the pore dimensions range from 2 to 5 µm^2, GM6001 leads to a migration stop and the formation of a nuclear prolapse in cancer cells, whereas the T-lymphocyte migration persists with a lower migration speed. Compared to cancer cells, T-lymphocytes can be distinguished by a two- to four-fold smaller nucleus and their inability to proteolytically degrade fibrillar collagen, thus their motility and their ability to cross barriers (such as an endothelial barrier) depends solely on shape changes (Wolf *et al* 2003b). However, all T-lymphocytes become unable to migrate through matrices when they have pore cross-sections of 1–2 µm^2, which is no longer in the spatial range for possible nuclear deformation. In summary, MMP-independent mononuclear cell migration uniformly depends on the ability to deform the nucleus in response to the lateral compression induced by structures of the connective tissue.

Similar behavior has been observed for PMN. PMN migration is not influenced by the absence or presence of GM6001 in low- to intermediate-density collagen matrices. In contrast to the homogeneous deformation of mononuclear nuclei, the segmented nucleus of PMNs is characterized by interconvertible folding states, such as compact configuration or pearl-chain-like complete and partial unfolding. The different folding states lead to migration speed changes in the PMWs. When the nucleus is compact, the cells are nearly immobile. By contrast, when the nucleus is unfolded, the cells are motile. In high-density rat-tail collagen matrices (3.3 mg ml^{-1}, pore cross-section 2–3 µm^2), PMN migration is still observed independently of MMP activity, however, the migration speed is reduced and the cells are immobile at higher collagen density (6.6 mg ml^{-1}), with pore cross-sections in the range of 0.5 to 1.5 µm^2, which is below that of the nuclear cross-sections. Similarly to cancer cells, immobilized PMNs display collapsed and spherical nuclei with occasional single-segment prolapse in the direction of the leading oscillating pseudopod. In summary, shape change patterns such as the hourglass-like compression of mononuclear nuclei or the unfolding of polymorphonuclear nuclei are needed for cellular migration through interstitial matrices. In high-density fibrillar 3D collagen matrices with no collagenolysis, the nuclear deformability determines the migration rates as a function of pore size. The cellular motility limit for cancer cells is at a cross-section of 7 µm^2 and for PMNs and T-lymphocytes it is at a cross-section of 2–3 µm^2. Taken together, inhibited cell migration is linearly correlated with pore size, but is independent of the matrix scaffold's stiffness.

Physical confinement of cell migration through non-degradable substrates
Migration rates through dense 3D collagen matrices and broad-spectrum MMP inhibitor treatment can also be regulated by additional parameters that have not yet been addressed, such as residual low-level MMP-independent collagenolysis performed by other classes of proteases that are not inhibited by GM6001. In addition, also following guidance by occasional gaps present in assembled matrix scaffolds and mechanical rupture of very small or incompletely polymerized collagen fibrils by migrating cells may alter the motility of cancer cells. In order to exclude such side-effects from interfering with the parameters, a polycarbonate filter model can be used for cell trafficking through non-degradable and hence

non-deformable barriers. In this 'stable' system, a subtotal inhibition (90%) of cell migration was detected at pore cross-sections of 7–$10\,\mu m^2$ for all mononuclear cell types and $4\,\mu m^2$ for PMN. Complete inhibition (>99%) was found at pore dimensions below 5–$6\,\mu m^2$ for mononuclear cells and $1\,\mu m^2$ for PMNs. Similarly to 3D collagen matrices, efficient transmigration through polycarbonate filter pores is facilitated through the cytoplasmic protrusion, which is followed by a nuclear deformation within a pore cross-sectional range of 7–$50\,\mu m^2$. The morphology of cells that adhere and are immobilized above small pores ($0.8\,\mu m^2$ cross-section) displays a long cytoplasmic protrusion extending into the pore, whereas the nucleus is still above the pore. This is very similar to the arrested phenotypes of cells observed in high-density fibrillar 3D collagen matrices. In summary, cell migration through non-degradable pores has certain physical limits that depend on pore size and nuclear deformability. Moreover, the physical limits of 3D fibrillar collagen matrices are confirmed by these cell migrations through non-degradable and non-deformable pores.

The rate-limiting physicochemical parameters for cell migration can be obtained through analyzing the change between MMP-dependent and MMP-independent cell motility with decreasing matrix scaffold porosity and by quantitatively determining the migration speed and the shape of the cell's body and nucleus. The first of these parameters is the space available between extracellular matrix fibrils or within the filter pore that can afford the movement of the cell's body, and the second is the deformability of the cell's nucleus due to the requirements of the matrix's or the filter pore's confinement. With decreasing cross-section, interfibrillar pores mechanically hinder cellular migration and evoke a deformation in a cell-type-specific manner until the limit for deformability is reached and the nucleus is mechanically trapped within the matrix's confinement. The balance between cell translocation and arrest is regulated by processes that control either the pore size or nucleus deformation. The cellular translocation is supported firstly by effectors modulating fibril remodeling through MT1-MMP to create 'neo-space', and secondly by mechanocoupling through the integrins and the cell's actomyosin contractility to push the nucleus through confined matrices.

How does the extracellular matrix porosity in vivo and in vitro
affect cancer cell motility?
Fibrillar collagen is organized *in vivo* as fibrils and fibers of diverse thickness, orientation and interfibrillar spacing (Starborg *et al* 2008, Wolf *et al* 2009). It has been reported that the interfibrillar space varies largely *in vivo* in loose and dense interstitial tissues, ranging between 2 and $30\,\mu m$ (Stoitzner *et al* 2002, Wolf *et al* 2009, Weigelin *et al* 2012). In order to have an appropriate model for cellular invasiveness, we varied the porosity of 3D collagen matrices similarly. This large range of porosity serves to both guide and mechanically challenge migrating cells that commonly display cross-sections of 30–$100\,\mu m^2$ in a cell-type–dependent manner.

What are the limits of cell migration in 3D?
It has been shown that migration speed reduces linearly with decreasing pore size in a cell-type-specific manner. The reduction in migration speed in response to

increasing mechanical confinement is a gradual process that largely depends upon cellular deformability, which is a strategy used by numerous migrating cells (Lautenschläger *et al* 2009, Wolf and Friedl 2011, Tong *et al* 2012, Mierke *et al* 2011). The maximum speed is reached at pore sizes that approximate the cell's body sizes, where the matrix substrate guides without physically hindering migration (Jacobelli *et al* 2010). With decreasing porosity of the matrices, cell migration decreases inverse proportional to the increased capability to deform, leading to a slower migration, which stops until the subtotal limit of cellular deformability is reached. Migration may be abrogated when a cell-type-specific maximum of deformation is reached; this is called the absolute limit. As the cell's shape is highly adaptable to microenvironmental conditions, pores of defined cross-sections differing only in their geometry (such as discontinuous polygonal-shaped pores in 3D fibrillar collagen matrices, flat and broad cleft-like spaces, evenly shaped cylindrical pores of polycarbonate membranes and elongated continuous channels of microdevices) are equally well suited for cell migration (Lautenschläger *et al* 2009, Jacobelli *et al* 2010, Rolli *et al* 2010, Ilina *et al* 2011, Balzer *et al* 2012, Tong *et al* 2012). It has been reported that a cross-section of 20–30 μm^2 is near-optimal for the interstitial migration of HT1080 and MDA-MB-231 cells through complex-shaped spaces in 3D collagen matrices and monomorphic transwell membrane filter pores, or through the 30 μm^2 cross-sectional areas of engineered rectangular microchannels, promoting cell migration with high speed (Tong *et al* 2012). Keratinocytes are a good example of cell shape adaptation as they can traverse gaps of 500 nm height and unrestricted width (Brunner *et al* 2006). This result confirms that it is only the cross-section of the transmigrated space and not the diameter that regulates the efficacy of cell migration in a confined matrix. Hence, the limits of cell migration depend on a two-parameter function of the matrix's or substrate's porosity and the cell's deformability.

What role does the nucleus play in the process of cell motility through dense extracellular matrices?

The nucleus is the largest and most rigid cell organelle, because of the chromatin content and the stabilizing nuclear lamina consisting of the intermediate filament lamin A/C (Dahl *et al* 2004, Gerlitz and Bustin 2011, Chow *et al* 2012). Migrating cells display two mechanically distinct forms of nuclear deformation: cell types with mononuclear nuclei show global deformation, resulting in transient hourglass-like or cigar-shaped elongated morphologies; and cell types with polymorphonuclear nuclei show the unfolding of polymorphonuclear nuclei in order to display pearl-chain-like configurations without major deformations in the individual nuclear segments. These reversible migration-associated nuclear morphologies are frequently detected *in vivo* during the dissemination of cancer cells (Yamauchi *et al* 2005, 2006, Alexander *et al* 2008, Beadle *et al* 2008, Friedl *et al* 2011) and their transendothelial migration (Feng *et al* 1998, Voisin *et al* 2009). The major restriction for cellular deformability is the stiff nucleus, because the deformability of the nucleus in dense spaces is finite, whereas the cytoplasm is able to squeeze through almost any pore size, including 1 μm^2 gaps in 3D collagen matrices and 0.8 μm^2 pores in the polycarbonate membranes used for transwell membrane assays (Schoumacher *et al* 2010, Shankar *et al* 2010).

Taken together, cell movement through dense 3D extracellular matrices is regulated by at least three properties of the nucleus—size, rigidity and shape (structure)—which regulate an adaptation range of factor two to five when mononuclear cancer cells are compared with PMN. A consistent ratio of nearly 1/10 has been found for minimum nuclear cross-sections relative to the non-deformed state, suggesting that the maximal compressibility is an absolute value, independent of cell type, basal nuclear shape and nuclear rigidity. The nuclear lamina seems to regulate the nuclear shape and/or deformability rather than the absolute compression limit, which leads to the suggestion that a non-compressible intranuclear component such as chromatin defines the compression maximum and, consequently, the physical restriction of cell migration.

How does proteolysis alter the porosity of the extracellular matrix?
By degrading fibrillar collagen at the cell–matrix interface, MT1-MMP–dependent proteolysis enlarges the collagen pore cross-sections to facilitate cell migration through degradable substrates such as 3D extracellular matrices (Wolf *et al* 2007, Fisher *et al* 2009, Sabeh *et al* 2009). It has been shown that MT1-MMP increases the migration at confining porosity above critical limits and hence provides slow and persistent migration even in very dense 3D extracellular matrices. However, with porosity high enough to perform migration through the deforming of the cell body, pericellular proteolysis is unnecessary, as the migration persists despite pharmacological or molecular targeting of MMP activity in either cancer cell or leukocyte populations. This suggests that MMPs regulate pore dimensions, thereby enhancing cell speed, deformation and the limits of cell migration in degradable matrix scaffolds. There has been confusion about the role of MMPs in pore size control (Wolf *et al* 2003a, Sabeh *et al* 2004, 2009), but another report using different collagen preparations side by side resolved the different results regarding the role of MMPs for cell invasion through 3D collagen scaffolds of different origin and porosity (Wolf *et al* 2013).

The impact of mechanocoupling on 3D cellular motility
Forces generated between the cell and 3D matrix scaffold may have an effect on cellular deformation and, consequently, on migration through a confined matrix. In particular, β1 integrins act as both the main adhesion receptors to fibrillar collagen and as mechanotransducers in migrating cells (Huttenlocher *et al* 1995, Puklin-Faucher and Sheetz 2009), hence providing the traction forces necessary to push the nucleus through narrow pores in dense 3D matrices. Rho kinase-mediated actomyosin contractility reinforces integrin-facilitated cell adhesion and regulates the retraction of the cell's rear (Vicente-Manzanares *et al* 2008), leading to the translocation of the cell's nucleus through small spaces (Lämmermann *et al* 2008). Taken together, MMP-dependent extracellular matrix degradation and integrin- and Rho-mediated force transmission may use complementary mechanisms in order to secure cellular migration, particularly when space is confined.

What affects the limits of cell migration in an indirect manner?

Additional physicochemical properties of the matrix's substrate may alter migration rates, such as the extracellular matrix's telopeptide status and stiffness (Sabeh *et al* 2009, Miron-Mendoza *et al* 2010, Ehrbar *et al* 2011). It has been shown that high telopeptide content in extracted rat-tail collagen accelerates fibril polymerization and increases mechanical strength, whereas the fibril thickness and the network porosity are reduced (Helseth and Veis 1981, Elbjeirami *et al* 2003, Sabeh *et al* 2009, Wolf *et al* 2009), which hinders non-proteolytic cell migration. By contrast, with slowed-down collagen fibrillogenesis, larger pore sizes are generated, which match nuclear size and deformability better, and hence support non-proteolytic (MMP-independent) cell migration. Finally, independently of the telopeptide content of the collagen preparation used, small pores cause increased nuclear deformability and exclude a MMP-independent migration mode. No influence on cell immobilization through matrix restrictions is observed to come from different substrate stiffness. A stiffness range of 20–700 Pa and larger than 10^6 Pa was investigated for reconstituted fibrillar collagen and polycarbonate membranes, respectively. The elastic modules of *in vivo* tissues range from 200 to 1,000 Pa for mammary interstitium and from 1,000 to 10,000 Pa for adipocytes and myofibers, and they can thus be mimicked by the two different artificial microenvironments (Stein *et al* 2008, Butcher *et al* 2009, Levental *et al* 2009, Buxboim *et al* 2010). Finally, when these results are normalized to the pore size, both soft and rigid substrates lead to nearly identical subtotal and absolute migration restrictions, excluding substrate stiffness larger than 20 Pa. These stiff substrates possess an independent mechanism for regulating cellular deformability and, consequently, physical migration arrest within these matrices.

Taken together, this multi-scale analysis of cell–matrix geometries and migration kinetics may help to reveal the biophysical processes involved in the regulation of cellular motility in confined microenvironment, such as dense 3D extracellular matrices. These biophysical parameters may include the elasticity of the nucleus by regulation of the laminA/C expression, nucleus–cytoskeleton linkage, intracellular and intranuclear pressure and hydration, chromatin organization and physical sensitivity to repetitive or long-lasting mechanical stresses (such as cellular stress-stiffening behavior). The deformability of the cell's body and nucleus is suitable for migration through 2D and 3D tracks, as well as through gaps that are constitutively present in healthy connective tissues (Wolf *et al* 2009, Weigelin *et al* 2012). However, it is suggested that the penetration of tumor-associated collagen-rich desmoplastic tissue, scarred stroma and the basement membranes involves even more complex invasion signaling pathways (Provenzano *et al* 2006, Rowe and Weiss 2009, Tanaka *et al* 2010, Salmon *et al* 2012). Finally, multi-scale wet-laboratory work and analyses of cell migration will together reveal the strategies used by cells to migrate through extracellular matrix barriers encountered *in vitro* and *in vivo*.

8.2 Matrix stiffness

The mechanical properties of tissues in which cells are embedded play an important role in cellular functions such as proliferation, survival, development of tissues, tissue homeostasis, vascularization and organ function. Thus, the impact of matrix stiffness on cellular motility has been investigated in reductionist *in vitro* microenvironment model systems such as artificial extracellular matrices, which mimic specific extracellular matrix functions under highly controlled conditions. In particular, they have frequently served to elucidate the role of cell–extracellular matrix interactions in regulating cell fate. To reveal the interplay of biophysical and biochemical effectors in controlling 3D cell migration, a poly(ethylene glycol)-based artificial extracellular matrix platform was used. The influence of the matrix cross-linking density, represented by the matrix stiffness, on cell migration *in vitro* and *in vivo* was investigated. The migration capacity of single preosteoblastic cells within hydrogels of varying stiffness and susceptibility to degradation by MMPs was analyzed *in vitro* using time-lapse video microscopy. Indeed, the motility of the cells was strongly dependent on matrix stiffness. In more detail, two migration regimes were identified: a non-proteolytic migration mode used at a relatively low matrix stiffness and a proteolytic migration mode performed at higher matrix stiffness. In line with this, *in vivo* experiments revealed a similar stiffness dependence for matrix remodeling by invasive cells; however, the dependence was less sensitive to the MMP sensitivity. Thus, this artificial extracellular matrix model system is indeed well suited to reveal the role of biophysical and biochemical parameters for physiologically relevant cell migration phenomena.

In tissue engineering, the regulation of 3D cell migration within and into biomaterial scaffolds plays an important role. In particular, biomaterials can serve as implants and can be designed to guide endogenous stem or progenitor cells to the site of a tissue defect to facilitate tissue regeneration by promoting repopulation and remodeling of an implant by host cells, which is a process that is regulated by large-scale cell migration (Lutolf and Hubbell 2005). The absence of sufficient cell migration is the most prominent limitation in creating large tissue-engineered constructs. For example, impaired endothelial cell invasion into connective tissue may evoke a lack of vascularization and may then, ultimately, lead to necrosis (Phelps and Garcia 2010). By contrast, in smart biomaterials designed to be carriers for cell delivery to targeted sites, encapsulated cells must be able to leave their delivery system and migrate extensively into 3D extracellular matrices (Mooney and Vandenburgh 2008). Thus, these critical biological and biomechanical requirements for engineered biomaterials containing cells need to be optimized for 3D cell migration properties and they have thus become the focus of biomedical research for cell carrier systems. For this optimization of the cell migration behavior within biomaterials it is necessary to reveal the 3D cell motility mechanisms for physiological and pathological situations as well as proteolytic (mesenchymal) or non-proteolytic (amoeboid) migration strategies (Zaman *et al* 2007, Ilina and Friedl 2009). By contrast to migration in 2D, cells in 3D have to overcome the sterical hindrances and confinements of their surrounding microenvironment. In the proteolytic migration mode, cells secrete active proteases in order to break down macromolecules of the extracellular matrix and thus create macroscopic cavities such as migration tunnels, which facilitate their movement. Matrix degradation takes place in the local microenvironment of the cell. In particular it is possible to degrade the extracellular matrix in two ways by using membrane-bound such as membrane-type (MT) MMPs or secreted proteases such as collagenases.

As an alternative to proteolytic migration through 3D extracellular matrices, a number of inflammatory cell types such as lymphocytes and dendritic cells or cancer cells are known to utilize migration strategies in order to overcome biophysical matrix resistance by squeezing through the extracellular matrix or by even deforming it without any proteolysis (the amoeboid migration mode). However, the same cell type can use both mechanisms, depending on the specific extracellular matrix context, such as constituents and confinement (Friedl and Wolf 2010).

Thus, several approaches for generating synthetic biomaterials that support both types of cellular migration have been reported (Lutolf and Hubbell 2005). The main focus is on designing biomaterials that are beneficial for cell migration through interconnected and preexisting pores (Moroni *et al* 2008). Indeed, biomaterials have been constructed as artificial extracellular matrices that are sensitive to cell-derived proteases degrading the artificial extracellular matrices. In more detail, peptidic substrates that are to be cleaved by MMPs or plasmin are placed in the backbone of cross-linked hydrophilic polymer chains, which results in the degradation of the resulting copolymer hydrogels (figure 8.1). Artificial extracellular matrices are based on a simple design strategy, which includes the incorporation of an integrin-binding peptide such as RGD or other protein components that promote cell adhesion, and these then serve as synthetic extracellular matrix analogs through which the cells can migrate using proteolytic mechanisms. Although artificial extracellular matrices are frequently used in tissue engineering and 3D cell culture to complement naturally derived extracellular matrices, such as collagen matrices or Matrigel (Lutolf 2009, Tibbitt and Anseth 2009), it is not yet clear whether there is an interaction of the various biochemical and biophysical artificial extracellular matrices characteristics in controlling 3D cell motility (Zaman *et al* 2007). A key question that has remained unanswered is whether certain artificial extracellular matrices aspects support proteolytically or non-proteolytically facilitated cell migration modes. However, this has been addressed using an engineered artificial extracellular matrices model system formed from poly(ethylene glycol)(PEG)-based macromers through the addition of transglutaminase (TG) factor XIII (Ehrbar *et al* 2007a, Ehrbar *et al* 2007b) (figure 8.1).

Unlike existing 3D extracellular matrices produced from naturally derived extracellular matrix components such as collagen matrices, these molecularly engineered TG-PEG gels possess no microstructure and are therefore non-porous. Indeed, these synthetic TG-PEG gels are composed of a molecular meshwork consisting of flexible cross-linked polymers with a mesh size of approximately tens of nanometers (Ehrbar *et al* 2007b). As with existing artificial extracellular matrix models (Ehrbar *et al* 2007b, Gobin and West 2002, Halstenberg *et al* 2002, Bryant and Anseth 2002, Shu *et al* 2003, Kim *et al* 2005, Raeber *et al* 2005, Almany and Seliktar 2005, Peyton *et al* 2006), the biochemical and biophysical properties of TG-PEG hydrogel networks are determined by incorporating protease-sensitive peptide domains and cell-adhesive ligands, or by tuning the matrix stiffness through the modification of the precursor polymer architecture or concentration. In order to investigate the modularity of this

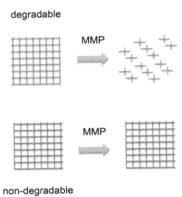

Figure 8.1. Schematic image of the design of PEG-based artificial extracellular matrices. Stoichiometrically balanced (the ratio of lysine to glutamine is 1 to 4) 8-arm PEG macromers in a buffer solution are enzymatically cross-linked through their pending glutamine acceptor [Gln] and lysine-donor [Lys] FXIIIa substrate sequences to assemble a hydrogel. Through the variation of the linker sequence and the initial precursor concentration, artificial extracellular matrices with different stiffness and different MMP sensitivities in presence of constant RGD concentrations can be obtained.

system, a systematic analysis was carried out to assess how 3D migration of single cells depends on the gel's stiffness, which was adjusted independently of the matrix's degradability.

Surprisingly, it was found that in artificial extracellular matrices with low stiffness, single cells can overcome the confinement of the matrix by performing a degradation-independent 3D migration mode, suggesting that these cells use preexisting or *de novo*-formed macroscopic gel defects. This particular finding highlights that the engineering of gel defects in seemingly homogeneous polymer gels is a powerful tool for increasing cellular permissiveness in dense artificial extracellular matrices.

Knowledge and optimization of 3D cell migration within and into biomaterial scaffolds is an advantage for tissue engineering. TG-PEG gels serve as a suitable model system for examining the impact of matrix stiffness and proteolytic remodeling on 3D cell migration *in vitro* and *in vivo*. The structural components of the native extracellular matrix (such as fibrillar collagen) ensure an interconnected microporous network that contains other matrix components such as glycosaminoglycans. 3D cell motility within this complex microenvironment (for example, TG-PEG) relies on proteolysis-dependent and proteolysis-independent mechanisms (Ilina and Friedl 2009), but to distinguish between both modes is not easy. Indeed, 3D cell migration data obtained from *in vivo* extracellular matrices or *in vitro* models derived from natural extracellular matrix components are difficult to interpret in terms of extracellular matrix characteristics and it is not easy to uncover the various cell–matrix interactions that determine 3D cell migration. Thus, many reports seem to be inconsistent; for example, Sabeh *et al* (Sabeh *et al* 2009) showed that cancer cells migrate within non-covalently cross-linked collagen gels independently of proteases, while in covalently cross-linked collagen gels, MT1-MMP activity was essential for movement.

Engineering approaches have led to the formation of artificial extracellular matrices that offer high control over the material microstructure, and in some cases scaffolds have been built that are essentially amorphous and pore-free. Indeed, by using protease inhibitors, it has been demonstrated that 3D migration in dense chemically cross-linked PEG hydrogels is strongly dependent on proteolysis (Raeber *et al* 2005, 2007). Using enzymatically cross-linked artificial PEG-based matrices, we expected a similar protease-dependence across the entire stiffness range analyzed. From the swelling ratios of all gel conditions (approximately tens of nanometers), the molecular mesh sizes can be calculated. As expected, the molecular mesh size is far below the threshold that a cell would be able to breach and penetrate the surrounding physical matrix barrier. In line with this, cells encapsulated within relatively stiff gels, polymerized at 2 or 2.5% solid content, were unable to migrate without proteolytic activity, as observed by blocking proteolysis. A slight decrease in cross-linking density, induced by changing the solid content to 1.5%, led to matrices through which cells could efficiently migrate using a protease-independent mechanism.

In addition to affecting confined migration in the absence of a microscopic porosity, the viscoelastic matrix properties seem to have a direct impact on the migratory behavior of cells in 3D matrices (Zaman *et al* 2006, 2007, Pedersen and Swartz 2005, Dikovsky *et al* 2008). It has been shown that certain cell types can adopt an amoeboid migration mode that is independent of matrix remodeling and is characterized by localized extracellular matrix deformation as well as cell shape alterations (Wolf *et al* 2003). Indeed, such a mechanism seems to be dependent on the mechanical properties of the microenvironment. Fibroblast-like preosteoblastic cells are expected to migrate using a mesenchymal mode (Friedl and Bröcker 2000). An amoeboid migration mode has only been observed in physically cross-linked matrices, which support material displacement by cellular forces. In particular, these PEG-based gels are cross-linked by strong covalent bonds. Alternatively, cells may use existing, macroscopic gel defects or exert sufficiently strong force to induce local propagating cracks in the hydrogels; however, this is difficult to investigate. Unfortunately, it is well known that single-component hydrogels—especially those formed from synthetic polymers such as PEG—are mechanically very fragile, which is a problem that can be overcome by choosing more sophisticated multicomponent design strategies (Gong *et al* 2003). However, such cracks within hydrogels seem to be accessible for migrating cells and could result in a migration speed that exceeds that of cells that rely purely on proteolytic migration in higher density hydrogels.

The hydrogel matrices used *in vitro* under well-defined culture conditions were transferred to the more complex *in vivo* situation to assess migratory events during the onset of bone regeneration. Similarly to the *in vitro* case, more densely cross-linked matrices led to less migration to the inside of the gel and thus to bone formation solely on the outside of the gel. Although the differences were significant, the sensitivity of the implanted materials to MMPs is not thought to play a major role during *in vivo* remodeling compared to the *in vitro* 3D migration model.

Many more factors *in vivo* influence the gel performance and thus cannot be ignored, such as the amount and range of proteases, which may be differ considerably between *in vitro* cell culture conditions and a wound-healing environment, due to the interaction of the implant with a larger number of different cells, with these specialized to remodel rapidly under *in vivo* conditions, for example in the presence of inflammatory cell types. Nevertheless, the good correlation between *in vivo* and *in vitro* behavior indicates that the *in vitro* migration model can be successfully applied to predict the response of modulated material characteristics for specific tissue regeneration.

Precise control over *in vivo* cell fates using engineered, artificial extracellular matrices would be an extremely useful capability (Chan and Mooney 2008). Through the variation of the network's stiffness or proteolytic susceptibility and hence specificity, recruited cells may be confined to the tissue-material interface or directed to migrate into the material. By comparison to wound healing processes, which depend on surface erosion, the inside–out generation of new tissue from an implant might be considerably faster. It has generally been found that the most efficient bone reformation in long-term experiments occurs with gel compositions that provide efficient cell migration. Taken together, soft gels, which are subject to very fast gel degradation, are not helpful in supporting complete wound healing, as the implant stability cannot sustain the mechanical properties needed for the formation of bone tissue.

8.3 Matrix composition

This subchapter describes the effect of matrix components such as proteins on the structure and the mechanical properties of the matrix. The effect of embedded cells is excluded here, but is addressed in detail in chapters 11 and 12. The extracellular matrix, which is a non-cellular component, is still present in all kinds of tissues and organs. Moreover, it provides essential physical scaffolding for the storage of cellular constituents, while it is also responsible for some crucial biochemical and mechanical properties of tissues, required for their morphogenesis, differentiation and homeostasis. The importance of the extracellular matrix is mirrored in the wide range of syndromes that arise from genetic abnormalities in extracellular matrix proteins (Jarvelainen *et al* 2009).

The extracellular matrix is fundamental in providing organ and tissue functions, and is composed of water, proteins and polysaccharides. Due to the tissue type, the extracellular matrix displays a unique composition and topology, which is built up during tissue development through a dynamic and reciprocal biochemical and biophysical remodeling between the various cellular components, such as epithelial cells, fibroblast, adipocytes and the endothelial cell linings of blood or lymphoid vessels. This all leads to a cellular and protein microenvironment that surrounds cells, tissues and organs. Although the physical, topological and biochemical composition of the extracellular matrix has been reported to be tissue-specific, it is also pronouncedly heterogeneous within the same tissue or even in the same tissue of different individual organisms.

The extracellular matrix is a network of secreted extracellular macromolecules (Lu *et al* 2012). In more detail, this consists of a variety of proteins and polysaccharides which are secreted locally and assembled into an organized network in close neighborhood to the surface of the cell that produced them. The extracellular matrix of connective tissue is frequently more profuse than the cells it surrounds and thus it regulates the overall tissue's mechanical properties. Connective tissues form the framework of the organism, but the quantities vary due to their location in organs such as cartilage and bone. In these two tissues, the connective tissues are the major component, but in the brain and spinal cord they are only minor constituents (Galtrey and Fawcett 2007). The cells that secrete the surrounding extracellular matrix can also help to organize the extracellular matrix structurally. In particular, the orientation of the cytoskeleton inside the cell regulates the orientation of the extracellular matrix produced outside (Mierke et al 2011). In most connective tissues, the extracellular matrix's macromolecules are mainly produced and secreted by fibroblasts. In several specialized types of connective tissues, such as cartilage and bone, the matrix's secreting cells are chondroblasts, which build cartilage, and osteoblasts, which assemble bone. However, both cell types still belong to the family of fibroblasts.

Two main classes of extracellular macromolecules assemble the matrix. The first consists of polysaccharide chains called glycosaminoglycans (GAGs). In more detail, these are usually covalently linked to proteins in the form of proteoglycans. The second class consists of fibrous proteins such as collagen, elastin, fibronectin and laminin. All these proteins possess both structural and adhesive functions. In particular, the proteoglycan molecules of the connective tissue build a highly hydrated, gel-like basal substance, in which the extracellular matrix proteins, growth factors, cytokines and chemokines can be embedded. Moreover, the polysaccharide gel can withstand compressive forces on the extracellular matrix while allowing the diffusion of nutrients, metabolites and hormones between the white blood cells and the connective tissue cells. The collagen fibers strengthen the matrix scaffold and thus structurally organize the matrix. The rubber-like elastin fibers provide the linear elasticity of the matrix. Besides a large number of proteins, the connective tissue also contains cells and hence it provides the morphological structure and stability to support cell survival, proliferation and differentiation, and it stores nutrients for cells (Adams and Watts 1993, Hay 1993). In addition, the extracellular matrix also plays an important role in wound-healing processes (Schultz and Wysocki 2009).

The most abundant protein in the extracellular matrix of connective tissue is collagen (Shoulders and Raines 2009). Collagen monomers can assemble into fibrils and networks and are thus involved in providing cell–matrix adhesion through binding to collagen receptors on the cells. Moreover, they give structure and stability to the extracellular matrix (Halper and Kjaer 2014). There are other proteins embedded in the extracellular matrix, such as fibronectin, laminin and elastin (Halper and Kjaer 2014). Fibronectin and laminin are cell-adhesion proteins that can bind to cell surface receptors such as integrins and thus they play a role in cell migration, differentiation, primary tumor formation and malignant cancer progression, such as cancer metastasis (Mierke 2014, Wess 2005). Elastin is a structural protein that affects the architecture of the extracellular matrix scaffold, matrix stability and elasticity (Halper and Kjaer 2014). Cell adhesion to the extracellular matrix is facilitated by several groups of extracellular matrix receptors, such as integrins, discoidin domain receptors and syndecans (Harburger and Calderwood 2009, Humphries *et al* 2006, Leitinger and Hohenester 2007, Xian *et al* 2010). The adhesion of cells facilitates the connection of the cell's cytoskeleton to the extracellular matrix (mechano-coupling) and is a major driving factor in cell migration through the extracellular matrix (Schmidt and Friedl 2010). Moreover, the extracellular matrix seems to be a very dynamic structure that is constantly remodeled and restructured enzymatically or non-enzymatically, and its molecular constituents are also changed by numerous post-translational modifications. Through these physical and biochemical characteristics, the extracellular matrix generates the biochemical and mechanical properties of each organ, such as tensile and compressive strength and linear elasticity, and it also provides protection through a buffering action that maintains extracellular homeostasis and water retention.

Proteoglycans are also an important component of the extracellular matrix. In particular, they are associated with collagen fibrils and other proteins such as elastin, which influence the extracellular matrix's structure and regulate water distribution within it (Halper and Kjaer 2014). In addition, proteoglycans play a role during the assembly of collagen fibers from collagen fibrils (Kalamajski and Oldberg 2010).

Cell–matrix adhesions are facilitated by extracellular matrix receptors to ensure cell survival and enable proliferation, differentiation and motility (Wess 2005). Integrins are the most common extracellular matrix receptors. They are associated with proteins of the cytoskeleton, such as actin, through the focal adhesion and mechanocoupling proteins such as vinculin and talin, which support cell adhesion by binding integrins to their ligands, such as the extracellular matrix proteins fibronectins, laminins and collagens (Ginsberg 2014).

Finally, water is a key component of the extracellular matrix, influencing processes such as osmosis, filtration swelling and diffusion. Its distribution through the extracellular matrix is modulated by proteoglycans and glycoproteins (Kalamajski and Oldberg 2010). In addition, the extracellular matrix regulates essential morphological organization and physiological functions through binding of growth factors (GFs) which binds to cell-surface receptors, inducing signal trans-duction as well as gene transcription regulation. The biochemical and mechanical protective and organizational properties of the extracellular matrix tissues can vary between different types, for example between lung, skin and bone tissue, and even within a single tissue, as in the renal cortex and medulla; they also differ between different physiological states, for example, between normal and cancerous states.

It has been suggested that cancer cells are influenced by their surrounding microenvironment, such as the extracellular matrix of connective tissue. Alterations in protein distributions (such as in the cell–matrix adhesion proteins of cancer cells and the structural proteins within the extracellular matrix) have an effect on cellular behavior and function (Wolf *et al* 2009). In addition, growth factors regulate the expression and suppression of genes, which can lead to alterations in several cellular processes (Normanno *et al* 2006). The extracellular matrix affects cancer disease by stimulation through proteins or molecules. The structural and mechanical properties of the extracellular matrix are also known to alter the properties of cancer cells and to promote malignant cancer progression. Indeed, there are studies reporting that matrix stiffness and structure have an effect on cellular processes such as cell migration (Wolf *et al* 2009). In turn, cells have the ability to restructure the extracellular matrix by expressing an enzyme that can degrade the extracellular matrix. MMPs (Fingleton 2005), a disintegrin and metalloproteinases (ADAMs), a disintegrin and metalloproteinases with thrombospondin motifs (ADAMTS) (Rocks *et al* 2008) and other proteolytic enzymes expressed by the cells can alter the extracellular matrix. Moreover, cancer cells can also express proteins such as fibronectin and laminin to cross-link extracellular matrix structures and provide ligands for their own cell–matrix adhesion molecules, thus restructuring the extracellular matrix by embedding additional extracellular matrix proteins (Huang and Charrabarty 1994, Mierke *et al* 2011, Sporn and Roberts 1985).

In the following, the main molecular components of the extracellular matrix are described and compared. For example, the differences between the properties of the extracellular matrices of a normal simple epithelial tissue and those found within a pathologically modified tissue (such as aged tissue, wounded or fibrotic tissue and tumors) are examined. The focus is mainly on the composition and architecture of the extracellular matrix and interactions with its cellular constituents, and also on common post-translational modifications that facilitate defined topological and viscoelasticity alterations in the tissue. In addition, the functional consequences of extracellular matrix remodeling on cellular behaviors such as altered GF sensitivity evoked by alterations in the tension of the extracellular matrix are discussed. The particular focus is on the interstitial stroma of simple glandular epithelial tissues; the basement membranes are excluded here.

The role of the composition of the extracellular matrix in cell motility
The extracellular matrix is composed of two main classes of macromolecules: proteoglycans (PGs) and fibrous proteins (Jarvelainen *et al* 2009, Schaefer and Schaefer 2010).

Structural and functional properties of proteoglycans
PGs consist of glycosaminoglycan (GAG) chains that are covalently linked to a specific protein core (one exception is of hyaluronic acid) (Iozzo and Murdoch (1996), Schaefer and Schaefer (2010)). In more detail, PGs have been classified according to their core proteins, localization and GAG composition. Thus, the three main families are: small leucine-rich proteoglycans (SLRPs), modular proteoglycans and cell-surface expressed proteoglycans (Schaefer and Schaefer 2010). The GAG chains on the protein core are unbranched polysaccharide ones composed of repeating disaccharide units, such as sulfated N-aceltylglucosamine or N-acetylgalactosamine, D-glucuronic or L-iduronic acid and galactose (-4 N-acetylglucosamine-β1, 3-galactose-β1), which can be subdivided into sulfated (chondroitin sulfate, heparan sulfate and keratan sulfate) and non-sulfated (hyaluronic acid) GAGs (Schaefer and Schaefer 2010). All these molecules are extremely hydrophilic and hence adopt highly extended conformations that are necessary for the formation of hydrogels and lead to matrices with the ability to

withstand high compressive forces. It has been demonstrated that many genetic diseases have been connected to mutations in PG genes (Jarvelainen *et al* 2009). In more detail, SLRPs have been involved in many signaling pathways as through their binding they activate the epidermal growth factor receptor (EGFR), insulin-like growth factor 1 receptor (IGFIR) and low-density lipoprotein-receptor-related protein 1 (LRP1), thus regulating the inflammatory response reaction through the binding and activation of TGFβ (Goldoni and Iozzo 2008, Schaefer and Schaefer 2010). Modular PGs can regulate many important cellular processes, such as cell adhesion, migration and proliferation (Schaefer and Schaefer 2010). Basement membrane modular PGs such as perlecan, agrin and collagen type XVIII fulfill a dual function as pro- and anti-angiogenic factors (Iozzo *et al* 2009). Cell surface expressed PGs such as syndecans and glypicans are able to act as co-receptors mediating ligands binding to signaling receptors (Schaefer and Schaefer 2010).

The synthesis of the extracellular matrix proteins collagen and fibronectin
To date, 28 different types of collagen have been identified in vertebrates (Gordon and Hahn 2010). The majority of collagen molecules assemble a triple-stranded helix that can subsequently assemble into supramolecular complexes, such as fibrils and networks. Whether fibrils or networks are formed depends on the type of collagen. For example, fibrous collagens build the backbone of the collagen fibril bundles found within the interstitial tissue stroma, while network collagens are detected within the basal membrane. The synthesis of collagen type I includes several enzymatic post-translational modifications (Gordon and Hahn 2010, Myllyharju and Kivirikko 2004), such as the hydroxylation of proline and lysine residues, glycosylation of lysine and N- and C-terminal cleavage of propeptides. After cleavage, collagen fibrils are strengthened through covalent cross-linking between the lysine residues of constituent collagen molecules mediated by lysyl oxidases (LOX) (Myllyharju and Kivirikko 2004, Robins 2007).

Fibronectin is secreted by cells as a dimer connected by two C-terminal disulfide bonds. In addition, it has several binding sites to other fibronectin dimers, collagen, heparin and cell surface expressed matrix adhesion molecules, such as integrins (Pankov and Yamada 2002). In particular, cell surface binding of the soluble fibronectin dimer is essential for its assembly into longer fibronectin fibrils. Moreover, cell contraction provided by the actomyosin cytoskeleton and the resulting clustering of integrins induces fibronectin–collagen fibril assembly by exposing cryptic binding sites, which bind one another (Leiss *et al* 2008, Mao and Schwarzbauer 2005, Vakonakis and Campbell 2007).

The main fibrous proteins of the extracellular matrix are collagens, elastins, fibronectins and laminins (Alberts *et al* 2007). By contrast, PGs fill the majority of the extracellular interstitial space within the tissue by forming a hydrated gel (Jarvelainen *et al* 2009). In addition, PGs have a wide variety of functions, mirroring their unique buffering, hydration, binding and force-resistance properties. In particular, in the basement membrane of the kidney glomerular, perlecan plays a prominent role in glomerular filtration (Harvey and Miner 2008, Morita *et al* 2005).

However, in ductal epithelial tissues, decorin, biglycan and lumican can bind to collagen fibers and consequently generate a molecular structure within a certain region of the extracellular matrix that is essential for mechanical buffering and hydration. Moreover, the binding makes GFs serve as an easily accessible storage sink within this substructure (Iozzo and Murdoch 1996).

Fibroblasts are a major source of the bulk of the secreted interstitial collagen resident in the stroma or recruited to it from neighboring tissues (De Wever et al 2008). Through the exertion tension on the extracellular matrix, fibroblasts are help to organize collagen fibrils into sheets and cables and, consequently, determine the alignment of collagen fibers. Although there is a heterogeneous mixture of different collagen types within a tissue, one type of collagen usually predominates.

Collagen interacts with elastin, which is a second major extracellular matrix fiber. Elastin fibers provide elastic repulsion to tissues that undergo repeated stretch. However, the elastin stretch is crucially limited by tight binding to collagen fibrils (Wise and Weiss 2009). Secreted tropoelastin (precursor of elastin) molecules can assemble into fibers and are strongly cross-linked to one another through their lysine residues by members of the lysyl oxidase (LOX) enzyme family, such as LOX and LOXL (Lucero and Kagan 2006). In more detail, elastin fibers are covered by glycoprotein microfibrils (such as fibrillins), which are fundamental for the integrity of the elastin fibers (Wise and Weiss 2009).

A third fibrous protein is fibronectin and this is important in directing the organization of the interstitial extracellular matrix. Additionally, fibronectin is crucial in facilitating cell adhesion and function. Indeed, fibronectin can be stretched repeatedly over its resting (persistence) length by cellular traction forces (Smith et al 2007). This force-dependent unfolding of fibronectin leads to the exposure of cryptic binding sites for integrins and, consequently, pleiotrophic alterations in cellular behavior. Due to these results, it has been suggested that fibronectin acts as an extracellular mechano-regulator (Smith et al 2007). As expected, 'tensed' fibronectin affects the catch bond 'force activation' and adhesion assembly of the $\alpha 5 \beta 1$ integrin through exposure of its synergy-binding site (Friedland et al 2009). Fibronectin is a major driving factor for cell migration and tissue invasion during development and has been implicated in cardiovascular disease and malignant cancer progression, such as tumor metastasis (Rozario and DeSimone 2010, Tsang et al 2010, Mierke et al 2011). As with fibronectin, other proteins of the extracellular matrix such as tenascin exert multiple unrelated effects on cellular behavior, such as the promotion of fibroblast migration during wound healing (Trebaul et al 2007, Tucker and Chiquet-Ehrismann 2009). Indeed, levels of tenascins C and W increase in the stroma of certain transformed tissues where they can block the interaction between syndecan4 and fibronectin in order to facilitate tumor growth and cancer metastasis (Tucker and Chiquet-Ehrismann 2009).

The extracellular matrix and tissue homeostasis under normal conditions
Normal glandular epithelial tissues are composed of a simple layer of epithelial cells displaying apical–basal polarity, where the basal side adheres to the basement membrane and the apical side is exposed to the fluid-filled lumen. In some glandular

epithelium there is a basal or myoepithelial cell layer, which separates the luminal epithelium from the interstitial extracellular matrix (Barsky and Karlin 2005). The homeostasis of epithelial tissue depends on the maintenance of tissue organization and provides a dynamic exchange with the surrounding stroma, mainly embedding non-activated fibroblasts and adipocytes and a stable unaltered population of transiting, non-stimulated leukocytes (Ronnov-Jessen *et al* 1996). Thus, non-activated tissue fibroblasts secrete and organize type I and III collagens, elastin, fibronectin, tenascin and a specific set of PGs, such as hyaluronic acid and decorin, which all keep the structural and functional integrity of the interstitial extracellular matrix in constant balance. Most of the glandular epithelial tissues, such as the breast, saliva gland, lung and prostate, are in tensional homeostasis and their normal state is reported to be highly mechanically compliant (Paszek and Weaver 2004). In a compliant tissue, the extracellular matrix is composed of a relaxed meshwork of collagens type I and III and elastin, which together with fibronectin build a relaxed network of fibers. These fibers are surrounded by a hydrogel of glycosaminoglycan-chain-containing PGs (Bosman and Stamenkovic 2003). Hence, the relaxed collagen and elastin fiber network allows the healthy extracellular matrix to resist a wide range of tensile stresses.

In addition, a functionally competent normal tissue can also easily withstand compressive stresses due to the binding of the hydrated glycosaminoglycan (GAG) network to the fibrous extracellular matrix proteins (Scott 2003). The tissue extracellular matrix is consequently a highly dynamic scaffold that is continuously remodeled, as the precise microstructure is crucial for the maintenance of its normal function (Egeblad *et al* 2010, Kass *et al* 2007). Tissue homeostasis is facilitated by the coordinated secretion of fibroblast's MMPs (Mott and Werb 2004), which is also counterbalanced by tissue inhibitors of metalloproteinases (TIMPs) located in the extracellular matrix (Cruz-Munoz and Khokha 2008) and the controlled activity of other enzymes, such as LOX and transglutaminases, which can cross-link and hence increase the stiffness of the extracellular matrix (Lucero and Kagan 2006). Numerous GFs are bound to the extracellular matrix and regulate these processes (Friedl 2010, Hynes 2009, Macri *et al* 2007, Murakami *et al* 2008, Oehrl and Panayotou 2008). These extracellular matrix-bound GFs are able to differentially affect cell growth and migration and upon release they are a part of a tightly controlled feedback cycle that is essential for the homeostasis of normal tissue (Hynes 2009).

The extracellular matrix under tissue aging conditions
When a tissue ages, the levels of junctional proteins such as cadherin, catenin and occludin decrease in their expression and subsequently this reduction can compromise junctional integrity, as gaps between the adjacent epithelial cells can occur frequently (Akintola *et al* 2008, Bolognia 1995). Moreover, old tissue is characterized by a thinning of the basement membrane, which may be caused by elevated MMP-driven degradation, as well as reduced synthesis of basement membrane proteins (Callaghan and Wilhelm 2008). In particular, the resident fibroblasts in aged tissues are non-proliferating and can even resist apoptotosis,

which is indicative of senescence (Campisi and d'Adda di Fagagna 2007). As expected, senescent fibroblasts express increased amounts of fibronectin, MMPs, GFs, interleukins, cytokines, plasminogen activator inhibitor (PAI) (Coppe *et al* 2010) and mitochondrial-related reactive oxygen species (ROS) (Untergasser *et al* 2005). From this, the extracellular matrix is frequently in a state similar to that of chronic inflammation. Indeed, the combined action of chronic inflammation and elevated levels of MMPs, PAI and ROS may destroy the integrity of the elastin network and cause alterations to the collagen fiber network, while reduced levels of tissue-associated GAGs also compromise the integrity of the basement membrane (Callaghan and Wilhelm 2008, Calleja-Agius *et al* 2007, Nomura 2006). Paradoxically, in an aging tissue the collagen fibers are frequently inappropriately cross-linked by glycation, byproducts of lipid oxidation, and upon exposure to UV light (Robins 2007). Glycation increases with age and several advanced glycation end products additionally act as cross-linkers, which contribute to the progressive insolubilization of the extracellular matrix and the enhanced stiffness of collagens in aged tissues (Avery and Bailey 2006). The combination of elevated and inappropriate collagen cross-linking causes tissue stiffening, which results in a mechanically weaker and less elastic but also more rigid aged tissue compared to a young tissue (Calleja-Agius *et al* 2007, Robins 2007, Schulze *et al* 2012). This aberrant mechanical state can severely compromise organization of the extracellular matrix, modify the epithelial organization and function, and also potentially support age-related diseases such as cancer (Coppe *et al* 2010, Freund *et al* 2010, Sprenger *et al* 2008).

The role of tensional homeostasis and fibrosis

The acute injury of tissues activates the fibrogenic processes and induces wound healing. An early event, which characterizes a wound-healing response, is vascular damage and the formation of a fibrin clot, which then induces the infiltration of monocytes to the damaged sites of the extracellular matrix. When monocytes bind to extracellular matrix-degradation products and cytokines, they rapidly differentiate into macrophages (Clark 2001). These activated macrophages in turn secrete and release multiple GFs, MMPs and cytokines, which induce angiogenesis and stimulate the motility of fibroblasts and induce their proliferation (Schultz and Wysocki 2009). Thereafter, recruited fibroblasts start to synthesize and deposit large quantities of extracellular matrix proteins, such as collagen type I and III, fibronectin and hyaluronic acid, which all affect the mechanical and structural properties of the extracellular matrix. Moreover, the increased mechanical stress associated with this profound extracellular matrix deposition can lead to the transdifferentiation of fibroblasts and other tissue-resident cells, which means they switch from an epithelial to a mesenchymal phenotype (the epithelial–mesenchymal transition (EMT)). Another example is circulating bone-marrow-derived mesenchymal stem cells, which switch into myofibroblasts (Schultz and Wysocki 2009, Velnar *et al* 2009). These cells possess a high capacity to synthesize extracellular matrix components and are highly contractile. Moreover, they promote the formation of large, rigid collagen bundles that, when cross-linked by LOX enzymes, mechanically strengthen and stiffen the connective tissue (Szauter *et al* 2005). This wounded and stiffened microenvironment disrupts the basement membrane that surrounds the epithelium and compromises the integrity of the tissue, as indicated by the loss of apical–basal polarity and destabilized cell–cell adhesions. The remodeled extracellular matrix additionally promotes the directional migration of cells within the tissue to the site of the inflammatory wound (Schafer and Werner 2008). In some cases, the release of transforming growth factor β (TGF-β) caused by tension and secretion of MMPs induces the EMT of the resident surrounding epithelium (Schultz and Wysocki 2009, Wipff *et al* 2007, Xu *et al* 2009). In a healthy tissue, once the wound has been repopulated strict feedback mechanisms are induced that lead to the reestablishment of tissue homeostasis and the dissolution of fibrosis (Schultz and Wysocki 2009, Velnar *et al* 2009). Under extreme conditions, such as repeated injury or when normal feedback mechanisms are compromised, continuous extracellular matrix synthesis, deposition and remodeling becomes permanent and myofibroblasts remain in the tissue, where TIMP production prevails over MMP synthesis. However, these aberrant conditions may lead to chronic vascular remodeling and elevated extracellular matrix cross-linking, which may cause aberrant fibrosis. Finally, the connective tissue will not be able to heal properly. This aberrant wound-healing scenario leads to altered mechanical stability and decreased elasticity, which is a typical feature of scarred tissue (Kisseleva and Brenner 2008). In special, extreme cases, a chronic wound is able to promote a tumor phenotype (De Wever *et al* 2008).

How is the extracellular matrix altered near primary tumors?
Cancer can be described as the loss of tissue organization and the abnormal behavior of cellular components. Thus, cell transformation caused by genetic mutations and epigenetic alterations further supports the malignant progression of cancer. In addition, primary tumors seem to be similar to wounds that fail to heal (Schafer and Werner 2008). The tumor stroma only exhibits certain characteristics found in an unresolved wound (Bissell and Radisky 2001). One characteristic feature of tumors is that they are stiffer than the surrounding normal healthy tissue. The stress stiffening of primary tumors is induced by extracellular matrix deposition and remodeling through resident tumor-associated fibroblasts and through the enhanced contractility of the transformed epithelium (Butcher *et al* 2009, Levental *et al* 2009). Moreover, chemokines and GFs (De Wever *et al* 2008) can also induce inflammation and thereby modify the number of infiltrating T lymphocytes (Tan and Coussens 2007). Tissue inflammation leads to stromal fibroblast activation and induces the transdifferentiation of fibroblasts into myofibroblasts, thus increasing and promoting tissue desmoplasia (De Wever *et al* 2008, Desmouliere *et al* 2004). Myofibroblasts deposit large quantities of extracellular matrix proteins, secrete GFs and exert strong contractile forces on the extracellular matrix (De Wever *et al* 2008, Desmouliere *et al* 2004). Thus, newly deposited and remodeled collagen as well as elastin fibers are reoriented and remodeled after cross-linking by LOX and transglutaminase. Finally, larger and more rigid fibrils are present that further stiffen the tissue extracellular matrix (Butcher *et al* 2009, Erler and Weaver 2009, Levental *et al*

2009, Lucero and Kagan 2006, Payne *et al* 2007, Rodriguez *et al* 2008). MMPs can be secreted and activated by cancer cells and myofibroblasts (De Wever *et al* 2008, Kessenbrock *et al* 2010) and are then able to remodel the basement membrane surrounding the primary tumor, releasing and activating extracellular matrix embedded GFs (Bosman and Stamenkovic 2003, Kessenbrock *et al* 2010). The release of GFs such as vascular endothelial growth factor (VEGF) enhances the vascular permeability of blood vessels and thus induces the growth of new vessels, which then generate additional interstitial tissue pressure. There is an amplifying cycle involving cancer-associated extracellular matrix stiffening, reciprocal extracellular matrix resistance induced by resident cancer cells and myoepithelial cells, and cell-generated contractility, which act as a positive-feedback loop to potentiate tumor growth and survival. This induces neoangiogenesis and cancer cell invasion, which may eventually foster cancer metastasis (Butcher *et al* 2009, Erler and Weaver 2009, Paszek and Weaver 2004, Paszek *et al* 2005).

What are the challenges encountered with natural and synthetic extracellular matrices?

As the extracellular matrix influences and regulates numerous fundamental cellular processes, many tissue-culture models have been developed to investigate the impact of the biochemical and biophysical properties from these microenvironments and to obtain insights into the molecular origins of cellular behaviors controlled by extracellular matrix ligation. In order to assess the fundamental process of cell adhesion and its impact on cell behavior, the majority of cancer research studies have been based on coated tissue culture dishes (plastic or glass), with purified preparations or mixtures of extracellular matrix proteins, in order to obtain 2D monolayers (Kuschel *et al* 2006). However, to analyze extracellular matrix rigidity, functionalized polyacrylamide (PAA) gels cross-linked with reconstituted basement membrane generated from Engelbreth–Holm–Swarm mouse carcinoma (Matrigel™), collagen type I, fibronectin or extracellular matrix peptides such as RGD peptide, have become the general approach (Johnson *et al* 2007, Pelham and Wang 1997). These experimental strategies do not mirror the behavior of cells within tissues under *in vivo* microenvironmental conditions in an appropriate manner, as no real 3D situation can be investigated, but the extracellular matrix can be readily restructured. To reveal the impact of 3D and extracellular matrix remodeling, natural extracellular matrix and reconstituted extracellular matrix gels can be analyzed in terms of tissue-specific differentiation and architecture. The reconstituted basement membrane, which mimics several of the biochemical and biophysical properties of endogenous epithelial basement membranes, has been used in 3D organotypic culture assays, for xenograft manipulations or tissue engineering and investigating tissue-specific morphogenesis, such as branching and acini formation and differentiation (Kleinman and Martin 2005, Kleinman *et al* 1986). However, basement membrane preparations such as Matrigel™ are well suited for determining normal epithelial or endothelial behavior and distinguishing between the normal and malignant behavior of certain tissues, although they have a complex and rudimentarily defined composition that does not reconstruct the physical state of the *in vivo* interstitial extracellular matrix. In addition, fibrin has been used as a natural biodegradable scaffold in vascular tissue engineering. A disadvantage of fibrin is that it lacks the mechanical strength and durability of native interstitial extracellular matrix (Blomback and Bark 2004, Shaikh *et al* 2008). By contrast, collagen type I is useful and can be combined with reconstituted basement membrane (such as purified laminin or fibronectin) to determine certain biological aspects of normal and diseased interstitial extracellular matrices (Friess 1998, Gudjonsson *et al* 2002). In particular, collagen type I can assemble into a mechanically tense network of fibrils that can be oriented, functionally modified and enzymatically as well as chemically cross-linked and hence stiffened. Thus collagen I gels are suitable substrates for determining the role of collagen and fibronectin stiffness and organization in the pathogenesis of tumor progression and invasion (Levental *et al* 2009, Provenzano *et al* 2009). Nevertheless, collagen gels have been reported to be quite heterogeneous and modifying their architecture alters their organization, pore size and ligand concentration, thereby complicating the interpretation of the data in terms of differentiating between the effects of each component and generalizing the results resulting from using this natural scaffold (Johnson *et al* 2007). To overcome this, denuded extracellular matrix scaffolds from various tissues have been isolated (Macchiarini *et al* 2008). These scaffolds contain colonies of stem cells which may reconstitute normal tissues with reasonable fidelity (Lutolf *et al* 2009). Indeed, extracellular matrices have been isolated and extracted from various tissues, including from the small intestine, skin, pancreas and breast (Rosso *et al* 2005), and these extracellular matrices have been used to engineer skin grafts (Badylak 2007), increase the healing processes of wounds and investigate tumor progression. An example is porcine-derived small intestinal submucosa, which is a proven clinical success for treating patients with hernias (Franklin *et al* 2002, Badylak 2007). Although these purified extracellular matrices have certainly been successful in special applications, they are difficult to obtain, as they need well-defined micro-environments in tissue regeneration and stem cell transplantation, in which animal byproducts and contaminants are eliminated. In order to reveal the molecular and biophysical mechanisms through which the extracellular matrix supports special effects in cellular differentiation and morphogenesis, it is important to use chemically and physically defined modular extracellular matrices, which have been reproduced reliably. In this respect, it has been demonstrated that synthetically produced matrices have defined and tunable composition, organization, mechanics and extracellular matrix remodeling capabilities (Ayres *et al* 2009, Dutta and Dutta 2009, Lutolf and Hubbell 2005, McCullen *et al* 2009, Rosso *et al* 2005, Zisch *et al* 2003). For example, polyethylene glycol (PEG) hydrogels are used frequently as biologically compatible synthetic matrices facilitating cell adhesion, supporting cell viability and growth (Lutolf and Hubbell 2005). Indeed, these matrices can be covalently cross-linked with extracellular matrix ligands, and can bind collagenase-degradable peptides and store GFs (Ehrbar *et al* 2007b, Zisch *et al* 2003), although they do not mirror the structural and architectural features of native collagen gels and their pore size often impedes cell migration. By contrast, peptide-based hydrogels such as peptideamphiphiles assemble into secondary structures that rebuild the collagen triple helix and subsequently facilitate stem cell growth and viability as well as direct multicellular morphogenesis (Hauser and Zhang 2010, Sieminski *et al* 2008, Smith and Ma 2004, Ulijn and Smith 2008). In more detail, these peptide-amphiphiles are modified by covalent binding of native proteins and MMP-degradable extracellular matrix peptides. As an alternative matrix, poly (lactic-co-glycolic acid) (PLGA), a copolymer of glycolic acid and lactic acid (McCullen *et al* 2009), which is biodegradable because it can be hydrolyzed into lactic and glycolic acid, has been conjugated to various extracellular matrix ligands and peptides, or coated with collagen or chitosan, in order to induce cell adhesion and provide viability and growth. One development is modular biocompatible extracellular matrices, which contain ligand-binding cassettes and have tunable stiffness features supporting a precise patterning of cell adhesion in 2D and 3D microenvironments (Serban and Prestwich 2008). The organization of the extracellular matrix's structure is crucial for cellular functions and hence has led to the development of new methodologies that lead to extracellular matrices whose fiber size, orientation, stiffness, ligand-binding function and remodeling potential can be controlled and varied (Zhang *et al* 2009). Anisotropically nanofabricated substrates assembled from scalable biocompatible PEG (Kim *et al* 2010, Smith *et al* 2009) are a novel development in the biomaterials field, and their only major advantage seems to be their ability to address the lack of functional assessment in physiological culture assays and animal models.

The main constituent of the extracellular matrix: collagen
Collagen is the most abundant fibrous protein within the interstitial extracellular matrices and provides up to 30% of the total protein mass of a multicellular animal. Collagens constitute the main structural element of the extracellular matrix and provide tensile strength, regulating cell adhesion, chemotaxis and migration, and consequently direct tissue development (Rozario and DeSimone 2010). In more detail, collagens are major proteins in the extracellular matrix of connective tissue and are the most prevalent component of mammalians that is mainly produced and altered by fibroblasts. These structural proteins are part of tissues in bones, tendons, ligaments, skin, blood vessels and the cornea of the eye. Hence, their broad distribution displays the wide range of properties that comes from their functional requirements (Wess 2005). Minerals are included in bones to stiffen and stabilize the matrix. In tendons and ligaments the collagen tissue needs to be more elastic and store energy. In the cornea of the eye, visible light should be able to pass through the collagen matrix that needs to be transparent. All these different properties are caused by special structural arrangements of collagen monomers, fibrils, fibers and interaction with other molecules, such as minerals, proteoglycans and glycosami-noglycans. All collagen types have a common primary structure: collagen monomers can assemble to three left-handed helical polypeptides (called α-chains), which then associate to form a right-handed helical fibrillar collagen monomer (Ramachandran and Kartha 1954, Rich and Crick 1955a). There are at least 40 genes encoding various α-chains, which assemble into different collagen types (Stamov and

Pompe 2012). In more detail, the α-chains consist of triplets made up of a glycine molecule and two non-equivalent amino acids Gly-X-Y (Rich and Crick 1955a). These amino acids are often (2S)-proline (28%) and (2S, 4R)-4-hydroxyproline (38%) (Shoulders and Raines 2009). Both N-terminal and C-terminal ends of an α-chain terminate in a propeptide, thus this α-chain is also called the protocollagen chain. Three protocollagen chains bind together through the establishment of hydrogen bonds (Rich and Crick 1955b). The resulting right-handed chain is called procollagen. After intracellular assembly, these chains are transported out of the cell. Finally, parts of the propeptide ends are cut off by specific MMPs, leaving telopeptide ends. The propeptides prevent fibrillation inside the cells, whereas the N- and C-terminal telopeptides support it. Thus, these telopeptides play an important role in the extracellular assembly of collagen fibrils and network structures. The resulting chain, tropocollagen, is the monomer of collagen structures.

Collagen classification
At least 29 different collagen types (and other proteins consisting of a collagen-like structure) have been identified in vertebrates (Ramachandran and Kartha 1954) (Ricard-Blum 2011). There are five main classes into which they can be divided (Birk and Bruckner 2005): fibrillar collagens, network forming collagens, fibril-associated collagens with interrupted triple helices (FACITs), membrane associated collagens with interrupted triple helices (MACITs) and collagens with multiple triple-helix domains with interruptions (MULTIPLEXINs) (table 8.1).

Fibrillar collagens
The monomers of fibrillar collagens aggregate to fibrils, which further assemble into fibers, bundles and whole networks. The most important members of this class of fibrillar collagens are types I and II. Collagen I is more widely distributed than collagen II, as it is found in the dermis, bone, tendon and ligaments. By contrast, collagen II is found in the cartilage and vitreous of the eye. As mentioned above, a fibrillar collagen monomer is a right-handed helical chain assembled by three left-handed polypeptides. Each fibrillar collagen polypeptide chain consists of 338–343 Gly-X-Y triplets (Ramshaw *et al* 1998). Collagen monomers have a length of approximately 300 nm and a thickness of up to 1.5 nm. They aggregate by establishing covalent bonds between the C-terminal telopeptide and the helical domain within fibrils, which show a repeating banding pattern with a periodicity of

Table 8.1. The five main collagen classes.

Type I. Skin, tendon, vascular ligature, organs and bone.
Type II. Cartilage.
Type III. Reticulate, commonly found alongside type I.
Type IV. Forms basal lamina, the epithelium-secreted layer of the basement membrane.
Type V. Cell surfaces, hair and placenta.

64–67 nm (named D periodicity). This banding pattern can be seen in high-resolution images, and is caused by the gap-overlap structure of regions with high and low electron densities within collagen monomers.

The composition of the three α-chains differs between the fibrillar collagen types: the monomers of collagen I are heterotrimer (one α-chain has another triplet composition), whereas collagen II is a homotrimer (α-chains have the same triplet composition). In addition, the number of amino acid triplets within the protocollagen chain can change.

Collagen fibrils are able to aggregate into fibers or networks with different structural and mechanical properties. Other molecules and proteins also play a role in fiber aggregation, including collagen III, collagen V, fibronectin, glycosaminoglycans, glycoproteins and proteoglycans (Di Lullo et al 2002). For example, interactions between proteoglycans or between a proteoglycan molecule and other glycosaminoglycans regulates the interfibrillar interaction (Wess 2005). In addition, collagen V is important for collagen I fibrillation and hence matrix organization (Wenstrup et al 2004), and inclusion of collagen III leads to alterations in the network structure (Lapiere et al 1977).

Network-forming collagens

This class of collagens assembles into network scaffolds. Important members of the class of network-building collagens are collagen IV located in the basement membrane (Glentis et al 2014) and collagen VI.

There are different aggregation modes of network-forming collagens, such as rectangular networks (collagen IV) and hexagonal lattice structures (collagen VI, VIII, V), which are formed by head-to-head interactions among and also between N-terminal and C-terminal domains. In some cases they can even form fibril-like structures (collagen VI), where two monomers assemble into a dimer through the interaction of the C-terminal domain with the helical structure of the second monomer via the formation of hydrogen bonds. Then, two dimers form a tetramer, which is the building block for further, more complex structures. Moreover, they can assemble into beaded microfibrils, broad-banded fibrils or hexagonal lattice structures by head-to-head interactions of the N-terminal regions. Indeed, interruptions in the triplet sequence of the polypeptide chains are observed and can lead to more flexibility in the overall structure. Interactions between collagens such as the fibrillar (collagen I and II) are possible. In more detail, collagen types such as collagen I and collagen IV interact in the basement membrane (Glentis et al 2014).

Fibril-associated collagens with interrupted triple helices (FACITs), MACITs and MULTIPLEXIN collagens are associated collagens. In particular, FACITs occupy the surface of fibrils such as collagen I or collagen II. The structural difference of several triple-helical domains interrupted by non-helical sequences supports the modification of the connective tissue. They lead to bridges in the extracellular matrix, inducing organization and stabilization of the extracellular matric (Shaw and Olsen 1991). MACITs also possess helical structure domains, which are interrupted by non-helical domains like FACITs and can bind to network-forming collagens such as collagen VI. They can also be found in the basement membrane associated with collagen IV, which alters the structural properties. FACITs cannot form fibrils by themselves, but they are able to associate with the surface of collagen fibrils. In particular, collagen IX is covalently linked to the surface of cartilage collagen fibrils composed mostly of collagen II (Olsen 1997) and collagens XII and XIV are found to be associated with collagen-I-containing fibrils. In addition, collagen XV is located in close association with collagen fibrils at the basement membrane and thus forms a bridge that links large, banded fibrils such as those containing collagens I and III (Amenta *et al* 2005). Similarly, MULTIPLEXIN collagens are frequently associated with membranes and consist of several interrupted collagenous domains.

Collagen type I
Collagen I is the most abundant collagen in mammalian tissue, as it accounts for more than 90% of the total collagen found in nature (Gobeaux *et al* 2008). Moreover, collagen I is the main protein of the extracellular matrix and it occurs in tendons, ligaments and the dermis. Monomers of the fibril forming collagen type I are assembled to heterotrimers consisting of two identical $\alpha 1$ chains and a $\alpha 2$ chain, which differs slightly from the $\alpha 1$ chain. The $\alpha 1$ and $\alpha 2$ chains are encoded by the COL1A1 and COL1A2 genes, respectively. They differ solely in their triplet structure Gly-X-Y. Collagen chains can vary in size from 662 to 3152 amino acids for the human $\alpha 1(X)$ and $\alpha 3(VI)$ chains, respectively (Ricard-Blum *et al* 2000, Gordon and Hahn 2010). In more detail, each α-chain consists of 338–343 uninterrupted amino acid triplets (Ramshaw *et al* 1998). However, the total number of the triplets depends on the collagen source. Beyond the existence of 28 different collagen types, further diversity occurs in the collagen family due to the existence of several molecular isoforms for the same collagen type, such as for collagens IV and VI and hybrid isoforms composed of chains belonging to two different collagen types, for example type V/XI hybrid molecules (Richard-Blum 2011).

The fibrillation behavior of single collagen I fibrils is well understood. The aggregation of monomers and the lateral growth of fibrils is temperature-dependent (Gelman *et al* 1979, Wolf *et al* 2013). Increasing temperature causes a decrease in the fibril diameter and length (Liu *et al* 2005). In addition, the pH of the non-polymerized collagen solution influences the structure of the fiber and the whole network. Thus it has been reported that with increasing pH the fibril diameter increases (Christiansen *et al* 2000). Furthermore, the collagen concentration influences fibril formation, as with increasing collagen concentration, the fibrillation rate and fibril diameter increase (Williams *et al* 1978, Gobeaux *et al* 2008). However, what impact does this have on the mechanical properties of 3D collagen matrices and on cancer cell migration through these special matrices?

8.4 The impact of fiber thickness, connection points and polymerization dynamics on cancer cell invasion

It has been shown that the fiber thickness can be modulated by polymerization of the collagen matrices at different temperatures, which influences the polymerization kinetics. Indeed, it has been reported that the polymerization process decreases with lowering temperature, leading to thicker collagen fibrils and larger pore sizes (Wolf *et al* 2013). Recently, the impact of the fiber thickness on cancer cell migration was determined (Sapudom *et al* 2005). However, the role played by the connection points within a 3D matrix is still elusive and must be investigated further. The connection points may contribute to the anisotropy of the 3D collagen matrices and may represent a greater confinement for cancer cell invasion than parallel collagen fibers and bundles.

Polymerization dynamics

The polymerization dynamics have a major impact on the structure and morphology and hence on the mechanical properties of the extracellular matrix. As is obvious, the temperature during the polymerization of collagen matrices is a critical factor for regulating the fiber thickness and the pore size or cross-sectional area. In addition, the pH value for the polymerization is important for the structure and morphology and consequently the mechanical properties of the extracellular matrix. However, the pH values can only be varied over a small range, that in which the buffer has its highest impact. There are still many points to investigate in terms of controlled collagen polymerization dynamics.

8.5 The role of a matrix stiffness gradient in cancer cell invasion

Cancer cells are tuned to the mechanical properties of the 3D microenvironment (for example, a 3D extracellular matrix). All cells—not only cancer cells, and including those in traditionally mechanically static tissues, such as those from the breast and brain—are exposed to isometric force or tension. The isometric force or tension is generated locally at the nanoscale level by cell–cell or cell–matrix interactions and it alters the cell's function through regulation of the actomyosin contractility and actin dynamics (Álvarez-González *et al* 2015). In more detail, each cell type is specifically adapted to the specific 'home' tissue in which it is embedded. In particular, the brain is pronouncedly softer than bone tissue (figure 8.2). Thus, neural cell growth, survival and differentiation are supported by a highly compliant matrix. By contrast, osteoblast differentiation and survival is favored on stiffer

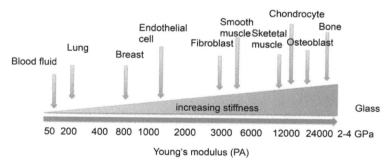

Figure 8.2. Mechanical properties such as stiffness (elastic modulus) of different tissue cell types. In breast tumors is the stiffness increased as indicated by the blue bar.

extracellular matrices with mechanical properties that are more similar to newly formed bone tissue. Normal mammary epithelial cell growth, survival, differentiation and morphogenesis are regulated by the interaction of the cells with a relatively soft matrix. Upon transformation of healthy tissue to cancerous tissue, breast tissue becomes progressively stiffer and consequently cancer cells become significantly more contractile and hyper-responsive to matrix compliance changes (Lopez *et al* 2011). In addition, it has been reported that normalizing the tensional homeostasis of cancer cells reverts them toward a non-malignant phenotype (Lühr *et al* 2012), thereby suggesting a functional link between the matrix's mechanical properties, cellular tension and normal tissue behavior. Although primary breast tumors are much stiffer than the normal healthy breast, the mechanical properties of a breast tumor are still pronouncedly softer than those of muscle or bone tissue tumors, highlighting the precise association between the tissue phenotype and the matrix's rigidity.

Further reading

Deryugina E I, Luo G X, Reisfeld R A, Bourdon M A and Strongin A 1997 Tumor cell invasion through matrigel is regulated by activated matrix metalloproteinase-2 *Anticancer Res.* **17**(5A) 3201–10

Friedl P and Bröcker E B 2004 Reconstructing leukocyte migration in 3D extracellular matrix by time-lapse videomicroscopy and computer-assisted tracking *Methods Mol. Biol.* **239** 77–90

Friedl P, Maaser K, Klein C E, Niggemann B, Krohne G and Zänker K S 1997 Migration of highly aggressive MV3 melanoma cells in three-dimensional collagen lattices results in local matrix reorganization and shedding of alpha2 and beta1 integrins and CD44 *Cancer Res.* **57** 2061–70

Hotary K B, Allen E D, Brooks P C, Datta N S, Long M W and Weiss S J 2003 Membrane type I matrix metalloproteinase usurps tumor growth control imposed by the three-dimensional extracellular matrix *Cell* **114** 33–45

Hotary K, Allen E, Punturieri A, Yana I and Weiss S J 2000 Regulation of cell invasion and morphogenesis in a three-dimensional type I collagen matrix by membrane-type matrix metalloproteinases 1, 2, and 3 *J. Cell Biol.* **149** 1309–23

Lin D C, Dimitriadis E K and Horkay F 2007 Robust strategies for automated AFM force curve analysis—I. Non-adhesive indentation of soft, inhomogeneous materials *J. Biomech. Eng.* **129** 430–40

Moss N M, Wu Y I, Liu Y, Munshi H G and Stack M S 2009 Modulation of the membrane type 1 matrix metalloproteinase cytoplasmic tail enhances tumor cell invasion and proliferation in three-dimensional collagen matrices *J. Biol. Chem.* **284** 19791–9

te Riet J, Katan A J, Rankl C, Stahl S W, Avan Buul A M and Phang I Y *et al* 2011 Interlaboratory round robin on cantilever calibration for AFM force spectroscopy *Ultramicroscopy* **111** 1659–69

van Spriel A B, Leusen J H, van Egmond M, Dijkman H B, Assmann K J, Mayadas T N and van de Winkel J G 2001 Mac-1 (CD11b/CD18) is essential for Fc receptor-mediated neutrophil cytotoxicity and immunologic synapse formation *Blood* **97** 2478–86

Yamamoto N, Jiang P, Yang M, Xu M, Yamauchi K, Tsuchiya H, Tomita K, Wahl G M, Moossa A R and Hoffman R M 2004 Cellular dynamics visualized in live cells *in vitro* and *in vivo* by differential dual-color nuclear-cytoplasmic fluorescent-protein expression *Cancer Res.* **64** 4251–6

Zaman M H, Trapani L M and Matsudaira P 2006 Migration of tumor cells in 3D matrices is governed by matrix stiffness along with cell–matrix adhesion and proteolysis *Proc. Natl Acad. Sci. USA* **103** 10889–94

References

Álvarez-González B, Meili R, Bastounis E, Firtel RA, Lasheras J C and Del Álamo J C 2015 Three-dimensional balance of cortical tension and axial contractility enables fast amoeboid migration *Biophys. J.* **108** 821–32

Adams J E and Watt F 1993 Regulation of development and differentiation by the extracellular matrix *Development* **117** 1183–98

Akintola A D, Crislip Z L, Catania J M, Chen G, Zimmer W E, Burghardt R C and Parrish A R 2008 Promoter methylation is associated with the age-dependent loss of N-cadherin in the rat kidney *Am. J. Physiol. Renal Physiol.* **294** F170–6

Alberts B, Johnson A, Lewis J, Raff M, Roberts K and Walter P 2007 *Mol. Biol. Cell* (London: Garland)

Alexander S, Koehl G E, Hirschberg M, Geissler E K and Friedl P 2008 Dynamic imaging of cancer growth and invasion: a modified skin-fold chamber model *Histochem. Cell Biol.* **130** 1147–54

Almany L and Seliktar D 2005 Biosynthetic hydrogel scaffolds made from fibrinogen and polyethylene glycol for 3D cell cultures *Biomaterials* **26** 2467–77

Amenta P S, Scivoletti N A, Newman M D, Sciancalepore J P, Li D and Myers J C 2005 Proteoglycan-collagen XV in human tissues is seen linking banded collagen fibers subjacent to the basement membrane *J. Histochem. Cytochem.* **53** 165–76

Avery N C and Bailey A J 2006 The effects of the Maillard reaction on the physical properties and cell interactions of collagen *Pathol. Biol.* **54** 387–95

Ayres C E, Jha B S, Sell S A, Bowlin G L and Simpson D G 2009 Nanotechnology in the design of soft tissue scaffolds: innovations in structure and function *Interdiscip. Rev. Nanomed. Nanobiotechnol.* **2** 20–34

Badylak S F 2007 The extracellular matrix as a biologic scaffold material *Biomaterials* **28** 3587–93

Balzer E M, Tong Z, Paul C D, Hung W C, Stroka K M, Boggs A E, Martin S S and Konstantopoulos K 2012 Physical confinement alters tumor cell adhesion and migration phenotypes *FASEB J.* **26** 4045–56

Barsky S H and Karlin N J 2005 Myoepithelial cells: autocrine and paracrine suppressors of breast cancer progression *J. Mammary Gland Biol. Neoplasia* **10** 249–60

Beadle C, Assanah M C, Monzo P, Vallee R, Rosenfeld S S and Canoll P 2008 The role of myosin II in glioma invasion of the brain *Mol. Biol. Cell* **19** 3357–68

Berton S, Belletti B, Wolf W, Canzonieri V, Lovat F, Vecchione A, Colombatti A, Friedl P and Baldassarre G 2009 The tumor suppressor functions of p27(kip1) include control of the mesenchymal/amoeboid transition *Mol. Cell Biol.* **29** 5031–45

Birk D E and Bruckner P 2005 *The extracellular matrix: an overview* (New York: Springer) pp 185–205

Bissell M J and Radisky D 2001 Putting tumours in context *Nat. Rev. Cancer* **1** 46–54

Blombäck B and Bark N 2004 Fibrinopeptides and fibrin gel structure *Biophys. Chem.* **112** 147–51

Bolognia J L 1995 Aging skin *Am. J. Med.* **98** 99S–103S

Bosman F T and Stamenkovic I 2003 Functional structure and composition of the extracellular matrix *J. Pathol.* **200** 423–8

Brunner C A, Ehrlicher A, Kohlstrunk B, Knebel D, Käs J A and Goegler M 2006 Cell migration through small gaps *Eur. Biophys. J.* **35** 713–9

Bryant S J and Anseth K S 2002 Hydrogel properties influence ECM production by chondrocytes photoencapsulated in poly(ethylene glycol) hydrogels *J. Biomed. Mater. Res.* **59** 63–72

Butcher D T, Alliston T and Weaver V M 2009 A tense situation: forcing tumour progression *Nat. Rev. Cancer* **9** 108–22

Buxboim A, Ivanovska I L and Discher D E 2010 Matrix elasticity, cytoskeletal forces and physics of the nucleus: how deeply do cells 'feel' outside and in? *J. Cell Sci.* **123** 297–308

Callaghan T M and Wilhelm K P 2008 A review of aging and an examination of clinical methods in the assessment of ageing skin. Part 2. Clinical perspectives and clinical methods in the evaluation of ageing skin *Int. J. Cosmet. Sci.* **30** 323–32

Calleja-Agius J, Muscat-Baron Y and Brincat M P 2007 Skin ageing *Menopause Int.* **13** 60–4

Campisi J and d'Adda di Fagagna F 2007 Cellular senescence: when bad things happen to good cells *Nat. Rev. Mol. Cell Biol.* **8** 729–40

Chan G and Mooney D J 2008 New materials for tissue engineering: towards greater control over the biological response *Trends Biotechnol.* **26** 382–92

Chow K H, Factor R E and Ullman K S 2012 The nuclear envelope environment and its cancer connections *Nat. Rev. Cancer* **12** 196–209

Christiansen D L, Huang E K and Silver F H 2000 Assembly of type I collagen: fusion of fibril subunits and the influence of fibril diameter on mechanical properties *Matrix Biol.* **19** 409–20

Clark R A 2001 Fibrin and wound healing *Ann. NY Acad. Sci.* **936** 355–67

Coppe J P, Desprez P Y, Krtolica A and Campisi J 2010 The senescence-associated secretory phenotype: the dark side of tumor suppression *Annu. Rev. Pathol.* **5** 99–118

Cruz-Munoz W and Khokha R 2008 The role of tissue inhibitors of metalloproteinases in tumorigenesis and metastasis *Crit. Rev. Clin. Lab. Sci.* **45** 291–338

Dahl K N, Kahn S M, Wilson K L and Discher D E 2004 The nuclear envelope lamina network has elasticity and a compressibility limit suggestive of a molecular shock absorber *J. Cell Sci.* **117** 4779–86

De Wever O, Demetter P, Mareel M and Bracke M 2008 Stromal myofibroblasts are drivers of invasive cancer growth *Int. J. Cancer* **123** 2229–38

Desmouliere A, Guyot C and Gabbiani G 2004 The stroma reaction myofibroblast: a key player in the control of tumor cell behavior *Int. J. Dev. Biol.* **48** 509–17

Di Lullo G A, Sweeney S M, Körkkö J, Ala-Kokko L and San Antonio J D 2002 Mapping the ligand-binding sites and disease-associated mutations on the most abundant protein in the human, type I collagen *J. Biol. Chem.* **277** 4223–31

Dikovsky D, Bianco-Peled H and Seliktar D 2008 Defining the role of matrix compliance and proteolysis in three-dimensional cell spreading and remodeling *Biophys. J.* **94** 2914–25

Dutta R C and Dutta A K 2009 Cell-interactive 3D scaffold: advances and applications *Biotechnol. Adv.* **27** 334–9

Egeblad M, Rasch M G and Weaver V M 2010 Dynamic interplay between the collagen scaffold and tumor evolution *Curr. Opin. Cell Biol.* **22** 697–706

Ehrbar M, Rizzi S C and Lutolf M P 2007a Enzymatic formation of modular cell-instructive fibrin analogs for tissue engineering *Biomaterials* **28** 3856–66

Ehrbar M, Rizzi S C and Lutolf M P 2007b Biomolecular hydrogels formed and degraded via site-specific enzymatic reactions *Biomacromolecules* **8** 3000–7

Ehrbar M, Sala A, Lienemann P, Ranga A, Mosiewicz K, Bittermann A, Rizzi S C, Weber F E and Lutolf M P 2011 Elucidating the role of matrix stiffness in 3D cell migration and remodeling *Biophys. J.* **100** 284–93

Elbjeirami W M, Yonter E O, Starcher B C and West J L 2003 Enhancing mechanical properties of tissue-engineered constructs via lysyl oxidase cross-linking activity *J. Biomed. Mater. Res.* A **66** 513–21

Erler J T and Weaver V M 2009 Three-dimensional context regulation of metastasis *Clin. Exp. Metastasis* **26** 35–49

Feng D, Nagy J A, Pyne K, Dvorak H F and Dvorak A M 1998 Neutrophils emigrate from venules by a transendothelial cell pathway in response to FMLP *J. Exp. Med.* **187** 903–15

Fingleton B 2005 Matrix metalloproteinases: roles in cancer and metastasis *Frontiers in bioscience* **11** 479–91

Fisher K E, Sacharidou A, Stratman A N, Mayo A M, Fisher S B, Mahan R D, Davis M J and Davis G E 2009 MT1-MMP- and Cdc42-dependent signaling co-regulate cell invasion and tunnel formation in 3D collagen matrices *J. Cell Sci.* **122** 4558–69

Franklin M E Jr, Gonzalez J J Jr, Michaelson R P, Glass J L and Chock D A 2002 Preliminary experience with new bioactive prosthetic material for repair of hernias in infected fields *Hernia* **6** 171–4

Freund A, Orjalo A V, Desprez P Y and Campisi J 2010 Inflammatory networks during cellular senescence: causes and consequences *Trends Mol. Med.* **16** 238–46

Friedl A 2010 Proteoglycans: master modulators of paracrine fibroblast-carcinoma cell interactions *Semin. Cell Dev. Biol.* **21** 66–71

Friedl P and Alexander S 2011 Cancer invasion and the microenvironment: plasticity and reciprocity *Cell* **147** 992–1009

Friedl P and Bröcker E B 2000 The biology of cell locomotion within three-dimensional extracellular matrix *Cell. Mol. Life Sci.* **57** 41–64

Friedl P and Wolf K 2009 Proteolytic interstitial cell migration: a five-step process *Cancer Metastasis Rev.* **28** 129–35

Friedl P and Wolf K 2010 Plasticity of cell migration: a multiscale tuning model *J. Cell Biol.* **188** 11–9

Friedl P, Wolf K and Lammerding J 2011 Nuclear mechanics during cell migration *Curr. Opin. Cell Biol.* **23** 55–64

Friedland J C, Lee M H and Boettiger D 2009 Mechanically activated integrin switch controls alpha5beta1 function *Science* **323** 642–4

Friess W 1998 Collagen-biomaterial for drug delivery *Eur. J. Pharm. Biopharm.* **45** 113–36

Galtrey C M and Fawcett J W 2007 The role of chondroitin sulfate proteoglycans in regeneration and plasticity in the central nervous system *Brain Res. Rev.* **54** 1–18

Gelman R A, Williams B R and Piez K A 1979 Collagen fibril formation. Evidence for a multistep process *Journal of Biological Chemistry* **254**(1) 180–6 PMID: 758319

Gerlitz G and Bustin M 2011 The role of chromatin structure in cell migration *Trends Cell Biol.* **21** 6–11

Ginsberg M H 2014 Integrin activation *BMB Rep.* **47** 655–9

Glentis A, Gurchenkov V and Matic Vignjevic D 2014 Assembly, heterogeneity and breaching of the basement membranes *Cell Adh. Migr.* **8** 236–45

Gobeaux F, Mosser G, Anglo A, Panine P, Davidson P, Giraud-Guille M-M and Belamie E 2008 Fibrillogenesis in dense collagen solutions: a physicochemical study *J. Mol. Biol.* **376** 1509–22

Gobin A S and West J L 2002 Cell migration through defined, synthetic ECM analogs *FASEB J.* **16** 751–3 PMID: 11923220

Goldberg M W, Huttenlauch I, Hutchison C J and Stick R 2008 Filaments made from A- and B-type lamins differ in structure and organization *J. Cell Sci.* **121** 215–25

Goldoni S and Iozzo R V 2008 Tumor microenvironment: modulation by decorin and related molecules harboring leucine-rich tandem motifs *Int. J. Cancer* **123** 2473–9

Gong J P, Katsuyama Y and Osada Y 2003 Double-network hydrogels with extremely high mechanical strength *Adv. Mater.* **15** 1155–8

Gordon M K and Hahn R A 2010 Collagens *Cell Tissue Res.* **339** 247–57

Grinnell F and Petroll W M 2010 Cell motility and mechanics in three-dimensional collagen matrices *Annu. Rev. Cell Dev. Biol.* **26** 335–61

Gudjonsson T, Ronnov-Jessen L, Villadsen R, Rank F, Bissell M J and Petersen O W 2002 Normal and tumor-derived myoepithelial cells differ in their ability to interact with luminal breast epithelial cells for polarity and basement membrane deposition *J. Cell Sci.* **115** 39–50 PMID: 11801722

Halper J and Kjaer M 2014 Basic components of connective tissues and extracellular matrix: elastin, fibrillin, fibulins, fibrinogen, fibronectin, laminin, tenascins and thrombospondins *Adv. Exp. Med. Biol.* **802** 31–47 PMID: 24443019

Halstenberg S, Panitch A and Hubbell JA 2002 Biologically engineered protein-graft-poly (ethylene glycol) hydrogels: a cell adhesive and plasmin-degradable biosynthetic material for tissue repair *Biomacromolecules* **3** 710–23

Harburger D S and Calderwood D A 2009 Integrin signalling at a glance *J. Cell Sci.* **122** 159–63

Harvey S J and Miner J H 2008 Revisiting the glomerular charge barrier in the molecular era *Curr. Opin. Nephrol. Hypertens.* **17** 393–8

Hauser C A and Zhang S 2010 Designer self-assembling peptide nanofiber biological materials *Chem. Soc. Rev.* **39** 2780–90

Hay E D 1993 Extracellular matrix alters epithelial differentiation *Curr. Opin. Cell Biol.* **5** 1029–35

Helseth D L Jr and Veis A 1981 Collagen self-assembly *in vitro*. Differentiating specific telopeptide-dependent interactions using selective enzyme modification and the addition of free amino telopeptide *J. Biol. Chem.* **256** 7118–28 PMID: 7251588

Huang S and Chakrabarty S 1994 Regulation of fibronectin and laminin receptor expression, fibronectin and laminin secretion in human colon cancer cells by transforming growth factor-β1 *Int. J. Cancer* **57** 742–6

Humphries J D, Byron A and Humphries M J 2006 Integrin ligands at a glance *J. Cell Sci.* **119** 3901–3

Huttenlocher A, Sandborg R R and Horwitz A F 1995 Adhesion in cell migration *Curr. Opin. Cell Biol.* **7** 697–706

Hynes R O 2009 The extracellular matrix: not just pretty fibrils *Science* **326** 1216–9

Ilina O and Friedl P 2009 Mechanisms of collective cell migration at a glance *J. Cell Sci.* **122** 3203–8

Ilina O, Bakker G J, Vasaturo A, Hofmann R M and Friedl P 2011 Two-photon laser-generated microtracks in 3D collagen lattices: principles of MMP dependent and independent collective cancer cell invasion *Phys. Biol.* **8** 015010

Iozzo R V and Murdoch A D 1996 Proteoglycans of the extracellular environment: clues from the gene and protein side offer novel perspectives in molecular diversity and function *FASEB J.* **10** 598–614 PMID: 8621059

Iozzo R V, Zoeller J J and Nystrom A 2009 Basement membrane proteoglycans: modulators *par excellence* of cancer growth and angiogenesis *Mol. Cells* **27** 503–13

Jacobelli J, Friedman R S, Conti M A, Lennon-Dumenil A M, Piel M, Sorensen C M, Adelstein R S and Krummel M F 2010 Confinement-optimized three-dimensional T cell amoeboid motility is modulated via myosin IIA-regulated adhesions *Nat. Immunol.* **11** 953–61

Jarvelainen H, Sainio A, Koulu M, Wight T N and Penttinen R 2009 Extracellular matrix molecules: potential targets in pharmacotherapy *Pharmacol. Rev.* **61** 198–223

Jawerth L M, Münster S, Vader D A, Fabry B and Weitz D A 2010 A blind spot in confocal reflection microscopy: the dependence of fiber brightness on fiber orientation in imaging biopolymer networks *Biophys. J.* **98** L1–3

Johnson K R, Leight J L and Weaver V M 2007 Demystifying the effects of a three-dimensional microenvironment in tissue morphogenesis *Methods Cell Biol.* **83** 547–83 PMID: 17613324

Kalamajski S and Oldberg 2010 The role of small leucine-rich proteoglycans in collagen fibrillogenesis *Matrix Biol.* **29** 248–53

Kass L, Erler J T, Dembo M and Weaver V M 2007 Mammary epithelial cell: influence of extracellular matrix composition and organization during development and tumorigenesis *Int. J. Biochem. Cell Biol.* **39** 1987–94

Kessenbrock K, Plaks V and Werb Z 2010 Matrix metalloproteinases: regulators of the tumor microenvironment *Cell* **141** 52–67

Kim D H, Lipke E A, Kim P, Cheong R, Thompson S, Delannoy M, Suh K Y, Tung L and Levchenko A 2010 Nanoscale cues regulate the structure and function of macroscopic cardiac tissue constructs *Proc. Natl Acad. Sci. USA* **107** 565–70

Kim S, Chung E H and Healy K E 2005 Synthetic MMP-13 degradable ECMs based on poly(n-isopropylacrylamide-co-acrylic acid) semi-interpenetrating polymer networks. I. Degradation and cell migration *J. Biomed. Mater. Res.* A **75** 73–88

Kisseleva T and Brenner D A 2008 Mechanisms of fibrogenesis *Exp. Biol. Med. (Maywood)* **233** 109–22

Kleinman H K and Martin G R 2005 Matrigel: basement membrane matrix with biological activity *Semin. Cancer Biol.* **15** 378–86

Kleinman H K, McGarvey M L, Hassell J R, Star V L, Cannon F B, Laurie G W and Martin G R 1986 Basement membrane complexes with biological activity *Biochemistry* **25** 312–8

Kuschel C, Steuer H, Maurer A N, Kanzok B, Stoop R and Angres B 2006 Cell adhesion profiling using extracellular matrix protein microarrays *Biotechniques* **40** 523–31

Lämmermann T *et al* 2008 Rapid leukocyte migration by integrin-independent flowing and squeezing *Nature* **453** 51–5

Lühr I *et al* 2012 Mammary fibroblasts regulate morphogenesis of normal and tumorigenic breast epithelial cells by mechanical and paracrine signals *Cancer Lett.* **325** 175–88

Lapiere C M, B Nusgens B and Pierard G E 1977 Interaction between collagen type I and type III in conditioning bundles organization *Connect. Tissue Res.* **5** 21–9

Lautenschläger F, Paschke S, Schinkinger S, Bruel A, Beil M and Guck J 2009 The regulatory role of cell mechanics for migration of differentiat- ing myeloid cells *Proc. Natl Acad. Sci. USA* **106** 15696–701

Leiss M, Beckmann K, Giros A, Costell M and Fassler R 2008 The role of integrin binding sites in fibronectin matrix assembly *in vivo Curr. Opin. Cell Biol.* **20** 502–7

Leitinger B and Hohenester E 2007 Mammalian collagen receptors *Matrix Biol.* **26** 146–55

Leventual K R *et al* 2009 Matrix cross-linking forces tumor progression by enhancing integrin signaling *Cell* **139** 891–906

Liu M-Y, Yeh M-L and Luo Z-P 2005 *In vitro* regulation of single collagen fibril length by buffer compositions and temperature *Biomed. Mater. Eng.* **15** 413–20 PMID: 16308457

Lopez J I, Kang I, You W K, McDonald D M and Weaver V M 2011 In situ force mapping of mammary gland transformation *Integr. Biol. (Camb.)* **3** 910–21

Lu P, Weaver V M and Werb Z 2012 The extracellular matrix: a dynamic niche in cancer progression *J. Cell Biol.* **196** 395–406

Lucero H A and Kagan H M 2006 Lysyl oxidase: an oxidative enzyme and effector of cell function *Cell Mol. Life Sci.* **63** 2304–16

Lutolf M P 2009 Integration column: artificial ECM: expanding the cell biology toolbox in 3D *Integr. Biol. (Camb).* **1** 235–41

Lutolf M P and Hubbell J A 2005 Synthetic biomaterials as instructive extracellular micro-environments for morphogenesis in tissue engineering *Nat. Biotechnol.* **23** 47–55

Lutolf M P, Gilbert P M and Blau H M 2009 Designing materials to direct stem-cell fate *Nature* **462** 433–41

Macchiarini P *et al* 2008 Clinical transplantation of a tissue-engineered airway *Lancet* **372** 2023–30

Macri L, Silverstein D and Clark R A 2007 Growth factor binding to the pericellular matrix and its importance in tissue engineering *Adv. Drug Deliv. Rev.* **59** 1366–81

Mao Y and Schwarzbauer J E 2005 Fibronectin fibrillogenesis, a cell-mediated matrix assembly process *Matrix Biol.* **24** 389–99

Matthew D Shoulders and Ronald T Raines 2009 Collagen structure and stability *Annu. Rev. Biochem.* **78** 929–58

McCullen S D, Ramaswamy S, Clarke L I and Gorga R E 2009 Nanofibrous composites for tissue engineering applications *Wiley Interdiscip. Rev. Nanomed. Nanobiotechnol.* **1** 369–90

Mierke C T, Frey B, Fellner M, Herrmann M and Fabry B 2011 Integrin α5β1 facilitates cancer cell invasion through enhanced contractile forces *J. Cell Sci.* **124** 369–83

Mierke CT 2013 The integrin alphav beta3 increases cellular stiffness and cytoskeletal remodeling dynamics to facilitate cancer cell invasion *New J. Phy.* **15** 015003

Mierke C T 2014 The fundamental role of mechanical properties in the progression of cancer disease and inflammation *Rep. Prog. Phys.* **77** 076602

Miron-Mendoza M, Seemann J and Grinnell F 2010 The differential regulation of cell motile activity through matrix stiffness and porosity in three dimensional collagen matrices *Biomaterials* **31** 6425–35

Mooney D J and Vandenburgh H 2008 Cell delivery mechanisms for tissue repair *Cell Stem Cell* **2** 205–13

Morita H, Yoshimura A, Inui K, Ideura T, Watanabe H, Wang L, Soininen R and Tryggvason K 2005 Heparan sulfate of perlecan is involved in glomerular filtration *J. Am. Soc. Nephrol.* **16** 1703–10

Moroni L, de Wijn J R and van Blitterswijk C A 2008 Integrating novel technologies to fabricate smart scaffolds *J. Biomater. Sci. Polym. Ed.* **19** 543–72

Mott J D and Werb Z 2004 Regulation of matrix biology by matrix metalloproteinases *Curr. Opin. Cell Biol.* **16** 558–64

Murakami M, Elfenbein A and Simons M 2008 Non-canonical fibroblast growth factor signalling in angiogenesis *Cardiovasc. Res.* **78** 223–31

Myllyharju J and Kivirikko K I 2004 Collagens, modifying enzymes and their mutations in humans, flies and worms *Trends Genet.* **20** 33–43

Nomura Y 2006 Structural change in decorin with skin aging *Connect. Tissue Res.* **47** 249–55

Normanno N, De Luca A, Bianco C, Strizzi L, Mancino M, Maiello M R, Carotenuto A, De Feo G, Caponigro F and Salomon D S 2006 Epidermal growth factor receptor (EGFR) signaling in cancer *Gene* **366** 2–16

Oehrl W and Panayotou G 2008 Modulation of growth factor action by the extracellular matrix *Connect. Tissue Res.* **49** 145–8

Olsen B R 1997 Collagen IX *Int. J. Biochem. Cell Biol.* **29** 555–8

Packard B Z, Artym V V, Komoriya A and Yamada K M 2009 Direct visualization of protease activity on cells migrating in three-dimensions *Matrix Biol.* **28** 3–10

Pankov R and Yamada K M 2002 Fibronectin at a glance *J. Cell Sci.* **115** 3861–3

Paszek M J and Weaver V M 2004 The tension mounts: mechanics meets morphogenesis and malignancy *J. Mammary Gland Biol. Neoplasia* **9** 325–42

Paszek M J et al 2005 Tensional homeostasis and the malignant phenotype *Cancer Cell* **8** 241–54

Payne S L, Hendrix M J and Kirschmann D A 2007 Paradoxical roles for lysyl oxidases in cancer- a prospect *J. Cell Biochem.* **101** 1338–54

Pedersen J A and Swartz M A 2005 Mechanobiology in the third dimension *Ann. Biomed. Eng.* **33** 1469–90

Pelham R J Jr and Wang Y 1997 Cell locomotion and focal adhesions are regulated by substrate flexibility *Proc. Natl Acad. Sci. USA* **94** 13661–5

Peyton S R, Raub C B and Putnam A J 2006 The use of poly (ethylene glycol) hydrogels to investigate the impact of ECM chemistry and mechanics on smooth muscle cells *Biomaterials* **27** 4881–93

Phelps E A and Garcia A J 2010 Engineering more than a cell: vascularization strategies in tissue engineering *Curr. Opin. Biotechnol.* **21** 704–9

Provenzano P P, Eliceiri K W and Keely P J 2009 Shining new light on 3D cell motility and the metastatic process *Trends Cell. Biol.* **19** 638–48

Provenzano P P, Eliceiri K W, Campbell J M, Inman D R, White J G and Keely P J 2006 Collagen reorganization at the tumor-stromal interface facilitates local invasion *BMC Med.* **4** 38

Puklin-Faucher E and Sheetz M P 2009 The mechanical integrin cycle *J. Cell Sci.* **122** 179–86

Raeber G P, Lutolf M P and Hubbell J A 2005 Molecularly engineered PEG hydrogels: a novel model system for proteolytically mediated cell migration *Biophys. J.* **89** 1374–88

Raeber G P, Lutolf M P and Hubbell J A 2007 Mechanisms of 3-D migration and matrix remodeling of fibroblasts within artificial ECMs *Acta Biomater.* **3** 615–29

Ramachandran G N and Kartha G 1954 Structure of collagen *Nature* **174** 269–70

Ramshaw J A, Shah N K and Brodsky B 1998 Gly-X-Y tripeptide frequencies in collagen: a context for host-guest triple-helical peptides *J. Struct. Biol.* **122** 86–91

Raub C B, Suresh V, Krasieva T, Lyubovitsky J, Mih J D, Putnam A J, Tromberg B J and George S C 2007 Noninvasive assessment of collagen gel microstructure and mechanics using multiphoton microscopy *Biophys. J.* **92** 2212–22

Ren X D, Wang R, Li Q, Kahek L A, Kaibuchi K and Clark R A 2004 Disruption of Rho signal transduction upon cell detachment *J. Cell Sci.* **117** 3511–8

Ricard-Blum S 2011 the collagen family *Cold Spring Harb Perspect Biol.* **3** a004978

Ricard-Blum S, Dublet B and van der Rest M 2000 *Unconventional Collagens: Types VI, VII, VIII, IX, X, XII, XIV, XVI and XIX (Protein Profile Series)* (Oxford: Oxford University Press)

Rich A and Crick F H 1955a Structure of polyglycine II *Nature* **176** 780

Rich A, Francis H and Crick F H 1955b The structure of collagen *Nature* **4489** 915–6

Ridley A J, Schwartz M A, Burridge K, Firtel R A, Ginsberg M H, Borisy G, Parsons J T and Horwitz A R 2003 Cell migration: integrating signals from front to back *Science* **302** 1704–9

Robins S P 2007 Biochemistry and functional significance of collagen cross-linking *Biochem. Soc. Trans.* **35** 849–52

Rocks N, Paulissen G, El Hour M, Quesada F, Crahay C, Gueders M, Foidart J-M, Noel A and Cataldo D 2008 Emerging roles of ADAM and ADAMTS metalloproteinases in cancer *Biochimie* **90** 369–79

Rodriguez C, Rodriguez-Sinovas A and Martinez-Gonzalez J 2008 Lysyl oxidase as a potential therapeutic target *Drug News Perspect.* **21** 218–24 PMID: 18560621

Rolli C G, Seufferlein T, Kemkemer R and Spatz J P 2010 Impact of tumor cell cytoskeleton organization on invasiveness and migration: a microchannel-based approach *PLoS ONE* **5** e8726

Ronnov-Jessen L, Petersen O W and Bissell M J 1996 Cellular changes involved in conversion of normal to malignant breast: importance of the stromal reaction *Physiol. Rev.* **76** 69–125 PMID: 8592733

Rosso F, Marino G, Giordano A, Barbarisi M, Parmeggiani D and Barbarisi A 2005 Smart materials as scaffolds for tissue engineering *J. Cell Physiol.* **203** 465–70

Rowe R G and Weiss S J 2009 Navigating ECM barriers at the invasive front: the cancer cell–stroma interface. *Annu. Rev. Cell Dev. Biol.* **25** 567–95

Rozario T and DeSimone D W 2010 The extracellular matrix in development and morphogenesis: a dynamic view *Dev. Biol.* **341** 126–40

Sabeh F *et al* 2004 Tumor cell traffic through the extracellular matrix is controlled by the membrane-anchored collagenase MT1-MMP *J. Cell Biol.* **167** 769–81

Sabeh F, Shimizu-Hirota R and Weiss S J 2009 Protease dependent versus independent cancer cell invasion programs: three-dimensional amoeboid movement revisited *J. Cell Biol.* **185** 11–9

Salmon H, Franciszkiewicz K, Damotte D, Dieu-Nosjean M C, Validire P, Trautmann A, Mami-Chouaib F and Donnadieu E 2012 Matrix architecture defines the preferential localization and migration of T cells into the stroma of human lung tumors *J. Clin. Invest.* **122** 899–910

Sanz-Moreno V, Gadea G, Ahn J, Paterson H, Marra P, Pinner S, Sahai E and Marshall C J 2008 Rac activation and inactivation control plasticity of tumor cell movement *Cell* **135** 510–23

Sapudom J, Rubner S, Martin S, Kurth T, Riedel S, Mierke C T and Pompe T 2005 The phenotype of cancer cell invasion controlled by fibril diameter and pore size of 3D collagen networks *Biomaterials* **52** 367–75

Schaefer L and Schaefer R M 2010 Proteoglycans: from structural compounds to signaling molecules *Cell Tissue Res.* **339** 237–46

Schafer M and Werner S 2008 Cancer as an overhealing wound: an old hypothesis revisited *Nat. Rev. Mol. Cell Biol.* **9** 628–38

Schmidt S and Friedl P 2010 Interstitial cell migration: integrin-dependent and alternative adhesion mechanisms *Cell Tissue Res.* **339** 83–92

Schoumacher M, Goldman R D, Louvard D and Vignjevic D M 2010 Actin, microtubules, and vimentin intermediate filaments cooperate for elongation of invadopodia *J. Cell Biol.* **189** 541–56

Schultz G S and Wysocki A 2009 Interactions between extracellular matrix and growth factors in wound healing *Wound Repair Regen.* **17** 153–62

Schulze C, Wetzel F, Kueper T, Malsen A, Muhr G, Jaspers S, Blatt T, Wittern K P, Wenck H and Käs J A 2012 Stiffening of human skin fibroblasts with age *Clin. Plast. Surg.* **39** 9–20

Scott J E 2003 Elasticity in extracellular matrix 'shape modules' of tendon, cartilage, etc. A sliding proteoglycan-filament model *J. Physiol.* **553** 335–43

Serban M A and Prestwich G D 2008 Modular extracellular matrices: solutions for the puzzle *Methods* **45** 93–8

Shaikh F M, Callanan A, Kavanagh E G, Burke P E, Grace P A and McGloughlin T M 2008 Fibrin: a natural biodegradable scaffold in vascular tissue engineering *Cells Tissues Organs* **188** 333–46

Shankar J, Messenberg A, Chan J, Underhill T M, Foster L J and Nabi I R 2010 Pseudopodial actin dynamics control epithelial-mesenchymal transition in metastatic cancer cells *Cancer Res.* **70** 3780–90

Shaw L M and Olsen B R 1991 FACIT collagens: diverse molecular bridges in extracellular matrices *Trends Biochem. Sci.* **16** 191–4

Shu X Z, Liu Y and Prestwich G D 2003 Disulfide-cross-linked hyaluronan-gelatin hydrogel films: a covalent mimic of the extracellular matrix for *in vitro* cell growth *Biomaterials* **24** 3825–34

Sieminski A L, Semino C E, Gong H and Kamm R D 2008 Primary sequence of ionic self-assembling peptide gels affects endothelial cell adhesion and capillary morphogenesis *J. Biomed. Mater. Res.* A **87** 494–504

Smith I O, Liu X H, Smith L A and Ma P X 2009 Nanostructured polymer scaffolds for tissue engineering and regenerative medicine *Interdiscip. Rev. Nanomed. Nanobiotechnol.* **1** 226–36

Smith L A and Ma P X 2004 Nano-fibrous scaffolds for tissue engineering *Colloids Surf. B: Biointerfaces* **39** 125–31

Smith M L, Gourdon D, Little W C, Kubow K E, Eguiluz R A, Luna-Morris S and Vogel V 2007 Force-induced unfolding of fibronectin in the extracellular matrix of living cells *PLoS Biol.* **5** e268

Sodek K L, Brown T J and Ringuette M J 2008 Collagen I but not Matrigel matrices provide an MMP-dependent barrier to ovarian cancer cell penetration *BMC Cancer* **8** 223

Sodek K L, Ringuette M J and Brown T J 2007 MT1-MMP is the critical determinant of matrix degradation and invasion by ovarian cancer cells *Br. J. Cancer* **97** 358–67

Sporn M B and Roberts A B 1985 Autocrine growth factors and cancer *Nature* **313** 745–7

Sprenger C C, Plymate S R and Reed M J 2008 Extracellular influences on tumour angiogenesis in the aged host *Br. J. Cancer* **98** 250–5

Stamov D R and Pompe T 2012 Structure and function of ECM-inspired composite collagen type I scaffolds *Soft Matter* **8** 10200–12

Starborg T, Lu Y, Kadler K E and Holmes D F 2008 Electron microscopy of collagen fibril structure *in vitro* and *in vivo* including three-dimensional reconstruction *Methods Cell Biol.* **88** 319–45

Stein A M, Vader D A, Jawerth L M, Weitz D A and Sander L M 2008 An algorithm for extracting the network geometry of three-dimensional collagen gels *J. Microsc.* **232** 463–75

Stoitzner P, Pfaller K, Stössel H and Romani N 2002 A close-up view of migrating Langerhans cells in the skin *J. Invest. Dermatol.* **118** 117–25

Szauter K M, Cao T, Boyd C D and Csiszar K 2005 Lysyl oxidase in development, aging and pathologies of the skin *Pathol. Biol.* **53** 448–56

Tan T T and Coussens L M 2007 Humoral immunity, inflammation and cancer *Curr. Opin. Immunol.* **19** 209–16

Tanaka Y, Matsuo K and Yuzuriha S 2010 Long-term histological compari- son between near-infrared irradiated skin and scar tissues *Clin. Cosmet. Investig. Dermatol.* **3** 143–9

Tibbitt M W and Anseth K S 2009 Hydrogels as extracellular matrix mimics for 3D cell culture *Biotechnol. Bioeng.* **103** 655–63

Tong Z, Balzer E M, Dallas M R, Hung W C, Stebe K J and Konstantopoulos K 2012 Chemotaxis of cell populations through confined spaces at single-cell resolution *PLoS ONE* **7** e29211

Trebaul A, Chan E K and Midwood K S 2007 Regulation of fibroblast migration by tenascin-C *Biochem. Soc. Trans.* **35** 695–7

Tsang K Y, Cheung M C, Chan D and Cheah K S 2010 The developmental roles of the extracellular matrix: beyond structure to regulation *Cell Tissue Res.* **339** 93–110

Tucker R P and Chiquet-Ehrismann R 2009 The regulation of tenascin expression by tissue microenvironments *Biochim. Biophys. Acta* **1793** 888–92

Ulijn R V and Smith A M 2008 Designing peptide based nanomaterials *Chem. Soc. Rev.* **37** 664–75

Untergasser G, Madersbacher S and Berger P 2005 Benign prostatic hyperplasia: age-related tissue-remodeling *Exp. Gerontol.* **40** 121–8

Vakonakis I and Campbell I D 2007 Extracellular matrix: from atomic resolution to ultrastructure *Curr. Opin. Cell Biol.* **19** 578–83

Velnar T, Bailey T and Smrkolj V 2009 The wound healing process: an overview of the cellular and molecular mechanisms *J. Int. Med. Res.* **37** 1528–42

Versaevel M, Grevesse T and Gabriele S 2012 Spatial coordination between cell and nuclear shape within micropatterned endothelial cells *Nat. Commun.* **3** 671

Vicente-Manzanares M, Koach M A, Whitmore L, Lamers M L and Horwitz A F 2008 Segregation and activation of myosin IIB creates a rear in migrating cells *J. Cell Biol.* **183** 543–54

Voisin M B, Woodfin A and Nourshargh S 2009 Monocytes and neutrophils exhibit both distinct and common mechanisms in penetrating the vascular basement membrane *in vivo* *Arterioscler. Thromb. Vasc. Biol.* **29** 1193–9

Weigelin B, Bakker G J and Friedl P 2012 Intravital third harmonic generation microscopy of collective melanoma cell invasion: principles of interface guidance and microvesicle dynamics *Intra Vital* **1** 32–43

Wenstrup R J, Florer J B, Brunskill E W, Bell S M, Inna Chervoneva I, David E and Birk D E 2004 Type V collagen controls the initiation of collagen fibril assembly *J. Biol. Chem.* **279** 53331–7

Wess T J 2005 Collagen fibril form and function *Adv. Protein Chem.* **70** 341–74 PMID: 15837520

Williams B R, Gelman R A, Poppke D C, Karl A and Piez K A 1978 Collagen fibril formation. Optimal in vitro conditions and preliminary kinetic results *J. Biol. Chem.* **253**(18) 6578–85 PMID: 28330

Wipff P J, Rifkin D B, Meister J J and Hinz B 2007 Myofibroblast contraction activates latent TGF-beta1 from the extracellular matrix *J. Cell Biol.* **179** 1311–23

Wise S G and Weiss A S 2009 Tropoelastin *Int. J. Biochem. Cell Biol.* **41** 494–7

Wolf K and Friedl P 2011 Extracellular matrix determinants of proteolytic and non-proteolytic cell migration *Trends Cell Biol.* **21** 736–44

Wolf K, Alexander S, Schacht V, Coussens L M, von Andrian U H, van Rheenen J, Deryugina E and Friedl P 2009 Collagen-based cell migration models *in vitro* and *in vivo* *Sem. Cell Deve. Biol.* **20** 931–41

Wolf K, Mazo I and Friedl P 2003 Compensation mechanism in tumor cell migration: mesen-chymal-amoeboid transition after blocking of pericellular proteolysis *J. Cell Biol.* **160** 267–77

Wolf K, Mazo I, Leung H, Engelke K, von Andrian U H, Deryugina E I, Strongin A Y, Bröcker E B and Friedl P 2003a Compensation mechanism in tumor cell migration: mesenchymal-amoeboid transition after blocking of pericellular proteolysis *J. Cell Biol.* **160** 267–77

Wolf K, Müller R, Borgmann S, Bröcker E B and Friedl P 2003b Amoeboid shape change and contact guidance: T-lymphocyte crawling through fibrillar collagen is independent of matrix remodeling by MMPs and other proteases *Blood* **102** 3262–9

Wolf K, Te Lindert M, Krause M, Alexander S, Te Riet J, Willis L, Hoffman M, Figdor G, Weiss J and Friedl P 2013 Physical limits of cell migration: control by ECM space and nuclear deformation and tuning by proteolysis and traction force *J. Cell Biol.* **201** 1069–84

Wolf K, Wu Y I, Liu Y, Geiger J, Tam E, Overall C, Stack M S and Friedl P 2007 Multi-step pericellular proteolysis controls the transition from individual to collective cancer cell invasion *Nat. Cell Biol.* **9** 893–904

Xian X, Gopal S and Couchman J R 2010 Syndecans as receptors and organizers of the extracellular matrix *Cell Tissue Res.* **339** 31–46

Xu J, Lamouille S and Derynck R 2009 TGF-beta-induced epithelial to mesenchymal transition *Cell Res.* **19** 156–72

Yamauchi K, Yang M, Jiang P, Xu M, Yamamoto N, Tsuchiya H, Tomita K, Moossa A R, Bouvet M and Hoffman R M 2006 Development of real-time subcellular dynamic multicolor imaging of cancer-cell trafficking in live mice with a variable-magnification whole-mouse imaging system *Cancer Res.* **66** 4208–14

Yamauchi K *et al* 2005 Real-time *in vivo* dual-color imaging of intracapillary cancer cell and nucleus deformation and migration *Cancer Res.* **65** 4246–52

Yang Y L and Kaufman L J 2009 Rheology and confocal reflectance microscopy as probes of mechanical properties and structure during collagen and collagen/hyaluronan self-assembly *Biophys. J.* **96** 1566–85

Yang Y L, Motte S and Kaufman L J 2010 Pore size variable type I collagen gels and their interaction with glioma cells *Biomaterials* **31** 5678–88

Zaman M H, Matsudaira P and Lauffenburger D A 2007 Understanding effects of matrix protease and matrix organization on directional persistence and translational speed in three-dimensional cell migration *Ann. Biomed. Eng.* **35** 91–100

Zaman M H, Trapani L M, Sieminski A L, Mackellar D, Gong H, Kamm R D, Wells A, Lauffenburger D A and Matsudaira P 2006 Migration of tumor cells in 3D matrices is governed by matrix stiffness along with cell–matrix adhesion and proteolysis *Proc. Natl Acad. Sci. USA* **103** 10889–94

Zhang X, Reagan M R and Kaplan D L 2009 Electrospun silk biomaterial scaffolds for regenerative medicine *Adv. Drug Deliv. Rev.* **61** 988–1006

Zisch A H, Lutolf M P and Hubbell J A 2003 Biopolymeric delivery matrices for angiogenic growth factors *Cardiovasc. Pathol.* **12** 295–310

IOP Publishing

Physics of Cancer

Claudia Tanja Mierke

Chapter 9

The impact of cells and substances within the extracellular matrix tissue on mechanical properties and cell invasion

Summary

In this chapter the role of extracellular matrix embedded cells such as tumor-associated fibroblasts in the matrix's mechanical properties and cancer cell motility in 3D microenvironments is discussed. The impact of substances such as hyaluronan (which is associated with the extracellular matrix network) is also discussed, including its interaction with the surface receptors of cancer cells (such as CD44) and its role as a key regulator of cancer cell migration and involvement in the malignant progression of cancer. The impact that the mechanical properties of the extracellular matrix have and how these properties contribute to the invasiveness of cancer cells and consequently cancer metastasis is described.

9.1 The impact of tumor-associated fibroblasts on matrix mechanical properties

Stromal fibroblasts that locally surround breast carcinomas often express the cell surface proteoglycan syndecan-1 (Sdc1). In human breast carcinoma samples, the stromal Sdc1 expression correlates with an organized and parallel extracellular matrix fiber architecture. In order to reveal a possible link between stromal Sdc1 and the fiber architecture of the extracellular matrix, bioactive cell-free 3D extracellular matrices were prepared from cultures of Sdc1-positive and Sdc1-negative murine and human mammary fibroblasts (called extracellular matrix-Sdc1 and extracellular matrix-mock, respectively). Indeed, extracellular matrix-Sdc1 led to a parallel fiber architecture, whereas the extracellular matrix-mock provided a random fiber arrangement. When breast carcinoma cells were embedded into the fibroblast-free extracellular matrices, extracellular matrix-Sdc1, but not extracellular matrix-mock, supported their adhesion, invasion and directional migration through the matrix. Moreover, the contribution of the structural or compositional alterations in extracellular matrix-Sdc1 was investigated in respect of cancer cell behavior. By microcontact printing of cell culture surfaces, the Sdc1-negative fibroblasts were forced to produce an extracellular matrix with parallel fiber organization, mimicking the architecture observed in extracellular matrix-Sdc1. However, it was seen that the fiber topography governs the directionality of cancer cell migration. By contrast, an elevated fibronectin level in extracellular matrix-Sdc1 was responsible for the enhanced adhesiveness of the breast cancer cells. These findings suggest that Sdc1 expression in breast carcinoma associated stromal fibroblasts supports the assembly of an architecturally abnormal extracellular matrix, which induces the breast carcinoma cell's directional migration and invasion.

Epithelial–stromal interactions seem to be crucial in directing mammary gland development and in maintaining normal tissue homeostasis. By contrast, during tumorigenesis, the stroma dramatically increases carcinoma growth and progression. In more detail, the predominant cell type within the stromal microenvironment is the fibroblast, which can synthesize, organize and maintain a 3D extracellular matrix network of glycoproteins and proteoglycans. It has been suggested that normal healthy stromal fibroblasts and their extracellular matrix provide an inhibitory constraint on tumor growth and progression (Bauer 1996, Kuperwasser et al 2004). In more detail, major alterations occur in the stromal fibroblasts and extracellular matrix during neoplastic transformation, indicating a permissive and supportive microenvironment for the development of carcinomas. By comparison with their quiescent normal fibroblast counterpart, carcinoma-associated fibroblasts exhibit an activated phenotype, which is characterized by the expression of smooth muscle markers, an enhanced proliferative and migratory capacity and altered gene expression profiles. Carcinoma-associated fibroblasts produce and deposit elevated quantities and abnormal varieties of extracellular matrix components (Barsky et al 1984, Schor et al 2003, Tuxhorn et al 2002). Recent evidence (Provenzano et al 2006, 2008) indicates that the extracellular matrix composition and architecture are both altered in close proximity to carcinomas and that these alterations may lead to tumor progression. However, the contribution of these stromal modifications to the tumor development, the molecular mechanisms and the signal transduction events underlying these alterations is not yet well understood.

Syndecans (Sdcs) belong to a family of transmembrane heparan sulfate proteoglycans with four identified members, for example Sdc1-4. Through their heparan sulfate glycosaminoglycan (HS-GAG) chains, Sdcs can interact with a wide variety of proteins, such as growth factors and extracellular matrix constituents (Lopes et al 2006, Tkachenko et al 2005, Zimmermann and David 1999). Thus, they have a role in cell growth, adhesion, migration and morphogenesis. In more detail, it has been suggested that Sdc2 is required for the assembly of laminin and fibronectin into a fibrillar matrix (Klaas et al 2000). Moreover, syndecan-4 has also been shown to participate in fibronectin matrix assembly. In line with this, concomitant engagement of Sdc4 and integrins triggers Rho GTPase and focal adhesion kinase (FAK) activity, which is crucial for efficient initiation of fibronectin matrix assembly (Saoncella *et al* 1999, Wilcox-Adelman *et al* 2002, Ilic *et al* 2004, Wierbicka-Patynowski *et al* 2002). However, Sdc1 is expressed primarily by the epithelial and plasma cells of healthy adult tissue (Sanderson *et al* 1992). An induction of Sdc1 expression in stromal fibroblasts of invasive breast carcinomas has been observed (Maeda *et al* 2004, Stanley *et al* 1999). Sdc1, which is aberrantly expressed by stromal fibroblasts in breast carcinomas, participates in a reciprocal carcinoma growth, which means that it promotes the feedback loop, requiring proteolytic shedding of its ectodomain (Maeda *et al* 2004, 2006, Su *et al* 2007). Although the role of Sdc1 in the assembly of the extracellular matrix has not yet been revealed, Sdc1 has been shown to interact with several extracellular matrix components, such as fibronectin, fibrillar collagens, laminin, vitronectin, thrombospondin and tenascin (Lopes *et al* 2006, Tkachenko *et al* 2005, Zimmermann and David 1999).

Thus, the role of Sdc1 expression by stromal fibroblasts has been explored and it has been suggested that Sdc1 may be functionally involved in the altered matrix production present around tumors, the so-called tumor stroma. It has been observed that in mammary stromal fibroblasts, Sdc1 facilitates extracellular matrix assembly and thus determines extracellular matrix fiber architecture (Yang *et al* 2011). Moreover, it has been shown that cell-free 3D extracellular matrices produced by Sdc1-expressing fibroblasts provide the directional migration of mammary carcinoma cells and link this activity to the parallel fiber architecture (Yang *et al* 2011).

It has been observed that Sdc1 is aberrantly expressed by stromal fibroblasts in most infiltrating breast carcinomas (Maeda *et al* 2004, Stanley *et al* 1999). It has been shown that Sdc1 expression in stromal fibroblasts induces and promotes breast carcinoma growth and angiogenesis (Maeda *et al* 2004, 2006, Su *et al* 2007). How stromal Sdc1 alters the extracellular matrix composition and architecture *in vivo* and *in vitro*, with the altered extracellular matrix fiber architecture subsequently supporting the directional migration of breast carcinoma cells, is discussed.

The extracellular matrix provides a complex macromolecular network of glycoproteins and proteoglycans that is necessary for cell survival, proliferation, migration and differentiation. During cancer cell invasion, the extracellular matrix may be subject to extensive alterations due to the abnormal synthesis of extracellular matrix components and their proteolytic remodeling (Schor *et al* 2003, Wilhelm *et al* 1988). Extensive accumulation of the extracellular matrix protein fibronectin has been detected in the stroma of a variety of solid human tumors (Wilhelm *et al* 1988, Moro *et al* 1992) and a linear function between the fibronectin content and tumor stage or adverse outcome has been demonstrated (Yang *et al* 2011). Indeed, these results support the hypothesis that the fibronectin production in stromal fibroblasts seems to be regulated by Sdc1. Fibronectin facilitates the attachment of breast carcinoma cells to the extracellular matrix, while it does not stimulate their migratory behavior. However, this observation in 3D matrices is in contrast to the biphasic relationship between migration velocity and adhesion molecule (such as fibronectin and laminin concentration observed under traditional 2D conditions) and may thus reflect the importance of the topographic presentation of adhesion ligands to the cell in the 3D microenvironment (DiMilla *et al* 1993, Goodman *et al* 1989, Palecek *et al* 1997).

By contrast to the intersecting meshwork of the extracellular matrix produced by Sdc1-negative fibroblasts, fibronectin and collagen I fibers in extracellular matrix-Sdc1 are organized in parallel patterns. Indeed, it has been observed that a parallel fiber arrangement is a characteristic feature of an extracellular matrix evoked by primary carcinoma-associated fibroblasts of the skin (Amatangelo *et al* 2005). However, the consequences of this parallel fiber arrangement for cancer cell behavior are elusive and require further investigation. In particular, the extracellular matrix architecture has been characterized as a functional determinant of the directional cell migration. The parallel fiber architecture produced by Sdc1-positive fibroblasts tends to reflect on the collagen fiber signature identified in the transgenic Wnt-1 mouse mammary tumor model (Provenzano *et al* 2006). In this study, the parallel collagen fibers perpendicular to the advancing edge of the tumors were described, and these were spatially associated with carcinoma cells invading singly or collectively into the extracellular matrix of connective tissue. Because of the inherent limitations of the model system, it is not possible to judge whether the invasion is a consequence or the cause of the parallel fiber arrangement.

Thus, the molecular mechanism through which stromal Sdc1 affects extracellular matrix assembly needs to be elucidated. An intact fibronectin matrix seems to be essential for the assembly and stability of a mature collagen-containing extracellular matrix (Sottile *et al* 2002, Velling *et al* 2002). Thus, knowledge of the regulation of the fibronectin fibril assembly may lead to the precise understanding of the extracellular matrix organization. Fibronectin fibrillogenesis is a complex cell-facilitated process that engages the fibronectin binding to cell surface receptors, fibronectin–fibronectin self-association and its interaction with the actin cytoskeleton (Wierzbicka-Patynowski and Schwarzbauer 2003, Mao *et al* 2005). Fibronectin is secreted as tightly folded, disulfide-bonded dimers consisting of three types of repeating modules, such as types I, II and III. In particular, these fibronectin dimers are initially inactive until they interact with their specific integrins and other receptors displayed at the cell surface. In more detail, this binding interaction induces intracellular signal transduction pathways, supports rearrangements of the actin cytoskeleton and may cause conformational alterations in fibronectin that change the inactive fibronectin molecule into an extended and active form. Stromal-derived Sdc1 seems to regulate the serial steps of the fibronectin fibrillogenesis, such as fibronectin fibril initiation and elongation. In more detail, the Sdc1 regulates the activity of several integrins, including $\alpha v \beta 3$, $\alpha v \beta 5$ and $\beta 4$ (Beauvais *et al* 2004, 2009, McQuade *et al* 2006, Ogawa *et al* 2007). Although the fibronectin matrix assembly appears to be induced mainly by Arg–Gly–Asp-binding (RGD-peptide-binding), integrin $\alpha 5 \beta 1$ (Fogerty *et al* 1990, Wu *et al* 1993) and alternative fibronectin-binding integrin receptors, such as $\alpha v \beta 3$, $\alpha 4 \beta 1$ and $\alpha v \beta 1$, can also support this process when properly activated (Wu *et al* 1995, Yang and Hynes 1996, Wu *et al* 1996, Wennerberg *et al* 1996, Sechler *et al* 2000). However, it is possible that Sdc1 induction activates integrins other than α5β1 and hence initiates an alternative assembly pathway, thus resulting in an extracellular matrix that is structurally and compositionally different from an extracellular matrix assembled under the stringent control of α5β1. Moreover, it seems to be possible that stromal Sdc1 directly facilitates the fibronectin matrix assembly. The compact conformation of fibronectin dimers is provided by intramolecular interactions involving the type III12–14 repeats of fibronectin (Johnson *et al* 1999, Hynes 1999). In particular, the type III12–14 repeats (termed heparin II binding domain) are able to interact with the HS chains from various members of the Sdc family (Tumova *et al* 2000). In turn, the binding of Sdc1 to heparin II can facilitate the unfolding of dimeric fibronectins and hence expose fibronectin self-assembly (fibronectin-binding) sites, consequently promoting fibronectin deposition and fibrillogenesis.

There is strong evidence supporting the importance of the mechanical properties of the extracellular matrix in the behavior of breast cancer cells. It has been shown that dense rigid mechanical properties suppress tubulogenesis and possibly stimulate invasion of well-differentiated breast carcinoma cells in collagen gels by inducing the activity of the small GTPase Rho (Wozniak *et al* 2003, Paszek *et al* 2005). In addition, the cell-derived extracellular matrices seem to be more suited than the basement membrane or collagen gels to mimic 3D matrix effects on the behavior of breast cancer cells (Green and Yamada 2007). Thus, a model is favored in which the fiber topography rather than the matrix rigidity governs the cell invasion in fibroblast-derived matrices. A dissection of the molecular mechanism of cell migration regulation by parallel extracellular matrix fibers and their orientation may involve Rac1, which is a member of the Rho family of GTPases. Indeed, the inhibition of Rac1 leads to a switch in the migration mode of fibroblasts and epithelial cells from a random to a more directionally persistent migration mode (Pankov *et al* 2005). The induction of cell migration is needed for cancer cell invasion and metastasis (Vicente-Manzanares *et al* 2005, Ridley *et al* 2003, Zijlstra *et al* 2008) and thus it seems plausible that extracellular matrices with parallel fiber organization facilitate cancer cell spread and invasiveness. Taken together, a novel pathway has been discovered for how aberrant expression of Sdc1 in stromal fibroblasts is able to increase cancer progression.

The components that comprise the extracellular matrix are integral to normal tissue homeostasis and the development and progression of breast tumors. In particular, the secretion, construction and remodeling of the extracellular matrix are regulated by a complex interplay between at least three different cell types: cancer cells, fibroblasts and macrophages. Transforming growth factor-β (TGF-β) is an essential molecule in facilitating the cellular production of extracellular matrix molecules and moreover, providing the adhesive interactions of cells with the extracellular matrix. In more detail, hypoxic cell signals (caused by oxygen deprivation), the presence of additional metabolic factors and receptor activation are associated with extracellular matrix structural architecture and consequently the progression of breast cancer. However, it has been suggested that both TGF-β and hypoxic cell signals play a key role in the functional and morphological alterations of cancer-associated fibroblasts and tumor-associated macrophages. In line with this, the increased recruitment of tumor and stromal cells in response to hypoxia-induced chemokines leads to enhanced deposition and remodeling processes for the extracellular matrix, elevated formation of new blood vessels through the induction of neoangiogenesis within the endothelial cell lining, and increased migration of cancer cells. Thus, greater knowledge of the collaborative interactive networks between cancer and stromal cells in response to the combined signals of TGF-β and hypoxia may reveal insights into the treatment parameters for targeting both cancer and stromal cells.

The extracellular matrix consists of approximately 300 proteins that facilitate organogenesis, tissue homeostasis and the progression of inflammation and disease (Hynes and Naba 2012, Tlsty and Coussens 2006). In human cancers, tumor initiation, proliferation, migration and metastasis are associated with the composition of the matrix (Egeblad *et al* 2010, Tlsty and Coussens 2006). However, increased production and the deposition of extracellular matrix proteins within the extracellular microenvironment are identified risk factors in human breast cancers (Keely 2011). This dysregulated activity is caused by deviant tumor and/or stromal cell function and finally leads to the formation of an altered matrix displaying fibrotic, stiff and dense properties (Levental *et al* 2009).

The extracellular matrix in normal mammary development and breast cancer tumorigenesis is highly regulated by cytokine TGF-β (Moses and Barcellos-Hoff 2011, Silberstein *et al* 1992). In particular, TGF-β also facilitates mammary morphogenesis by inhibiting mammary lateral branching and ductal growth (Lanigan *et al* 2007) and increases tumor growth via dysregulated cell signal pathways inhibiting then cell cycle arrest (Donovan and Slingerland 2000). However, the aberrant functions of TGF-β during breast cancer development seems to be linked to the disease progression into various metastatic sites, which additionally require TGF-β during their formation, such as the brain (Dobolyi *et al* 2012), liver (Karkampouna *et al* 2012), lung (Bartram and Speer 2004) and skeletal bone (Janssens *et al* 2005). However, site-specific organotropism is a complex process involving not only TGF-β, but also the tumor cell genotype and tumor–stroma interactions at the primary site, as well as at the targeted organ (Eckhardt *et al* 2012; Ganapathy *et al* 2012, Lu and Kang 2007). The various phenotypes of each metastatic site in combination with TGF-β have been investigated (Drabsch and ten Dijke 2011, Eckhardt *et al* 2012, Lu and Kang 2007, Nishizuka *et al* 2002).

In particular, TGF-β ligands are members of the TGF-β superfamily, consisting of more than 25 closely related proteins, such as growth differentiation factors (GDFs), bone morphogenetic proteins (BMPs), activins and inhibins (Kingsley 1994). The three isoforms of TGF-β—TGF-β1, TGF-β2 and TGF-β3—have been revealed in humans, and each of these molecules as well as their associated receptors have been characterized in human breast tissue cancer and stromal cells (Chakravarthy *et al* 1999). More precisely, TGF-β ligands are secreted from cells as an inactive homodimer bound non-covalently to a latency associated peptide (LAP) that is mediated by a disulfide bound to the latent TGF-β binding protein (LTBP) (Horiguchi *et al* 2012). After the release of this large latent complex, the tissue transglutaminase-2 (TG2) enzymatically cross-links fibrillar proteins in the extracellular matrix (such as fibrillin and fibronectin) to LTBP and the associated LAP. The TGF-β complex then becomes bound to the extracellular matrix in a still-inactive form (Nurminskaya and Belkin 2012, Zilberberg et al 2012). The active form of TGF-β can be released through integrin-facilitated mechanical deformation of the extracellular matrix and/or through the degradation of LAP using cellular proteases such as MMPs, thrombospondin-1 and plasmin. However, the release of active TGF-β in turn induces adhesive interactions via the cell signals that regulate integrin expression and the production of additional extracellular matrix proteins and TGF-β molecules (Chandramouli et al 2011, Horiguchi et al 2012).

Cell surface integrins are trans-membrane receptors that facilitate cell-to-extracellular matrix interactions in tissue homeostasis, disease formation and immunity (Luo et al 2007). Each integrin receptor consists of a heterodimer, composed of one α and one β subunit and the extracellular and intracellular microenvironments are connected by binding to special ligands outside the cell and cytoskeletal components underneath the integrins assembled in focal adhesions, which facilitate this link (such as vinculin and focal adhesion kinase) (Mierke et al 2008, 2010, Mierke 2013, Berman et al 2003, Luo et al 2007). Regarding the 24 known integrin heterodimers, it has been observed that the extracellular matrix ligands predominantly bind to the α subunit and thus activate intracellular signaling events via the β subunit (Hehlgans et al 2007). Indeed, subsequent conformational alterations and integrin clustering, which may even include cell surface TG2-to-integrin binding interactions, build a 3D matrix adhesion signaling complex that seems to be involved in tumor proliferation and migration, as well as matrix deposition and remodeling (Keely 2011, Nurminskaya and Belkin 2012, Provenzano et al 2009, Wozniak et al 2003). The expression of the integrin subunits (such as α5, αv, β1, β3 and β5) that bind the extracellular matrix is also enhanced by TGF-β cell signals, and ligation of these integrins by their specific ligands (α2β1:collagen, α5β1: fibronectin, αvβ3 or αvβ5:periostin) in turn induces the production of TGF-β, leading to a feed-forward loop between cancer cells and the extracellular matrix (Bianchi-Smiraglia et al 2012, Garamszegi et al 2009, Kudo 2011, Margadant and Sonnenberg 2010, Soikkeli et al 2010).

The extracellular matrix in human primary breast cancers (such as ductal and non-ductal) includes higher protein levels of collagen I, III and IV, fibronectin, periostin, tenascin-C and vitronectin compared to normal healthy breast tissue (Aaboe et al 2003, al Adnani et al 1987, Gould et al 1990, Guttery et al 2010, Kadowaki et al 2011, Kharaishvili et al 2011, Vasaturo et al 2005, Zhang et al 2010). Indeed, these proteins are all regulated by TGF-β (Garamszegi et al 2009, Grande et al 1997, Guttery et al 2010, Koli et al 1991, Kudo 2011, Margadant and Sonnenberg 2010, Soikkeli et al 2010) and are produced by various cell types, such as breast cancer cells (collagen I and IV, fibronectin, periostin and tenascin-C), fibroblasts (collagen I and III, fibronectin, periostin and tenascin-C), endothelial cells (collagen IV), macrophages (fibronectin and tenascin-C) and hepatocytes (vitronectin) (Arancibia et al 2013, Goh et al 2010, Guttery et al 2010, Hielscher et al 2012, Kleinman et al 1981, Philippeaux et al 2009, Preissner 1991, Shao et al 2004,

Taylor-Papadimitriou *et al* 1981). The adhesive interactions among these extracellular matrix proteins favor their co-localization (Kudo 2011).

The matrix is further altered by enzymatic cross-linking of collagen through molecules commonly expressed in breast cancer, such as TG2 and lysyl oxidase (LOX), which finally generates a stiff matrix (Barker *et al* 2012, Jiang *et al* 2003, Levental *et al* 2009, Nurminskaya and Belkin 2012, Taylor *et al* 2011). These molecules are additionally regulated by TGF-β, as well as hypoxic cell signals (Barker *et al* 2012, Nurminskaya and Belkin 2012). Interestingly, in a murine model of breast cancer metastases to the bone, small molecule inhibition of hypoxia (2-methoxyestradiol) or TGF-β (SD-208) reduced osteolytic lesions and increased the survival of mice compared to their healthy controls. The combined inhibition of both factors induced a synergistic response that was potentially regulated by cancer cell vascular endothelial growth factor (VEGF) production and the CXCR4 chemokine receptor expression, as these two end-points synergistically decreased in response to both inhibitors *in vitro* (Dunn *et al* 2009). Taken together, TG2, LOX and additional TGF-β cell signal end-points associated with the extracellular matrix and tumorigenesis may also be synergistically increased by hypoxia.

The reduction of oxygen/perfusion within the local tumor microenvironment produces hypoxia and consequently activates the hypoxia inducible factors (HIFs) (Porporato *et al* 2011). It has been suggested that these factors are also induced by receptor-facilitated cell signals, such as insulin, growth factors as well as cytokines, increased free radical production and cellular alterations in iron and/or metabolic homeostasis (Cascio *et al* 2008, Knowles and Harris 2001, Lopez-Lazaro 2009, Schulze and Downward 2011, Selak *et al* 2005, Spangenberg *et al* 2006, Thornton *et al* 2000). However, increased activation of HIF genes redirects the cellular metabolism away from oxidative phosphorylation to aerobic glycolysis and finally leads to the production of lactate (Porporato *et al* 2011). These metabolic alterations may help to fully understand the increased breast cancer risks evoked by premenopausal iron deficiency, postmenopausal obesity, hyper-insulinemia and even iron overload (Braun *et al* 2011, Jian *et al* 2011, Rose and Vona-Davis 2012).

HIFs are activated in metabolic disorders such as obesity and it has been revealed that obesity is a poor prognostic indicator in patients diagnosed with breast cancer (Braun *et al* 2011, von Drygalski *et al* 2011). As a characteristic feature of various solid tumors such as breast cancer, the expression of HIFs strongly increases metastasis, resists radiation/chemotherapeutic therapy and offers poor patient prognosis (Charpin *et al* 2012, Fokas *et al* 2012). The tumorigenic potential of HIFs involves the cancer and stromal cell production of hypoxia-induced growth factors (Krock *et al* 2011). The elevated production of growth factors and their associated receptor-facilitated cell signals alter cancer and stromal cell affinity and avidity for the extracellular matrix and provide immune tolerance, induce angio-genesis and finally support metastatic disease (Chouaib *et al* 2012, Hood and Cheresh 2002). Hence, tumor progression is a collaborative effort between cancer and stromal cells, such as fibroblasts and macrophages, which are located within a hypoxic microenvironment.

The effect of hypoxia on breast cancer cells

The hypoxia marker HIF-1α has been detected and characterized in primary human ductal carcinomas and elevated levels of HIF-1α correlate significantly with an unfavorable outcome for patients (Brito *et al* 2011, Charpin *et al* 2012). The additional identification of HIF-1α in circulating cancer cells isolated from the peripheral blood of metastatic breast cancer patients (Kallergi *et al* 2009) further supports the idea of an association between hypoxic marker concentration and cancer cell migration. Nonetheless, a migratory mechanism can involve the action of the fibronectin-associated integrin α5β1. Indeed, the increased expression of this dimer pair has been observed to occur in response to oxygen deprivation and human epidermal growth factor receptor-2 (HER-2)-induced HIF activation (Spangenberg *et al* 2006). Increased expression of α5β1 integrin compared to other integrins could thus fundamentally alter cancer cell-to-extracellular matrix interactions and finally alter the associated cancer cell migratory patterns (Mierke *et al* 2011, Mierke 2013).

In addition, the production of TGF-β and fibronectin is also elevated by interleukin-19 (IL-19), which is a cytokine that can be induced by hypoxia in the 4T1 murine mammary tumor cell line (Hsing *et al* 2012). Moreover, similarly to HIF-1α, the expression of IL-19 has been identified as a poor prognostic indicator in invasive ductal carcinoma patients (Hsing *et al* 2012). There is indeed evidence that hypoxia and TGF-β independently regulate α5β1 expression (Bianchi-Smiraglia *et al* 2012, Margadant and Sonnenberg 2010, Spangenberg *et al* 2006) and this suggests that the HIF transcriptional factors may cooperate together synergistically with TGF-β (or other associated mediators, such as IL-19, HER-2 ligands and estrogen), which was also reported for the chemokine receptor, CXCR4 (Dunn *et al* 2009).

Although the activation of HER-2 has been shown to elevate the expression of α5β1 (Spangenberg *et al* 2006), similar findings were not seen for the expression of the α2 component of the collagen/laminin receptor, α2β1 (Ye *et al* 1996). Indeed, the α2 integrin has been identified as a tumor metastasis suppressor in a murine model of breast cancer (Ramirez *et al* 2011). *In vitro* examination of human breast cancer cell lines indicates that the expression of α2 seems to be dependent on estrogen and progesterone cell signals (Lanzafame *et al* 1996), suggesting that the phenotype may be common to HER-2+, estrogen receptor (ER)- and progesterone receptor (PR)-cancers. Moreover, hypoxia-related breast cancer studies regarding α2 and other integrins of interest (αv, β3 and β5) have not changed significantly at the present time. However, it has been revealed that in human mesenchymal stem cells hypoxia induces the expression of various integrins (α1, α3, α5, α6, α11, αv, β1 and β3), but notably not the collagen/laminin integrin, α2 (Saller *et al* 2012). These data suggest that hypoxia may differentially direct migration, depending on the individual genotype and possibly via substrates other than collagen. Indeed, this is supported by a study that examined human breast cancer cells (triple-negative (HER-2-, ER-, PR-) MDA-MB-231) cultured in the mammary fat pad of severe combined immunodeficient mice, in which identified hypoxic regions displayed significantly fewer and less dense collagen type I fibers (Kakkad *et al* 2010).

By contrast, other studies noted increased collagen surrounding mammary tumors overall (Iacobuzio-Donahue *et al* 2002). Alterations to the tumor-associated matrix, including increased alignment of collagen fibers and matrix stiffness identified in murine models and human breast cancer tissue (Conklin *et al* 2011, Provenzano *et al* 2006), may be due to the hypoxic induction of MMPs, TG2 and LOX by cancer cells (Barker *et al* 2012, Choi *et al* 2011, Munoz-Najar *et al* 2006, Nurminskaya and Belkin 2012). Additionally, it should be noted that these studies differ from that of Kakkad and colleagues in their assessment of collagen deposition in immunocompetent animals (Kakkad *et al* 2010). Thus, the effect of hypoxia on collagen deposition seems to be facilitated by immune cells.

As breast cancer cell triggered production of MMP-9 is additionally regulated by fibronectin adhesion and this interaction is enhanced by LOX cell signals (Maity *et al* 2011, Zhao *et al* 2009), hypoxia supports forward processes initiated by oncogenes. Additional cancer cell facilitated production of chemokines such as endothelin-2 (ET2), chemokine C-C motif ligand 5 (CCL5) (Grimshaw *et al* 2002a, Lin *et al* 2012) and growth factors such as vascular endothelial growth factor (VEGF), basic fibroblast growth factor (bFGF) and connective tissue growth factor (CTGF) (Dunn *et al* 2009, Kondo *et al* 2002, Le and Corry 1999) in response to hypoxia alter the recruitment and activity of stromal cells that subsequently affect tumorigenesis.

The effect of hypoxia on fibroblasts
Fibroblasts are a pervasive and diverse population of cells that produce extracellular matrix proteins and maintain the extracellular matrix in normal tissue homeostasis, playing an active role in the wound-healing response and tumorigenesis (Sorrell and Caplan 2009). Comparison of fibroblasts isolated from normal tissue with human breast cancer tissue revealed that the cancer-associated fibroblasts (CAFs) contained a subpopulation of fibroblasts and myofibroblasts which could react to increase tumor cell growth, encourage tumor vascularization, exhibit increased collagen contractility and release higher levels of the chemokine stromal cell-derived factor-1 (SDF-1) compared to normal fibroblasts (NFs) (Orimo and Weinberg 2006, Shimoda *et al* 2010). Autocrine TGF-β and SDF-1, which are both cytokines found to increase in close proximity to tumors, evoked cell signals that support the differentiation of primary human breast NFs to CAFs (Kojima *et al* 2010). In human synovial fibroblasts, hypoxic conditions can facilitate the release of SDF-1 and the cytokine interleukin-1β (IL-1β), which activates both the SDF-1 and HIF-1α genes in these same cells (Hitchon *et al* 2002, Thornton *et al* 2000). Indeed, the expression of IL- 1β has been detected in human breast cancer tissue (Jin *et al* 1997, Kurtzman *et al* 1999) and linked to production by either cancer cells or macrophages located in the tumor microenvironment (Jin *et al* 1997). Thus, cytokines such as TGF-β and IL-1β that are secreted by cancer cells and cancer-associated macro-phages may encourage NF to CAF differentiation, which is then further enhanced by hypoxia.

In a murine xenograft model involving human breast cancer cells (MDA-MB-231) and immortalized fibroblast (hTERT-BJ1) cell lines, the ectopic expression of HIF-1α in fibroblasts dramatically enhances the tumor growth through a mechanism that may support the transport of fibroblast metabolites such as lactate and pyruvate to the adjacent cancer cells (Chiavarina *et al* 2010). Similar results have been revealed in comparable studies involving MDA-MB-231 and the CL4 human foreskin mutant fibroblast cell line, which has previously been characterized as favoring aerobic glycolysis compared to the CL3 variant that favors the oxidative metabolism (Migneco *et al* 2010). However, utilizing the same murine model, the constitutive activation of HIF-2α in fibroblasts did not lead to a shift toward aerobic glycolysis or increased tumor growth (Chiavarina *et al* 2012). Furthermore, in the transgenic mouse mammary tumor virus model of breast cancer involving the polyoma virus middle T transgene (MMTV-pyMT), targeted fibroblast deletion of HIF-1α or VEGF increases tumor growth, whereas HIF-2α does not alter it (Kim *et al* 2012). These differences may possibly be explained by the varying genotype and/or mutations within the cancer or fibroblast cells used in these research studies. In addition, there may be an interplay between hypoxic targets such as HIF-1α, HIF-2α and HIF-3α, differences between primary murine cells and human cell lines, and/or differences in the mediators or cell–cell contacts in the local tumor micro-environment. Thus, the presence of paradoxical and convergent findings suggests there are still aspects of the regulation of hypoxic responses that are not yet fully understood.

It has been reported that human dermal fibroblast production of collagen, fibronectin, MMP-1, MMP-2 and MMP-3 is induced in co-cultures involving the estrogen receptor positive human breast cancer cell MCF-7 cell line where cell-to-cell contact is required and dependent solely on the fibroblast population for maximal stimulation (Ito *et al* 1995, Noel *et al* 1992a, 1992b). An additional study demonstrated that suspension of human dermal fibroblasts and MCF-7 cells in Matrigel™ and subsequent implantation of the cell/matrix mix into nude mice induces tumor growth that can be inhibited by MMP inhibitors (TIMPs) (Noel *et al* 1998). Thus, this tumor growth mechanism seems to involve paracrine interactions between MMP-2 bound to fibroblasts and cancer cell surface membrane-type 1 MMP (MT1-MMP), which is able to activate MMP-2 (Saad *et al* 2002). Subsequently, MMP-2 and MT1-MMP have also been implicated in the proteolytic degradation of the matrix cross-linker TG2, which additionally facilitates crosslinks between integrins (Nurminskaya and Belkin 2012). Thus, the interactions between cancer cells and fibroblasts may affect the expression of TG2 at the cell surface or within the extracellular matrix and facilitate cell adhesion and migration. Moreover, these interactions may then be further supported by hypoxia, which increases the expression of MT-MMP-1 in MDA-MB-231 breast cancer cells and in addition enhances the production of an additional factor, CTGF (Kondo *et al* 2002).

The mediator CTGF is produced by both cancer cells and fibroblasts and is thus implicated in fibrosis, metastatic disease and chemotherapeutic resistance (Chien *et al* 2011, Shi-Wen *et al* 2008, Wang *et al* 2009). Fibroblast production of CTGF is found to begin in response to TGF-β, MMP-2 and collagen ligation (Grotendorst *et al* 2004, Tall *et al* 2010). Fibroblast treatment with CTGF may then facilitate fibroblast-to-myofibroblast trans-differentiation, LOX activity and collagen deposition (Droppelmann *et al* 2009, Grotendorst *et al* 2004, Hong *et al* 1999). Interestingly, in a xenograft model overexpression of CTGF in hTERT-BJ1 immortalized fibroblasts supported the growth of co-injected MDA-MB-231 cells through a mechanism that included the induction of autophagy and HIF-1α dependent metabolic alterations, independent of the matrix deposition (Capparelli *et al* 2012). Thus, the deposition of a stiff matrix may be in response to TGF-β, CGTF and/or hypoxia stimulation of either cancer cells or CAFs that then start to produce collagen, LOX and TG2.

The effect of hypoxia on macrophages

Macrophages are sentinel cells in innate and adaptive immunity that fulfill additional roles in tissue homeostasis and the wound-healing response (Murray and Wynn 2011). In particular, macrophages are found during normal mammary gland development, the process of involution and breast cancer progression (Laoui *et al* 2011, Schwertfeger *et al* 2006). The localization of macrophages within the stromal tissue locally surrounding human breast tumors, but not directly within the growing primary tumors, serves as a prognostic indicator of a poor patient outcome (Medrek *et al* 2012). As a result, macrophages at the tumor invasive front are thought to increase tumorigenesis, angiogenesis and the promotion of cancer metastatic (Hao *et al* 2012). However, this may occur partly in a paracrine-fashion interaction, where cancer cells secrete colony-stimulating factor-1 (CSF-1) and sense epidermal growth factor (EGF), whereas macrophages secrete EGF and sense CSF-1 (Patsialou *et al* 2009). Additional chemotactic signals that facilitate the recruitment of macrophages are substances released from hypoxic endothelial cells (SDF-1), fibroblasts (SDF-1) and/or breast cancer cells (ET-2, CCL5) (Grimshaw *et al* 2002a, Grimshaw *et al* 2002b, Jin *et al* 2012, Lin *et al* 2012, Schmid *et al* 2011, Soria and Ben-Baruch 2008). There is some evidence to suggest that hypoxia inhibits macrophage recruitment and decreases CSF-1 production, which would explain why macrophages are located around the peripheral edge of tumors rather than directly within the primary tumor complex where hypoxia is most pronouncedly identified (Green *et al* 2009; Hockel and Vaupel 2001, Turner *et al* 1999).

The directed migration of macrophages toward the tumor is facilitated by the binding of the blood monocyte integrin α4β1 to the endothelium via the vascular cell adhesion molecule-1 (VCAM-1), diapedesis through the blood vessel wall, monocyte-to-macrophage differentiation within the extracellular matrix microenvironment and chemokine-facilitated migration into mammary tissue (Jin *et al* 2006, Schmid *et al* 2011, Stewart *et al* 2012). The increased deposition of type I collagen in human breast cancer tissue (Guo *et al* 2001, Ramaswamy *et al* 2003) suggests that macrophages may migrate through the α2β1 integrin, as has been demonstratd previously with mouse peritoneal macrophages (Philippeaux *et al* 2009). Collagen-binding interactions with murine macrophages also trigger the production of fibronectin, which indeed increases the cellular adhesion to collagen that is partially inhibited by the addition of competitive RGD peptides known to block integrin binding sites to fibronectin by binding to them (Philippeaux *et al* 2009). In more detail, macrophages bind to fibronectin via α4β1 and α5β1, and it has been proposed that α5β1 is the primary integrin involved in fibronectin-induced human or murine macrophage MMP-9 production (Xie *et al* 1998).

Increased production of MMP-9 and vascular endothelial growth factor (VEGF) has been detected in human breast cancers (Vinothini *et al* 2011), where the release of VEGF is triggered by the hypoxic-induced activation of both cancer cells and tumor-associated macrophages (Harmey *et al* 1998). Additional growth factors such as platelet derived growth factor (PDGF) and bFGF are produced at higher levels in breast cancer and are additionally produced and secreted by hypoxic human macrophages in order to induce and enhance endothelial cell migration and proliferation (Kuwabara *et al* 1995, Rykala *et al* 2011). Thus, breast cancer associated macrophages seem to increase matrix deposition and remodeling as well as supporting the formation/recruitment of new blood vessels (neoangiogenesis).

Moreover, macrophages regulate the innate and adaptive immune responses via various mechanisms, such as phagocytosis, free radical production and the professional presentation of antigen to T cell subsets (Laskin 2009). The cytokine TGF-β, which is produced by cancer cells and fibroblasts (Kojima *et al* 2010, Margadant and Sonnenberg 2010), down-regulates the free radical production in macrophages and reduces the expression of co-stimulatory molecules required for antigen presentation (Li *et al* 2006). These immuno-suppressive functions of TGF-β are supposed to be enhanced by breast tumor hypoxic induction of the anti-inflammatory cytokine IL-19 (Azuma *et al* 2011, Hoffman *et al* 2011, Hsing *et al* 2012). The identified expression of IL-19 in murine macrophages may also suggest that macrophages produce this cytokine in response to hypoxia, as has been identified for 4T1 cancer cells (Azuma *et al* 2011, Hsing *et al* 2012). Moreover, it has also been indicated that hypoxia increases phagocytosis and enhances antigen presentation in the RAW 264.7 cell line and murine peritoneal macrophages in a HIF-1α-dependent fashion (Acosta-Iborra *et al* 2009, Anand *et al* 2007). The expression of HIF-1α has been detected in the differentiation of murine myeloid-derived suppressor cells, such as macrophages (Corzo *et al* 2010). Conditional deletion of HIF-1α in these cells within MMTV-pyMT mice showed a loss of T cell suppression and hence decreased tumor growth that was dependent on a lack of inducible nitric oxide synthase (iNOS) and arginase 1 (Arg1) and independent of the VEGF production (Doedens *et al* 2010). However, it has been reported that TGF-β cell signals in murine macrophages antagonize the expression of iNOS (Sugiyama *et al* 2012), but induce the expression of Arg1 (Li *et al* 2012), suggesting a possible cell signal synergy between TGF-β and HIF-1α in regulating Arg1 expression.

Interestingly, macrophage TGF-β production is induced through a process called efferocytosis, which involves the phagocytosis of apoptotic cells (Fadok *et al* 1998, Korns *et al* 2011). The process of efferocytosis occurs via various mechanisms, such as macrophage integrin (αvβ3, αvβ5)-facilitated interactions with apoptotic cells (Korns *et al* 2011). Molecules that bind these integrins and confine them, such as blocking antibodies or the high mobility group box 1, have been shown to eliminate efferocytosis (Friggeri *et al* 2010, Stern *et al* 1996), suggesting that macrophage αvβ3 and αvβ5 interactions with known extracellular matrix ligands such as vitronectin, fibronectin and periostin (Kudo 2011, Nemeth *et al* 2007) may indeed impede efferocytosis and thereby foster chronic inflammation. In addition, the expression of TG2 on the cell surface of murine macrophages induces the phagocytosis of apoptotic cells by creating a bridge between the β3 integrin and milk fat globulin epidermal growth factor (EGF) factor 8 (MFG-E8), which contains the MGF-E8 RGD motif, which can bind to αvβ3 and the MGF-E8 factor VIII-homologous domain, which can bind to apoptotic cell phosphatidylserine (Toda *et al* 2012, Toth *et al* 2009). In particular, TG2 is produced in response to hypoxia and the bacterial cell wall component, lipopolysaccharide (LPS) (Ghanta *et al* 2011, Nurminskaya and Belkin 2012), indicating that a hypoxic microenvironment and/or infection may lead to macrophage phagocytosis or TG2 matrix deposition.

What is the future direction of cancer research?
The etiology of human breast cancer is still elusive but it may involve endogenous (hereditary genes, hormonal/reproductive alterations, breast density, obesity) and/or exogenous (carcinogens, radiation, hormone replacement therapy, alcohol consumption) risk factors, which may even act synergistically in promoting the formation of tumors (Mansfield 1993). Genetic and epigenetic alterations involved in oncogenesis also trigger the conversion of TGF-β cell signals from cytostatic to tumorigenic in an uncharacterized process known as the TGF-β paradox (Schiemann 2007). The tumor-promoting properties of TGF-β not only affect the cancer cell, but also the stromal cell functions associated with angiogenesis and immunosuppression. Thus, the blockade of TGF-β cell signals via various inhibitors, such as neutralizing antibodies, decoy receptors, antisense oligonucleotides, small molecule receptor kinase inhibitors and peptide aptamers, is being assessed as a therapeutic target (Connolly *et al* 2012). Indeed, the function of MMPs in the formation of active TGF-β and tumor progression suggest that the inhibition of MMPs may also reduce tumor viability and cancer cell migration. Thus, previous studies in murine models using MMP inhibitors were found to pronouncedly reduce tumor growth, however, clinical trials have not yet been similarly successful (Coussens *et al* 2002). This may be due to MMP production coming from primarily stromal cells in human cancers or the lack of drug specificity to a particular MMP expressed at a certain tumor stage or within certain patient subsets (Coussens *et al* 2002, Zucker and Cao 2009). However, perhaps a more refined therapy directed against specific MMPs will lead to more specificity and success. In line with this, this type of drug selectivity has been described in studies that require HER-2 (Herceptin) or estrogen (Tamoxifen) receptor expression for efficacy (Khasraw and Bell 2012, Pearson *et al* 1982). In particular, studies *in vitro* and in an *in vivo* murine xenograft model of breast cancer showed that a soluble synthetic cytoplasmic tail peptide of MT1-MMP significantly inhibits HIF-1α, lactate production and tumor growth (Sakamoto *et al* 2011). Similar studies were performed in murine macrophages that revealed a mechanism involving MT1-MMP cytoplasmic tail induced localization of the factor inhibiting HIF-1 (FIH-1) with an inhibitor of FIH-1 (Mint3), which then provided the transcriptional activity of HIF-1α (Sakamoto and Seiki 2010). Moreover, it has been suggested that these associated functions also affect fibroblast MMP-2 production, which relies on MT1-MMP function for activation, in breast tumorigenesis (Saad *et al* 2002). As MT1-MMP and additional MMPs

are induced by hypoxia (Harmey *et al* 1998, Munoz-Najar *et al* 2006), mechanisms to abrogate the hypoxic response may also be therapeutically useful.

However, receptor activation by HER-2 ligands and estrogen seems to be implicated in the regulation of the hypoxic response. In more detail, HER-2 ligands induce the activation of HIF-1α, suggesting that in HER-2+ tumors Herceptin may be able to decrease the expression of the poor prognostic indicator HIF-1α (Charpin *et al* 2012, Spangenberg *et al* 2006). With respect to the estrogen receptor, the functions of HIF-1α appear to be more complex. In more detail, it has been indicated that hypoxia acts synergistically with estrogen in the activation of estrogen response elements (ERE) and the transcription of genes supporting tumorigenesis, angiogenesis and metabolism (Seifeddine *et al* 2007, Yi *et al* 2009). This synergy has led to increased clinical research in this field indicating that HIF-1α expression is a poor response predictor for chemoendocrine (Tamoxifen, Epirubicin) therapy (Generali *et al* 2006), suggesting that the examination of additional treatment approaches for hypoxia is necessary.

Alternative approaches for blocking hypoxic cell signals may include 2-deoxy-D-glucose (2-DG), antioxidants and aryl hydrocarbon receptor (AHR) ligands. In more detail, 2-DG is a stable glucose analogue, which can be actively taken up by hexose transporters and is phosphorylated by a hexokinase. This phosphorylated molecule (2-DG6P) then becomes a non-competitive inhibitor to hexokinase and a competitive inhibitor to glucose phosphoisomerase in the glycolytic pathway (Chen and Gueron 1992, Sols and Crane 1954, Wick *et al* 1975). Typical use of 2-DG involves F-18 labeling and non-invasive detection and staging of human tumors via positron emission tomography (PET) (Dwarakanath and Jain 2009). Renewed interest in 2-DG as a therapeutic agent has been reported in a combined treatment of 2-DG and a mitochondrial inhibitor (Mito-CP), which significantly decreased tumor weight without detriment to vital organs in a breast cancer xenograft model (Cheng *et al* 2012). In addition, it has also been reported that the cytotoxic effects of 2-DG increase under hypoxic conditions, thus enhancing the sensitivity of HIF-1 positive tumors to radiation and drug therapies (Aghaee *et al* 2012). Moreover, the functional mechanisms of 2-DG are not limited to the reduction of energy (ATP), because the molecule blocks protein glycosylation (Kurtoglu *et al* 2007). As the formation of collagen relies on glycosylation (Gelse *et al* 2003), fibroblast extra-cellular matrix deposition and remodeling seem to be altered by 2-DG. There is evidence that integrins also rely on glycosylation in establishing conformational structures and activation (Janik *et al* 2010), thus linking the migratory functions of cancer and stromal cells to potential inhibition by 2-DG.

In addition, it has also been suggested that antioxidants regulate integrin expression, particularly in subsets of immune cells, in which the molecule N-acetyl-L-cysteine (NAC) dramatically decreases the expression of the integrin associated with macrophage tissue invasion, alpha4 (Curran and Bertics 2012, Jin *et al* 2006, Laragione *et al* 2003, Puig-Kroger *et al* 2000). In a variety of cancer cell lines and murine models, NAC has been seen to decrease tumor growth and reduce HIF-1α levels by stimulating the HIF-1α degradation in a prolyl hydroxylase- and von Hippel–Lindau-dependent fashion (Gao *et al* 2007).

The activity of HIF-1α may also be inhibited by the cross-talk with the aryl hydrocarbon receptor (AHR), which shares a dimerization partner, HIF-1β (also called the AHR nuclear translocator, or ARNT), with HIF-1α for transcriptional activation of responsive genes (Chan *et al* 1999, Zhang and Walker 2007). In particular, the AHR is also able to exhibit positive and negative cross-talk with the estrogen receptors after activation by xenobiotic ligands, such as dioxin, or endogenous ligands, such as tryptophan metabolites (Denison and Nagy 2003). However, increased AHR cell signaling downstream of CXCR4 has been characterized as a biomarker of Tamoxifen resistant MCF-7 cell lines, suggesting that AHR antagonists may offer therapeutic advantages for Tamoxifen-resistant patient subsets (Dubrovska *et al* 2012, Ohtake *et al* 2011). Alternatively, in a murine model involving mitoxantrone-selected MDA-MB-231 cells injected IV into NOD scid gamma mice, gavage treatment with the non-toxic AHR ligand, named Tranilast, significantly reduced tumor growth and lung metastases (Prud'homme *et al* 2010). This result could be in response to direct tumor cytotoxicity, Tranilast-induced TGF-β inhibition, (Prud'homme *et al* 2010), reduced differentiation of macrophages (van Grevenynghe *et al* 2003), an absence of AHR-induced T regulatory cells (Schulz *et al* 2012) and/or altered fibroblast extracellular matrix production (Lehmann *et al* 2011).

As TGF-β and hypoxia cell signals lead to elevated extracellular matrix deposition, the cellular responses to the extracellular matrix may also be of interest as therapeutic targets. In particular, reduced breast tumor growth and metastases *in vitro* and *in vivo* has been found to involve small peptides (ATN-161, Cilengitide) or blocking antibodies (Abegrin) that antagonize integrin (α5β1, αvβ3, αvβ5) activation (Bauerle *et al* 2011, Khalili *et al* 2006, Mulgrew *et al* 2006). The force associated with breast density and tumorigenesis also increases integrin expression and activity (Paszek and Weaver 2004, Sawada *et al* 2006). Cross-linker molecules such as TG2 and LOX ensure the formation of a stiff matrix and the generation of force (Nurminskaya and Belkin 2012, Paszek and Weaver 2004). In a murine model of breast cancer, tumorigenesis is connected to the increased expression of stromal cell LOX, and inhibition of LOX with antibodies or β–aminopropionitrile (BAPN) increased tumor latency and subsequently decreased tumor incidence (Levental *et al* 2009). Additional studies with cancer cell TG-2 expression also revealed reduced tumor growth in the absence of TG-2 (Oh *et al* 2011), suggesting that continued research regarding the particular intracellular and extracellular functions of extra-cellular matrix cross-linkers (which, it has been suggested, play a role in both cancer and stromal cell function) is warranted.

The downstream targets of the integrin activation are expressed at enhanced levels in response to a dense collagen matrix such as focal adhesion kinase (FAK) and RhoA (Heck *et al* 2012, Keely 2011, Provenzano *et al* 2009). Interestingly, the expression of FAK in the mammary epithelium is essential for murine MMTV-PyMT breast tumor progression, suggesting that FAK inhibitors may act as potential inhibitors of breast cancer (Provenzano *et al* 2008, Schultze and Fiedler 2010). The molecule RhoA not only coordinates cell migration and survival, it has also been reported to be an inhibitor to macrophage efferocytosis (Korns *et al* 2011, Ridley 2004).

The identification of RhoA in human tumors is linearly correlated with lymph node metastasis and tumor invasion and it may be mediated through a mechanism involving hypoxia-induced RhoA, as has been demonstrated in the human MCF-7 cell line (Ma *et al* 2012). In the orthotopic breast cancer model involving MDA-MB-231 cells, the tumor burden was drastically reduced upon treatment with an oral inhibitor of Rho kinases, Fasudil (Ying *et al* 2006). Thus, FAK and RhoA, as well as additional downstream signals of the integrin receptor (such as extracellular matrix interactions) may serve as effectual therapeutic targets in order to treat breast cancer. However, understanding how these pathways agree or differ in cancer and stromal cells would provide therapeutic advantages for certain tumors or patient subsets.

Taken together, breast cancer is a heterogeneous disease disparately characterized by receptor status (HER2, ER, PR) and regulated by the cytokine TGF-β and the properties of the extracellular matrix that locally surrounds the primary tumor. Cell-facilitated adhesive interactions induce the production and mediate the release of TGF-β, which in turn supports cell signals associated with the expression of integrins, the deposition and remodeling of the extracellular matrix. Moreover, the formation of an extracellular matrix that serves as a bridge to blood vessels and metastatic sites involves contributions from both cancer and stromal cells. The introduction of hypoxia to the microenvironment differentially affects each parti-cular cell type and may even depend additionally on cell-specific mutations, cell–cell contact, extracellular matrix interactions and the production of effectors. Moreover, cells not only reveal specific signs of hypoxia in response to oxygen deprivation, but also from additional metabolic factors, cytokines and growth factors. The subsequent activation of HIFs induces metabolic shifts and evokes the production of chemokines and adhesive factors involved in tumor survival, recruitment and migration. However, these transcription factors are also implicated in fibroblast and macrophage differentiation and in the production of proteins that remodel the extracellular matrix, such as the enzyme cross-linking molecules of LOX and TG2. Due to their complexity, LOX and TG2 are also pleiotropic proteins involved in cell surface receptor-facilitated functions and intracellular cell signaling. These various interactions display and illuminate the diverse nature of the hypoxic tumor micro-environment and the complex highly integrated networks between cancer and stromal cells in tumor development and the immune response. However, the elucidation of these networks may lead to improved treatments that target both cancer and stromal cells in breast cancer.

9.2 The role of substances and growth factors within the extracellular matrix in cancer cell's mechanical properties

Low molecular weight hyaluronan (LMW-HA), which is generated by degradation of the extracellular matrix component hyaluronan (HA), has been identified as being crucial to cancer progression. However, no systematic clinical study of breast cancer has been performed in order to reveal a correlation between LMW-HA levels and cancer metastasis. In was found that in 176 serum specimens the serum LMW-HA (but not total HA) level correlated significantly with lymph node metastasis, suggesting that serum LMW-HA represents a better prognostic indicator of breast cancer progression than HA (Wu *et al* 2014). Similarly, it was observed that breast cancer cells displaying higher invasive potential had a higher LMW-HA concentration than less-invasive cancer cells. Indeed, this higher LMW-HA level was accompanied by the overexpression of hyaluronan synthase (HAS2) and hyaluronidase (both HYAL1 and HYAL2) (Wu *et al* 2014). Importantly, decreasing LMW-HA production significantly inhibited breast cancer cell migration and invasion (Wu *et al* 2014). Taken together, these results suggest that during cancer progression cancer cells may actively remodel their microenvironment through an autocrine/paracrine-like process, which results in elevated LMW-HA levels that in turn promote cancer progression by supporting the migration and invasion of cancer cells into the extracellular matrix of connective tissue. Thus, cancer-associated LMW-HA seems to be a more promising molecular biomarker than total HA for detecting metastasis and it may provide further applications in the treatment of breast cancer.

Hyaluronan is known to be a prominent component of the local microenvironment in most malignant tumors and can be used as a prognostic factor for tumor progression. Extensive experimental evidence in animal models indicates that hyaluronan interactions in tumor growth and metastasis are essential, but it is also known that a balance of synthesis and turnover by hyaluronidases is critical. CD44 is a major hyaluronan receptor, and is commonly but not uniformly associated with malignancy and hence frequently serves as a marker for cancer stem cells in human carcinomas. Multivalent interactions of hyaluronan with CD44 collaborate in inducing numerous tumor-promoting signaling pathways and transporter activities. It is widely accepted that hyaluronan-CD44 interactions are crucial in both malignancy and resistance to therapy, while investigating the mechanism for the activation of hyaluronan-CD44 signaling in cancer cells, the relative importance of variant forms of CD44 and other hyaluronan receptors (such as Rhamm) in different tumor contexts and the role of stromal versus cancer cell driven hyalronan and its turnover are some of the major challenges for future research. Despite these, it is clear that hyaluronan–CD44 interactions are an important target for translation into the clinic. Among the approaches that seem to be promising are antibodies and vaccines for specific variants of CD44 that are uniquely expressed at critical progression stages of a particular cancer, hyaluronidase-mediated reduction of barriers to drug access and small hyaluronan oligosaccharides, which attenuate constitutive hyaluronan–receptor signaling and enhance chemosensitivity. In addition, as a novel development in drug delivery, hyaluronan is being used to tag drugs and delivery vehicles for targeting of anti-cancer agents toward CD44-expressing cancer cells.

The importance of the microenvironment in cancer progression has been demonstrated in numerous studies (Nelson and Bissell 2006, Polyak *et al* 2009, Joyce and Pollard 2009, Kalluri and Weinberg 2009). Hyaluronan is a prominent and well-known constituent of the local microenvironment in most malignant tumors. In particular, hyaluronan immediately surrounds the pericellular milieu around cancer cells and is located in the tumor stroma, but its association with either compartment can be prognostic for tumor progression (Knudson *et al* 1989, Tammi *et al* 2008). A major receptor for hyaluronan, CD44, is currently much investigated, as it is a common marker for 'tumor-initiating cells/cancer stem cells' (CSCs) in human carcinomas. Although the nature of these CSCs is highly controversial in the literature, there is a reasonable consensus that CD44-expressing sub-fractions of many human carcinomas are highly malignant and resistant to therapy, properties that are also associated with CSCs (Visvader and Lindeman 2008, Polyak and Weinberg 2009). However, the functions of CD44 and its hyaluronan ligand for the properties of these particular cells are still elusive and need further investigation. The functional dynamics of hyaluronan and its receptors such as CD44 were recently investigated in more detail with respect to cancer (Tammi *et al* 2008, Stern 2008, 2009, Maxwell *et al* 2008, Bourguignon 2008, Lokeshwar and Selzer 2008, Itano and Kimata 2008, Naor *et al* 2008, Toole *et al* 2008, Simpson and Lokeshwar 2008). The current state of knowledge for the functions of hyaluronan-CD44 interactions in cancer and the mechanisms through which these interactions influence a large number of signaling pathways and cellular behaviors was summarized (Toole 2009). Some of these activities are summarized in figure 9.1.

Hyaluronan synthases produce and extrude hyaluronan, which may be retained by the synthase or released into the pericellular milieu. The extruded hyaluronan has been shown to interact multivalently with CD44 to induce and/or stabilize signaling domains within the cell membrane. These signaling domains contain receptor tyrosine kinases such as ErbB2 and EGFR, other signaling receptors (TGFβR1) and non-receptor kinases (Src family) that are able to drive oncogenic pathways, for instance, the MAP kinase and PI3 kinase/Akt cell proliferation and survival pathways, as well as various transporters that provide drug resistance and induce malignant cell properties (Toole *et al* 2008, Bourguignon 2009). Various adaptor proteins, such as Vav2, Grb2 and Gab-1, facilitate the interaction of CD44 with upstream effectors, such as RhoA, Rac1 and Ras, which can drive these pathways (Bourguignon 2008, 2009). In other cases, carbohydrate side groups on variant regions of CD44, such as heparan sulfate chains, bind additional regulatory

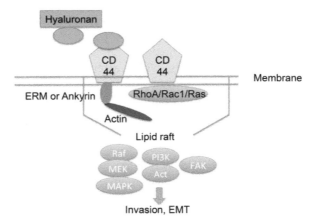

Figure 9.1. Regulation of signal transduction cascades by hyaluronan-CD44 interaction.

factors and co-activate receptor tyrosine kinases, such as the c-Met receptor (Ponta *et al* 2003). Hyaluronan-CD44 interactions also facilitate cytoskeletal alterations that induce cell motility and invasion. In this case, actin filaments are recruited to the cytoplasmic tail of CD44 through members of the ezrin–radixin–moesin (ERM) family or ankyrin (Bourguignon 2008, Ponta *et al* 2003). Proteoglycans and associated factors attached to pericellular hyaluronan are also able to influence these activities (Itano and Kimata 2008, Evanko *et al* 2007) (figure 9.1). However, hyaluronan produced by stromal cells may have overlapping or different activities, while the relative contributions of stromal and tumor-derived hyaluronan are still elusive (Tammi *et al* 2008).

Hyaluronan

Hyaluronan (also termed hyaluronic acid or hyaluronate) is a very large, linear, negatively charged polysaccharide, which consists of repeating disaccharides of glucuronate and N-acetylglucosamine. Hyaluronan is produced by three hyaluronan synthases (Has1/Has2/Has3), which are integral cell membrane proteins whose active sites are located at the intracellular site of the cell membrane (Weigel and DeAngelis 2007). In particular, newly synthesized hyaluronan is extruded as it is elongated and then targeted to the cell surface or to pericellular extracellular matrices. However, hyaluronan is widely distributed in vertebrate tissues, but it is especially concentrated in regions of cell division and cell invasion (Toole 2001). In adult tissues such as synovial fluid, cartilage and dermis, hyaluronan plays a pronounced structural role, based on its unique hydrodynamic properties and its interactions with other extracellular matrix components. On the other hand, hyaluronan has an instructive cell signaling role during dynamic restructuring cell processes such as morphogenesis, inflammation, wound repair and cancer, in which hyaluronan–receptor interactions are activated and induce numerous signaling pathways (Bourguignon 2008, Turley *et al* 2002, Toole 2001). In addition to the signal transduction, the hyaluronan–receptor interactions act in at least two other important physiological processes: endocytosis of hyaluronan and assembly of pericellular matrices (Toole 2001, Knudson *et al* 2002, Evanko *et al* 2007).

Hyaluronan receptors

Hyaluronan interacts with several cell surface receptors, such as CD44, Rhamm, LYVE-1, HARE/stabilin-2 and Toll-like receptors-2 and 4 (Turley *et al* 2002, Jiang *et al* 2007, Jackson 2009). CD44 is widely distributed, but particularly important in the immune system and inflammatory processes (Jiang *et al* 2007, Johnson and Ruffell 2009), and in diseases such as atherosclerosis and cancer (Toole 2001, 2002). In contrast to CD44, LYVE-1 and HARE, Rhamm does not belong to the linkage module family of hyaluronan-binding proteins. Instead, Rhamm is located either in the cytoplasm or on the cell surface and is an important factor in cell motility in wound-healing processes and diseases such as cancer (Maxwell *et al* 2008). LYVE-1 is closely related to CD44 and is mainly restricted to lymphatic vessel and lymph node endothelia, whereas its function is not well understood (Jackson 2009). In addition, HARE/stabilin-2 is a scavenging receptor that is able to clear hyaluronan and other glycosaminoglycans from the blood vessel circulation (Pandey et al 2008). Moreover, the Toll-like receptors recognize hyaluronan fragments during inflammatory reactions (Jiang et al 2007).

The major receptors implicated in cancer are CD44 and Rhamm. Thus, the focus is on CD44 during the investigation of cancer progression and metastasis, while it is still important to note that CD44 and Rhamm can exhibit both cooperative and interchangeable signaling functions. Moreover, it has been revealed that interactions at the cell membrane between CD44 and Rhamm activate CD44 signaling through ERK1/2 and subsequently promote cancer cell motility (Maxwell et al 2008). In some cases, for instance, in animal models with autoimmune diseases, Rhamm is able to compensate for CD44, which is a very important consideration when interpreting experiments in CD44-null mice (Naor et al 2007).

CD44 is a single-chain, single-pass and transmembrane glycoprotein, which is widely expressed in physiological and pathological systems. In more detail, CD44 was first characterized as playing a role in hyaluronan–cell interactions, lymphocyte homing and cell adhesion (Toole 1990) and its role in these phenomena seems to be well established (Johnson and Ruffell 2009, Ponta et al 2003). Although CD44 arises from a single gene, numerous transcripts are provided by alternative splicing, in which different (alternative) splice sites are used in order to splice out the introns. The standard and normal CD44 is comprised of the constant, non-variant exon products, whereas the so-called variant isoforms are generated by alternative splicing of numerous additional exon products into a single site within the membrane-proximal region of the ectodomain (Ponta et al 2003). In more detail, cancer cells typically produce several variant forms of CD44 and the standard CD44, whereas some cancer types such as gliomas, mainly produce the standard form of CD44 (Heider et al 2004). All forms of CD44 include an N-terminal, membrane-distal, hyaluronan-binding domain, which has significant homology with the hyaluronan-binding region, such as the link module of several other proteins and proteoglycans. Hyaluronan is the most widely studied ligand for CD44, but other ligands are similarly important. The best characterized among these are osteopontin and factors such as FGF and selectin ligands, which are able to recognize carbohydrate side chains covalently bound to CD44 (Ponta et al 2003, Sackstein 2009). One of the most interesting aspects of CD44 is its activation upon hyaluronan binding and subsequent signaling. Possible factors contributing to the activation include post-translational modifications of CD44 such as glycosylation, CD44-cytoskeletal interactions, localization of CD44 within specialized domains such as lipid raft domains in the cell membrane, as well as the mode of pericellular organization and presentation of hyaluronan. However, which of these activation factors or steps is the most important remains elusive and will have to be worked out in more detail. Nevertheless, it is common knowledge that numerous cytokines, growth factors and alterations in the cell context can induce the events that result in CD44 activation (Ponta et al 2003).

Hyaluronan–CD44 signal transduction processes

In several types of cancer cells, binding of hyaluronan to CD44 leads to a direct or indirect interaction of CD44 with signaling receptors such as ErbB2, EGFR and TGF-beta receptor type I and thus impacts on the activity of these receptors (Toole *et al* 2008, Toole 2004, Bourguignon 2009). However, it can subsequently promote interaction with and the altered activity of non-receptor kinases of the Src family or Ras family GTPases (Bourguignon 2008, Bourguignon 2009). The complex formation with adaptor proteins such as Vav2, Grb2 and Gab-1 facilitates the interaction of CD44 with upstream effectors such as RhoA, Rac1 and Ras, which all induce intracellular signaling pathways (Bourguignon 2008, Bourguignon 2009). Thus, hyaluronan-CD44 binding influences the activity of a variety of downstream signaling pathways, especially the MAP kinase and PI3 kinase/Akt pathways and consequently supports cancer cell proliferation, survival, motility, invasiveness and chemoresistance (Bourguignon 2008, Toole *et al* 2004, 2008, Turley *et al* 2002, Ponta *et al* 2003, Bourguignon 2009). In addition, binding of hyaluronan to CD44 activates multidrug and metabolic transporters that are important in providing therapy resistance (Toole *et al* 2008, Bourguignon *et al* 2004, Miletti-Gonzalez *et al* 2005, Colone *et al* 2008, Slomiany *et al* 2009a, 2009b) and induces the presentation of proteases that promote cancer cell invasion (Ponta *et al* 2003, Stamenkovic and Yu 2009). Most of the current evidence indicates that these interactions involve specific variants of CD44, while the particular variant almost certainly depends on the type of cancer cell as well as on the stage of malignant progression and in some cases even standard CD44 rather than variant CD44 is critically involved. However, the regulation mechanisms of these various interactions in different cancer cell types and stages are not yet well understood and need further investigation, while the widespread deregulation of many normal pathways in cancer cells also contributes to the induction of anomalous involvement of hyaluronan-CD44 interactions that operate normally in other contexts, such as embryonic development (Toole 2001) and inflammation (Jiang *et al* 2007). A related possibility is that deregulated splicing in cancer cells (Skothein and Nees 2007) causes the different CD44 variants that support oncogenic events such as inappropriate Ras signaling (Cheng *et al* 2006) and binding of osteopontin, stromal growth factors and proteases (Ponta *et al* 2003).

In addition to its function as a co-receptor or co-activator of membrane-associated signaling molecules, CD44 may regulate other cellular, disease-associated events, such as cancer cell proliferation and motility through cross-linking to the actin cytoskeleton via ankyrin or members of the ezrin–radixin–moiesin family (Bourguignon 2008, Ponta *et al* 2003, Stamenkovic and Yu 2009). The tumor suppressor merlin has been shown to act by blocking the hyaluronan–CD44 interaction and in addition dissociating the ezrin–radixin–moiesin proteins from the cytoplasmic tail of CD44. In turn, the release of merlin suppression may trigger the activation of hyaluronan–CD44 binding, which then supports the formation of signaling complexes in order to facilitate tumor progression and cancer cell invasiveness (Stamenkovic and Yu 2009). Another regulatory mechanism whereby hyaluronan–CD44 interactions induce intracellular signal transduction pathways functions is through the intracellular cleavage of CD44, its translocation of the cytoplasmic part into the nucleus and subsequently its activation of gene transcription (Nagano and Saya 2004).

Many studies indicate that CD44 is localized at least in part to the lipid micro-domains with the properties of so-called lipid rafts and that it associates indirectly or directly therein with signaling proteins and transporters. Moreover, most of these studies also demonstrate that CD44 is recruited into these lipid raft domains in response to ligand interactions (Bourguignon 2008, Bourguignon et al 2004, Ghatak et al 2005, Lee et al 2008). In particular, endocytosis of hyaluronan and CD44 occurs from these lipid raft domains (Thankamony and Knudson 2006). Because of the large variety of signal transduction pathways regulated by CD44, there is a strong possibility that indirect and direct interactions with a wide variety of effectors occur within such lipid rafts domains and, consequently, that these domains are induced and/or stabilized by the multivalent interactions of hyaluronan with CD44; this offers an interesting hypothesis to guide current and future investigations.

A puzzling aspect of many studies investigating hyaluronan-induced oncogenic signal transduction is that they are based on exogenous hyaluronan, which is simply added to cultured cancer cells. Although these studies have led to robust data on the pathways under investigation, it is difficult to take these investigations further, even though there is a long history of safe utilization of hyaluronan in numerous reconstructive or regenerative experiments performed on human patients. For instance, hyaluronan is used successfully widely in eye and knee surgeries and in the prevention of adhesions (Balazs and Denlinger 1989, Prestwich and Kuo 2008). Moreover, hyaluronan-based hydrogels have also been developed for a variety of purposes, such as drug delivery, encapsulation of progenitor cells and tissue engineering (Allison and Grande-Allen 2006, Prestwich 2008). Can hyaluronan still be used or is the risk of developing cancer indeed too high? However, such studies suggest that the oncogenic effects of hyaluronan only occur in the context of the tumor micro-environment and that stromal hyaluronan, as well as tumor-cell-produced hyaluronan, play important roles in tumorigenesis, which is supported by correlative studies of numerous human tumor types (Tammi et al 2008). Strong evidence for the tumor-promoting effects of hyaluronan have been derived from studies in which tumor hyaluronan levels and interactions with receptors have been manipulated in vivo.

Hyaluronan–CD44 interactions
Strong experimental evidence for the involvement of hyaluronan in tumor growth and cancer metastasis has been gained from animal models of several tumor types. The approaches performed include the manipulation of the hyaluronan levels and the perturbation of endogenous hyaluronan–receptor interactions using a number of methods (Itano and Kimata 2008, Toole 2004). In line with this, it has become evident that the turnover of hyaluronan by hyaluronidases is indeed an essential aspect of the promotion of cancer progression by hyaluronan and that the balance of synthesis and degradation is critical (Lokeshwar and Selzer 2008, Simpson and Lokeshwar 2008). Hyaluronan synthesis has been reported to be conditionally increased expressed in mammary tumors that arise spontaneously in MMTV-Neu

mice, which highlights the importance of hyaluronan in cancer promotion, especially through the enhanced recruitment of stromal cells and *de novo* induced angiogenesis (Itano and Kimata 2008). As expected, numerous studies have demonstrated that hyaluronan-CD44 interactions have an important role in the recruitment or homing of various cell types, including circulating immune cells and precursor cells (Johnson and Ruffell 2009, Haylock and Nilsson 2006). The MMTV-Neu studies (Itano and Kimata 2008) also revealed the importance of hyaluronan in EMT. It was found that there are cells that are not able to undergo EMT. In particular, a major defect in the Has2-null mouse is its inability to undergo EMT during early cardiac development (Camenisch *et al* 2000). In turn, the increased expression of Has2 in phenotypically normal epithelium induces EMT characteristics, including anchorage-independent growth and enhanced invasiveness (Zoltan-Jones *et al* 2003), which are two of the major capabilities of malignant cells.

The evidence for the involvement of CD44 in cancer progression is strong, but seems to be very complex. Studies of tumorigenesis in CD44-null mice and manipulation of CD44 levels in various tumor systems have produced contradictory results, while treatments with CD44 antibodies and vaccines have demonstrated the key regulatory role of CD44 in tumor growth and metastasis in mouse models of leukemias and carcinomas (Naor *et al* 2008, Jin *et al* 2006, Krause *et al* 2006, Wallach-Dayan 2008). Several studies have implicated variants of CD44 rather than standard CD44 in cancer progression, however, this is not a unique effect of the CD44 variants for all cancer types, rather it depends on the stage of progression and type of the tumor (Naor *et al* 2008, Ponta *et al* 2003). A special feature of CD44 is its emergence as a marker for sub-populations of several types of human carcinomas, termed CSCs, which exhibit highly malignant and chemoresistant properties (Visvader and Lindeman 2008, Polyak and Weinberg 2009). In line with this, the characteristics of EMT have recently been connected to the properties of these cell sub-populations. For instance, a CD44+/CD24-sub-population exhibiting CSC properties was induced through the up-regulation of EMT-associated transcription factors in the primary human breast epithelium and a similar sub-population exhibiting both EMT and CSC properties was isolated from transformed epithelial cells (Polyak and Weinberg 2009, Hollier *et al* 2009). These cells exhibited anchorage-independent growth of colonies in soft agar, which is a property that usually reflects resistance to apoptosis and in turn seems to be linked to chemo-resistance. Numerous studies have revealed that the CSC sub-population of carcinomas and other tumor types is indeed resistant to chemotherapeutic agents, which is most likely attributable to enhanced anti-apoptotic pathway activity and enrichment of multidrug transporters (Polyak and Weinberg 2009, Toole *et al* 2008, Hollier *et al* 2009). Another important feature of EMT is invasiveness (Kalluri and Weinberg 2009, Turley *et al* 2008) and thus CSCs have been connected to invasiveness and metastasis (Visvader and Lindeman 2008, Polyak and Weinberg 2009, Sleeman *et al* 2007). As noted previously, hyaluronan is closely associated with EMT and in addition these same properties of anchorage-independent growth, resistance to apoptosis, drug resistance and invasiveness are induced or enhanced by up-regulation of hyaluronan synthesis and can be even reversed by antagonists of hyaluronan-CD44 interactions (Toole *et al* 2008, Toole 2004). In more detail, there is strong evidence that hyaluronan-dependent interaction of CD44 with receptor kinases (Toole 2004, Ponta *et al* 2003, Bourguignon 2009) and transporters (Toole *et al* 2008, Slomiany *et al* 2009a, 2009b, Bourguignon *et al* 2004, Colone *et al* 2008) plays a prominent role in drug resistance and malignancy. The question of whether there are hyaluronan-CD44 interactions in a CSC-like subpopulation of cells isolated from human patient's ovarian carcinoma ascites has been investigated. It has been demonstrated that CSCs are enriched in receptor tyrosine kinases and ABC-family drug transporters and in addition that these proteins are present in close proximity to CD44 in the cell membrane of the CSCs, while this association depends on constitutive hyaluronan interactions.

Hyaluronan–CD44 interactions in cancer
Although the studies on hyaluronan-CD44 interactions in cancer disease are contradictory and contain paradoxes, agreement has been reached in the field that these interactions are an important target for translation into the clinic. A frequently expressed concern regarding these studies is the widespread expression and the broad variety of cellular functions of hyaluronan and CD44 under normal physiological conditions. However, two observations seem to promise that therapeutic interventions can be developed that target oncogenic events with some degree of specificity or differential sensitivity. First, there is the finding that many of the interactions involve variants of CD44, which are amplified greatly in many tumor types in comparison to normal processes (Naor *et al* 2008, Heider *et al* 2004, Skotheim and Nees 2007). Second, there is the nature of the activation process that leads to hyaluronan–CD44 interactions in malignant cancer cells. Although these processes may have some overlapping features with immune and inflammatory pathways, there are also clear differences, and it may be possible to selectively target them. Some of the studies have utilized antagonists that may ultimately have therapeutic value, but these have not yet reached the clinic.

CD44 antibodies and vaccines
Several studies have reported that the administration of antibodies against CD44 reduces tumor growth and restricts tumor progression. For instance, injection of monoclonal antibodies directed against CD44 that inhibit the binding of hyaluronan eliminated the invasion of mouse lymphoma cells into local lymph nodes (Naor *et al* 2008). Moreover, blocking antibodies directed against CD44 have been shown to inhibit homing and promote differentiation of acute myeloid leukemic stem cells, consequently eliminating tumor-initiating cells (Jin *et al* 2006). In line with this, prolonged survival also occurred in mice with leukemic stem cells expressing BCR-ABL after treatment with an inhibitory CD44 antibody (Krause *et al* 2006). Indeed, a CD44 variant-based vaccine was designed to decrease mouse mammary carcinoma tumor growth and metastases (Wallach-Dayan *et al* 2008). Moreover, it is recognized in this medical field that these approaches will be greatly improved by tailoring antibodies and vaccines to specific variants of CD44 that are then uniquely expressed during the critical stages of cancer progression (Naor *et al* 2008,

Ponta *et al* 2003). However, phase I trials in breast as well as in head and neck carcinoma patients with an antibody directed against CD44v6 have been discontinued due to toxicity (Rupp *et al* 2007, Riechelmann *et al* 2008).

Hyaluronidases

Although constitutive hyaluronidase may support the pro-oncogenic functions of hyaluronan, over-expression or exogenous administration of large amounts of hyaluronidase is usually inhibitory to malignant cancer progression (Lokeshwar and Selzer 2008, Simpson and Lokeshwar 2008, Stern 2008). Hence, hyaluronidase has been used in the clinic for several years as an adjunct to chemotherapy, because it has been supposed to improve the access of drugs to cancer cells through the interference with cancer cell adhesion and their capacity to break through matrix barriers (Lokeshwar and Selzer 2008, Baumgartner *et al* 1998). Indeed, highly purified recombinant hyaluronidase (Frost 2007) is currently being used in a phase I trial for patients with advanced solid tumors. Interestingly, it was also demonstrated that hyaluronidase sensitizes mouse mammary carcinoma cells to chemotherapeutic drugs, if the cells are cultured as drug-resistant spheroids (St Croix *et al* 2000), which is a technique reported to enrich for CSCs. Although hyaluronidase may indeed function in part by reducing barriers to drug diffusion, it is also able to act through its oligosaccharide products, which have been found to inhibit constitutive hyaluronan–CD44 signal transduction processes, resulting in reduced cell survival and induced chemoresistance (Toole *et al* 2008).

Small hyaluronan oligosaccharides

Small oligomers of hyaluronan can suppress anti-apoptotic signaling pathways in cancer cells and thus inhibit the activity of transporters that increase resistance to therapeutic agents (Toole *et al* 2008). Initially, the use of these was based on certain findings that oligomers consisting of three–nine disaccharides bind CD44 monovalently (Lesley *et al* 2000) and displace the hyaluronan polymer from membrane-bound receptors (Underhill and Toole 1979), while another study showed that these oligomers even reduce hyaluronan synthesis (Slomiany *et al* 2009a). Indeed, the treatment of cancer cells with these oligomers leads to disassembly of CD44-transporter and CD44-receptor tyrosine kinase complexes, internalization of the disassembled components and attenuation of CD44 function (Slomiany *et al* 2009a, Slomiany *et al* 2009b, Ghatak *et al* 2005). In addition, treatment *in vivo* with small hyaluronan oligomers suppresses tumor growth and/or even triggers tumor regression in experiments using xenografts of various tumor types, such as melanoma, carcinomas, glioma, osteosarcoma and malignant peripheral nerve sheath tumors (Toole *et al* 2008, Toole 2001, Slomiany *et al* 2009a, Hosono *et al* 2007, Gilg *et al* 2008). One of these studies reported significant effects on cancer metastasis (Hosono *et al* 2007). In addition, significant effects on tumor growth and invasion were demonstrated when CSC-like sub-populations were obtained from a glioma cell line (Gilg *et al* 2008) or from human patient ovarian carcinoma ascites and used for co-culture or supernatant experiments. Moreover, systemic administration of sub-optimal doses of hyaluronan oligomers were shown to sensitize highly resistant, malignant peripheral nerve sheath tumors to doxorubicin treatment *in vivo* (Slomiany *et al* 2009a). Although this approach might be expected to interfere with all activated hyaluronan–CD44 interactions, malignant tumors appear to be far more sensitive than normal physiological processes.

Targeting drugs to cancer cell CD44
In addition to targeting hyaluronan–CD44 interactions directly, the interactions can be used for the targeted delivery of chemotherapeutic drugs and other anti-cancer agents to cancer cells. It has been reported that increased efficacy in cell and animal tumor models has been obtained through conjugating drugs to hyaluronan, a CD44 antibody, incorporating drugs or siRNAs (gene knock-down) into vehicles such as liposomes, hydrogels and nanoparticles, which are masked with hyaluronan or antibodies against CD44 in order to target them to cancer cells and primary tumors (Platt and Szoka 2008). Indeed, the first human-patient trials with drugs conjugated to CD44 antibody have shown some promising results, although there were complications regarding various toxicities (Platt and Szoka 2008). However, specific targeting to the relevant variants of CD44 is a crucial point of this drug approach. In particular, it has also been revealed that the enormous hydrodynamic domain encompassed by hyaluronan is well suited to entrap drugs, without any need for chemical conjugation, and finally to target them to CD44-expressing tumors (Brown 2008). In line with this, when connected to hyaluronan, irinotecan has demonstrated increased safety and efficacy, and this has been shown to be a promising approach for colorectal carcinoma patients in a pilot trial (Gibbs *et al* 2009).

Further reading

Baba F, Swartz K, van Buren R, Eickhoff J, Zhang Y, Wolberg W and Friedl A 2006 Syndecan-1 and syndecan-4 are overexpressed in an estrogen receptor-negative, highly proliferative breast carcinoma subtype *Breast Cancer Res. Treat.* **98** 91–8

Bauer M, Eickhoff J C, Gould M N, Mundhenke C, Maass N and Friedl A 2008 Neutrophil gelatinase-associated lipocalin (NGAL) is a predictor of poor prognosis in human primary breast cancer *Breast Cancer Res. Treat.* **108** 389–97

Bernard A, Delamarche E, Schmid H, Michel B, Bosshard H R and Biebuyck H 1998 Printing patterns of proteins *Langmuir* **14** 2225–9

Cukierman E 2002a *Preparation of extracellular matrices produced by cultured fibroblasts* ed J B Harford and K M Yamada (Philadelphia, PA: Lippincott-Schwartz)

Cukierman E 2002b *Preparation of extracellular matrices produced by cultured fibroblasts* ed J S Bonifacino, M Dasso, J Lippincott-Schwartz, J B Harford and K M Yamada (New York: Wiley)

Cukierman E 2005 Cell migration analyses within fibroblast-derived 3D matrices *Methods Mol. Biol.* **294** 79–93

Harvey J M, Clark G M, Osborne C K and Allred D C 1999 Estrogen receptor status by immunohistochemistry is superior to the ligand-binding assay for predicting response to adjuvant endocrine therapy in breast cancer *J. Clin. Oncol.* **17** 1474–81 PMID: 10334533

Jin L, Hope K J, Zhai Q, Smadja-Joffe F and Dick J E 2006 Targeting of CD44 eradicates human acute myeloid leukemic stem cells *Nat. Med.* **12** 1167–74

Junqueira L C, Bignolas G and Brentani R R 1979 Picrosirius staining plus polarization microscopy, a specific method for collagen detection in tissue sections *Histochem. J.* **11** 447–55

Provenzano P P, Eliceiri K W, Campbell J M, Inman D R, White J G and Keely P J 2006 Collagen reorganization at the tumor stromal interface facilitates local invasion *BMC Med.* **4** 38

Slomiany M G, Grass G D, Robertson A D, Yang X Y, Maria B L, Beeson C and Toole B P 2009a Hyaluronan, CD44 and emmprin regulate lactate efflux and membrane localization of monocarboxylate transporters in human breast carcinoma cells *Cancer Res.* **69** 1293–301

Slomiany MG, Dai L, Bomar P A, Knackstedt T J, Kranc D A, Tolliver L, Maria B L and Toole B P 2009b Abrogating drug resistance in malignant peripheral nerve sheath tumors by disrupting hyaluronan-CD44 interactions with small hyaluronan oligosaccharides *Cancer Res.* **69** 4992–8

St Croix B, Man S and Kerbel R S 2000 Reversal of intrinsic and acquired forms of drug resistance by hyaluronidase treatment of solid tumors *Cancer Lett.* **131** 35–44

Thompson E W *et al* 1992 Association of increased basement membrane invasiveness with absence of estrogen receptor and expression of vimentin in human breast cancer cell lines *J. Cell Physiol.* **150** 534–44

Weigelt B and Bissell M J 2008 Unraveling the microenvironmental influences on the normal mammary gland and breast cancer *Semin. Cancer Biol.* **18** 311–21

Wierzbicka-Patynowski I and Schwarzbauer J E 2002 Regulatory role for SRC and phosphatidylinositol 3-kinase in initiation of fibronectin matrix assembly *J. Biol. Chem.* **277** 19703–8

Zoltan-Jones A, Huang L, Ghatak S and Toole B P 2003 Elevated hyaluronan production induces mesenchymal and transformed properties in epithelial cells *J. Biol. Chem.* **278** 45801–10

References

Aaboe M, Offersen B V, Christensen A and Andreasen P A 2003 Vitronectin in human breast carcinomas *Biochim. Biophys. Acta* **1638** 72–82

Acosta-Iborra B, Elorza A, Olazabal I M, Martin-Cofreces N B, Martin-Puig S, Miro M, Calzada M J, Aragones J, Sanchez-Madrid F and Landazuri M O 2009 Macrophage oxygen sensing modulates antigen presentation and phagocytic functions involving IFN-gamma production through the HIF-1 alpha transcription factor *J. Immunol.* **182** 3155–64

Aghaee F, Pirayesh Islamian J and Baradaran B 2012 Enhanced radiosensitivity and chemosensitivity of breast cancer cells by 2-deoxy-d-glucose in combination therapy *J. Breast Cancer* **15** 141–7

al Adnani M S, Taylor S, al-Bader A A, al-Zuhair A G and McGee J O 1987 Immunohistochemical localization of collagens and fibronectin in human breast neoplasms *Histol. Histopathol.* **2** 227–38 PMID: 2980725

Allison D D and Grande-Allen K J 2006 Hyaluronan: a powerful tissue engineering tool *Tissue Eng.* **12** 2131–40

Amatangelo M D, Bassi D E, Klein-Szanto A J and Cukierman E 2005 Stroma-derived three-dimensional matrices are necessary and sufficient to promote desmoplastic differentiation of normal fibroblasts *Am. J. Pathol.* **167** 475–88

Anand R J, Gribar S C, Li J, Kohler J W, Branca M F, Dubowski T, Sodhi C P and Hackam D J 2007 Hypoxia causes an increase in phagocytosis by macrophages in a HIF-1alpha-dependent manner *J. Leukoc Biol.* **82** 1257–65

Arancibia R, Oyarzun A, Silva D, Tobar N, Martinez J and Smith P C 2013 Tumor necrosis factor-α inhibits transforming growth factor-β-stimulated myofibroblastic differentiation and extracellular matrix production in human gingival fibroblasts *J. Periodontol.* **84**(5) 683–93

Azuma Y T, Matsuo Y, Nakajima H, Yancopoulos G D, Valenzuela D M, Murphy A J, Karow M and Takeuchi T 2011 Interleukin-19 is a negative regulator of innate immunity and critical for colonic protection *J. Pharmacol. Sci.* **115** 105–11

Balazs E A and Denlinger J L 1989 Clinical uses of hyaluronan *Ciba Found Symp.* **143** 265–80 PMID: 2680347

Barker H E, Cox T R and Erler J T 2012 The rationale for targeting the LOX family in cancer *Nat. Rev. Cancer* **12** 540–52

Barsky S H, Green W R, Grotendorst G R and Liotta L A 1984 Desmoplastic breast carcinoma as a source of human myofibroblasts *Am. J. Pathol.* **115** 329–33 PMID: 6329001

Bartram U and Speer C P 2004 The role of transforming growth factor beta in lung development and disease *Chest* **125** 754–65

Bauer G 1996 Elimination of transformed cells by normal cells: a novel concept for the control of carcinogenesis *Histol. Histopathol.* **11** 237–55 PMID: 8720467

Bauerle T, Komljenovic D, Merz M, Berger M R, Goodman S L and Semmler W 2011 Cilengitide inhibits progression of experimental breast cancer bone metastases as imaged noninvasively using VCT, MRI and DCE-MRI in a longitudinal *in vivo* study *Int. J. Cancer* **128** 2453–62

Baumgartner G, Gomar-Hoss C, Sakr L, Ulsperger E and Wogritsch C 1998 The impact of extracellular matrix on the chemoresistance of solid tumors—experimental and clinical results of hyaluronidase as additive to cytostatic chemotherapy *Cancer Lett.* **131** 85–99

Beauvais D M, Burbach B J and Rapraeger A C 2004 The syndecan-1 ectodomain regulates alphavbeta3 integrin activity in human mammary carcinoma cells *J. Cell Biol.* **167** 171–81

Beauvais D M, Ell B J, McWhorter A R and Rapraeger A C 2009 Syndecan-1 regulates alphavbeta3 and alphavbeta5 integrin activation during angiogenesis and is blocked by synstatin, a novel peptide inhibitor *J. Exp. Med.* **206** 691–705

Berman A E, Kozlova N I and Morozevich G E 2003 Integrins: structure and signaling *Biochemistry (Mosc)* **68** 1284–99

Bianchi-Smiraglia A, Paesante S and Bakin A V 2012 Integrin beta5 contributes to the tumorigenic potential of breast cancer cells through the Src-FAK and MEK-ERK signaling pathways *Oncogene* **32** (25) 3049–58

Bourguignon L Y 2008 Hyaluronan-mediated CD44 activation of RhoGTPase signaling and cytoskeleton function promotes tumor progression *Semin. Cancer Biol.* **18** 251–9

Bourguignon L Y 2009 Hyaluronan-mediated CD44 interaction with receptor and non-receptor kinases promotes oncogenic signaling, cytoskeleton activation and tumor progression *Hyaluronan in Cancer Biology* ed R Stern (San Diego, CA: Academic) pp 89–107

Bourguignon L Y, Singleton P A, Diedrich F, Stern R and Gilad E 2004 CD44 interaction with Na+-H+ exchanger (NHE1) creates acidic microenvironments leading to hyaluronidase-2 and cathepsin B activation and breast tumor cell invasion *J. Biol. Chem.* **279** 26991–7007

Braun S, Bitton-Worms K and LeRoith D 2011 The link between the metabolic syndrome and cancer *Int. J. Biol. Sci.* **7** 1003–15

Brito L G, Schiavon V F, Andrade J M, Tiezzi D G, Peria F M and Marana H R 2011 Expression of hypoxia-inducible factor 1-alpha and vascular endothelial growth factor-C in locally advanced breast cancer patients *Clinics (Sao Paulo)* **66** 1313–20 PMID: 21915477

Brown T J 2008 The development of hyaluronan as a drug transporter and excipient for chemotherapeutic drugs *Curr. Pharm. Biotechnol.* **9** 253–60

Camenisch T D, Spicer A P, Brehm-Gibson T, Biesterfeldt J, Augustine M L, Calabro A Jr, Kubalak S, Klewer S E and McDonald J A 2000 Disruption of hyaluronan synthase-2 abrogates normal cardiac morphogenesis and hyaluronan-mediated transformation of epithelium to mesenchyme *J. Clin. Invest.* **106** 349–60

Capparelli C *et al* 2012 CTGF drives autophagy, glycolysis and senescence in cancer-associated fibroblasts via HIF1 activation, metabolically promoting tumor growth *Cell Cycle* **11** 2272–84

Cascio S, Bartella V, Auriemma A, Johannes G J, Russo A, Giordano A and Surmacz E 2008 Mechanism of leptin expression in breast cancer cells: role of hypoxia-inducible factor-1alpha *Oncogene* **27** 540–7

Chakravarthy D, Green A R, Green V L, Kerin M J and Speirs V 1999 Expression and secretion of TGF-beta isoforms and expression of TGF-beta-receptors I, II and III in normal and neoplastic human breast *Int. J. Oncol.* **15** 187–94

Chan W K, Yao G, Gu Y Z and Bradfield C A 1999 Cross-talk between the aryl hydrocarbon receptor and hypoxia inducible factor signaling pathways. Demonstration of competition and compensation *J. Biol. Chem.* **274** 12115–23

Chandramouli A, Simundza J, Pinderhughes A and Cowin P 2011 Choreographing metastasis to the tune of LTBP *J. Mammary Gland Biol. Neoplasia* **16** 67–80

Charpin C *et al* 2012 Validation of an immunohistochemical signature predictive of 8-year outcome for patients with breast carcinoma *Int. J. Cancer* **131** E236–43

Chen W and Gueron M 1992 The inhibition of bovine heart hexokinase by 2-deoxy-D-glucose-6-phosphate: characterization by 31 P NMR and metabolic implications *Biochimie* **74** 867–73

Cheng C, Yaffe M B and Sharp P A 2006 A positive feedback loop couples Ras activation and CD44 alternative splicing *Genes Dev.* **20** 1715–20

Cheng G, Zielonka J, Dranka B P, McAllister D, Mackinnon A C Jr, Joseph J and Kalyanaraman B 2012 Mitochondria-targeted drugs synergize with 2-deoxyglucose to trigger breast cancer cell death *Cancer Res.* **72** 2634–44

Chiavarina B, Martinez-Outschoorn U E, Whitaker-Menezes D, Howell A, Tanowitz H B, Pestell R G, Sotgia F and Lisanti M P 2012 Metabolic reprogramming and two-compartment tumor metabolism: Opposing role(s) of HIF1alpha and HIF2alpha in tumor-associated fibroblasts and human breast cancer cells *Cell Cycle* **11** 3280–9

Chiavarina B *et al* 2010 HIF1-alpha functions as a tumor promoter in cancer associated fibroblasts, and as a tumor suppressor in breast cancer cells: Autophagy drives compartment-specific oncogenesis *Cell Cycle* **9** 3534–51

Chien W, O'Kelly J, Lu D, Leiter A, Sohn J, Yin D, Karlan B, Vadgama J, Lyons K M and Koeffler H P 2011 Expression of connective tissue growth factor (CTGF/CCN2) in breast cancer cells is associated with increased migration and angiogenesis *Int. J. Oncol.* **38** 1741–7

Choi J Y, Jang Y S, Min S Y and Song J Y 2011 Overexpression of MMP-9 and HIF-1alpha in breast cancer cells under hypoxic conditions *J. Breast Cancer* **14** 88–95

Chouaib S, Messai Y, Couve S, Escudier B, Hasmim M and Noman M Z 2012 Hypoxia promotes tumor growth in linking angiogenesis to immune escape *Front Immunol.* **3** 21

Colone M *et al* 2008 The multidrug transporter P-glycoprotein: a mediator of melanoma invasion? *J. Invest. Dermatol.* **128** 957–71

Conklin M W, Eickhoff J C, Riching K M, Pehlke C A, Eliceiri K W, Provenzano P P, Friedl A and Keely P J 2011 Aligned collagen is a prognostic signature for survival in human breast carcinoma *Am. J. Pathol.* **178** 1221–32

Connolly E C, Freimuth J and Akhurst R J 2012 Complexities of TGF-beta targeted cancer therapy *Int. J. Biol. Sci.* **8** 964–78

Corzo C A *et al* 2010 HIF-1alpha regulates function and differentiation of myeloid-derived suppressor cells in the tumor microenvironment *J. Exp. Med.* **207** 2439–53

Coussens L M, Fingleton B and Matrisian L M 2002 Matrix metalloproteinase inhibitors and cancer: trials and tribulations *Science* **295** 2387–92

Curran C S and Bertics P J 2012 Lactoferrin regulates an axis involving CD11b and CD49d integrins and the chemokines MIP-1alpha and MCP-1 in GM-CSF-treated human primary eosinophils *J. Inter. Cytok. Res.* **32** 450–61

Denison M S and Nagy S R 2003 Activation of the aryl hydrocarbon receptor by structurally diverse exogenous and endogenous chemicals *Annu. Rev. Pharmacol. Toxicol.* **43** 309–34

DiMilla P A, Stone J A, Quinn J A, Albelda S M and Lauffenburger D A 1993 Maximal migration of human smooth muscle cells on fibronectin and type IV collagen occurs at an intermediate attachment strength *J. Cell Biol.* **122** 729–37

Dobolyi A, Vincze C, Pal G and Lovas G 2012 The neuroprotective functions of transforming growth factor Beta proteins *Int. J. Mol. Sci.* **13** 8219–58

Doedens A L, Stockmann C, Rubinstein M P, Liao D, Zhang N, DeNardo D G, Coussens L M, Karin M, Goldrath A W and Johnson R S 2010 Macrophage expression of hypoxia-inducible factor-1 alpha suppresses T-cell function and promotes tumor progression *Cancer Res.* **70** 7465–75

Donovan J and Slingerland J 2000 Transforming growth factor-beta and breast cancer: Cell cycle arrest by transforming growth factor-beta and its disruption in cancer *Breast Cancer Res.* **2** 116–24

Drabsch Y and ten Dijke P 2011 TGF-beta signaling in breast cancer cell invasion and bone metastasis *J. Mammary Gland Biol. Neoplasia* **16** 97–108

Droppelmann C A, Gutierrez J, Vial C and Brandan E 2009 Matrix metalloproteinase-2-deficient fibroblasts exhibit an alteration in the fibrotic response to connective tissue growth factor/ CCN2 because of an increase in the levels of endogenous fibronectin *J. Biol. Chem.* **284** 13551–61

Dubrovska A, Hartung A, Bouchez L C, Walker J R, Reddy V A, Cho C Y and Schultz P G 2012 CXCR4 activation maintains a stem cell population in tamoxifen-resistant breast cancer cells through AhR signalling *Br. J. Cancer* **107** 43–52

Dunn L K, Mohammad K S, Fournier P G, McKenna C R, Davis H W, Niewolna M, Peng X H, Chirgwin J M and Guise T A 2009 Hypoxia and TGF-beta drive breast cancer bone metastases through parallel signaling pathways in tumor cells and the bone microenvironment *PLoS One* **4** e6896

Dwarakanath B and Jain V 2009 Targeting glucose metabolism with 2-deoxy-D-glucose for improving cancer therapy *Future Oncol.* **5** 581–5

Eckhardt B L, Francis P A, Parker B S and Anderson R L 2012 Strategies for the discovery and development of therapies for metastatic breast cancer *Nat. Rev. Drug Discov.* **11** 479–97

Egeblad M, Rasch M G and Weaver V M 2010 Dynamic interplay between the collagen scaffold and tumor evolution *Curr. Opin. Cell Biol.* **22** 697–706

Evanko S P, Tammi M I, Tammi R H and Wight T N 2007 Hyaluronan-dependent pericellular matrix *Adv. Drug Deliv. Rev.* **59** 1351–65

Fadok V A, Bratton D L, Konowal A, Freed P W, Westcott J Y and Henson P M 1998 Macrophages that have ingested apoptotic cells *in vitro* inhibit proinflammatory cytokine production through autocrine/paracrine mechanisms involving TGF-beta, PGE2, and PAF *J. Clin. Invest.* **101** 890–8

Fogerty F J, Akiyama S K, Yamada K M and Mosher D F 1990 Inhibition of binding of fibronectin to matrix assembly sites by anti-integrin (alpha 5 beta 1) antibodies *J. Cell Biol.* **111** 699–708

Fokas E, McKenna W G and Muschel R J 2012 The impact of tumor microenvironment on cancer treatment and its modulation by direct and indirect antivascular strategies *Cancer Metast. Rev.* **31** 823–42

Friggeri A, Yang Y, Banerjee S, Park Y J, Liu G and Abraham E 2010 HMGB1 inhibits macrophage activity in efferocytosis through binding to the alphavbeta3-integrin *Am. J. Physiol. Cell Physiol.* **299** C1267–76

Frost G I 2007 Recombinant human hyaluronidase (rHuPH20): an enabling platform for subcutaneous drug and fluid administration *Expert Opin. Drug Deliv.* **4** 427–40

Ganapathy V, Banach-Petrosky W, Xie W, Kareddula A, Nienhuis H, Miles G and Reiss M 2012 Luminal breast cancer metastasis is dependent on estrogen signaling *Clin. Exp. Metast.* **29** 493–509

Gao P *et al* 2007 HIF-dependent antitumorigenic effect of antioxidants *in vivo Cancer Cell* **12** 230–8

Garamszegi N, Garamszegi S P, Shehadeh L A and Scully S P 2009 Extracellular matrix-induced gene expression in human breast cancer cells *Mol. Cancer Res.* **7** 319–29

Gelse K, Poschl E and Aigner T 2003 Collagens—structure, function, and biosynthesis *Adv. Drug Deliv. Rev.* **55** 1531–46

Generali D *et al* 2006 Hypoxia-inducible factor-1alpha expression predicts a poor response to primary chemoendocrine therapy and disease-free survival in primary human breast cancer *Clin. Cancer Res.* **12** 4562–8

Ghanta K S, Pakala S B, Reddy S D, Li D Q, Nair S S and Kumar R 2011 MTA1 coregulation of transglutaminase 2 expression and function during inflammatory response *J. Biol. Chem.* **286** 7132–8

Ghatak S, Misra S and Toole B P 2005 Hyaluronan regulates constitutive ErbB2 phosphorylation and signal complex formation in carcinoma cells *J. Biol. Chem.* **280** 8875–83

Gibbs P, Brown T J, Ng R, Jennens R, Cinc E, Pho M, Michael M and Fox R M 2009 A pilot human evaluation of a formulation of irinotecan and hyaluronic acid in 5-fluorouracil-refractory metastatic colorectal cancer patients *Chemotherapy* **55** 49–59

Gilg A G, Tye S L, Tolliver L B, Wheeler W G, Visconti R P, Duncan J D, Kostova F V, Bolds L N, Toole B P and Maria B L 2008 Targeting hyaluronan interactions in malignant gliomas and their drug-resistant multipotent progenitors *Clin. Cancer Res.* **14** 1804–13

Goh F G, Piccinini A M, Krausgruber T, Udalova I A and Midwood K S 2010 Transcriptional regulation of the endogenous danger signal tenascin-C: a novel autocrine loop in inflammation *J. Immunol.* **184** 2655–62

Goodman S L, Risse G and von der Mark K 1989 The E8 subfragment of laminin promotes locomotion of myoblasts over extracellular matrix *J. Cell Biol.* **109** 799–809

Gould V E, Koukoulis G K and Virtanen I 1990 Extracellular matrix proteins and their receptors in the normal, hyperplastic and neoplastic breast *Cell Differ. Dev.* **32** 409–16

Grande J P, Melder D C and Zinsmeister A R 1997 Modulation of collagen gene expression by cytokines: stimulatory effect of transforming growth factor-beta1, with divergent effects of epidermal growth factor and tumor necrosis factor-alpha on collagen type I and collagen type IV *J. Lab Clin. Med.* **130** 476–86

Green C E, Liu T, Montel V, Hsiao G, Lester R D, Subramaniam S, Gonias S L and Klemke R L 2009 Chemoattractant signaling between tumor cells and macrophages regulates cancer cell migration, metastasis and neovascularization *PLoS One* **4** e6713

Green J A and Yamada K M 2007 Three-dimensional microenvironments modulate fibroblast signaling responses *Adv. Drug Deliv. Rev.* **59** 1293–8

Grimshaw M J, Naylor S and Balkwill F R 2002a Endothelin-2 is a hypoxia-induced autocrine survival factor for breast tumor cells *Mol. Cancer Ther.* **1** 1273–81

Grimshaw M J, Wilson J L and Balkwill F R 2002b Endothelin-2 is a macrophage chemo-attractant: implications for macrophage distribution in tumors *Eur. J. Immunol.* **32** 2393–400

Grotendorst G R, Rahmanie H and Duncan M R 2004 Combinatorial signaling pathways determine fibroblast proliferation and myofibroblast differentiation *Faseb J.* **18** 469–79

Guo Y P, Martin L J, Hanna W, Banerjee D, Miller N, Fishell E, Khokha R and Boyd N F 2001 Growth factors and stromal matrix proteins associated with mammographic densities *Cancer Epidemiol. Biomarkers Prev.* **10** 243–8 PMID: 11303594

Guttery D S, Shaw J A, Lloyd K, Pringle J H and Walker R A 2010 Expression of tenascin-C and its isoforms in the breast *Cancer Metast. Rev.* **29** 595–606

Hao N B, Lu M H, Fan Y H, Cao Y L, Zhang Z R and Yang S M 2012 Macrophages in tumor microenvironments and the progression of tumors *Clin. Dev. Immunol.* **2012** 948098

Harmey J H, Dimitriadis E, Kay E, Redmond H P and Bouchier-Hayes D 1998 Regulation of macrophage production of vascular endothelial growth factor (VEGF) by hypoxia and transforming growth factor beta-1 *Ann. Surg. Oncol.* **5** 271–8

Haylock D N and Nilsson S K 2006 The role of hyaluronic acid in hemopoietic stem cell biology *Regen. Med.* **1** 437–45

Heck J N, Ponik S M, Garcia-Mendoza M G, Pehlke C A, Inman D R, Eliceiri K W and Keely P J 2012 Microtubules regulate GEF-H1 in response to extracellular matrix stiffness *Mol. Biol. Cell* **23** 2583–92

Hehlgans S, Haase M and Cordes N 2007 Signalling via integrins: implications for cell survival and anticancer strategies *Biochim. Biophys. Acta* **1775** 163–80

Heider K H, Kuthan H, Stehle G and Munzert G 2004 CD44v6: a target for antibody-based cancer therapy *Cancer Immunol. Immunother.* **53** 567–79

Hielscher A C, Qiu C and Gerecht S 2012 Breast cancer cell-derived matrix supports vascular morphogenesis *Am. J. Physiol. Cell Physiol.* **302** C1243–56

Hitchon C, Wong K, Ma G, Reed J, Lyttle D and El-Gabalawy H 2002 Hypoxia-induced production of stromal cell-derived factor 1 (CXCL12) and vascular endothelial growth factor by synovial fibroblasts *Arthritis Rheum.* **46** 2587–97

Hockel M and Vaupel P 2001 Tumor hypoxia: definitions and current clinical, biologic, and molecular aspects *J. Natl Cancer Inst.* **93** 266–76

Hoffman C, Park S H, Daley E, Emson C, Louten J, Sisco M, de Waal Malefyt R and Grunig G 2011 Interleukin-19: a constituent of the regulome that controls antigen presenting cells in the lungs and airway responses to microbial products *PLoS One* **6** e27629

Hollier B G, Evans K and Mani S A 2009 The epithelial-to-mesenchymal transition and cancer stem cells: a coalition against cancer therapies *J. Mammary Gland Biol. Neoplasia* **14** 29–43

Hong H H, Uzel M I, Duan C, Sheff M C and Trackman P C 1999 Regulation of lysyl oxidase, collagen, and connective tissue growth factor by TGF-beta1 and detection in human gingiva *Lab. Invest.* **79** 1655–67 PMID: 10616214

Hood J D and Cheresh D A 2002 Role of integrins in cell invasion and migration *Nat. Rev. Cancer* **2** 91–100

Horiguchi M, Ota M and Rifkin D B 2012 Matrix control of transforming growth factor-beta function *J. Biochem.* **152**(4) 321–9

Hosono K, Nishida Y, Knudson W, Knudson C B, Naruse T, Suzuki Y and Ishiguro N 2007 Hyaluronan oligosaccharides inhibit tumorigenicity of osteosarcoma cell lines MG-63 and LM-8 *in vitro* and *in vivo* via perturbation of hyaluronan-rich pericellular matrix of the cells *Am. J. Pathol.* **171** 274–86

Hsing C H, Cheng H C, Hsu Y H, Chan C H, Yeh C H, Li C F and Chang M S 2012 Upregulated IL-19 in breast cancer promotes tumor progression and affects clinical outcome *Clin. Cancer Res.* **18** 713–25

Hynes R O 1999 The dynamic dialogue between cells and matrices: implications of fibronectin's elastici *Proc. Natl Acad. Sci. USA* **96** 2588–90

Hynes R O and Naba A 2012 Overview of the matrisome—an inventory of extracellular matrix constituents and functions *Cold Spring Harb Perspect. Biol.* **4** a004903

Iacobuzio-Donahue C A, Argani P, Hempen P M, Jones J and Kern S E 2002 The desmoplastic response to infiltrating breast carcinoma: gene expression at the site of primary invasion and implications for comparisons between tumor types *Cancer Res.* **62** 5351–7 PMID: 12235006

Ilic D *et al* 2004 FAK promotes organization of fibronectin matrix and fibrillar adhesions *J. Cell. Sci.* **117** 177–87

Itano N and Kimata K 2008 Altered hyaluronan biosynthesis in cancer progression *Semin. Cancer Biol.* **18** 268–74

Ito A, Nakajima S, Sasaguri Y, Nagase H and Mori Y 1995 Co-culture of human breast adenocarcinoma MCF-7 cells and human dermal fibroblasts enhances the production of matrix metalloproteinases 1, 2 and 3 in fibroblas *Br. J. Cancer* **71** 1039–45

Jackson D G 2009 Immunological functions of hyaluronan and its receptors in the lymphatics *Immunol. Rev.* **230** 216–31

Janik M E, Litynska A and Vereecken P 2010 Cell migration—the role of integrin glycosylation *Biochim. Biophys. Acta* **1800** 545–55

Janssens K, ten Dijke P, Janssens S and Van Hul W 2005 Transforming growth factor-beta1 to the bone *Endocr. Rev.* **26** 743–74

Jian J, Yang Q, Dai J, Eckard J, Axelrod D, Smith J and Huang X 2011 Effects of iron deficiency and iron overload on angiogenesis and oxidative stress—a potential dual role for iron in breast cancer *Free Radic. Biol. Med.* **50** 841–7

Jiang D, Liang J and Noble P W 2007 Hyaluronan in tissue injury and repair *Annu. Rev. Cell Dev. Biol.* **23** 435–61

Jiang W G, Ablin R, Douglas-Jones A and Mansel R E 2003 Expression of transglutaminases in human breast cancer and their possible clinical significance *Oncol. Rep.* **10** 2039–44

Jin F, Brockmeier U, Otterbach F and Metzen E 2012 New insight into the SDF-1/CXCR4 axis in a breast carcinoma model: hypoxia-induced endothelial SDF-1 and tumor cell CXCR4 are required for tumor cell intravasation *Mol. Cancer Res.* **10** 1021–31

Jin H, Su J, Garmy-Susini B, Kleeman J and Varner J 2006 Integrin alpha4beta1 promotes monocyte trafficking and angiogenesis in tumors *Cancer Res.* **66** 2146–52

Jin L *et al* 1997 Expression of interleukin-1beta in human breast carcinoma *Cancer* **80** 421–34

Johnson K J, Sage H, Briscoe G and Erickson H P 1999 The compact conformation of fibronectin is determined by intramolecular ionic interactions *J. Biol. Chem.* **274** 15473–9

Johnson P and Ruffell B 2009 CD44 and its role in inflammation and inflammatory diseases *Inflamm. Allergy Drug Targets* **8** 208–20

Joyce J A and Pollard J W 2009 Microenvironmental regulation of metastasis *Nat. Rev. Cancer* **9** 239–52

Kadowaki M *et al* 2011 Identification of vitronectin as a novel serum marker for early breast cancer detection using a new proteomic approach *J. Cancer Res. Clin. Oncol.* **137** 1105–15

Kakkad S M, Solaiyappan M, O'Rourke B, Stasinopoulos I, Ackerstaff, Raman V, Bhujwalla Z M and Glunde K 2010 Hypoxic tumor microenvironments reduce collagen I fiber density *Neoplasia* **12** 608–17

Kallergi G, Markomanolaki H, Giannoukaraki V, Papadaki M A, Strati A, Lianidou E S, Georgoulias V, Mavroudis D and Agelaki S 2009 Hypoxia-inducible factor-1alpha and vascular endothelial growth factor expression in circulating tumor cells of breast cancer patients *Breast Cancer Res.* **11** R84

Kalluri R and Weinberg R A 2009 The basics of epithelial-mesenchymal transition *J. Clin. Invest.* **119** 1420–8

Karkampouna S, Ten Dijke P, Dooley S and Kruithof-de Julio M 2012 TGFbeta signaling in liver regeneration *Curr. Pharm. Des.* **18** 4103–13

Keely P J 2011 Mechanisms by which the extracellular matrix and integrin signaling act to regulate the switch between tumor suppression and tumor promotion *J. Mammary Gland Biol. Neoplasia* **16** 205–19

Khalili P, Arakelian A, Chen G, Plunkett M L, Beck I, Parry G C, Donate F, Shaw D E, Mazar A P and Rabbani S A 2006 A non-RGD-based integrin binding peptide (ATN-161) blocks breast cancer growth and metastasis *in vivo Mol. Cancer Ther.* **5** 2271–80

Kharaishvili G, Cizkova M, Bouchalova K, Mgebrishvili G, Kolar Z and Bouchal J 2011 Collagen triple helix repeat containing 1 protein, periostin and versican in primary and metastatic breast cancer: an immunohistochemical study *J. Clin. Pathol.* **64** 977–82

Khasraw M and Bell R 2012 Primary systemic therapy in HER2-amplified breast cancer: a clinical review *Exp. Rev. Anticancer Ther.* **12** 1005–13

Kim J W *et al* 2012 Loss of fibroblast HIF-1alpha accelerates tumorigenesis *Cancer Res.* **72** 3187–95

Kingsley D M 1994 The TGF-beta superfamily: new members, new receptors, and new genetic tests of function in different organisms *Genes Dev.* **8** 133–46

Klass C M, Couchman J R and Woods A 2000 Control of extracellular matrix assembly by syndecan-2 proteoglycan *J. Cell Sci.* **113** 493–506 PMID: 10639336

Kleinman H K, Klebe R J and Martin G R 1981 Role of collagenous matrices in the adhesion and growth of cells *J. Cell Biol.* **88** 473–85

Knowles H J and Harris A L 2001 Hypoxia and oxidative stress in breast cancer. Hypoxia and tumourigenesis *Breast Cancer Res.* **3** 318–22

Knudson W, Biswas C, Li XQ, Nemec R E and Toole B P 1989 The role and regulation of tumour-associated hyaluronan *Ciba Found Symp.* **143** 150–9 PMID: 2680343

Knudson W, Chow G and Knudson C B 2002 CD44-mediated uptake and degradation of hyaluronan *Matrix Biol.* **21** 15–23

Kojima Y *et al* 2010 Autocrine TGF-beta and stromal cell-derived factor-1 (SDF-1) signaling drives the evolution of tumor-promoting mammary stromal myofibroblasts *Proc. Natl Acad. Sci. USA* **107** 20009–14

Koli K, Lohi J, Hautanen A and Keski-Oja J 1991 Enhancement of vitronectin expression in human HepG2 hepatoma cells by transforming growth factor-beta 1 *Eur. J. Biochem.* **199** 337–45

Kondo S, Kubota S, Shimo T, Nishida T, Yosimichi G, Eguchi T, Sugahara T and Takigawa M 2002 Connective tissue growth factor increased by hypoxia may initiate angiogenesis in collaboration with matrix metalloproteinases *Carcinogenesis* **23** 769–76

Korns D, Frasch S C, Fernandez-Boyanapalli R, Henson P M and Bratton D L 2011 Modulation of macrophage efferocytosis in inflammation *Front Immunol.* **2** 57

Krause D S, Lazarides K, von Andrian U H and Van Etten R A 2006 Requirement for CD44 in homing and engraftment of BCR-ABL-expressing leukemic stem cells *Nat. Med.* **12** 1175–80

Krock B L, Skuli N and Simon M C 2011 Hypoxia-induced angiogenesis: good and evil *Genes Cancer* **2** 1117–33

Kudo A 2011 Periostin in fibrillogenesis for tissue regeneration: periostin actions inside and outside the cell *Cell Mol. Life Sci.* **68** 3201–7

Kuperwasser C, Chavarria T, Wu M, Magrane G, Gray J W, Carey L, Richardson A and Weinberg R A 2004 Reconstruction of functionally normal and malignant human breast tissues in mice *Proc. Natl Acad. Sci. USA* **101** 4966–71

Kurtoglu M, Gao N, Shang J, Maher J C, Lehrman M A, Wangpaichitr M, Savaraj N, Lane A N and Lampidis T J 2007 Under normoxia, 2-deoxy-D-glucose elicits cell death in select tumor types not by inhibition of glycolysis but by interfering with N-linked glycosylation *Mol. Cancer Ther.* **6** 3049–58

Kurtzman S H, Anderson K H, Wang Y, Miller L J, Renna M, Stankus M, Lindquist R R, Barrows G and Kreutzer D L 1999 Cytokines in human breast cancer: IL-1alpha and IL-1beta expression *Oncol. Rep.* **6** 65–70

Kuwabara K *et al* 1995 Hypoxia-mediated induction of acidic/basic fibroblast growth factor and platelet-derived growth factor in mononuclear phagocytes stimulates growth of hypoxic endothelial cells *Proc. Natl Acad. Sci. USA* **92** 4606–10

Lanigan F, O'Connor D, Martin F and Gallagher W M 2007 Molecular links between mammary gland development and breast cancer *Cell Mol. Life Sci.* **64** 3159–84

Lanzafame S, Emmanuele C and Torrisi A 1996 Correlation of alpha 2 beta 1 integrin expression with histological type and hormonal receptor status in breast carcinomas *Pathol. Res. Pract.* **192** 1031–8

Laoui D, Movahedi K, Van Overmeire E, Van den Bossche J, Schouppe E, Mommer C, Nikolaou A, Morias Y, De Baetselier P and Van Ginderachter J A 2011 Tumor-associated macrophages in breast cancer: distinct subsets, distinct functions *Int. J. Dev. Biol.* **55** 861–7

Laragione T, Bonetto V, Casoni F, Massignan T, Bianchi G, Gianazza E and Ghezzi P 2003 Redox regulation of surface protein thiols: identification of integrin alpha-4 as a molecular target by using redox proteomics *Proc. Natl Acad. Sci. USA* **100** 14737–41

Laskin D L 2009 Macrophages and inflammatory mediators in chemical toxicity: a battle of forces *Chem. Res. Toxicol.* **22** 1376–85

Le Y J and Corry P M 1999 Hypoxia-induced bFGF gene expression is mediated through the JNK signal transduction pathway *Mol. Cell Biochem.* **202** 1–8

Lee J L, Wang M J, Sudhir P R and Chen J Y 2008 CD44 engagement promotes matrix-derived survival through the CD44-SRC-integrin axis in lipid rafts *Mol. Cell Biol.* **28** 5710–23

Lehmann G M *et al* 2011 The aryl hydrocarbon receptor ligand ITE inhibits TGFbeta1-induced human myofibroblast differentiation *Am. J. Pathol.* **178** 1556–67

Lesley J, Hascall V C, Tammi M and Hyman R 2000 Hyaluronan binding by cell surface CD44 *J. Biol Chem.* **275** 26967–75

Levental K R *et al* 2009 Matrix crosslinking forces tumor progression by enhancing integrin signaling *Cell* **139** 891–906

Li M O, Wan Y Y, Sanjabi S, Robertson A K and Flavell R A 2006 Transforming growth factor-beta regulation of immune responses *Annu. Rev. Immunol.* **24** 99–146

Li Z, Pang Y, Gara S K, Achyut B R, Heger C, Goldsmith P K, Lonning S and Yang L 2012 Gr -1+CD11b+ cells are responsible for tumor promoting effect of TGF-beta in breast cancer progressi *Int. J. Cancer* **131** 2584–95

Lin S, Wan S, Sun L, Hu J, Fang D, Zhao R, Yuan S and Zhang L 2012 Chemokine C-C motif receptor 5 and C-C motif ligand 5 promote cancer cell migration under hypoxia *Cancer Sci.* **103** 904–12

Lokeshwar V B and Selzer M G 2008 Hyaluronidase: both a tumor promoter and suppressor *Semin Cancer Biol.* **18** 281–7

Lopes C C, Dietrich C P and Nader H B 2006 Specific structural features of syndecans and heparan sulfate chains are needed for cell signaling *Braz. J. Med. Biol. Res.* **39** 157–67

Lopez-Lazaro M 2009 Role of oxygen in cancer: looking beyond hypoxia *Anticancer Agents Med. Chem.* **9** 517–25 PMID: 19519293

Lu X and Kang Y 2007 Organotropism of breast cancer metastasis *J. Mammary Gland Biol. Neoplasia* **12** 153–62 PMID: 17566854

Luo B H, Carman C V and Springer T A 2007 Structural basis of integrin regulation and signaling *Annu. Rev. Immunol.* **25** 619–47

Ma J et al 2012 Ras homolog gene family, member A promotes p53 degradation and vascular endothelial growth factor-dependent angiogenesis through an interaction with murine double minute 2 under hypoxic conditions *Cancer* **118** 4105–16

Maeda T, Alexander C M and Friedl A 2004 Induction of syndecan-1 expression in stromal fibroblasts promotes proliferation of human breast cancer cells *Cancer Res.* **64** 612–21

Maeda T, Desouky J and Friedl A 2006 Syndecan-1 expression by stromal fibroblasts promotes breast carcinoma growth *in vivo* and stimulates tumor angiogenesis *Oncogene* **25** 1408–12

Maity G, Choudhury P R, Sen T, Ganguly K K, Sil H and Chatterjee A 2011 Culture of human breast cancer cell line (MDA-MB-231) on fibronectin-coated surface induces pro-matrix metalloproteinase-9 expression and activity *Tumour Biol.* **32** 129–38

Mansfield C M 1993 A review of the etiology of breast cancer *J. Natl Med. Assoc.* **85** 217–21 PMID: 8474136

Mao Y and Schwarzbauer J E 2005 Fibronectin fibrillogenesis, a cell-mediated matrix assembly process *Matrix Biol.* **24** 389–99

Margadant C and Sonnenberg A 2010 Integrin-TGF-beta crosstalk in fibrosis, cancer and wound healing *EMBO Rep.* **11** 97–105

Maxwell C A, McCarthy J and Turley E 2008 Cell-surface and mitotic-spindle RHAMM: moonlighting or dual oncogenic functions? *J. Cell Sci.* **121** 925–32

McQuade K J, Beauvais D M, Burbach B J and Rapraeger A C 2006 Syndecan-1 regulates alphavbeta5 integrin activity in B82L fibroblasts *J. Cell Sci.* **119** 2445–56

Medrek C, Ponten F, Jirstrom K and Leandersson K 2012 The presence of tumor associated macrophages in tumor stroma as a prognostic marker for breast cancer patients *BMC Cancer* **12** 306

Mierke C T 2011 Cancer cells regulate biomechanical properties of human microvascular endolethial cells *J. Biol. Chem.* **286** 40025–37

Mierke C T 2013 The role of focal adhesion kinase in the regulation of cellular mechanical properties *Phys. Biol.* **10** 065005

Mierke C T, Kollmannsberger P, Paranhos-Zitterbart D, Diez G, Koch T M, Marg S, Ziegler W H, Goldmann W H and Fabry B 2010 Vinculin facilitates cell invasion into 3D collagen matrices *J. Biol. Chem.* **285** 13121–30

Mierke C T, Kollmannsberger P, Paranhos-Zitterbart D, Smith J, Fabry B and Goldmann W H 2008 Mechano-coupling and regulation of contractility by the vinculin tail domain *Biophys. J.* **94** 661–70

Migneco G *et al* 2010 Glycolytic cancer associated fibroblasts promote breast cancer tumor growth, without a measurable increase in angiogenesis: evidence for stromal–epithelial metabolic coupling *Cell Cycle* **9** 2412–22

Miletti-Gonzalez K E, Chen S, Muthukumaran N, Saglimbeni G N, Wu X, Yang J, Apolito K, Shih W J, Hait W N and Rodríguez-Rodríguez L 2005 The CD44 receptor interacts with P-glycoprotein to promote cell migration and invasion in cancer *Cancer Res.* **65** 6660–7

Moro L, Colombi M, Molinari Tosatti M P and Barlati S 1992 Study of fibronectin and mRNA in human laryngeal and ectocervical carcinomas by *in situ* hybridization and image analysis *Int. J. Cancer* **51** 692–7

Moses H and Barcellos-Hoff M H 2011 TGF-beta biology in mammary development and breast cancer *Cold Spring Harb Perspect. Biol.* **3** a003277

Mulgrew K *et al* 2006 Direct targeting of alphavbeta3 integrin on tumor cells with a monoclonal antibody, Abegrin *Mol. Cancer Ther.* **5** 3122–9

Munoz-Najar U M, Neurath K M, Vumbaca F and Claffey K P 2006 Hypoxia stimulates breast carcinoma cell invasion through MT1-MMP and MMP-2 activation *Oncogene* **25** 2379–92

Murray P J and Wynn T A 2011 Protective and pathogenic functions of macrophage subsets *Nat. Rev. Immunol.* **11** 723–37

Nagano O and Saya H 2004 Mechanism and biological significance of CD44 cleavage *Cancer Sci.* **95** 930–5

Naor D *et al* 2007 CD44 involvement in autoimmune inflammations: the lesson to be learned from CD44-targeting by antibody or from knockout mice *Ann. N Y Acad. Sci.* **1110** 233–47

Naor D, Wallach-Dayan S B, Zahalka M A and Sionov R V 2008 Involvement of CD44, a molecule with a thousand faces, in cancer dissemination *Semin. Cancer Biol.* **18** 260–7

Nelson C M and Bissell M J 2006 Of extracellular matrix, scaffolds, and signaling: tissue architecture regulates development, homeostasis, and cancer *Annu. Rev. Cell Dev. Biol.* **22** 287–309

Nemeth J A, Nakada M T, Trikha M, Lang Z, Gordon M S, Jayson G C, Corringham R, Prabhakar U, Davis H M and Beckman R A 2007 Alpha-v integrins as therapeutic targets in oncology *Cancer Invest.* **25** 632–46

Nishizuka I *et al* 2002 Analysis of gene expression involved in brain metastasis from breast cancer using cDNA microarray *Breast Cancer* **9** 26–32

Noel A *et al* 1998 Inhibition of stromal matrix metalloproteases: effects on breast-tumor promotion by fibroblasts *Int. J. Cancer* **76** 267–73

Noel A, Munaut C, Boulvain A, Calberg-Bacq C M, Lambert C A, Nusgens B, Lapiere C M and Foidart J M 1992a Modulation of collagen and fibronectin synthesis in fibroblasts by normal and malignant cells *J. Cell Biochem.* **48** 150–61

Noel A, Munaut C, Nusgens B, Foidart J M and Lapiere C M 1992b The stimulation of fibroblasts' collagen synthesis by neoplastic cells is modulated by the extracellular matrix *Matrix* **12** 213–20

Nurminskaya M V and Belkin A M 2012 Cellular functions of tissue transglutaminase *Int. Rev. Cell Mol. Biol.* **294** 1–97

Ogawa T, Tsubota Y, Hashimoto J, Kariya Y and Miyazaki K 2007 The short arm of laminin gamma2 chain of laminin-5 (laminin-332) binds syndecan-1 and regulates cellular adhesion and migration by suppressing phosphorylation of integrin beta4 chain *Mol. Biol. Cell* **18** 1621–33

Oh K, Ko E, Kim H S, Park A K, Moon H G, Noh D Y and Lee D S 2011 Transglutaminase 2 facilitates the distant hematogenous metastasis of breast cancer by modulating interleukin-6 in cancer cells *Breast Cancer Res.* **13** R96

Ohtake F, Fujii-Kuriyama Y, Kawajiri K and Kato S 2011 Cross-talk of dioxin and estrogen receptor signals through the ubiquitin system *J. Steroid Biochem. Mol. Biol.* **127** 102–7

Orimo A and Weinberg R A 2006 Stromal fibroblasts in cancer: a novel tumor-promoting cell type *Cell Cycle* **5** 1597–601

Palecek S P, Loftus J C, Ginsberg M H, Lauffenburger D A and Horwitz A F 1997 Integrin-ligand binding properties govern cell migration speed through cell-substratum adhesiveness *Nature* **385** 537–40

Pandey M S, Harris E N, Weigel J A and Weigel P H 2008 The cytoplasmic domain of the hyaluronan receptor for endocytosis (HARE) contains multiple endocytic motifs targeting coated pit-mediated internalization *J. Biol. Chem.* **283** 21453–61

Pankov R, Endo Y, Even-Ram S, Araki M, Clark K, Cukierman E, Matsumoto K and Yamada K M 2005 A Rac switch regulates random versus directionally persistent cell migration *J. Cell Biol.* **170** 793–802

Paszek M J and Weaver V M 2004 The tension mounts: mechanics meets morphogenesis and malignancy *J. Mammary Gland Biol. Neoplasia* **9** 325–42

Paszek M J *et al* 2005 Tensional homeostasis and the malignant phenotype *Cancer Cell* **8** 241–54

Patsialou A, Wyckoff J, Wang Y, Goswami S, Stanley E R and Condeelis J S 2009 Invasion of human breast cancer cells *in vivo* requires both paracrine and autocrine loops involving the colony-stimulating factor-1 receptor *Cancer Res.* **69** 9498–506

Pearson O H, Manni A and Arafah B M 1982 Antiestrogen treatment of breast cancer: an overview *Cancer Res.* **42** 3424s–9s PMID: 7044524

Philippeaux M M, Bargetzi J P, Pache J C, Robert J, Spiliopoulos A and Mauel J 2009 Culture and functional studies of mouse macrophages on native-like fibrillar type I collagen *Eur. J. Cell Biol.* **88** 243–56

Platt V M and Szoka F C Jr 2008 Anticancer therapeutics: targeting macromolecules and nanocarriers to hyaluronan or CD44, a hyaluronan receptor *Mol. Pharm.* **5** 474–86

Polyak K and Weinberg R A 2009 Transitions between epithelial and mesenchymal states: acquisition of malignant and stem cell traits *Nat. Rev. Cancer* **9** 265–73

Polyak K, Haviv I and Campbell I G 2009 Co-evolution of tumor cells and their microenvironment *Trends Genet.* **25** 30–8

Ponta H, Sherman L and Herrlich P 2003 CD44: from adhesion molecules to signalling regulators *Nat. Rev. Mol. Cell Biol.* **4** 33–45

Porporato P E, Dhup S, Dadhich R K, Copetti T and Sonveaux P 2011 Anticancer targets in the glycolytic metabolism of tumors: a comprehensive review *Front Pharmacol.* **2** 49

Preissner K T 1991 Structure and biological role of vitronectin *Annu. Rev. Cell Biol.* **7** 275–310

Prestwich G D 2008 Engineering a clinically-useful matrix for cell therapy *Organogenesis* **4** 42–7

Prestwich G D and Kuo J W 2008 Chemically-modified HA for therapy and regenerative medicine *Curr. Pharm. Biotechnol.* **9** 242–5

Provenzano P P, Eliceiri K W, Campbell J M, Inman D R, White J G and Keely P J 2006 Collagen reorganization at the tumor–stromal interface facilitates local invasion *BMC Med.* **4** 38

Provenzano P P, Inman D R, Eliceiri K W and Keely P J 2009 Matrix density-induced mechanoregulation of breast cell phenotype, signaling and gene expression through a FAK-ERK linkage *Oncogene* **28** 4326–43

Provenzano P P, Inman D R, Eliceiri K W, Beggs H E and Keely P J 2008 Mammary epithelial-specific disruption of focal adhesion kinase retards tumor formation and metastasis in a transgenic mouse model of human breast cancer *Am. J. Pathol.* **173** 1551–65

Provenzano P P, Inman D R, Eliceiri K W, Knittel J G, Yan L, Rueden C T, White J G and Keely P J 2008 Collagen density promotes mammary tumor initiation and progression *BMC Med.* **6** 11

Prud'homme G J, Glinka Y, Toulina A, Ace O, Subramaniam V and Jothy S 2010 Breast cancer stem-like cells are inhibited by a non-toxic aryl hydrocarbon receptor agonist *PLoS One* **5** e13831

Puig-Kroger A, Sanz-Rodriguez F, Longo N, Sanchez-Mateos P, Botella L, Teixido J, Bernabeu C and Corbi A L 2000 Maturation-dependent expression and function of the CD49d integrin on monocyte-derived human dendritic cells *J. Immunol.* **165** 4338–45

Ramaswamy S, Ross K N, Lander E S and Golub T R 2003 A molecular signature of metastasis in primary solid tumors *Nat. Genet.* **33** 49–54

Ramirez N E, Zhang Z, Madamanchi A, Boyd K L, O'Rear L D, Nashabi A, Li Z, Dupont W D, Zijlstra A and Zutter M M 2011 The alpha(2)beta(1) integrin is a metastasis suppressor in mouse models and human cancer *J. Clin. Invest.* **121** 226–37

Ridley A J 2004 Rho proteins and cancer *Breast Cancer Res. Treat.* **84** 13–9

Ridley A J, Schwartz M A, Burridge K, Firtel R A, Ginsberg M H, Borisy G, Parsons J T and Horwitz A R 2003 Cell migration: integrating signals from front to back *Science* **302** 1704–9

Riechelmann H, Sauter A, Golze W, Hanft G, Schroen C, Hoermann K, Erhardt T and Gronau S 2008 Phase I trial with the CD44v6-targeting immunoconjugate bivatuzumab mertansine in head and neck squamous cell carcinoma *Oral Oncol.* **244** 823–9

Rose D P and Vona-Davis L 2012 The cellular and molecular mechanisms by which insulin influences breast cancer risk and progression *Endocr. Relat. Cancer* **19** (6) R225–41

Rupp U *et al* 2007 Safety and pharmacokinetics of bivatuzumab mertansine in patients with CD44v6-positive metastatic breast cancer: final results of a phase I study *Anticancer Drugs* **18** 477–85

Rykala J, Przybylowska K, Majsterek I, Pasz-Walczak G, Sygut A, Dziki A and Kruk-Jeromin J 2011 Angiogenesis markers quantification in breast cancer and their correlation with clinicopathological prognostic variables *Pathol. Oncol. Res.* **17** 809–17

Saad S, Gottlieb D J, Bradstock K F, Overall C M and Bendall L J 2002 Cancer cell-associated fibronectin induces release of matrix metalloproteinase-2 from normal fibroblasts *Cancer Res.* **62** 283–9 PMID: 11782389

Sackstein R 2009 Glycosyltransferase-programmed stereosubstitution (GPS) to create HCELL: engineering a roadmap for cell migration *Immunol. Rev.* **230** 51–74

Sakamoto T and Seiki M 2010 A membrane protease regulates energy production in macrophages by activating hypoxia-inducible factor-1 via a non-proteolytic mechanism *J. Biol. Chem.* **285** 29951–64

Sakamoto T, Niiya D and Seiki M 2011 Targeting the Warburg effect that arises in tumor cells expressing membrane type-1 matrix metalloproteinase *J. Biol. Chem.* **286** 14691–704

Saller M M *et al* 2012 Increased stemness and migration of human mesenchymal stem cells in hypoxia is associated with altered integrin expression *Biochem. Biophys Res. Commun.* **423** 379–85

Sanderson R D, Hinkes M T and Bernfield M 1992 Syndecan-1, a cell-surface proteoglycan, changes in size and abundance when keratinocytes stratify *J. Invest. Dermatol.* **99** 390–6

Saoncella S, Echtermeyer F, Denhez F, Nowlen J K, Mosher D F, Robinson S D, Hynes R O and Goetinck P F 1999 Syndecan-4 signals cooperatively with integrins in a Rho-dependent manner in the assembly of focal adhesions and actin stress fibers *Proc. Natl Acad. Sci. USA* **96** 2805–10

Sawada Y, Tamada M, Dubin-Thaler B J, Cherniavskaya O, Sakai R, Tanaka S and Sheetz M P 2006 Force sensing by mechanical extension of the Src family kinase substrate p130Cas *Cell* **127** 1015–26

Schiemann W P 2007 Targeted TGF-beta chemotherapies: friend or foe in treating human malignancies? *Expert Rev. Anticancer Ther.* **7** 609–11

Schmid M C, Avraamides C J, Foubert P, Shaked Y, Kang S W, Kerbel R S and Varner J A 2011 Combined blockade of integrin-alpha4beta1 plus cytokines SDF-1alpha or IL-1beta potently inhibits tumor inflammation and growth *Cancer Res.* **71** 6965–75

Schor S L *et al* 2003 Migration-stimulating factor: a genetically truncated onco-fetal fibronectin isoform expressed by carcinoma and tumor-associated stromal cells *Cancer Res.* **63** 8827–36

Schultze A and Fiedler W 2010 Therapeutic potential and limitations of new FAK inhibitors in the treatment of cancer *Expert Opin. Investig. Drugs* **19** 777–88

Schulz V J, Smit J J, Bol-Schoenmakers M, Duursen M B, van den Berg M and Pieters R H 2012 Activation of the aryl hydrocarbon receptor reduces the number of precursor and effector T cells, but preserves thymic CD4(+)CD25(+)Foxp3(+) regulatory T cells *Toxicol. Lett.* **215** (2) 100–9

Schulze A and Downward J 2011 Flicking the Warburg switch-tyrosine phosphorylation of pyruvate dehydrogenase kinase regulates mitochondrial activity in cancer cells *Mol. Cell* **44** 846–8

Schwertfeger K L, Rosen J M and Cohen D A 2006 Mammary gland macrophages: pleiotropic functions in mammary development *J. Mammary Gland Biol. Neoplasia* **11** 229–38

Sechler J L, Cumiskey A M, Gazzola D M and Schwarzbauer J E 2000 A novel RGD-independent fibronectin assembly pathway initiated by alpha4beta1 integrin binding to the alternatively spliced V region *J. Cell Sci.* **113** (Pt 8) 1491–8 PMID: 10725231

Seifeddine R, Dreiem A, Tomkiewicz C, Fulchignoni-Lataud M C, Brito I, Danan J L, Favaudon V, Barouki R and Massaad-Massade L 2007 Hypoxia and estrogen co-operate to regulate gene expression in T-47D human breast cancer cells *J. Steroid Biochem. Mol. Biol.* **104** 169–79

Selak M A, Armour S M, MacKenzie E D, Boulahbel H, Watson D G, Mansfield K D, Pan Y, Simon M C, Thompson C B and Gottlieb E 2005 Succinate links TCA cycle dysfunction to oncogenesis by inhibiting HIF-alpha prolyl hydroxylase *Cancer Cell* **7** 77–85

Shao R, Bao S, Bai X, Blanchette C, Anderson R M, Dang T, Gishizky M L, Marks J R and Wang X F 2004 Acquired expression of periostin by human breast cancers promotes tumor angiogenesis through up-regulation of vascular endothelial growth factor receptor 2 expression *Mol. Cell Biol.* **24** 3992–4003

Shimoda M, Mellody K T and Orimo A 2010 Carcinoma-associated fibroblasts are a rate-limiting determinant for tumour progression *Semin. Cell Dev. Biol.* **21** 19–25

Shi-Wen X, Leask A and Abraham D 2008 Regulation and function of connective tissue growth factor/CCN2 in tissue repair, scarring and fibrosis *Cytokine Growth Factor Rev.* **19** 133–44

Silberstein G B, Flanders K C, Roberts A B and Daniel C W 1992 Regulation of mammary morphogenesis: evidence for extracellular matrix-mediated inhibition of ductal budding by transforming growth factor-beta 1 *Dev. Biol.* **152** 354–62

Simpson M A and Lokeshwar V B 2008 Hyaluronan and hyaluronidase in genitourinary tumors *Front Biosci.* **13** 5664–80

Skotheim R I and Nees M 2007 Alternative splicing in cancer: noise, functional, or systematic? *Int. J. Biochem. Cell Biol.* **39** 1432–49

Sleeman J P and Cremers N 2007 New concepts in breast cancer metastasis: tumor initiating cells and the microenvironment *Clin. Exp. Metastasis* **24** 707–15

Soikkeli J *et al* 2010 Metastatic outgrowth encompasses COL-I, FN1, and POSTN up-regulation and assembly to fibrillar networks regulating cell adhesion, migration, and growth *Am. J. Pathol.* **177** 387–403

Sols A and Crane R K 1954 Substrate specificity of brain hexokinase *J. Biol. Chem.* **210** 581–95 PMID: 13211595

Soria G and Ben-Baruch A 2008 The inflammatory chemokines CCL2 and CCL5 in breast cancer *Cancer Lett.* **267** 271–85

Sorrell J M and Caplan A I 2009 Fibroblasts-a diverse population at the center of it all *Int. Rev. Cell Mol. Biol.* **276** 161–214 PMID: 19584013

Sottile J and Hocking D C 2002 Fibronectin polymerization regulates the composition and stability of extracellular matrix fibrils and cell-matrix adhesions *Mol. Biol. Cell* **13** 3546–59

Spangenberg C *et al* 2006 ERBB2-mediated transcriptional up-regulation of the alpha5beta1 integrin fibronectin receptor promotes tumor cell survival under adverse conditions *Cancer Res.* **66** 3715–25

Stamenkovic I and Yu Q 2009 CD44 meets merlin and ezrin: their interplay mediates the pro-tumor activity of CD44 and tumor-suppressing effect of merlin ed R Stern *Hyaluronan in Cancer Biology* (San Diego, CA: Academic) pp 71–87

Stanley M J, Stanley M W, Sanderson R D and Zera R 1999 Syndecan-1 expression is induced in the stroma of infiltrating breast carcinoma *Am. J. Clin. Pathol.* **112** 377–83 PMID: 10478144

Stern M, Savill J and Haslett C 1996 Human monocyte-derived macrophage phagocytosis of senescent eosinophils undergoing apoptosis. Mediation by alpha v beta 3/CD36/thrombo-spondin recognition mechanism and lack of phlogistic response *Am. J. Pathol.* **149** 911–21 PMID: 8780395

Stern R 2008 Hyaluronidases in cancer biology *Semin. Cancer Biol.* **18** 275–80

Stern R (ed) 2009 *Hyaluronan in Cancer Biology* (San Diego, CA: Academic)

Stewart D A, Yang Y, Makowski L and Troester M A 2012 Basal-like breast cancer cells induce phenotypic and genomic changes in macrophages *Mol. Cancer Res.* **10** 727–38

Su G, Blaine S A, Qiao D and Friedl A 2007 Shedding of syndecan-1 by stromal fibroblasts stimulates human breast cancer cell proliferation via FGF2 activation *J. Biol. Chem.* **282** 14906–15

Sugiyama Y, Kakoi K, Kimura A, Takada I, Kashiwagi I, Wakabayashi Y, Morita R, Nomura M and Yoshimura A 2012 Smad2 and Smad3 are redundantly essential for the suppression of iNOS synthesis in macrophages by regulating IRF3 and STAT1 pathways *Int. Immunol.* **24** 253–65

Tall E G, Bernstein A M, Oliver N, Gray J L and Masur S K 2010 TGF-beta-stimulated CTGF production enhanced by collagen and associated with biogenesis of a novel 31-kDa CTGF form in human corneal fibroblasts *Invest. Ophthalmol. Vis. Sci.* **51** 5002–11

Tammi R H, Kultti A, Kosma V M, Pirinen R, Auvinen P and Tammi M I 2008 Hyaluronan in human tumors: Pathobiological and prognostic messages from cell-associated and stromal hyaluronan *Semin. Cancer Biol.* **18** 288–95

Taylor M A, Amin J D, Kirschmann D A and Schiemann W P 2011 Lysyl oxidase contributes to mechanotransduction-mediated regulation of transforming growth factor-beta signaling in breast cancer cells *Neoplasia* **13** 406–18

Taylor-Papadimitriou J, Burchell J and Hurst J 1981 Production of fibronectin by normal and malignant human mammary epithelial cells *Cancer Res.* **41** 2491–500 PMID: 7016316

Thankamony S P and Knudson W 2006 Acylation of CD44 and its association with lipid rafts are required for receptor and hyaluronan endocytosis *J. Biol. Chem.* **281** 34601–09

Thornton R D, Lane P, Borghaei R C, Pease E A, Caro J and Mochan E 2000 Interleukin 1 induces hypoxia-inducible factor 1 in human gingival and synovial fibroblasts *Biochem. J.* **350**(Pt 1) 307–12

Tkachenko E, Rhodes J M and Simons M 2005 Syndecans: new kids on the signaling block *Circ. Res.* **96** 488–500

Tlsty T D and Coussens L M 2006 Tumor stroma and regulation of cancer development *Annu. Rev. Pathol.* **1** 119–50

Toda S, Hanayama R and Nagata S 2012 Two-step engulfment of apoptotic cells *Mol. Cell Biol.* **32** 118–25

Toole B P 1990 Hyaluronan and its binding proteins, the hyaladherins *Curr. Opin. Cell Biol.* **2** 839–44

Toole B P 2001 Hyaluronan in morphogenesis *Semin Cell Dev. Biol.* **12** 79–87

Toole B P 2004 Hyaluronan: from extracellular glue to pericellular cue *Nat. Rev. Cancer* **4** 528–39

Toole B P 2009 Hyaluronan-CD44 interactions in cancer: paradoxes and possibilities *Clin. Cancer Res.* **15** 7462–8

Toole B P and Slomiany M G 2008 Hyaluronan, CD44 and Emmprin: partners in cancer cell chemoresistance *Drug Resist. Updat.* **11** 110–21

Toole B P, Wight T N and Tammi M 2002 Hyaluronan–cell interactions in cancer and vascular disease *J. Biol. Chem.* **277** 4593–6

Toth B *et al* 2009 Transglutaminase 2 is needed for the formation of an efficient phagocyte portal in macrophages engulfing apoptotic cells *J. Immunol.* **182** 2084–92

Tumova S, Woods A and Couchman J R 2000 Heparan sulfate chains from glypican and syndecans bind the Hep II domain of fibronectin similarly despite minor structural differences *J. Biol. Chem.* **275** 9410–7

Turley E A, Noble P W and Bourguignon L Y 2002 Signaling properties of hyaluronan receptors *J. Biol. Chem.* **277** 4589–92

Turley E A, Veiseh M, Radisky D C and Bissell M J 2008 Mechanisms of disease: epithelial-mesenchymal transition—does cellular plasticity fuel neoplastic progression? *Nat. Clin. Pract. Oncol.* **5** 280–90

Turner L, Scotton C, Negus R and Balkwill F 1999 Hypoxia inhibits macrophage migration *Eur. J. Immunol.* **29** 2280–7

Tuxhorn J A, Ayala G E, Smith M J, Smith V C, Dang T D and Rowley D R 2002 Reactive stroma in human prostate cancer: induction of myofibroblast phenotype and extracellular matrix remodeling *Clin. Cancer Res.* **8** 2912–23 PMID: 14581350

Underhill C B and Toole B P 1979 Binding of hyaluronate to the surface of cultured cells *J. Cell Biol.* **82** 475–84

van Grevenynghe J, Rion S, Le Ferrec E, Le Vee M, Amiot L, Fauchet R and Fardel O 2003 Polycyclic aromatic hydrocarbons inhibit differentiation of human monocytes into macrophages *J. Immunol.* **170** 2374–81

Vasaturo F *et al* 2005 Comparison of extracellular matrix and apoptotic markers between benign lesions and carcinomas in human breast *Int. J. Oncol.* **27** 1005–11

Velling T, Risteli J, Wennerberg K, Mosher D F and Johansson S 2002 Polymerization of type I and III collagens is dependent on fibronectin and enhanced by integrins alpha 11beta 1 and alpha 2beta 1 *J. Biol. Chem.* **277** 37377–81

Vicente-Manzanares M, Webb D J and Horwitz A R 2005 Cell migration at a glance *J. Cell Sci.* **118** 4917–9

Vinothini G, Aravindraja C, Chitrathara K and Nagini S 2011 Correlation of matrix metalloproteinases and their inhibitors with hypoxia and angiogenesis in premenopausal patients with adenocarcinoma of the breast *Clin. Biochem.* **44** 969–74

Visvader J E and Lindeman G J 2008 Cancer stem cells in solid tumors: accumulating evidence and unresolved questions *Nat. Rev. Cancer* **8** 755–68

von Drygalski A, Tran T B, Messer K, Pu M, Corringham S, Nelson C and Ball E D 2011 Obesity is an independent predictor of poor survival in metastatic breast cancer: retrospective analysis of a patient cohort whose treatment included high-dose chemotherapy and autologous stem cell support *Int. J. Breast Cancer* **2011** 523276

Wallach-Dayan S B, Rubinstein A M, Hand C, Breuer R and Naor D 2008 DNA vaccination with CD44 variant isoform reduces mammary tumor local growth and lung metastasis *Mol. Cancer Ther.* **7** 1615–23

Wang M Y, Chen P S, Prakash E, Hsu H C, Huang H Y, Lin M T, Chang K J and Kuo M L 2009 Connective tissue growth factor confers drug resistance in breast cancer through concomitant up-regulation of Bcl-xL and cIAP1 *Cancer Res.* **69** 3482–91

Weigel P H and DeAngelis P L 2007 Hyaluronan synthases: a decade-plus of novel glycosyltransferases *J. Biol. Chem.* **282** 36777–81

Wennerberg K, Lohikangas L, Gullberg D, Pfaff M, Johansson S and Fässler R 1996 Beta 1 integrin-dependent and -independent polymerization of fibronectin *J. Cell Biol.* **132** 227–38

Wick A N, Drury D R, Nakada H I and Wolfe J B 1975 Localization of the primary metabolic block produced by 2-deoxyglucose *J. Biol. Chem.* **224** 963–9 PMID: 13405925

Wierzbicka-Patynowski I and Schwarzbauer J E 2003 The ins and outs of fibronectin matrix assembly *J. Cell Sci.* **116** 3269–76

Wilcox-Adelman S A, Denhez F and Goetinck P F 2002 Syndecan-4 modulates focal adhesion kinase phosphorylation *J. Biol. Chem.* **277** 32970–7

Wilhelm O, Hafter R, Coppenrath E, Pflanz M A, Schmitt M, Babic R, Linke R, Gossner W and Graeff H 1988 Fibrin–fibronectin compounds in human ovarian tumor ascites and their possible relation to the tumor stroma *Cancer Res.* **48** 3507–14 PMID: 3130986

Wozniak M A, Desai R, Solski P A, Der C J and Keely P J 2003 ROCK-generated contractility regulates breast epithelial cell differentiation in response to the physical properties of a three-dimensional collagen matrix *J. Cell Biol.* **163** 583–95

Wu C, Bauer J S, Juliano R L and McDonald J A 1993 The alpha 5 beta 1 integrin fibronectin receptor, but not the alpha 5 cytoplasmic domain, functions in an early and essential step in fibronectin matrix assembly *J. Biol. Chem.* **268** 21883–8 PMID: 7691819

Wu C, Hughes P E, Ginsberg M H and McDonald J A 1996 Identification of a new biological function for the integrin alpha v beta 3: initiation of fibronectin matrix assembly *Cell Adhes. Commun.* **4** 149–58

Wu C, Keivens V M, O'Toole T E, McDonald J A and Ginsberg M H 1995 Integrin activation and cytoskeletal interaction are essential for the assembly of a fibronectin matrix *Cell* **83** 715–24

Wu M, Cao M, He Y, Liu Y, Yang C, Du Y, Wang W and Gao F 2014 A novel role of low molecular weight hyaluronan in breast cancer metastasis *FASEB J* at press

Xie B, Laouar A and Huberman E 1998 Fibronectin-mediated cell adhesion is required for induction of 92-kDa type IV collagenase/gelatinase (MMP-9) gene expression during macrophage differentiation. The signaling role of protein kinase C-beta *J Biol Chem.* **273** 11576–82

Yang J T and Hynes R O 1996 Fibronectin receptor functions in embryonic cells deficient in alpha 5 beta 1 integrin can be replaced by alpha V integrins *Mol. Biol. Cell* **7** 1737–48

Yang N, Mosher R, Seo S, Beebe D and Friedl A 2011 Syndecan-1 in breast cancer stroma fibroblasts regulates extracellular matrix fiber organization and carcinoma cell motility *Am. J. Pathol.* **178**(1) 325–35

Ye J, Xu R H, Taylor-Papadimitriou J and Pitha P M 1996 Sp1 binding plays a critical role in Erb-B2- and v-ras-mediated downregulation of alpha2-integrin expression in human mammary epithelial cells *Mol. Cell Biol.* **16** 6178–89 PMID: 8887648

Yi J M, Kwon H Y, Cho J Y and Lee Y J 2009 Estrogen and hypoxia regulate estrogen receptor alpha in a synergistic manner *Biochem. Biophys. Res. Commun.* **378** 842–6

Ying H, Biroc S L, Li W W, Alicke B, Xuan J A, Pagila R, Ohashi Y, Okada T, Kamata Y and Dinter H 2006 The Rho kinase inhibitor fasudil inhibits tumor progression in human and rat tumor models *Mol. Cancer Ther.* **5** 2158–64

Zhang N and Walker M K 2007 Crosstalk between the aryl hydrocarbon receptor and hypoxia on the constitutive expression of cytochrome P4501A1 mRNA *Cardiovasc. Toxicol.* **7** 282–90

Zhang Y, Zhang G, Li J, Tao Q and Tang W 2010 The expression analysis of periostin in human breast cancer *J. Surg. Res.* **160** 102–6

Zhao Y, Min C, Vora S R, Trackman P C, Sonenshein G E and Kirsch K H 2009 The lysyl oxidase pro-peptide attenuates fibronectin-mediated activation of focal adhesion kinase and p130Cas in breast cancer cells *J. Biol. Chem.* **284** 1385–93

Zijlstra A, Lewis J, Degryse B, Stuhlmann H and Quigley J P 2008 The inhibition of tumor cell intravasation and subsequent metastasis via regulation of *in vivo* tumor cell motility by the tetraspanin CD151 *Cancer Cell* **13** 221–34

Zilberberg L, Todorovic V, Dabovic B, Horiguchi M, Courousse T, Sakai L Y and Rifkin D B 2012 Specificity of latent TGF-beta binding protein (LTBP) incorporation into matrix: Role of fibrillins and fibronectin *J. Cell Physiol.* **227** 3828–36

Zimmermann P and David G 1999 The syndecans, tuners of transmembrane signaling *FASEB J.* **13**(Suppl) S9–100

Zucker S and Cao J 2009 Selective matrix metalloproteinase (MMP) inhibitors in cancer therapy: ready for prime time? *Cancer Biol. Ther.* **8** 2371–3

IOP Publishing

Physics of Cancer

Claudia Tanja Mierke

Chapter 10

The role of endothelial cell–cell adhesions

Summary

During cancer metastasis, the transendothelial migration of cancer cells seems to be crucial in providing the formation of secondary tumors in targeted organs. Whether these cancer cells transmigrate transcellularly (through the living endothelial cell of the confluent endothelial monolayer lining blood vessels) or paracellularly (through the cell–cell adhesions of neighboring endothelial cells) is not yet clear. It seems to be the case that cancer cells are able to use both migration routes at different cellular loci. How these routes and loci are selected is discussed and hypothesized below.

10.1 The expression of cell–cell adhesion molecules

Invasive cancer cells regulate the expression of endothelial
cell–cell adhesion molecules

The expression of endothelial cell–cell adhesion molecules is important for inter-endothelial adhesion strength and hence for the integrity of the endothelial cell monolayer. However, it has not been shown how the expression of these molecules is regulated by aggressive and invasive cancer cells, but not by non-invasive cancer cells. There are still many questions that need to be raised and answered in order to understand the transendothelial migration step of the metastatic cascade. Do only aggressive and invasive cancer cells alter the expression of cell–cell adhesion molecules on endothelial cells? Nevertheless, the question of how specific cancer cells transmigrate through the endothelium remains a controversial one. However, the question of whether highly invasive cancer cells are able to regulate the cell–cell adhesion molecule expression on human microvascular endothelial cells during co-culture has been investigated. Primary human endothelial cells derived from the lung (called HPMECs) were co-cultured for 16 h with highly invasive MDA-MB-231 and weakly invasive MCF-7 cells. Using the flow cytometry technique, the co-culture of microvascular endothelial cells with highly invasive MDA-MB-231 breast cancer cells revealed that the platelet endothelial cell adhesion molecule-1 (PECAM-1) and the vascular endothelial-cadherin (VE-cadherin) were both down-regulated during co-culture with highly invasive MDA-MB-231 cells compared to mono-cultured endothelial cells, whereas the co-culture of endothelial cells with weakly invasive MCF-7 cells showed no effect on the expression of the two endothelial cell–cell adhesion proteins (Mierke *et al* 2011). These findings suggest that the down-regulation of endothelial cell–cell adhesion molecules seems to be cancer-cell-specific and may depend on their individual invasive potential as well as on their mechanical properties. Taken together, these results indeed show that the co-culture of endothelial cells with highly invasive cancer cells evoke a down-regulation in the cell surface expression of the cell–cell adhesion molecules, VE-cadherin and PECAM-1, thus altering the biomechanical properties of endothelial cells and effecting the break-down of the endothelial barrier function.

In order to investigate which particular mechanism facilitates the reduced endothelial cell–cell adhesion molecule expression during co-culture with MDA-MB-231 cells, the membrane shedding of these adhesion molecules was inhibited during the transendothelial migration of the MDA-MB-231 cells. In more detail, it was investigated whether the decreased expression of PECAM-1 and VE-cadherin receptors on human pulmonary microvascular endothelial cells (HPMECs) during co-culture with MDA-MB-231 cells is due to membrane shedding of these receptors. Thus the co-culture of cancer cells and endothelial cells was performed in the presence and absence of the broad matrix-metallo-proteinase inhibitor GM6001. Indeed, the reduction of the cell–cell adhesion receptors PECAM-1 and VE-cadherin on endothelial cells during the co-culture with MDA-MB-231 cells has been shown to be caused by increased membrane shedding of these cell–cell adhesion receptors.

10.2 The strength of cell–cell adhesions

Vascular leakage is a hallmark of many inflammatory and often life-threatening diseases and hence contributes to disease severity in disorders such as sepsis, cancer, diabetes and atherosclerosis (Weis and Cheresh 2005). Despite the tremendous medical importance of vascular leakage, only a few specific therapies are available in order to counteract it and current therapies often fail (Groeneveld 2002). However the *in vivo* molecular targets are not yet fully understood, although a wealth of data obtained from *in vitro* studies is available for the signal transduction pathways that regulate the vascular permeability (Mehta and Malik 2006, Jacobson and Garcia 2007). Among various new agents that potentially reduce endothelial hyperpermeability, such as cholesterol-lowering statin drugs, some have been proposed to reduce vascular leakage, as they are able to inhibit RhoA proteins (Jacobson *et al* 2005, van de Visse *et al* 2006). In a proof-of-principle it has been shown that increased RhoA activity fosters vascular hyperpermeability *in vivo* (Gorovoy *et al* 2007). In more detail, it has been found that an increase of RhoA activity by deletion of one of its inhibitory proteins such as RhoGDI leads to a reduction of endothelial junctional integrity and finally reduces the vascular endothelial cell barrier function.

Rho GTPases such as RhoA, Rac1 and Cdc42 have been revealed as key regulators of cell shape, movement and proliferation. However, *in vitro* studies have shown that the balance of activities of these small G proteins regulate the blocking potential of

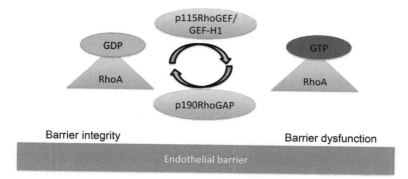

Figure 10.1. The endothelial barrier function is tightly regulated through a precise balance of the individual activities of small G proteins. In particular, vasoactive agents such as VEGF and thrombin, and interaction with leukocytes, can impair the endothelial barrier function through specific receptors. Indeed, several signal transduction mechanisms are simultaneously activated, such as the influx of calcium ions, the activation of small Rho GTPases and various other kinases and the phosphorylation of adherence junctional proteins. Among the small GTPases, RhoA seems to be mainly involved in inducing endothelial hyperpermeability, whereas Rac1, Cdc42 and Rap1 foster an intact endothelial cell barrier function. The activity of the small GTPases has been shown to be precisely determined by three classes of regulatory proteins, GDIs, GEFs and GAPs. In more detail, RhoGDI has been identified as a novel therapeutic target and evidence has been provided that in the healthy vasculature RhoGDI is a key regulatory point, restricting the RhoA activity to low levels.

endothelial barrier (Wojciak-Stothard and Ridley 2002) (figure 10.1): Cdc42 enhances the process of recovery of a disturbed barrier (Kouklis *et al* 2004), Rac1 is required to establish a tight barrier function (Wojciak-Stothard and Ridley 2002) and RhoA is involved in the induction of the endothelial hyperpermeability caused by various stimuli, such as thrombin, VEGF, angiopoietin-2 and LPA (Essler *et al* 1998, van Nieuw Amerongen *et al* 1998, Parikh *et al* 2006). In addition, these Rho GTPases regulate other vascular cells and leukocytes and hence facilitate other vascular functions.

The inhibition of the RhoA-target ROCK1/2 (Rho kinase) by inhibitors revealed the involvement of RhoA/ROCK activation in embryonic development/cytokinesis (Lai *et al* 2005) and in various vascular pathologies, such as (pulmonary) hypertension, atherosclerosis, stroke and even heart failure (Shimokawa *et al* 2005). Indeed the first evidence has been obtained that ROCK inhibition by Y-27632 can reduce pulmonary edema in animals after LPS stimulation or reexpansion of the lung (Tasaka *et al* 2005, Sawafuji *et al* 2005). Because of the central importance of ROCK in the regulation of many basal cellular functions, such as migration and proliferation, it has been suggested that the activity of Rho proteins is strongly controlled by regulatory proteins. In more detail, guanine dissociation inhibitors or GDIs hold Rho proteins in their inactivate GDP-bound mode (DerMardirossian *et al* 2005), while by contrast guanine exchange factors or GEFs activate Rho proteins by facilitating the exchange of GDP for GTP, and GTPase activating proteins or GAPs inactivate Rho proteins by inducing the conversion of Rho-bound GTP to GDP. These regulatory proteins act in close concert: the dissociation of RhoA from RhoGDI is a prerequisite for its activation by RhoGEF. Regulatory proteins of all three classes (RhoGDIs, RhoGEFs and RhoGAPs) are involved in the regulation of thrombin-enhanced *in vitro* endothelial permeability (Holinstat *et al* 2006, Mehta *et al* 2001, Birukova *et al* 2006). Epac, an analogous cAMP-activated exchange factor for Rap1 (a small GTPase not belonging to the Rho family of small GTPases), plays a crucial role in increasing vascular endothelial cadherin-facilitated cell–cell contacts (Cullere *et al* 2005, Fukuhara *et al* 2005).

In order to investigate the pulmonary vasculature, a model of isolated mouse lungs was selected. An increase in capillary permeability is the basic underlying abnormality of acute lung injury or acute respiratory distress syndrome (ALI/ARDS), which builds a continuum from mild to severe lung damage. ARDS often develops in septic patients or after trauma and thus is a major cause of death in the intensive care setting. It is thus important to investigate the effect of treatment with endotoxin, as a model for sepsis, inducing RhoA activity in the lung.

An important finding was that increased permeability in RhoGDI−/− mice was completely reversible by pharmacological inhibition of the Rho kinase. Firstly, this indicates that the hyperpermeability in RhoGDI−/− mice was indeed caused by enhanced RhoA/Rho kinase signaling, excluding possible side-effects. Secondly, this brings the concept of RhoGDI/RhoA-facilitated vascular leakage into the focus of medical research, as Rho kinase inhibitors with a reasonable safety profile (such as fasudil) are available.

However, no signs of edema were detected in the intact animals, which was attributed to the presence of safety factors such as lymphatic drainage. It remains to be investigated whether the vasculature in an intact animal is hyperpermeable, using appropriate dye extravasation experiments.

The data are challenging and should be interpreted with care. It is tempting to conclude that the absence of Rho-GDI and the accompanying effect of RhoA on vascular leakage was caused by activation of RhoA in endothelial cells. Indeed, it has been shown that the endothelial junctions in capillaries and postcapillary venules become disturbed, but the biochemical measurements were performed in whole lung homogenates and therefore are inconclusive. Although the siRNA approach in cultured endothelial cells confirmed that deletion of RhoGDI by itself is sufficient for barrier dysfunction, the present study does not exclude the possibility that non-endothelial effects in RhoGDI−/− mice may also contribute to the enhanced vascular permeability in the intact lung. In particular, future studies are needed to investigate whether rescuing of RhoGDI specifically in endothelial cells or leuko-cytes excludes the possibility that the effect on vascular junctions is not indirectly evoked by an alteration in leukocyte influx after LPS challenge, or an alteration in resident leukocytes and mast cells due to life-long depletion of RhoGDI. However, even if such indirect effects contribute to vascular leakage in RhoGDI−/− mice, this study provides fuel for the suggestion that inhibition of RhoA is a potential target for reducing vascular leakage. It has been reported for statins, which in addition to inhibiting cholesterol synthesis, inhibit the isoprenylation of proteins such as RhoA, required for their membrane anchoring, and it has been suggested that vascular leakage induced by sepsis may benefit more from treatment by statins than other, similar leakages caused by other stimuli in cardiovascular patients would (Jacobson *et al* 2005, van de Visse *et al* 2006). Due to the contribution of vascular leakage to many other non-pulmonary disorders, future studies are required to reveal whether similar mechanisms apply to other vascular beds and other disease states.

It remains a future challenge to develop therapies that increase RhoGDI activity. The answer may be found in the inhibition of the kinases that phosphorylate RhoGDI and stimulate the release of RhoA from RhoGDI. Two candidate kinases are Src and PKC (Holinstat *et al* 2006, DerMardirossian *et al* 2006). Interestingly, in line with this, it has been suggested that inhibition of p190RhoGAP by Angiopoietin-1 reduces endotoxin-enhanced vascular permeability in the mouse lung, indicating that targeting of Rho-regulatory proteins seems to be a feasible approach for reducing vascular leakage (Mammoto *et al* 2007). However, information about the activity status of the different Rho-regulatory proteins in (human) disease would provide valuable information for a directed therapy. In summary, the concept of RhoA being very important in the regulation of vascular endothelial leakage is approaching clinical applicability.

10.3 The cancer cell transmigration route

How can cancer cells transmigrate through an endothelium?
Most cancer-related deaths are caused by cancer metastasis, a process that starts with dissociation of cancer cells from the primary tumor and is followed by tissue invasion, entrance into blood or lymph vessels (intravasation) and transport to remote sites. The transmigration through the endothelial cell layer of blood or lymph vessels is a step in the metastatic cascade of malignant tumor progression. The transmigration of cancer cells through the endothelial cell lining of blood vessels is not yet understood in detail. It is assumed that cancer cells can then escape from the microvasculature (called extravasation), invade the target tissue and form secondary tumors in distant organs (Liotta *et al* 1991, Langley and Fidler 2007, Steeg 2006). A potentially rate-limiting step in the metastatic cascade, therefore, would be the extravasation process that involves adhesion of cancer cells to endothelial cells and their transmigration through the endothelial cell monolayer and finally the basement membrane (Steeg 2006, Nicolson 1989, Stetler-Stevenson *et al* 1993, Luzzi *et al* 1998). Indeed, specific cancer cell types have been demonstrated, both *in vitro* and *in vivo*, to be able to overcome the endothelial barrier (Luzzi *et al* 1998, Weis *et al* 2004, Voura *et al* 2001, Tremblay *et al* 2006, Sandig *et al* 1997, Fidler and Hart 1982). However, cancer cell extravasation need not be the only mechanism for metastasis formation, as has been pointed out by Al-Mehdi and colleagues (Al-Mehdi *et al* 2000), who reported that cancer cells can adhere and grow onto the endothelial layer and form a metastasis without ever leaving the blood or lymph vessel confinement. Either way, the role of the endothelial monolayer of blood or lymph vessels in this process seems to be crucial in that it can actively regulate metastasis formation by either allowing or blocking the adhesion and possibly transmigration, of cancer cells (Voura *et al* 2001, Tremblay *et al* 2006, Sandig *et al* 1997). The details for the endothelial cell functions in this process, however, are poorly understood and the extent to which the endothelium restricts or even promotes the process of metastasis is not yet clear.

Transmigrating cancer cells are thought to be able to overcome the endothelial barrier by inducing alterations within endothelial cells, such as the up-regulation of adhesion molecule receptor expression (Laferriere *et al* 2001), the reorganization of the cytoskeleton (Rousseau *et al* 1997), Src-mediated disruption of endothelial VE-cadherin-beta-catenin cell–cell adhesions (Weis *et al* 2004), the formation of 'holes' within the endothelial layer (Li and Zhu 1999) and even the induction of apoptosis (Heyder *et al* 2002). However, cancer cell invasion seems to be similar to leukocyte trafficking, for which the endothelium acts as a passive barrier, greatly reducing invasion rates (Wittchen *et al* 2005). For instance, the function of the endothelial cell barrier against both leukocyte trafficking and cancer cell trans-migration is reduced in the presence of inflammatory cytokines such as tumor necrosis factor-alpha and interleukin-1beta (Voura *et al* 2001, Laferriere *et al* 2001, Chandrasekharan *et al* 2006, McGettrick *et al* 2006). In more detail, these cytokines are known to trigger an up-regulation of the adhesion molecule E-selectin (Laferriere *et al* 2001). The subsequent adhesion of cancer cells to E-selectin leads in turn to an up-regulation of stress-activated protein kinase-2 (SAPK2/p38) in endothelial cells (Laferriere *et al* 2001), finally triggering actin polymerization and reorganization into stress fibers in endothelial cells (Rousseau *et al* 1997). These results indicate that the mechanical properties of endothelial cells may be altered by invasive cancer cells adhering and transmigrating through an endothelial cell monolayer, which is an endothelial cell barrier lining blood or lymphoid vessels.

Chemokines and their receptors are also important for leukocyte trafficking (Gallatin *et al* 1983, Hillyer *et al* 2003) and cancer cell invasion (Reiland *et al* 1999). In more detail, chemokines are a superfamily of small cytokine-like proteins that induce cytoskeletal rearrangements in endothelial cells and leukocytes, the firm adhesion of leukocytes to endothelial cells and the directional migration of leukocytes (Gallatin *et al* 1983). The involvement of chemokines in tumor–endothelial inter-actions and their effect on cancer cell mechanics during matrix invasion are considerably less well understood and hence require further investigation.

Thus, it has been investigated whether the endothelium is able to regulate the transmigration and invasion of cancer cells into an extracellular matrix. The invasion of human cancer cell lines into a 3D collagen gel matrix covered with an endothelial cell monolayer was performed. Interestingly, in the presence of an endothelium, the invasion of special cancer cell lines increased pronouncedly. Moreover, gene expression analysis of endothelial cells cocultured with invasive cancer cells revealed an up-regulation of Gro-beta and IL-8 chemokines compared with endothelial cells cocultured with non-invasive cancer cells. Finally, it was demonstrated that Gro-beta and IL-8 receptor (called CXCR2) expression on cancer cells serves as a key mediator responsible for the break-down of the endothelial barrier function through enhancing cancer cell force generation and cytoskeletal remodeling dynamics.

10.3.1 The paracellular transendothelial migration route

It has been hypothesized that the endothelial cell's actin cytoskeleton may provide a migration scaffold for transmigrating cancer cells. The endothelial cell lining of vessels represents a strong barrier against the invasion of specific cancer cells and is thus a key rate-limiting step against the transmigration, invasion and metastasis of aggressive and invasive cancer cells (Zijlstra *et al* 2008). In particular, the endothelial vessel wall has been commonly considered to be a strong tissue barrier against the dissemination of cancer cells through pronouncedly reducing their invasiveness and consequently eliminating their metastatic potential (Wittchen *et al* 2005). However, recent results have led to the establishment of a novel and unexpected role for the endothelial cell lining of vessels. In this, endothelial cells enhance the invasiveness of certain cancer cells. Firstly, breast cancer cells showed increased dispersion and clearance through hematogeneous dissemination adjacent to blood vessels (Kedrin *et al* 2008). Secondly, the invasiveness of special cancer cell lines is endothelial-cell-dependent and is thus enhanced in certain highly invasive cancer cells, whereas in weakly invasive cancer cells the endothelium acts as a classical barrier confinement for cancer cell invasion (Mierke *et al* 2008). Although the process of cancer cell invasion and metastasis has been the subject of numerous research papers, the molecular and mechanical mechanisms of cancer cell transendothelial migration are still not yet precisely understood and thus require further investigation.

The physical and biochemical aspects of the cancer cell intravasation process involve the interaction of at least three cell types: an invasive cancer cell, a macrophage and an opposing endothelial cell representing the barrier function. In more detail, all three cell types will engage the mechano- and biochemical-transduction properties of the cytoskeleton of all three neighboring cells. In order to reveal the cancer-cell-induced signals in endothelial cells, a 3D transmigration and invasion assay can be used in which the real-time intra-endothelial signaling events evoked by invasive cancer cells or macrophages are analyzed and compared to monocultured endothelial cells (Khuon *et al* 2010, Dovas *et al* 2013, Roh-Johnson *et al* 2014). In particular, this assay involves the assembly of a vasculature network within a 3D collagen matrix using endothelial cells that express a fluorescent resonant energy transfer-based biosensor reporting the activity of myosin light chain kinase (MLCK) in endothelial cells in real time (Chew *et al* 2002). As expected, endothelial cells react to mechano-sensing events in the 3D collagen matrix. For instance, the 3D microenvironment induces lumen formation of endothelial cells and endothelial cells show basal–apical polarity in the proper orientation indicated by $\alpha 4$ laminin deposition. As hypothesized before, it was confirmed that invasive cancer cells affect the MLCK-facilitated actomyosin function within the underlying endothelium. In addition, cancer cells can transmigrate through the endothelial barrier confinement in at least two different cellular ways: firstly, through transcellular routes (by transmigrating directly through the cytoplasm of an adjacent under lying endothelial cell) and secondly, through paracellular routes (by transmigrating between the endothelial cell–cell junctions of two neighboring individual endothelial cells) (Khuon *et al* 2010).

10.3.2 The transcellular transendothelial migration route

The transcellular transmigration route has for a long time been regarded solely as an artifact, but this mode has now been related in several reports, and it has even been presented in videos. Thus, this method of transedothelial migration is now established and seems to be real. However, the precise regulatory mechanisms are not yet well understood and require further investigation. Moreover, this transcellular mode of transendothelial migration of cancer cells seems to involve the generation of forces and utilizes the actomyosin cytoskeleton of the endothelial cell to migrate through its cytoplasm in a directed manner using the confinement of the endothelial cell as a migration grid.

However, when cancer cells use a transcellular invasion path, they trigger MLCK activation in this endothelial cell, which correlates with increased locationally and spatially restricted phosphorylation of the myosin-II regulatory light chain (RLC) and localized endothelial myosin contraction. Indeed, this has been functionally analyzed using endothelial cells expressing a RLC mutant that cannot be phosphorylated; the intravasation events of cancer cells migrating intracellularly (transcellularly) through the endothelial cell body are drastically reduced. In summary, (i) invasive cancer cells are capable of undergoing transcellular migration; (ii) cancer cells induce transient and local MLCK activation, as well as myosin contraction in adjacent endothelial cells at the site of transmigration and tissue invasion; and (iii) the transcellular invasion path through endothelial cells depends on the phosphorylation of myosin-II RLC. However, this result has to be confirmed through investigating more cancer cell types and cancer cells isolated from different stages of cancer disease. Nonetheless, all these findings demonstrate that the endothelium fulfills an exceptional and active role in cancer cells' intravasation—and possibly also in their extravasation.

10.4 The role of cancer cell exerted invadopodia during transendothelial migration

Cancer cell extravasation is a key step during cancer metastasis, yet the precise mechanisms that regulate this dynamic process are still unclear. A high-resolution time-lapse intravital imaging approach has been utilized to visualize the dynamics of cancer cell extravasation *in vivo*. During intravascular migration, cancer cells build protrusive structures identified as invadopodia through their enrichment of MT1-MMP, cortactin, Tks4 and, importantly, Tks5, which localizes exclusively to invadopodia. In more detail, cancer cells exert invadopodia through the endothelium into the extravascular stroma prior to their total extravasation at endothelial junctions. Genetic or pharmacological inhibition of invadopodia initiation (cortactin), maturation (Tks5) or function (Tks4) results in the elimination of cancer cell extravasation and metastatic colony formation in an experimental mouse lung metastasis model. Thus, this provides direct evidence for a functional role of invadopodia during the process of cancer cell extravasation and distant cancer metastasis and moreover reveals an opportunity for therapeutic intervention in this important process.

Metastasis is a complex scenario consisting of a multistep process that represents the most deadly aspect of cancer. Cancer cells that successfully disseminate from the primary tumor and survive in the vascular system eventually extravasate across the endothelium to colonize secondary sites. However, the process of cancer cell extravasation is the least understood step in the metastatic cascade, as it is difficult to investigate as no appropriate human microvascular endothelial cell lines exist and hence primary human endothelial cells have to used. Immune cell extravasation or diapedesis relies on ligand–receptor interactions for adhesion to the endothelium, assembling specialized structures called podosomes to promote their transmigration across the endothelial layer (Carman et al 2007). Morphologically, these podosomes have been described as invadosome-like protrusions that are regarded as normal counterparts of subcellular protrusions commonly referred to as invadopodia in cancer cells (Carman and Springer 2004, Carman et al 2007, Murphy and Courtneidge 2011). Indeed, podosomes have been observed in vivo during athero-genesis of the intimal layer of mouse aorta (Quintavalle et al 2010) and in neural crest migration during embryonic development (Murphy et al 2011), however a physiological role for invadopodia in cancer disease has not been revealed.

Invadopodia are cancer-specific protrusive and adhesive structures and were initially observed in vitro as flat protrusions on the basolateral side of cancer cells. Extensive efforts to characterize invadopodia and podosomes have shown that these two structures are composed of proteins such as cortactin, N-WASP, Tks4 and Tks5 (Clark et al 2007, Murphy and Courtneidge 2011, Oser et al 2009, Weaver et al 2002). However, Tks5 is exclusively present within podosomes, suggesting that it fulfills a specific role in invadopodia formation and maturation (Abram et al 2003, Seals et al 2005). Invadopodia contain proteases (such as MT1-MMP, MMP9 and MMP2) for local directed release and/or activity during extracellular matrix break-down (Clark et al 2007) and, in particular, Tks5 is required in both invadopodia and podosomes to degrade the extracellular matrix in vitro (Caldieri et al 2009, Furmaniak-Kazmierczak et al 2007, Magalhaes et al 2011, Pignatelli et al 2012, Seals et al 2005).

The visualization of cancer cell invadopodia in living organisms has been elusive due to the challenges associated with distinguishing them from other invasive structures, such as lamellipodia (Gligorijevic et al 2012). Established components of invadopodia, such as cortactin, MT1-MMP and Tks4, are also expressed in other protrusive structures, such as lamellipodia. Recent evidence suggests that invado-podia can indeed be specifically inhibited through loss of function of Tks5 (Burger et al 2014, Diaz et al 2009, Sharma et al 2013a). Moreover, cytoplasmic extensions assembled by cancer cells during intravascular arrest in capillary beds have been observed in zebra-fish models (Stoletov et al 2010, Yamauchi et al 2006), although a functional link to extravasation has to be established through further experiments. The visualization and characterization of invadopodia structures in models of breast cancer (Kedrin et al 2008, Roh-Johnson et al 2014) have suggested that invadopodia are key mediators of intravasation (Eckert et al 2011, Gligorijevic et al 2012). Indeed, invadopodia have been observed in ex vivo experiments (Schoumacher et al 2010, Weaver et al 2013), leading to additional support for a role in vivo. However, despite the evidence that invadopodia are prevalent in metastatic cancer cells and fulfill a major functional role in the invasion and metastasis of cancer, direct evidence for their contribution in vivo is still elusive and needs further investigation.

As podosomes are involved in immune cell extravasation and invadopodia are prevalent in metastatic cancer cells, the role of invadopodia should be investigated in cancer cell extravasation *in vivo*. In particular, to visualize these dynamic cell interactions in high-resolution detail in real time, an intravital microscopy platform specifically developed to investigate cancer cell migration *in vivo* has been utilized in the *ex ovo* chicken embryo model (Arpaia *et al* 2012, Leong *et al* 2010, 2012a, 2012b). Moreover, real-time 3D time-lapse intravital imaging was performed to visualize the behavior and dynamics of cancer cell extravasation *in vivo*. Indeed, direct evidence of the role of invadopodia during cancer cell extravasation has been provided. Moreover, it was demonstrated that disruption of invadopodia assembly via RNA interference with structural proteins (such as cortactin, Tks4 and Tk5), by either genetic or pharmacological means, results in reduced extravasation rates and the elimination of cancer metastasis.

Intravital imaging of human cancer cells and endothelium
The chorioallantoic membrane (CAM) of the chicken embryo, with its highly organized capillary bed network supported by arteries, veins and stromal cells, is an ideal model for visualizing the behavior of disseminating human cancer cells *in vivo* (Deryugina and Quigley 2008, Koop *et al* 1996). In order to investigate this, an intravital microscopy platform capable of capturing high-resolution 3D time-lapse imagery of human tumor growth, cell migration and extravasation using the *ex ovo* chicken embryo model was utilized (Arpaia *et al* 2012, Goulet *et al* 2011, Leong *et al* 2012a, 2012b, Palmer *et al* 2014, Zijlstra *et al* 2008). In more detail, visualization of the luminal surface of the CAM endothelium was achieved upon IV injection of fluorescent *Lens culinaris* agglutinin, which interacts specifically with the glycocalyx of avian endothelial cells (Jilani *et al* 2003). Moreover, the blood volume was visualized using IV fluorescent dextran. Extravascular regions of the CAM, marked by stromal cells were identified during intravital imaging through their lack of lectin–rhodamine staining. As has been reported previously (Arpaia *et al* 2012), a significant accumulation of lectin–rhodamine staining is located at endothelial junctions during intravital imaging experiments, allowing their precise localization within the tissue. In order to confirm this, the avian endothelium was transduced with cytoplasmic zsGreen *in vivo* using IV lentivirus prior to injection of lectin–rhodamine, which indeed resulted in cytoplasmic GFP expression in roughly 5% of the CAM endothelial cells. Thus, strong lectin staining corresponded precisely with the endothelial cell–cell junctions.

Cancer cells intravasate
In order to define the characteristics of cancer cell arrest and extravasation in this model, time-lapse intravital imaging was performed after the IV injection of fluorescent human epidermoid carcinoma (HEp3-GFP) cells into the vitelline vein of *ex ovo* chicken embryos, which were preinjected with lectin–rhodamine. Circulating cancer cells initially arrest at the distal end of CAM arterioles where they meet the capillary plexus. Cancer cells within the vascular lumen adopt an amoeboid morphology, maintaining close contact with the endothelium during

intravascular migration, before extravasating into the adjacent targeted stromal layer of the CAM. During this process, highly dynamic cytoplasmic protrusions are exerted at the leading edge of intravascular HEp3-GFP cells. Although the majority of these dynamic protrusions are intravascular, a proportion of them extend through the entire endothelium into the adjacent stroma.

Cancer cells exert invasive cytoplasmic extensions, breaching the endothelium
Intravital imaging of the early stages of cancer cell extravasation revealed that cancer cells form cytoplasmic protrusions that extend through the endothelial cell lining into the extravascular stroma. In particular, it has been observed that cells typically migrate intravascularly along the endothelium for 6–8 h and then cease migration over the endothelium. At this point, protrusions are observed that cross the endothelium into the extravascular stroma. Approximately 1 1/2 h after initial contact, the cancer cell has begun extravasation and over the next 3 h, the extravascular portion grows larger, while the intravascular portion gradually shrinks as the cell translocates from the vessel lumen into the stroma. It has been observed that even single cancer cells form two distinct protrusions into the extravascular stroma. However, in all the cases observed cancer cells follow a single protrusion in order to complete their extravasation. These invasive structures were observed in a substantial fraction of a panel of cancer cells, as quantified by intravital imaging.

In particular, the precise route of extravasation was also investigated using high-resolution intravital imaging. When individual planes of 3D image volumes from extravasating HEp3-GFP cells were examined, it was clear that the majority of the invasive protrusions extended through endothelial junctions have separated slightly, and moreover that these separations dissolve once extravasation is completed. Indeed, a panel of cancer cell lines including HT1080 (fibrosarcoma), B16F10 (melanoma), MDA-MB-231LN (breast) and T24 (bladder) were evaluated and also found to extravasate mainly at endothelial junctions using the paracellular trans-endothelial migration mode. In some cases, it was observed that cancer cells extensively remodeled the local endothelium, appearing to displace endothelial cells to breach the endothelium in order to access the extravascular stroma. However, in these cases, the endothelial cells remained viable and no evidence of interendothelial transit was revealed (Carman and Springer 2008).

Cancer cells exert invadopodia through the endothelium
It has been hypothesized that components of invadopodia are present within the cytoplasmic protrusions of extravasating cancer cells observed *in vivo*. Indeed, histology was performed on frozen cross-sectional slices of CAM containing extravasating HT1080 fibrosarcoma cells. Visualized by fluorescence microscopy, extravasating cells exerted protrusions projecting through the endothelial layer and extending into the extravascular stroma (ES). In more detail, these cell protrusions contain both F-actin and cortactin, indicating that they may indeed be invadopodia.

In order to further characterize the identity and dynamics of these protrusions during extravasation, structural components known to be localized or concentrated in invadopodia (such as cortactin, Tks4 and Tks5) (Buschman *et al* 2009,

Oser *et al* 2010, Sung *et al* 2011, Abram *et al* 2003, Seals *et al* 2005) were assessed in extravasating cancer cells using intravital imaging. Constructs encoding the fusion proteins cortactin-zsGreen (CTTN-zsG) and Tks4-zsGreen (Tks4-zsG) were stably transfected into three aggressive human cancer cell lines—epidermoid carcinoma HEp3, fibrosarcoma HT1080 (HT1080-tdT) and breast cancer MDA-MB-231LN (231LN-tdT). The localization of each of these cancer cells was confirmed using fluorescence imaging.

During extravasation of 231LN-tdT cells *in vivo*, cortactin-zsGreen was abundant in cytoplasmic protrusions extending through the endothelium. Although cortactin-zsGreen is also present in foci throughout the cell, the cytoplasmic extension exerted into the stromal layer contains organized structures of cortactin-zsGreen that would be expected in invadopodia. Indeed, localization of cortactin-zsGreen in HEp3 cells was present through diffuse puncta throughout the cell, as well as within thin invasive structures at the site of cancer cell extravasation. Indeed, a significant enrichment in cortactin-zsGreen was observed within protrusions that extended through the endothelium.

Moreover, the localization of Tks4-zsG in HEp3, HT1080-tdT and 231LN-tdT cell lines was also evaluated during extravasation, using intravital imaging. In intravascular 231LN-tdT breast cancer cells, the Tks4-zsGreen construct was present as a diffuse signal throughout the cytoplasm. During extravasation, however, Tks4-zsGreen localized to the apical tips of cytoplasmic protrusions, which are exerted into the CAM stroma. The localization of Tks5-GFP, by contrast, was concentrated at the base of the protrusion and even throughout the protrusion invading into the extravascular stroma. However, a similar concentration at the base of invadopodia was not observed in cells expressing Tks4-zsG. The proteolytic activity of MT1-MMP is a key marker of invadopodia and it has also been observed in protrusions formed by extravasating 231LN-tdT cells and HT1080 cancer cells expressing MT1-MMP-GFP. The enrichment of cortactin, Tks4, Tks5 and MT1-MMP in these invasive protrusions further suggests that they indeed display invadopodia and moreover the intravital imaging experiments suggest that cancer cells extend invadopodia between endothelial cells and into the extravascular stroma prior to and during extravasation.

Tks4, Tks5, and cortactin are required for invadopodia formation
To reveal the impact of invadopodia on extravasation *in vivo*, loss-of-function experiments that targeted Tks4 and Tks5 individually were performed; this was reported to inhibit invadopodia function *in vitro* (Buschman *et al* 2009, Diaz *et al* 2009). However, it has been hypothesized that RNAi knockdown of Tks4 or Tks5 can inhibit invadopodia assembly in cancer cells, thus inhibiting cancer cell extravasation and finally cancer metastasis. In particular, cancer cell lines with stable small hairpin RNA (shRNA) knockdowns of cortactin, Tks4 and Tks5 were evaluated for their ability to form invadopodia, to extravasate and to establish metastatic colonies *in vivo*. These results were compared with cells lacking the key cell migration regulator RhoA. When assessed for the cancer cell's ability to degrade extracellular matrix, loss of cortactin, but not loss of RhoA, reduced the fraction of cells able to degrade gelatin.

In more detail, HEp3-GFP cells expressing RhoA or cortactin shRNAs were assessed in extravasation assays with the chick embryo CAM using an intravital imaging approach. Thus, cancer cells arrested in the CAM were visualized and classified according to their location as intravascular, in the process of extravasating, and extravascular. The majority of shLuc HEp3-GFP cells (40%–50%) extravasate within 24 h, with very few cells present in the intravascular space. By contrast, a significantly greater proportion of shCTTN cells remained in the intravascular space at each time point, whereas very few shRhoA cells were present within the intravascular space. Moreover, very few shRhoA and shCTTN cells successfully extravasated 24 h after IV injection. Because the numbers of extravasating cells can be monitored over time in each embryo, the number of cells that have been lost or have died can also be determined. Cells expressing shRhoA exhibited the greatest loss between the 0 h to 6 h time points, indicating that most of these cells probably died in the intravascular space. Cell damage was much lower in the first 6 h for shLuc and shCTTN cells, indicating that the significant difference in the numbers of intravascular and extravascular cells after 24 h was not due to cell death, but to an inhibition of the shCTTN cells to extravasate.

The impact of cortactin on subsequent metastatic colony formation was then investigated using intravital imaging and an experimental metastasis approach. Two hundred thousand cells per embryo were injected IV and after seven days the number of metastases, micrometastases and single cancer cells present throughout the entire CAM organ was put in a line. In more detail, embryos injected with shRhoA and shCTTN cells displayed the fewest metastases, micrometastases and single cancer cells compared to both parental and shLuc control cells. In summary, these data indicate that RhoA depletion inhibits metastatic colony formation primarily through cell damage, whereas the inhibition of invadopodia through the depletion of cortactin inhibits metastasis as a result of inhibited cancer cell extravasation.

Although Tks4 is predominantly localized to invadopodia, it has also been observed within lamellipodia, whereas Tks5 has only been observed in invadopodia (Abram et al 2003, Buschman et al 2009, Seals et al 2005). To establish the requirement of invadopodia for extravasation and metastatic colony formation, 231LN-tdT cell lines with stable shRNA knock-downs for Tks4 and Tks5 were evaluated for their ability to extravasate in the CAM of avian embryos over a 24 h time period. In particular, 231LN-tdT cells lacking RhoA, CTTN, Tks4 or Tks5 had significantly reduced extravasation rates compared to control cells. Moreover, these cells were also observed to have decreased incidence of protrusions formed by cells arrested in the intravascular space 3 h after IV injection of cells. Taken together, this provides strong evidence that invadopodia are indeed required for efficient extravasation in the CAM. These in vivo loss-of-function experiments revealed a functional requirement for CTTN, Tks4 and Tks5 in the formation of invadopodia and finally cancer cell extravasation.

Src kinase inhibition eliminates invadopodia formation
Src kinase regulates invadopodia formation via phosphorylation of cortactin to its active state (Evans et al 2012, Mader et al 2011, Oser et al 2009), while also

phosphorylating a number of other targets (Ferrando *et al* 2012). The treatment of HEp3-GFP cells with the Src kinase inhibitor Saracatinib at 1.0 mM significantly altered *in vitro* cell morphology from a stellate to cobblestone morphology. A marked reduction of cortactin-rich invadopodia and the associated adhesion-type structures that are characteristic of invadopodia was detected. The impact of Src kinase inhibition on the extravasation kinetics of HEp3-GFP cells in the avian embryo CAM was analyzed *in vivo* over a 24 h time period, comparing Saracatinib at 1.0 mM versus vehicle. As expected, no *in vivo* or *in vitro* cytotoxic effects were observed when Saracatinib was administered to a final concentration of 1.0 mM. Saracatinib-treated cancer cells at 3–6 h after IV injection exhibited a significant reduction in invadopodia assembly compared to vehicle-treated cells. Over 24 h, Saracatinib treatment resulted in significantly decreased extravasation rates compared to vehicle control after 24 h and thus was accompanied by a retention of Saracatinib-treated cells in the intravascular space after 24 h. The interpretation of the data is complicated due to the fact that more Saracatinib-treated cells were lost or died in the first 3 h compared to vehicle-control-treated cells, resulting in significantly fewer cells successfully extravasating after 24 h. However, these data suggest that Saracatinib directly impacts on cancer cell extravasation, which indicates that it might be a useful therapeutic reagent.

Inhibition of Tks4 or Tks5 reduces metastatic colony formation in a mouse model

Following the observations in the chicken embryo CAM model, a mammalian adult cancer model was used to confirm them. Clonogenic MDA-MB-231LN-tdT cell lines were generated with stable shRNA knockdowns for Tks4 and Tks5 and evaluated for their ability to extravasate and form metastatic colonies in murine lungs after tail vein injection in nude beige mice. Extravasation efficiency was determined by comparing the number of cells that extravasated 24 h after IV injection to the number of cells that initially arrested in the lung 0 h after IV injection. According to the histological sections of injected murine lungs, the majority of cells were intravascular immediately after injection, and thus all cells that were present in the lungs after 24 h had successfully extravasated, as determined by confocal microscopy. The knockdown of Tks4 or Tks5 led to significant decreases in extravasation compared to the shLUC control. Indeed, the majority of the Tks4 and Tks5 shRNA clones exhibited negligible rates of extravasation, with no extravasated cancer cells after 24 h post-injection. Moreover, the metastatic burden was also evaluated 4 weeks post-injection, where knockdown of Tks4 or Tks5 resulted in a significant reduction in the number of macrometastases, micro-metastases and extravascular single cells. In summary, these data demonstrate that in two xenograft models of human cancer metastasis invadopodia are required for extravasation and metastatic colonization of distant sites.

The extravasation of cancer cells at distant sites occurs predominantly within capillary beds (Chambers *et al* 2002) and is thought to be a key step in the metastatic cascade, preceding metastatic colony formation. In the past, a putative role for invadopodia in cancer cell migration and invasion has been extensively reported using *in vitro* approaches (Artym *et al* 2006, Buschman *et al* 2009, Diaz *et al* 2009, Linder 2007, Mader *et al* 2011, Oser *et al* 2009), whereas a lack of direct *in vivo* evidence for invadopodia has raised questions regarding their physiological relevance to cancer disease. Using cortactin, MT1-MMP and Tks4/5 fusion expression constructs, and shRNA-facilitated knockdown, it was determined that invadopodia are formed early in the extravasation process, manifesting as protrusions that breach the endothelial layer, and that they are required for successful extravasation. Taken together, cancer cell extravasation is a highly coordinated and dynamic process that occurs within 24 h, consistent with other observations during the initial intravital imaging experiments (Koop *et al* 1996). However, the inhibition of the structural or functional components of invadopodia results in a pronounced reduction in metastatic colony formation in two different experimental models of metastasis. Finally, by providing direct evidence of the functional importance of invadopodia in cancer cell extravasation *in vivo*, these studies demonstrate that invadopodia are crucial in the metastatic cascade and represent a potential therapeutic target for antimetastasis strategies.

The assembly of invadopodia is a precisely regulated and sequential process that is characterized by the initial formation of the non-degradative invadopodium precursors that are enriched in actin regulators such as cortactin, Arp2/3 and cofilin (Clark *et al* 2007, Oser *et al* 2009). Next, these precursors mature through a sequence of events involving stabilization through Tks5 interactions (Blouw *et al* 2008), actin polymerization and the recruitment of matrix proteases such as MT1-MMP, whose localization and stability is in turn regulated by factors such as Tks4 (Buschman *et al* 2009). Moreover, this results in mature, matrix-degrading invadopodia that regulate the remodeling of the extracellular matrix to increase cell migration and translocation. Based on these observations regarding the ability of invadopodia to facilitate cancer cell transmigration through the endothelial layer in the early steps of cancer cell extravasation, it has been hypothesized that their inhibition may prevent extravasation and consequently inhibit the formation of metastatic colonies. Thus it has been investigated how the inhibition of distinct steps of invadopodia initiation (cortactin), maturation (Tks5) and function (Tks4) impacts on the extravasation of metastatic cancer cells. In addition these results have been compared to the results regarding the loss of RhoA, which, it has been suggested, may inhibit cell migration in an invadopodia-independent way. Indeed, the extravasation rates were significantly reduced in the CAM of the avian embryo and in mouse lungs when any of these contributors to invadopodia were depleted in cancer cells. The depletion of Tks4 or Tks5 in cancer cells also led to a reduction of micrometastases and single migratory cancer cells compared to control cancer cells, indicating that the inhibition of invadopodia has additional anti-metastatic effects that finally lead to a further reduction in metastatic efficiency. These observations are consistent with other studies that have established a role for invadopodia in other steps within the metastatic cascade, such as tumor growth and intravasation (Blouw *et al* 2008, Gligorijevic *et al* 2012, Sharma *et al* 2013b), and which implicate these structures in a general mechanism for cancer cell motility and translocation.

Taken together, it has been established that the inhibition of extravasation by targeting invadopodia seems to be a viable antimetastasis approach. Nonetheless, a clinical window of opportunity for antimetastatic therapies may not exist for all cancer patients, especially considering the primary tumor's ability to disseminate cancer cells into the circulation at early stages of progression. However, there may be possibilities for therapeutic opportunities at other distinct stages of cancer development and treatment, but this is not addressed here. A substantial body of clinical evidence suggests that cancers such as prostate cancer acquire metastatic potential during the course of progression. Thus, some benefit from antimetastasis agents might be provided for patients who are identified as being at increased risk of metastasis (Palmer *et al* 2014). Finally, based on the growing evidence that cancer cells are shed into the circulation after core-needle biopsy (Hansen *et al* 2004) or surgery (Juratli *et al* 2014), it would be worth exploring whether an antiextravasation approach might help to eliminate a potential metastasis risk from these procedures.

The understanding of cancer cell extravasation has been advanced to a significant extent by well-characterized mechanisms of immune cell transendothelial migration or diapedesis. In particular, this both clarifies and substantiates the postulated concepts of cancer cell extravasation through the dynamic visualization of individual cancer cells at high resolution. Although the processes of cancer cell extravasation and leukocyte diapedesis share several features in common (Carman and Springer 2008), there seems to be no evidence from intravital imaging experiments that cancer cells undergo transcellular migration, or migration through pores created in endothelial cells, as has been observed with leukocytes. This raises the question of whether this transcellular mode is indeed a commonly used transmigration mode for cancer cells. The examination of thousands of cells from a panel of cancer cell lines undergoing extravasation demonstrated clearly that cancer cells in this model system only use a paracellular mode of transendothelial migration into the extravascular stroma. The simplest explanation for this observation is that cancer cells are typically much larger in volume than leukocytes or endothelial cells, which may cause a large hole within the endothelial cell upon transcellular migration that might eventually destroy its cytoskeletal architecture. As the cytoskeletal architecture is required for the transcellular transmigration of cancer cells, this mode does not seem to be the most likely to be used. It has been observed that a minority of extravasating cells gained access to the extravascular stroma through a more pronounced displacement of endothelial cells. However, this did not appear to impact on the viability of the endothelial cells and is consistent with previous observations in an *in vivo* model (Stoletov *et al* 2010), suggesting that cancer cells can play an active role in remodeling the local endothelium, as has been observed in an *in vitro* transendothelial migration model (Mierke *et al* 2008, Mierke 2011). This is particularly interesting, as it has been suggested that invadopodia biogenesis is linked with the secretion of exosomes (Hoshino *et al* 2013). In the intravital imaging experiments, invadopodia have been revealed to be highly dynamic in morphology as they extend into the extravascular stroma and they are even sporadically associated with the transient release of microparticles. The idea that invadopodia utilize a tightly regulated microvesicle release mechanism in order to influence the local micro-environment is a compelling one and thus requires further detailed investigation. Taken together, a powerful model has been established to visualize and dissect the key functional and structural components of invadopodia and provide significant evidence for an invadopodia role during cancer cell extravasation.

Further reading

Martinelli R, Zeiger A S, Whitfield M, Sciuto T E, Dvorak A, Van Vliet K J, Greenwood J and Carman C V 2014 Probing the biomechanical contribution of the endothelium to lymphocyte migration: diapedesis by the path of least resistance *J. Cell Sci.* **127** 3720–34

Vantyghem S A, Allan A L, Postenka C O, Al-Katib W, Keeney M, Tuck A B and Chambers A F 2005 A new model for lymphatic metastasis: development of a variant of the MDA-MB-468 human breast cancer cell line that aggressively metastasizes to lymph nodes *Clin. Exp. Metastasis* **22** 351–61

References

Abram C L, Seals D F, Pass I, Salinsky D, Maurer L, Roth T M and Courtneidge S A 2003 The adaptor protein fish associates with members of the ADAMs family and localizes to podosomes of Src-transformed cells *J. Biol. Chem.* **278** 16844–51

Al-Mehdi A B, Tozawa K, Fisher A B, Shientag L, Lee A and Muschel R J 2000 Intravascular origin of metastasis from the proliferation of endothelium-attached tumor cells: a new model for metastasis *Nat. Med.* **6** 100–2

Arpaia E, Blaser H, Quintela-Fandino M, Duncan G, Leong H S, Ablack A, Nambiar S C, Lind E F, Silvester J and Fleming C K *et al* 2012 The interaction between caveolin-1 and Rho-GTPases promotes metastasis by controlling the expression of alpha5-integrin and the activation of Src, Ras and Erk *Oncogene* **31** 884–96

Artym V V, Zhang Y, Seillier-Moiseiwitsch F, Yamada K M and Mueller S C 2006 Dynamic interactions of cortactin and membrane type 1 matrix metalloproteinase at invadopodia: defining the stages of invadopodia formation and function *Cancer Res.* **66** 3034–43

Birukova A A, Adyshev D, Gorshkov B, Bokoch G M, Birukov K G and Verin A D 2006 GEF-H1 is involved in agonist-induced human pulmonary endothelial barrier dysfunction *Am. J. Physiol. Lung Cell Mol. Physiol.* **290** L540–8

Blouw B, Seals D F, Pass I, Diaz B and Courtneidge S A 2008 A role for the podosome/invadopodia scaffold protein Tks5 in tumor growth *in vivo Eur. J. Cell Biol.* **87** 555–67

Burger K L, Learman B S, Boucherle A K, Sirintrapun S J, Isom S, Diaz B, Courtneidge S A and Seals D F 2014 Src-dependent Tks5 phosphorylation regulates invadopodia-associated invasion in prostate cancer cells *Prostate* **74** 134–48

Buschman M D, Bromann P A, Cejudo-Martin P, Wen F, Pass I and Courtneidge S A 2009 The novel adaptor protein Tks4 (SH3PXD2B) is required for functional podosome formation *Mol. Biol. Cell* **20** 1302–11

Caldieri G, Giacchetti G, Beznoussenko G, Attanasio F, Ayala I and Buccione R 2009 Invadopodia biogenesis is regulated by caveolin-mediated modulation of membrane cholesterol levels *J. Cell Mol. Med.* **13** (8B) 1728–40

Carman C V and Springer T A 2004 A transmigratory cup in leukocyte diapedesis both through individual vascular endothelial cells and between them *J. Cell Biol.* **167** 377–88

Carman C V and Springer T A 2008 Transcellular migration: cell–cell contacts get intimate *Curr. Opin. Cell Biol.* **20** 533–40

Carman C V, Sage P T, Sciuto T E, de la Fuente M A, Geha R S, Ochs H D, Dvorak H F, Dvorak A M and Springer T A 2007 Transcellular diapedesis is initiated by invasive podosomes *Immunity* **26** 784–97

Chambers A F, Groom A C and MacDonald I C 2002 Dissemination and growth of cancer cells in metastatic sites *Nat. Rev. Cancer* **2** 563–72

Chandrasekharan U M *et al* 2006 TNF-a receptor-II is required for TNF-a-induced leukocyte-endothelial interaction *in vivo Blood* **109** 1938–44

Chew T L, Wolf W A, Gallagher P J, Matsumura F and Chisholm R L 2002 A fluorescent resonant energy transfer-based biosensor reveals transient and regional myosin light chain kinase activation in lamella and cleavage furrows *J. Cell Biol.* **156** 543–53

Clark E S, Whigham A S, Yarbrough W G and Weaver A M 2007 Cortactin is an essential regulator of matrix metalloproteinase secretion and extracellular matrix degradation in invadopodia *Cancer Res.* **67** 4227–35

Cullere X, Shaw S K, Andersson L, Hirahashi J, Luscinskas F W and Mayadas T N 2005 Regulation of vascular endothelial barrier function by Epac, a cAMP-activated exchange factor for Rap GTPase *Blood* **105** 1950–5

DerMardirossian C and Bokoch G M 2005 GDIs: central regulatory molecules in Rho GTPase activation *Trends Cell Biol.* **15** 356–63

DerMardirossian C, Rocklin G, Seo J Y and Bokoch G M 2006 Phosphorylation of RhoGDI by Src regulates Rho GTPase binding and cytosol-membrane cycling *Mol. Biol. Cell* **17** 4760–8

Deryugina E I and Quigley J P 2008 Chick embryo chorioallantoic membrane model systems to study and visualize human tumor cell metastasis *Histochem. Cell Biol.* **130** 1119–30

Diaz B, Shani G, Pass I, Anderson D, Quintavalle M and Courtneidge S A 2009 Tks5-dependent, nox-mediated generation of reactive oxygen species is necessary for invadopodia formation *Sci. Signal.* **2** ra53

Dovas A, Patsialou A, Harney A S, Condeelis J and Cox D 2013 Imaging interactions between macrophages and tumour cells that are involved in metastasis *in vivo* and *in vitro J. Microsc.* **251** 261–9

Eckert M A, Lwin T M, Chang A T, Kim J, Danis E, Ohno-Machado L and Yang J 2011 Twist1-induced invadopodia formation promotes tumor metastasis *Cancer Cell* **19** 372–86

Essler M, Amano M, Kruse H J, Kaibuchi K, Weber P C and Aepfelbacher M 1998 Thrombin inactivates myosin light chain phosphatase via Rho and its target Rho kinase in human endothelial cells *J. Biol. Chem.* **273** 21867–74

Evans J V, Ammer A G, Jett J E, Bolcato C A, Breaux J C, Martin K H, Culp M V, Gannett P M and Weed S A 2012 Src binds cortactin through an SH2 domain cystine-mediated linkage *J. Cell Sci.* **125** 6185–97

Ferrando I M *et al* 2012 Identification of targets of c-Src tyrosine kinase by chemical complementation and phosphoproteomics *Mol. Cell. Proteomics* **11** 355–69

Fidler I J and Hart I R 1982 Biological diversity in metastatic neoplasms: origins and implications *Science* **217** 998–1003

Fukuhara S, Sakurai A, Sano H, Yamagishi A, Somekawa S, Takakura N, Saito Y, Kangawa K and Mochizuki N 2005 Cyclic AMP potentiates vascular endothelial cadherin-mediated cell–cell contact to enhance endothelial barrier function through an Epac-Rap1 signaling pathway *Mol. Cell Biol.* **25** 136–46

Furmaniak-Kazmierczak E, Crawley S W, Carter R L, Maurice D H and Cote G P 2007 Formation of extracellular matrix-digesting invadopodia by primary aortic smooth muscle cells *Circ. Res.* **100** 1328–36

Gallatin W M, Weissman I L and Butcher E C 1983 A cell-surface molecule involved in organ-specific homing of lymphocytes *Nature* **304** 30–4

Gligorijevic B, Wyckoff J, Yamaguchi H, Wang Y, Roussos E T and Condeelis J 2012 N-WASP-mediated invadopodium formation is involved in intravasation and lung metastasis of mammary tumors *J. Cell Sci.* **125** 724–34

Gorovoy M *et al* 2007 RhoGDI-1 modulation of the activity of monomeric RhoGTPase RhoA regulates endothelial barrier function in mouse lungs *Circ. Res.* **101** 50–8

Goulet B, Kennette W, Ablack A, Postenka C O, Hague M N, Mymryk J S, Tuck A B, Giguere V, Chambers A F and Lewis J D 2011 Nuclear localization of maspin is essential for its inhibition of tumor growth and metastasis *Lab. Invest.* **91** 1181–7

Groeneveld A B 2002 Vascular pharmacology of acute lung injury and acute respiratory distress syndrome *Vascul. Pharmacol.* **39** 247–56

Hansen N M, Ye X, Grube B J and Giuliano A E 2004 Manipulation of the primary breast tumor and the incidence of sentinel node metastases from invasive breast cancer *Arch. Surg.* **139** 634–9

Heyder C, Gloria-Maercker E, Entschladen F, Hatzmann W, Niggemann B, Zanker K S and Dittmar T 2002 Real-time visualization of tumor cell/endothelial cell interactions during transmigration across the endothelial barrier *J. Cancer Res. Clin. Oncol.* **128** 533–8

Hillyer P, Mordelet E, Flynn G and Male D 2003 Chemokines, chemokine receptors and adhesion molecules on different human endothelia: discriminating the tissue-specific functions that affect leukocyte migration *Clin. Exp. Immunol.* **134** 431–41

Holinstat M, Knezevic N, Broman M, Samarel A M, Malik A B and Mehta D 2006 Suppression of RhoA activity by focal adhesion kinase-induced activation of p190RhoGAP: role in regulation of endothelial permeability *J. Biol. Chem.* **281** 2296–305

Hoshino D, Kirkbride K C, Costello K, Clark E S, Sinha S, Grega-Larson N, Tyska M J and Weaver A M 2013 Exosome secretion is enhanced by invadopodia and drives invasive behavior *Cell Rep.* **5** 1159–68

Jacobson J R and Garcia J G 2007 Novel therapies for microvascular permeability in sepsis *Curr. Drug Targets* **8** 509–14

Jacobson J R, Barnard J W, Grigoryev D N, Ma S F, Tuder R M and Garcia J G 2005 Simvastatin attenuates vascular leak and inflammation in murine inflammatory lung injury *Am. J. Physiol. Lung Cell Mol. Physiol.* **288** L1026–32

Jilani S M, Murphy T J, Thai S N M, Eichmann A, Alva J A and Iruela-Arispe M L 2003 Selective binding of lectins to embryonic chicken vasculature *J. Histochem. Cytochem.* **51** 597–604

Juratli M A, Sarimollaoglu M, Siegel E R, Nedosekin D A, Galanzha E I, Suen J Y and Zharov V P 2014 Real-time monitoring of circulating tumor cell release during tumor manipulation using *in vivo* photoacoustic and fluorescent flow cytometry *Head Neck* **36** 1207–15

Kedrin D, Gligorijevic B, Wyckoff J, Verkhusha V V, Condeelis J, Segall J E and van Rheenen J 2008 Intravital imaging of metastatic behavior through a mammary imaging window *Nat. Methods* **5** 1019–21

Kedrin D, Gligorijevic B, Wyckoff J, Verkhusha V V, Condeelis J, Segall J E and van Rheenen J 2008 Intravital imaging of metastatic behavior through a mammary imaging window *Nat. Methods* **5** 1019–21

Khuon S, Liang L, Dettman R W, Sporn P H, Wysolmerski R B and Chew T L 2010 Myosin light chain kinase mediates transcellular intravasation of breast cancer cells through the underlying endothelial cells: a three-dimensional FRET study *J. Cell Sci.* **123** 431–40

Koop S, Schmidt E E, MacDonald I C, Morris V L, Khokha R, Grattan M, Leone J, Chambers A F and Groom A C 1996 Independence of metastatic ability and extravasation: metastatic ras-transformed and control fibroblasts extravasate equally well *Proc. Natl Acad. Sci. USA* **93** 11080–4

Kouklis P, Konstantoulaki M, Vogel S, Broman M and Malik A B 2004 Cdc42 regulates the restoration of endothelial barrier function *Circ. Res.* **94** 159–66

Laferriere J, Houle F, Taher M M, Valerie K and Huot J 2001 Transendothelial migration of colon carcinoma cells requires expression of E-selectin by endothelial cells and activation of stress-activated protein kinase-2 (SAPK2/p38) in the tumor cells *J. Biol. Chem.* **276** 33762–72

Lai S L, Chang C N, Wang P J and Lee S J 2005 Rho mediates cytokinesis and epiboly via ROCK in zebrafish *Mol. Reprod. Dev.* **71** 186–96

Langley R R and Fidler I J 2007 Tumor cell-organ microenvironment interactions in the pathogenesis of cancer metastasis *Endocr. Rev.* **28** 297–321

Leong H S, Chambers A F and Lewis J D 2012b Assessing cancer cell migration and metastatic growth *in vivo* in the chick embryo using fluorescence intravital imaging *Methods Mol. Biol.* **872** 1–14

Leong H S, Lizardo M M, Ablack A, McPherson V A, Wandless T J, Chambers A F and Lewis J D 2012a Imaging the impact of chemically inducible proteins on cellular dynamics *in vivo PLoS One* **7** e30177

Leong H S, Steinmetz N F, Ablack A, Destito G, Zijlstra A, Stuhlmann H, Manchester M and Lewis J D 2010 Intravital imaging of embryonic and tumor neovasculature using viral nanoparticles *Nat. Protoc.* **5** 1406–17

Li Y H and Zhu C 1999 A modified Boyden chamber assay for tumor cell transendothelial migration *in vitro Clin. Exp. Metastasis* **17** 423–9

Linder S 2007 The matrix corroded: podosomes and invadopodia in extracellular matrix degradation *Trends Cell Biol.* **17** 107–17

Liotta L A, Steeg P S and Stetler-Stevenson W G 1991 Cancer metastasis and angiogenesis: an imbalance of positive and negative regulation *Cell* **64** 327–36

Luzzi K J, MacDonald I C, Schmidt E E, Kerkvliet N, Morris V L, Chambers A F and Groom A C 1998 Multistep nature of metastatic inefficiency: dormancy of solitary cells after successful extravasation and limited survival of early micrometastases *Am. J. Pathol.* **153** 865–73

Mader C C, Oser M, Magalhaes M A O, Bravo-Cordero J J, Condeelis J, Koleske A J and Gil-Henn H 2011 An EGFR-Src-Arg-cortactin pathway mediates functional maturation of invadopodia and breast cancer cell invasion *Cancer Res.* **71** 1730–41

Magalhaes M A O, Larson D R, Mader C C, Bravo-Cordero J J, Gil-Henn H, Oser M, Chen X, Koleske A J and Condeelis J 2011 Cortactin phosphorylation regulates cell invasion through a pH-dependent pathway *J. Cell Biol.* **195** 903–20

Mammoto T, Parikh S M, Mammoto A, Gallagher D, Chan B, Mostoslavsky G, Ingber D E and Sukhatme V P 2007 Angiopoietin-1 requires p190RhoGAP to protect against vascular leakage *in vivo J. Biol. Chem.* **282** 23910–8

McGettrick H M, Lord J M, Wang K Q, Rainger G E, Buckley C D and Nash G B 2006 Chemokine- and adhesion-dependent survival of neutrophils after transmigration through cytokine-stimulated endothelium *J. Leukoc. Biol.* **79** 779–88

Mehta D and Malik A B 2006 Signaling mechanisms regulating endothelial permeability *Physiol. Rev.* **286** 279–367

Mehta D, Rahman A and Malik A B 2001 Protein kinase C-alpha signals rho-guanine nucleotide dissociation inhibitor phosphorylation and rho activation and regulates the endothelial cell barrier function *J. Biol. Chem.* **276** 22614–20

Mierke C T 2011 Cancer cells regulate biomechanical properties of human microvascular endothelial cells *J. Biol. Chem.* **286** 40025–37

Mierke C T, Paranhos-Zitterbart D, Kollmannsberger P, Raupach C, Schlötzer-Schrehardt U, Goecke T W, Behrens J and Fabry B 2008 Break-down of the endothelial barrier function in tumor cell transmigration *Biophys. J.* **94** 2832–46

Murphy D A and Courtneidge S A 2011 The 'ins' and 'outs' of podosomes and invadopodia: characteristics, formation and function *Nat. Rev. Mol. Cell Biol.* **12** 413–26

Murphy D A, Diaz B, Bromann P A, Tsai J H, Kawakami Y, Maurer J, Stewart R A, Izpisu'a-Belmonte J C and Courtneidge S A 2011 A Src-Tks5 pathway is required for neural crest cell migration during embryonic development *PLoS One* **6** e22499

Nicolson G L 1989 Metastatic tumor cell interactions with endothelium, basement membrane and tissue *Curr. Opin. Cell Biol.* **1** 1009–19

Oser M, Mader C C, Gil-Henn H, Magalhaes M, Bravo-Cordero J J, Koleske A J and Condeelis J 2010 Specific tyrosine phosphorylation sites on cortactin regulate Nck1-dependent actin polymerization in invadopodia *J. Cell Sci.* **123** 3662–73

Oser M, Yamaguchi H, Mader C C, Bravo-Cordero J J, Arias M, Chen X, Desmarais V, van Rheenen J, Koleske A J and Condeelis J 2009 Cortactin regulates cofilin and N-WASp activities to control the stages of invadopodium assembly and maturation *J. Cell Biol.* **186** 571–87

Palmer T D *et al* 2014 Integrin-free tetraspanin CD151 can inhibit tumor cell motility upon clustering and is a clinical indicator of prostate cancer progression *Cancer Res.* **74** 173–87

Parikh S M, Mammoto T, Schultz A, Yuan H T, Christiani D, Karumanchi S A and Sukhatme V P 2006 Excess circulating angiopoietin-2 may contribute to pulmonary vascular leak in sepsis in humans *PLoS Med.* **3** e46

Pignatelli J, Tumbarello D A, Schmidt R P and Turner C E 2012 Hic-5 promotes invadopodia formation and invasion during TGF-b-induced epithelial–mesenchymal transition *J. Cell Biol.* **197** 421–37

Quintavalle M, Elia L, Condorelli G and Courtneidge S A 2010 MicroRNA control of podosome formation in vascular smooth muscle cells *in vivo* and *in vitro J. Cell Biol.* **189** 13–22

Reiland J, Furcht L T and McCarthy J B 1999 CXC-chemokines stimulate invasion and chemotaxis in prostate carcinoma cells through the CXCR2 receptor *Prostate* **41** 78–88

Roh-Johnson M, Bravo-Cordero J J, Patsialou A, Sharma V P, Guo P, Liu H, Hodgson L and Condeelis J 2014 Macrophage contact induces RhoA GTPase signaling to trigger tumor cell intravasation *Oncogene* **33** 4203–12

Rousseau S, Houle H, Landry J and Huot J 1997 p38 MAP kinase activation by vascular endothelial growth factor mediates actin reorganization and cell migration in human endothelial cells *Oncogene* **15** 2169–77

Sandig M, Voura E B, Kalnins V I and Siu C H 1997 Role of cadherins in the transendothelial migration of melanoma cells in culture *Cell Motil. Cytoskeleton* **38** 351–64

Sawafuji M, Ishizaka A, Kohno M, Koh H, Tasaka S, Ishii Y and Kobayashi K 2005 Role of Rho-kinase in reexpansion pulmonary edema in rabbits *Am. J. Physiol. Lung Cell Mol. Physiol.* **289** L946–53

Schoumacher M, Goldman R D, Louvard D and Vignjevic D M 2010 Actin, microtubules, and vimentin intermediate filaments cooperate for elongation of invadopodia *J. Cell Biol.* **189** 541–56

Seals D F, Azucena E F Jr, Pass I, Tesfay L, Gordon R, Woodrow M, Resau J H and Courtneidge S A 2005 The adaptor protein Tks5/Fish is required for podosome formation and function, and for the protease-driven invasion of cancer cells *Cancer Cell* **7** 155–65

Sharma V P, Eddy R, Entenberg D, Kai M, Gertler F B and Condeelis J 2013a Tks5 and SHIP2 regulate invadopodium maturation, but not initiation, in breast carcinoma cells *Curr. Biol.* **23** 2079–89

Sharma V P, Entenberg D and Condeelis J 2013b High-resolution live-cell imaging and time-lapse microscopy of invadopodium dynamics and tracking analysis *Methods Mol. Biol.* **1046** 343–57

Shimokawa H and Takeshita A 2005 Rho-kinase is an important therapeutic target in cardiovascular medicine *Arterioscler. Thromb. Vasc. Biol.* **25** 1767–75

Steeg P S 2006 Tumor metastasis: mechanistic insights and clinical challenges *Nat. Med.* **12** 895–904

Stetler-Stevenson W G, Aznavoorian S and Liotta L A 1993 Tumor cell interactions with the extracellular matrix during invasion and metastasis *Annu. Rev. Cell Biol.* **9** 541–73

Stoletov K, Kato H, Zardouzian E, Kelber J, Yang J, Shattil S and Klemke R 2010 Visualizing extravasation dynamics of metastatic tumor cells *J. Cell Sci.* **123** 2332–41

Sung B H, Zhu X, Kaverina I and Weaver A M 2011 Cortactin controls cell motility and lamellipodial dynamics by regulating ECM secretion *Curr. Biol.* **21** 1460–9

Tasaka S *et al* 2005 Attenuation of endotoxin-induced acute lung injury by the Rho-associated kinase inhibitor, Y-27632 *Am. J. Respir. Cell Mol. Biol.* **32** 504–10

Tremblay P L, Auger F A and Huot J 2006 Regulation of transendothelial migration of colon cancer cells by E-selectin-mediated activation of p38 and ERK MAP kinases *Oncogene* **25** 6563–73

van de Visse E P, van der H M, Verheij J, van Nieuw Amerongen G P, van Hinsbergh V W, Girbes A R and Groeneveld A B 2006 Effect of prior statin therapy on capillary permeability in the lungs after cardiac or vascular surgery *Eur. Respir. J.* **27** 1026–32 PMID: 16707397

van Nieuw Amerongen G P, Draijer R, Vermeer M A and van Hinsbergh V W 1998 Transient and prolonged increase in endothelial permeability induced by histamine and thrombin: role of protein kinases, calcium, and RhoA *Circ. Res.* **83** 1115–23

Voura E B, Ramjeesingh R A, Montgomery A M and Siu C H 2001 Involvement of integrin alpha(v)beta(3) and cell adhesion molecule L1 in transendothelial migration of melanoma cells *Mol. Biol. Cell* **12** 2699–710

Weaver A M, Heuser J E, Karginov A V, Lee W L, Parsons J T and Cooper J A 2002 Interaction of cortactin and N-WASp with Arp2/3 complex *Curr. Biol.* **12** 1270–8

Weaver A M, Page J M, Guelcher S A and Parekh A 2013 Synthetic and tissue-derived models for studying rigidity effects on invadopodia activity *Methods Mol. Biol.* **1046** 171–89

Weis S Ma and Cheresh D A 2005 Pathophysiological consequences of VEGF-induced vascular permeability *Nature* **437** 497–504

Weis S, Cui J, Barnes L and Cheresh D 2004 Endothelial barrier disruption by VEGF-mediated Src activity potentiates tumor cell extravasation and metastasis *J. Cell Biol.* **167** 223–9

Wittchen E S, Worthylake R A, Kelly P, Casey P J, Quilliam L A and Burridge K 2005 Rap1 GTPase inhibits leukocyte transmigration by promoting endothelial barrier function *J. Biol. Chem.* **280** 11675–82

Wittchen E S, Worthylake R A, Kelly P, Casey P J, Quilliam L A and Burridge K 2005 Rap1 GTPase inhibits leukocyte transmigration by promoting endothelial barrier function *J. Biol. Chem.* **280** 11675–82

Wojciak-Stothard B and Ridley A J 2002 Rho GTPases and the regulation of endothelial permeability *Vascul. Pharmacol.* **39** 187–99

Yamauchi K, Yang M, Jiang P, Xu M, Yamamoto N, Tsuchiya H, Tomita K, Moossa A R, Bouvet M and Hoffman R M 2006 Development of real-time subcellular dynamic multicolor imaging of cancer-cell trafficking in live mice with a variable-magnification whole-mouse imaging system *Cancer Res.* **66** 4208–14

Zijlstra A, Lewis J, Degryse B, Stuhlmann H and Quigley J P 2008 The inhibition of tumor cell intravasation and subsequent metastasis via regulation of *in vivo* tumor cell motility by the tetraspanin CD151 *Cancer Cell* **13** 221–34

IOP Publishing

Physics of Cancer

Claudia Tanja Mierke

Chapter 11

The mechanical properties of endothelial cells altered by aggressive cancer cells

Summary

The mechanical properties of cells have long been ignored and only the structural and the compositional aspects of cellular components have been analyzed. Mechanotransduction has become a research focus for biophysical research, and especially for cancer research, and it is still of high importance. Moreover, it has been reported that the mechanical properties of cancer cells support their invasive and aggressive potential in order to promote the malignant progression of cancers. However, the mechanical properties of the endothelial barrier of blood or lymph vessels has not yet become the focus of cancer metastasis research, and only a few papers have investigated the role they have in providing the passive endothelial barrier function or promoting the transmigration of special cancer cells.

11.1 The role of endothelial cell stiffness

How is endothelial cell stiffness impaired by substrate stiffness in 2D and 3D?
Over the last two decades, there has been growing experimental interest in the impact of passive mechanical properties—such as the viscosity (Edwards *et al* 1996), microstructure (Sieminski *et al* 2002), and especially the stiffness (Discher *et al* 2005, Peyton *et al* 2007)—of the local microenvironment on cellular functions relevant to development, homeostasis and disease. For instance, in fibroblasts the substrate stiffness has been shown to affect the rate (Pelham and Wang 1997) and direction-ality (Lo *et al* 2000) of cell migration, the assembly of focal adhesions (Pelham and Wang 1997) and the formation of actin stress fibers (Halliday and Tomasek 1995, Yeung *et al* 2005). In addition, neurons display increased branching densities when cultured on compliant substrates, whereas glia cells, which are normally co-cultured with neurons, do not even survive on deformable substrates (Flanagan *et al* 2002). Thus, these results show how important the stiffness of the substrate on which the

doi:10.1088/978-0-7503-1134-2ch11

cells are cultured or investigated on is. Moreover, the substrate stiffness can no longer be ignored and many 'old' experiments performed on flat 2D substrates need to be refined using 3D substrates. How can cells form stress fibers and focal adhesions in 3D extracellular matrices? Is the quantity of stress fibers and focal adhesion different to that for 2D substrates? What about structures such as filopodia and lamellipodia, do they also assemble in 3D matrices?

Substrate stiffness pronouncedly influences the differentiation of mesenchymal stem cells, with soft, intermediate and stiff materials being neurogenic, myogenic and osteogenic, respectively (Engler et al 2006). Moreover, it has been shown that sensitivity to substrate stiffness requires non-muscle myosin II activity, which applies tension to the actin cytoskeleton and hence stiffens cortical actin structures. It has been shown consistently that mesenchymal stem cells (such as C2C12 myoblasts and hFOB cells cultured on collagen-laminated polyacrylamide gels with a substrate stiffness of 1000, 10 000 and 40 000 Pa) have a matrix stiffness that is dependent on the increase in cellular stiffness (Engler et al 2006). It has been reported that in the case of fibroblasts grown on fibronectin-coated polyacrylamide gels, the cells' elastic moduli were equal to (or slightly lower than) those of their underlying substrates for a range of substrate stiffness up to 20 kPa (Solon et al 2007).

These studies used cells cultured on top of protein-laminated polyacrylamide gels, where the cells interact with a 2D surface. In particular, this system provides excellent control of the substrate stiffness, while it does not provide an option for the cells to be cultured within a 3D matrix, which for many cell types is more representative of their native microenvironment. For these cell types, cell culture within 3D extracellular matrix gels would provide a more realistic microenvironment, with the stiffness of the gels controlled by the gel concentration, crosslinker concentration or the mechanical boundary conditions of the gels (Nehls and Herrmann 1996, Roeder et al 2002, Sieminski et al 2004). For instance, the stiffness of collagen gels regulates the morphology of embedded fibroblasts, with compliant gels supporting elaborate dendritic extensions. Indeed, the proliferation and migration of mammary epithelial cells is regulated by the stiffness of the surrounding gels (Paszek et al 2005). In particular, endothelial cells within 3D collagen gels form microvascular networks, with the average length of the network and the average lumen area altered by the stiffness of the gel (Sieminski et al 2004). Moreover, endothelial cells display similar stiffness-induced alterations in network morphology when cultured in self-assembling peptide gels (Sieminski et al 2007).

While there is indeed evidence that substrate stiffness can regulate cellular functions, little is known about how a given response to alterations in 2D substrate stiffness might correlate with cellular responses to alterations in 3D substrate stiffness (Peyton et al 2007). Thus, the effects of 2D and 3D substrate stiffness on the regulation of endothelial cell stiffness can be investigated. In addition, the substrate stiffness effects on actin can be investigated, as actin is a major determinant of the cell's mechanical properties, with both the abundance of actin and its tension evoked by myosin playing important roles (Pourati et al 1998). Endothelial cells are an ideal cell type for these studies, as they reside in vivo in both types of microenvironments: the endothelial monolayers lining blood or lymph vessels resemble the 2D *in vitro* microenvironment and capillaries surrounded by the basement membrane resemble the 3D *in vitro* microenvironment.

Endothelial cells can alter their cell stiffness in response to the stiffness of the underlying substrate that they are adhered to. Indeed, these observations confirm the results of earlier studies of fibroblasts, mesenchymal stem cells and immortalized osteoblast and myoblast cell lines (Engler *et al* 2006, Solon *et al* 2007). In addition, it has been reported that the stiffness of the 3D matrix in which cells are cultured also alters the stiffness of the embedded cells. As technical restrictions prevent the measurement of cell stiffness within 3D gels, for these cells stiffness measurements can be performed directly after they have been isolated from gels of different stiffness. For the following reasons, it is reasonable to expect that the cell stiffness measured on cells directly isolated from gels reflects their stiffness within the gels. Firstly, quantification of the actin staining performed on cells still within the 3D gels reveals differences in the intensity of the actin staining consistent with the trends observed in the stiffness measurements. The knowledge that alterations in cell stiffness are accompanied by alterations in the actin cytoskeleton was gained from a recent study in which fibroblasts were cultured on fibronectin-coated polyacrylamide gels of different stiffness (Solon *et al* 2007). Secondly, cells isolated from gels of different stiffness had different initial stiffness and these differences remained stable for several hours, which is all that is required to perform microaspiration measurements. Finally, over a period of 8–16 h, the stiffness and actin of cells isolated from gels stabilized as they adapted to their new microenvironment. The timescale of this stiffness adaptation process is much slower than is required for the measurements of cell stiffness.

The choice of the stiffness range for the 2D polyacrylamide gels was guided by the work of Yeung and colleagues, which showed that cells cultured on matrices with a stiffness of 1600 Pa or less lacked stress fibers or other actin bundles, whereas cells on matrices with a Young's modulus E of 3200 Pa or greater displayed significant stress fibers (Yeung *et al* 2005). In more detail, it has been shown that endothelial cells attached to, but do not spread on top of, gels with a modulus less than approximately 1000 Pa. Thus a range of gel stiffness is selected that would affect endothelial cells while still ensuring that the cells can exhibit normal behavior (Yeung *et al* 2005) on top of the substrate. The results for endothelial cells on polyacrylamide gels confirm published results showing cells that are able to respond to alterations in stiffness on the order of 1000 or 10 000 Pa (Byfield *et al* 2009). Thus it may be suggested that alterations across a similar stiffness range would be required to stimulate alterations in endothelial cells within gels. However, there are several reasons for selecting a lower stiffness range for the 3D gels. Firstly, it has been demonstrated that endothelial cells within 3D gels can alter their morphology, which may lead to alterations in this range of substrate stiffness (Sieminski *et al* 2004, Kanzawa *et al* 1993). Secondly, it has been shown that endothelial cells in significantly stiffer 3D gels (~1000 Pa) cannot form microvascular networks and hence remain essentially spherical (Sieminski *et al* 2007). Thus, as in the 2D studies, a range of gel stiffness is selected that affects endothelial cell behavior, while still ensuring that the cells exhibit normal behavior.

While endothelial cells displayed similar alterations in their cellular stiffness in response to alterations in either 2D substrate (cells on top of a gel) stiffness or 3D substrate (cells embedded in a gel) stiffness, the magnitude of the 2D-substrate-stiffness-induced effects was about an order of magnitude greater than those from 3D substrate stiffness. The idea that matrix dimensionality may facilitate the matrix stiffness effects is consistent with other concepts. In particular, it has been suggested that the matrix dimensionality and matrix stiffness both work together to regulate assembly of cell–matrix adhesions (Cukierman *et al* 2001). Cellular responses to alterations in either 2D substrate stiffness or 3D substrate stiffness have been detected and hence support the conclusion that the dimensionality of the substrate may indeed regulate stiffness effects. For instance, there are several studies that have investigated the effects of 2D or 3D substrate stiffness on different endothelial cell functions, however, those analyzing the cells within 3D gels observed responses at lower stiffness compared to the 2D situation. Similarly, those studies investigating fibroblasts in 3D collagen matrices saw effects attributed to stiffness at collagen concentrations in the range of those used by Grinnell and colleagues (Grinnell 2003), whereas those studying matrix stiffness effects in fibroblasts on top of extracellular matrix-laminated polyacrylamide gels used higher stiffness levels, ranging from 1000 to 30 000 Pa (Lo *et al* 2000, Jiang *et al* 2006, Kostic and Sheetz 2006).

In response to the notion that dimensionality is a key variable, it should be said that other differences between the cells in the collagen gel system and the cells on the polyacrylamide gel system may account for the observed difference in effectual stiffness. For instance, the ligand density, which will vary between these two systems, and the stiffness are known to interact, regulating cellular responses in other systems (Engler *et al* 2004). However, the apparent differences in the effectual stiffness that were observed may not represent true differences in the endothelial cells' response to their local microenvironment, but may rather result from the difficulty of characterizing the stiffness of a non-linearly elastic material. Collagen gels, but not polyacrylamide gels, exhibit non-linear elasticity with increased stiffness at greater strains (Storm *et al* 2005). In addition, the structure of the collagen matrices cannot be regarded to be homogenous and isotropic. This special feature may also be concentration, temperature, pH-value and cross-linker dependent. Additionally, in strain-stiffening materials the stiffness of collagen gels is much higher at larger strains than at lower strains. Thus, it is possible that the stiffness used experimentally for collagen gels, based on the linear elastic region observed at relatively low strains, may be considerably lower than the actual stiffness in the pericellular microenvironment of tumors, when larger strains are present (Byfield *et al* 2009). By contrast, the viscoelastic property of collagen, which is opposed to the purely elastic nature of polyacrylamide, provides collagen with the ability to creep over longer times compared to purely elastic gels, which then decreases the stress and consequently lowers collagen's effective stiffness. Regardless of the reasons for the apparent differences in the effectual stiffness in the 2D and 3D systems, these data further support the notion that alterations in both 2D substrate stiffness or 3D substrate stiffness can have pronounced effects on cells and they show that cells alter their cellular stiffness due to alterations in the stiffness of their 3D microenvironment

(Byfield *et al* 2009). Taken together, these results demonstrate that the magnitude of the stiffness initiating a response depends pronouncedly on the specific features of the microenvironment, which potentially includes the dimensionality of the matrix. Thus, when extrapolating the effectual stiffness obtained in one environment to another one, one must be very careful (Byfield *et al* 2009).

11.2 The role of the endothelial contractile apparatus

Do transmigrating invasive cancer cells regulate the biomechanical properties of the endothelium?

The impact of endothelial cells on the regulation of cancer cell invasiveness into 3D extracellular matrices is not yet well understood and requires further investigation. In particular, the regulation of cancer cell transmigration seems to be a complex scenario that has not yet been fully characterized. In numerous studies, the endothelium acts as a functional and passive barrier against the invasion of cancer cells (Al-Mehdi *et al* 2000, Zijlstra *et al* 2008). Moreover, the endothelium reduces the invasion of cancer cells and hence metastasis (Van Sluis *et al* 2009). By contrast, several recent reports have proposed a novel paradigm in which endothelial cells are able to regulate the invasiveness of certain cancer cells by increasing their dissemination through vessels (Kedrin *et al* 2008) or by increasing the individual invasiveness of cancer cells (Mierke *et al* 2008a). Although several adhesion molecules have been identified as functioning in tumor–endothelial cell interactions, hence promoting cancer metastasis formation, the role of endothelial mechanical properties during cancer cell transmigration and invasion is not yet known. However, it has been suggested that the altered mechanical properties of endothelial cells may support one of the two main functions of the endothelium in cancer metastasis: firstly, they act as a passive barrier for cancer cell invasion and secondly, they serve as an active enhancer for cancer cell invasion. A main biochemical signal transduction pathway of the tumor–endothelial interaction has been shown to involve cell adhesion receptors and integrins, such as platelet endothelial cell adhesion molecule-1 (PECAM-1) and $\alpha v \beta 3$ integrins, respectively (Voura *et al* 2000). As integrins connect the extracellular matrix and the actomyosin cytoskeleton (Neff *et al* 1982, Damsky *et al* 1985, Riveline *et al* 2001), the link between both components is facilitated through the mechano-coupling focal adhesion and cytoskeletal adaptor protein vinculin (Mierke *et al* 2008b), which subsequently determines the quantity of cellular counter-forces maintaining cellular shape, morphology and cellular stiffness (Rape *et al* 2011). A biomechanical approach investigating the endothelial barrier break-down in the presence of co-cultured invasive cancer cells, in which both the cancer and endothelial cells' mechanical properties are determined, remains elusive. Microrheologic measurements (such as magnetic tweezer microrheology) have been adequate for analyzing the endothelial cell's mechanical properties, such as cellular stiffness during co-culture with invasive or non-invasive cancer cells compared to mono-cultured endothelial cells. As expected, it has been reported that the endothelial stiffness influences co-cultured invasive cancer cells. In more detail, highly invasive breast cancer cells alter the cellular mechanical properties of co-cultured microvascular endothelial cells, as they lower the stiffness of endothelial cells, whereas by contrast non-invasive cancer cells have no effect on endothelial cell stiffness (Mierke 2011). In addition, nanoscale-particle-tracking-method-based diffusion measurements of actomyosin cytoskeletal-bound microbeads (which serve as fiducial markers for structural alterations of the intercellular cytoskeletal scaffold) are suitable for revealing the actomyosin-driven cytoskeletal remodeling dynamics. Thus, the cytoskeletal-remodeling dynamics of endothelial cells have been found to increase during co-culture with highly invasive cancer cells, whereas endothelial cytoskeletal remodeling dynamics are not altered by non-invasive cancer cells (Mierke 2011). Taken together, these findings demonstrate that highly invasive breast cancer cells can actively alter the mechanical properties of nearby co-cultured endothelial cells. Thus, these results may provide a proper explanation for the break-down of the endothelial barrier function of vessel wall monolayers.

Further reading

Byfield F J, Aranda-Espinoza H, Romanenko V G, Rothblat G H and Levitan I 2004 Cholesterol depletion increases membrane stiffness of aortic endothelial cells *Biophys. J.* **87** 3336–43

References

Al-Mehdi A B, Tozawa K, Fisher A B, Shientag L, Lee A and Muschel R J 2000 Intravascular origin of metastasis from the proliferation of endothelium-attached tumor cells: a new model for metastasis *Nat. Med.* **6** 100–2

Byfield F J, Reen R K, Shentu T P, Levitan I and Gooch K J 2009 Endothelial actin and cell stiffness is modulated by substrate stiffness in 2D and 3D *J. Biomech.* **42** 1114–9

Cukierman E, Pankov R, Stevens D R and Yamada K M 2001 Taking cell–matrix adhesions to the third dimension *Science* **294** 1708–12

Damsky C H, Knudsen K A, Bradley D, Buck C A and Horwitz A F 1985 Distribution of the cell substratum attachment (CSAT) antigen on myogenic and fibroblastic cells in culture *J. Cell Biol.* **100** 1528–39

Discher D E, Janmey P and Wang Y L 2005 Tissue cells feel and respond to the stiffness of their substrate *Science* **310** 1139–43

Edwards D, Gooch K J, Zhang I, McKinley G H and Langer R 1996 The nucleation of receptor-mediated endocytosis *Proc. Natl Acad. Sci. USA* **93** 1786–91

Engler A J, Sen S, Sweeney H L and Discher D 2006 Matrix elasticity directs stem cell lineage specification *Cell* **126** 677–89

Engler A, Bacakova L, Newman C, Hategan A, Griffin M and Discher D 2004 Substrate compliance versus ligand density in cell on gel responses *Biophys. J.* **86** 617–28

Flanagan L A, Ju Y E, Marg B, Osterfield M and Janmey P A 2002 Neurite branching on deformable substrates *Neuroreport* **13** 2411–5

Grinnell F 2003 Fibroblast biology in three-dimensional collagen matrices *Trends Cell Biol.* **13** 264–9

Halliday N L and Tomasek J J 1995 Mechanical properties of the extracellular matrix influence fibronectin fibril assembly *in vitro Exp. Cell Res.* **217** 109–17

Jiang G, Huang A H, Cai Y, Tanase M and Sheetz M P 2006 Rigidity sensing at the leading edge through alphavbeta3 integrins and RPTPalpha *Biophys. J.* **90** 1804–9

Kanzawa S, Endo H and Shioya N 1993 Improved *in vitro* angiogenesis model by collagen density reduction and the use of type III collagen *Ann. Plast. Surg.* **30** 244–51

Kedrin D, Gligorijevic B, Wyckoff J, Verkhusha V V, Condeelis J, Segall J E and van Rheenen J 2008 Intravital imaging of metastatic behavior through a mammary imaging window *Nat. Methods* **5** 1019–21

Kostic A and Sheetz M P 2006 Fibronectin rigidity response through Fyn and p130Cas recruitment to the leading edge *Mol. Biol. Cell* **17** 2684–95

Lo C M, Wang H B, Dembo M and Wang Y L 2000 Cell movement is guided by the rigidity of the substrate *Biophys. J.* **79** 144–52

Mierke C T 2011 Cancer cells regulate biomechanical properties of human microvascular endothelial cells *J. Biol. Chem.* **286** 40025–37

Mierke C T, Kollmannsberger P, Paranhos-Zitterbart D, Smith J, Fabry B and Goldmann W H 2008b Mechano-coupling and regulation of contractility by the vinculin tail domain *Biophys. J.* **94** 661–70

Mierke C T, Zitterbart D P, Kollmannsberger P, Raupach C, Schlotzer-Schrehardt U, Goecke T W, Behrens J and Fabry B 2008a Breakdown of the endothelial barrier function in tumor cell transmigration *Biophys. J.* **94** 2832–46

Neff N T, Lowrey C, Decker C, Tovar A, Damsky C, Buck C and Horwitz A F 1982 A monoclonal antibody detaches embryonic skeletal muscle from extracellular matries *J. Cell Biol.* **95** 654–66

Nehls V and Herrmann R 1996 The configuration of fibrin clots determines capillary morphogenesis and endothelial cell migration *Microvasc. Res.* **51** 347–64

Paszek M J *et al* 2005 Tensional homeostasis and the malignant phenotype *Cancer Cell* **8** 241–54

Pelham R J Jr and Wang Y 1997 Cell locomotion and focal adhesions are regulated by substrate flexibility *Proc. Natl Acad. Sci. USA* **94** 13661–5

Peyton S R, Ghajar C M, Khatiwala C B and Putnam A J 2007 The emergence of ECM mechanics and cytoskeletal tension as important regulators of cell function *Cell Biochem. Biophys.* **47** 300–20

Pourati J, Maniotis A, Spiegel D, Schaffer J L, Butler J P, Fredberg J J, Ingber D E, Stamenovic D and Wang N 1998 Is cytoskeletal tension a major determinant of cell deformability in adherent endothelial cells? *Am. J. Physiol.* **274** C1283–9 PMID: 9612215

Rape A D, Guo W H and Wang Y L 2011 The regulation of traction force in relation to cell shape and focal adhesions *Biomaterials* **32** 2043–51

Riveline D, Zamir E, Balaban N Q, Schwarz U S, Ishizaki T, Narumiya S, Kam Z, Geiger B and Bershadsky A D 2001 Focal contacts as mechanosensors: externally applied local mechanical force induces growth of focal contacts by an mDia1-dependent and ROCK-independent mechanism *J. Cell Biol.* **153** 1175–86

Roeder B A, Kokini K, Sturgis J E, Robinson J P and Voytik-Harbin S L 2002 Tensile mechanical properties of three-dimensional type I collagen extracellular matrices with varied microstructure *J. Biomech. Eng.* **124** 214–22

Sieminski A L, Hebbel R P and Gooch K J 2004 The relative magnitudes of endothelial force generation and matrix stiffness modulate capillary morphogenesis *in vitro* *Exp. Cell Res.* **297** 574–84

Sieminski A L, Padera R, Padera R F, Blunk T and Gooch K J 2002 Systemic delivery of hGH using genetically modified tissue-engineered microvascular networks: prolonged delivery and endothelial survival with inclusion of non- endothelial cells *Tissue Eng.* **8** 1057–69

Sieminski A L, Was A S, Kim G, Gong H and Kamm R D 2007 The stiffness of three-dimensional ionic self-assembling peptide gels affects the extent of capillary-like network formation *Cell Biochem. Biophys.* **49** 73–83

Solon J, Levental I, Sengupta K, Georges P C and Janmey P A 2007 Fibroblast adaptation and stiffness matching to soft elastic substrates *Biophys. J.* **93** 4453–61

Storm C, Pastore J J, MacKintosh F C, Lubensky T C and Janmey P A 2005 Nonlinear elasticity in biological gels *Nature* **435** 191–4

Van Sluis G L, Niers T M, Esmon C T, Tigchelaar W, Richel D J, Buller H R, Van Noorden C J and Spek C A 2009 Endogenous activated protein C limits cancer cell extravasation through sphingosine-1-phosphate receptor 1-mediated vascular endothelial barrier enhancement *Blood* **114** 1968–73

Voura E B, Chen N and Siu C H 2000 Platelet-endothelial cell adhesion molecule-1 (CD31) redistributes from the endothelial junction and is not required for the transendothelial migration of melanoma cells *Clin. Exp. Metastasis* **18** 527–32

Yeung T, Georges P C, Flanagan L A, Marg B, Ortiz M, Funaki M, Zahir N, Ming W, Weaver V and Janmey P A 2005 Effects of substrate stiffness on cell morphology, cytoskeletal structure and adhesion *Cell Signal. Cytoskel.* **60** 24–34

Zijlstra A, Lewis J, Degryse B, Stuhlmann H and Quigley J P 2008 The inhibition of tumor cell intravasation and subsequent metastasis via regulation of *in vivo* tumor cell motility by the tetraspanin CD151 *Cancer Cell* **13** 221–34

IOP Publishing

Physics of Cancer

Claudia Tanja Mierke

Chapter 12

The role of macrophages during cancer cell transendothelial migration

Summary

This chapter discusses how tumor-associated macrophages guide cancer cells through the endothelial cell lining of blood or lymphoid vessels. Moreover, it covers the precise role of macrophages in the progress of cancer and examines whether they impact on the cellular functions and properties of cancer cells in order to provide cancer cell invasiveness into extracellular matrices of connective tissue. In addition, the function of macrophages during the intravasation of cancer cells into blood or lymphoid vessels through the endothelial vessel lining is discussed. Some open questions are also included in the discussion. Are macrophages necessary to provide cancer cell transendothelial migration and invasiveness? What is special about these tumor-associated macrophages? Do they also affect the properties of the endothelium in order to facilitate cancer cell transmigration and invasion?

What is the role of macrophages during cancer disease?
Macrophages are terminally differentiated cells of the mononuclear phagocytic lineage, which have developed under the influence of their primary growth and differentiation factor, the colony-stimulating factor-1 (CSF-1). Although the mononuclear phagocytic lineage differentiates into heterogeneous populations, depending upon the tissue of residence, motility is an important feature of its function. In order to mediate their migration and invasion through tissues, macrophages express a unique adhesion range and contain specific cytoskeletal proteins. In addition, macrophages do not assemble large, stable adhesions or actin stress fibers, but rely on small, short-lived point focal contacts, focal complexes and podosomes for generation and transmission traction. Thus, macrophages are present in their targeted tissue in order to respond rapidly to migratory stimuli. In addition to CSF-1 mediated increased growth and differentiation, it is also a chemokine

supporting macrophage migration through the activation of the CSF-1 receptor tyrosine kinase. In particular, CSF-1R autophosphorylation of several intracellular tyrosine residues leads to its association and the activation of many downstream signaling molecules. Phosphorylation of just one residue, Y721, facilitates the association of PI3K with the receptor to activate the major motility signal transduction pathways in macrophages. However, dissection of these pathways will lead to the identification of possible drug targets for the inhibition of diseases in which macrophages contribute to adverse outcomes.

Macrophages reside in almost every tissue of the body and thus their adaptation to the different tissue microenvironments is exceptional; they can adopt a diverse range of morphologies and carry out a variety of functions. Despite their heterogeneity, macrophages all originate from the same pluripotent hematopoietic stem cell and, under the influence of hematopoietic growth factors, differentiate through several multipotent progenitor stages to a lineage named mononuclear phagocytic precursors, which is located in the bone marrow (Pixley and Stanley 2004, Gordon and Taylor 2005, Pollard 2009). The mononuclear phagocyte system is composed of mononuclear phagocyte precursors (such as monoblasts and promonocytes), circulating monocytes and fully differentiated, tissue-resident macrophages (Pixley and Stanley 2004, Pollard 2009, Gordon and Taylor 2005, van Furth et al 1972). CSF-1 has long been recognized as the primary growth factor regulating the survival, proliferation and differentiation of cells of the mononuclear phagocytic lineage (Pixley and Stanley 2004, Pollard 2009, Stanley and Heard 1977). It is also an essential differentiation factor for the bone resorbing osteoclast (Asagiri and Takayanagi 2007). A spontaneously occurring inactivating mutation in the mouse CSF-1 gene (osteopetrotic, Csf-1op) has been shown to be associated with reduced tissue macrophage numbers and a marked reduction in osteoclasts, while it also causes osteopetrosis, together with other developmental defects (Pixley and Stanley 2004, Yoshida et al 1990, Wiktor-Jedrzejczak et al 1990, Cecchini et al 1994). The CSF-1 protein signals through the CSF-1 receptor tyrosine kinase (RTK), which is encoded by the c-fms proto-oncogene (Sherr et al 1985). In addition, RTK induces a series of phosphorylation cascades, facilitating cellular responses to CSF-1 (Pixley and Stanley 2004). While the phenotype of mice nullizygous for the CSF-1R (Csf1r-/Csf1r-) largely recapitulates that seen in the Csf1op/Csf1op mouse, it is even more severe and this discrepancy has been explained by the discovery of a second partially redundant ligand for the CSF-1R, the interleukin-34 (IL-34) (Dai et al 2002, Lin et al 2008, Wei et al 2010).

Macrophages are professionally motile cells, which perform a variety of roles in immune surveillance and normal tissue development by secreting cytokines and growth factors, and also phagocytosing foreign material and apoptotic cells. Transendothelial and interstitial motility is an essential part of their function as they must be able to migrate to specific sites upon stimulation. From studies in primary macrophages and in CSF-1 dependent macrophage cell lines, it has been reported that CSF-1 is not only a mononuclear phagocyte lineage growth factor, it is also an important inducer of macrophage motility (Pixley and Stanley 2004, Wang et al 1988, Webb et al 1996, Pixley et al 2001). In more detail, the depletion of specific subsets of tissue macrophages in the Csf1op/Csf1op mouse and their reconstitution upon restoration of CSF-1 expression revealed that CSF-1 facilitates the differentiation and migration of trophic and/or scavenger macrophages that are physiologically essential for normal development and tissue homeostasis rather than immune function (Pollard 2009, Cecchini *et al* 1994, Dai *et al* 2002, Ryan *et al* 2001). In more detail, CSF-1 or CSF-1R deficient mice demonstrate abnormal neural, skeletal and glandular development, not only due to reduced macrophage content and decreased osteoclast numbers, but more likely through impaired matrix remodeling (Pollard 2009). Thus, CSF-1-induced motility seems to be an important element of macrophage function in development. Beyond their critical physiological role, CSF-1 dependent macrophages can promote disease progression in conditions ranging from cancer to atherosclerosis and arthritis (Pixley and Stanley 2004, Pollard 2009, Chitu and Stanley 2006, Hamilton 2008). Reactivation of developmental macrophage functions seems to be possible and may underlie the progression of these pathologies (Pollard 2009).

In order to participate in the disease process, macrophages must initially migrate to the affected tissue. Moreover, in the case of an increase in tumor invasion, tumor-associated macrophages and mammary carcinoma cells have been reported to migrate away from the primary tumor site together (Wyckoff *et al* 2004). How the macrophage motility is facilitated remains elusive. How the motility machinery differs from other cell types and whether inhibition of macrophage motility may improve disease outcomes is not yet well understood. Moreover, CSF-1 can activate signal transduction pathways that activate molecules or protein isoforms selectively expressed in macrophages (Pixley and Stanley 2004); some of these may be attractive therapeutic targets to specifically inhibit macrophage infiltration into disease sites. When considering the contribution of macrophages and CSF-1 to tumor dissemination and the progression of several inflammatory disorders (Pollard 2009, Chitu and Stanley 2006, Hamilton 2008), the focus is on understanding macrophage migration and its facilitation by CSF-1.

It has been reported that distinct Mena isoforms are expressed in invasive and migratory cancer cells *in vivo* and that the invasion isoform (MenaINV) potentiates carcinoma cell metastasis in murine models of breast cancer. However, the specific step of metastatic progression affected by this isoform and the effects on metastasis of the Mena11a isoform, expressed in primary cancer cells, have not yet been fully revealed. Evidence has been provided that elevated MenaINV increases coordinated streaming motility and increases the transendothelial migration and intravasation of cancer cells. Indeed, the promotion of these early stages of metastasis by MenaINV is dependent on a macrophage–cancer cell paracrine loop. In addition, it has been shown that increased Mena11a expression correlates with decreased expression of CSF-1 and leads to a dramatically decreased ability to participate in paracrine-mediated invasion and intravasation. These results illustrate the importance of paracrine-mediated cell streaming and intravasation for cancer cell dissemination and moreover show that the relative abundance of MenaINV and Mena11a helps to regulate these key steps of metastatic progression in breast cancer cells.

Cellular motility is essential for many aspects of metastasis. However, there are only a few molecular markers that can indicate and even predict the migratory potential of a cancer cell *in vivo*. The intravital multiphoton imaging technique can be used to characterize carcinoma and stromal cell behavior in detail within intact primary tumors in living animal models (Condeelis and Segall 2003, Wang *et al* 2007, Egeblad *et al* 2008, Kedrin *et al* 2008, Andresen *et al* 2009, Perentes *et al* 2009). Indeed, these imaging approaches deliver direct information at single-cell resolution and even permit the quantification of cellular motility and interactions between cancer and stromal cells, as well as providing direct observation of the invasion, intravasation and extravasation steps that all play an important role in cancer metastasis. In mammary tumors, this technology was used to describe the microenvironments in which cancer cells can invade, migrate and intravasate, while also revealing essential roles for macrophages in these events (Condeelis and Segall 2003, Condeelis and Pollard 2006, Yamaguchi *et al* 2006, Kedrin *et al* 2007). In more detail, chemotaxis of cancer cells toward macrophages is an essential step for invasion in mouse primary mammary tumors (Wyckoff *et al* 2004, Goswami *et al* 2005), while chemotaxis of cancer cells toward perivascular macrophages is essential for their intravasation (Wyckoff *et al* 2007).

Expression profiling of invasive cancer cells isolated from the primary tumor was used to obtain molecular information regarding the pathways providing carcinoma cell invasion and intravasation (Wyckoff *et al* 2000a, Wang *et al* 2004, Wang *et al* 2007). The 'invasion signature' revealed by this profile provides sets of coordinated expression alterations associated with the enhanced invasive potential (Goswami *et al* 2004, Wang *et al* 2004, Wang *et al* 2006, Wang *et al* 2007, Goswami *et al* 2009). As suggested, Mena, a regulator of actin polymerization and cell motility, is up-regulated in invasive cancer cells obtained from rat, mouse and human tumors (Di Modugno *et al* 2006, Goswami *et al* 2009, Robinson *et al* 2009). However, the conservation of Mena up-regulation in invasive cancer cells across species indicates that it plays an important role in the metastatic progression of cancer disease.

In patients, Mena expression correlates with metastatic risk: for instance, relatively high Mena expression has been observed in patient samples from high-risk primary and metastatic breast tumors (Di Modugno *et al* 2006), as well as cervical, colorectal and pancreatic cancers (Gurzu *et al* 2008, Pino *et al* 2008, Gurzu *et al* 2009). Mena is also a component of a marker for metastatic risk called tumor microenvironment for metastasis (TMEM) (Robinson *et al* 2009). TMEMs are identified by co-localization of Mena-positive cancer cells, macrophages and endothelial cells and, in particular, the TMEM score predicts the risk independently of the clinical subtype of cancer (Robinson *et al* 2009). The contribution of Mena to metastasis is independent of clinical subtype.

These findings highlight the importance of the mechanism by which Mena and its isoforms differentially facilitate metastatic progression. In particular, Mena is a member of the Ena/VASP family of proteins and can bind actin in order to affect the geometry and assembly of filament networks through: (i) anti-capping protein activity (Bear *et al* 2002, Barzik *et al* 2005, Hansen and Mullins 2010), which includes binding to profilin and both G- and F-actin; (ii) Mena tetramerization; and

(iii) a decrease in the density of actin-related proteins 2 and 3 (Arp2/3)-facilitated branching (Gertler *et al* 1996, Barzik *et al* 2005, Ferron *et al* 2007, Pasic *et al* 2008, Bear and Gertler 2009, Hansen and Mullins 2010). Moreover, alternative splicing for the Mena gene has been shown: a 19 amino acid residue insertion just after the EVH1 domain leads to the Mena invasion isoform (named MenaINV, formerly Mena + + +) (Gertler *et al* 1996, Philippar *et al* 2008), whereas a 21 residue insertion in the EVH2 domain leads to the Mena11a isoform (Di Modugno *et al* 2007). A comparison of the invasive and migratory cancer cells collected *in vivo*, with primary cancer cells isolated from mouse, rat and human cell-line-derived mammary tumors, showed that MenaINV expression is up-regulated and Mena11a is down-regulated selectively in the invasive and migrating carcinoma cell population (Goswami *et al* 2009). Moreover, the differential regulation of Mena isoforms across species indicates that the two isoforms fulfill important roles in invasion and metastasis.

In previous studies, the expression of MenaINV in a xenograft mouse mammary model has been reported that the tumor promotes increased formation of spontaneous lung metastases from these orthotopic primary tumors and alters the sensitivity of cancer cells towards epidermal growth factor (EGF) (Philippar *et al* 2008). This study was performed in order to identify the step(s) in the metastatic cascade regulated by MenaINV expression and to analyze the effect of expression of the second regulated isoform, Mena11a, on metastatic progression. In more detail, each step of metastatic progression was investigated in order to determine which steps are regulated by expression of MenaINV, which finally enhances the metastatic dissemination, and whether the same steps are also regulated by Mena11a expression in cancer cells.

MTLn3 cells have been selected in order to investigate the impact of these two Mena isoforms, as these cells are well characterized regarding cancer cell invasion, migration and metastasis (Levea *et al* 2000, Sahai 2005, Le Devedec *et al* 2009, Le Devedec *et al* 2010); tumor–stromal cell interactions (Sahai 2005); TGFβ signaling in metastatic progression (Giampieri *et al* 2009); and the functional role of Mena in breast cancer metastasis (Philippar *et al* 2008, Goswami *et al* 2009). Moreover, MTLn3 cells were obtained from the clonal selection of metastatic lung lesions from rats with mammary tumors (Neri *et al* 1982). In more detail, these rat mammary tumors were characterized as estrogen-independent and were observed to metastasize in the lymph nodes and lungs (Neri *et al* 1982). The evaluation of vimentin and keratins in MTLn3 mammary tumors, associated lymph nodes and lung metastases demonstrated that MTLn3 cancer cells are comparable to a basal-like subtype of breast cancer (Lichtner *et al* 1989).

MenaINV promotes coordinated cell migration in the form of streams of single cells

It has been found that expression of Mena and MenaINV increases *in vivo* cell motility, and it has been hypothesized that this supports the increased lung metastasis detected in these cells (Philippar *et al* 2008). However, different types of motility are regarded as playing diverse roles during cancer cell invasion (Wolf *et al* 2003, Gaggioli *et al* 2007, Ilina and Friedl 2009, Friedl and Wolf 2010), thus it has been hypothesized that MenaINV expression supports a type of motility that provides enhanced cancer cell invasion. To test this hypothesis, a multiphoton-based intravital imaging (IVI) approach was used to examine the types of motility displayed by cancer cells expressing the different Mena isoforms. In all tumors, two patterns of movement were usually observed: coordinated cell movement, whereby the cancer cells align and move in an ordered single file line (called streaming); and random cellular movement, whereby cancer cells move independently of other cancer cells in a somewhat uncoordinated fashion. As expected, both MenaINV-expressing and Mena11a-expressing cancer cells exhibited streaming and random movement *in vivo*. However, streaming movement was significantly more common in MenaINV-expressing cancer cells *in vivo*. The quantification of cells moving within the primary tumor showed that MenaINV expression significantly elevated both random and streaming cancer cell movement compared to GFP-expressing control cells and Mena11a-expressing primary mammary tumors. Both movement types were only slightly increased in Mena11a-expressing cancer cells. In addition, to characterize streaming motility more precisely, time-lapse images were recorded in order to investigate cell crawling, and photoconversion from green to red was performed to measure the stability of the streams over a 24 h period (Kedrin *et al* 2008). Carcinoma cell streams have been observed over 24 h after photoconversion, indicating that streaming is a long-lived behavior involving crawling cells. The results from the co-injection of MTLn3 cells expressing GFP–Mena11a–GFP or Cerulean–MenaINV confirmed that the tissue architecture specific to each Mena isoform type was even now preserved, compared with the injection of either Mena11a- or MenaINV-expressing cells separately.

Streaming cell movement is more efficient than random cell movement
The underlying motility parameters contributing to streaming and random movement were investigated by determining cell speed, net path length, directionality of the motion and the turning frequency *in vivo*. *In vivo*, streaming cells indeed migrated pronouncedly faster compared to randomly moving cells, regardless of Mena isoform expression, by exhibiting average speeds of greater than 1.9 μm min^{-1}. All cell types participating in this random movement displayed a narrow distribution of velocities, whereas cells that participated in the streaming movement showed a broad distribution of velocities. However, this suggests that random cell movement is autonomous, whereas streaming cells are restricted in their velocities due to their multiple cell–cell signaling interactions.

The directionality, net path length and turning frequency of a cell are all measures of its locomotion efficiency. In general, the net path length and directionality of streaming carcinoma cells *in vivo* were elevated, whereas the turning frequency was reduced compared to randomly moving cells. Thus, these results suggest that the streaming cell movement is more efficient, as it has been seen that streaming cells move faster and further, and turn less frequently. The enhanced streaming found *in vivo* for cells expressing MenaINV means that MenaINV-expressing cells migrate more efficiently *in vivo*. In steady-state conditions *in vitro*, these cells only move randomly and the Mena isoform-expressing cells do not differ pronouncedly from each other in their speed and the directionality of their movement. Moreover, this suggests that additional factors are essential for streaming that are not present *in vitro* and which must be revealed in *in vivo* experiments.

In vivo invasion is enhanced by MenaINV and suppressed by Mena11a

In order to analyze whether the enhanced streaming exhibited by MenaINV-expressing cancer cells correlates with chemotaxis-dependent invasion, the *in vivo* invasion assay was performed to evaluate the capacity of cancer cells to migrate and invade toward EGF *in vivo* (Wyckoff *et al* 2000a). MTLn3 cells were shown to display a characteristic biphasic response to EGF, whereby maximal chemotactic invasion was achieved in response to 25 nM EGF (Segall *et al* 1996, Wyckoff *et al* 2000a). Moreover, expression of MenaINV shifts this biphasic response and hence maximal invasion was achieved in response to 1 nM EGF (Philippar *et al* 2008). This result demonstrates that sensitivity to EGF chemotaxis is increased *in vivo* and is indeed consistent with the increased sensitivity of MenaINV-expressing cells to EGF *in vitro* (Philippar *et al* 2008). *In vitro*, MenaINV-expressing cells exhibit protrusive activity in response to EGF concentrations as low as 0.1 nM, whereas cells expressing GFP and Mena11a-expressing cells do not respond to stimulation with either 0.1 or even 0.5 nM EGF. Importantly, MenaINV expression not only sensitizes cancer cells to EGF, it also significantly increases the number of invasive cells collected with the peak concentration of EGF, which in general indicates more efficient cell migration. However, Mena11a-expressing tumors did not invade significantly above background levels in response to a broad range of EGF concentration. Taken together, Mena11a and MenaINV seem to have opposite effects on chemotaxis-dependent invasion *in vivo*.

Mena isoforms alter paracrine loop signaling with macrophages during invasion

Using the *in vivo* invasion assay, cancer cells have been observed to chemotax into needles containing either EGF or CSF-1 (Wyckoff *et al* 2004, Patsialou *et al* 2009). This can only be detected if paracrine signaling is established with macrophages, because, in the absence of macrophages, the chemostatic signal cannot be transmitted over long distances and only a few cells are collected (Wyckoff *et al* 2004). Therefore, the question of whether the cancer cell–macrophage paracrine loop is involved in the increased *in vivo* invasion of MenaINV-expressing cells and the suppression of invasion in Mena11a-expressing cells was addressed. Both macrophages and cancer cells enter collection needles during *in vivo* invasion (Wyckoff *et al* 2004) and typing of cells collected following *in vivo* invasion confirm the presence of both cancer cells and macrophages in tumors expressing GFP, MenaINV and Mena11a.

To assess the impact of paracrine signaling between macrophages and MenaINV-expressing carcinoma cells during *in vivo* cancer cell invasion, the *in vivo* invasion assay was performed in the presence of either 6.25 nM Erlotinib (Tarceva), an EGF receptor (EGFR) tyrosine kinase inhibitor, or a mouse CSF1 receptor-blocking antibody (α-CSF1R) (Wyckoff *et al* 2004). *In vivo* invasion of both MenaINV- and GFP-expressing cells was reduced to background levels in assays with Erlotinib as compared with invasion toward needles containing EGF + DMSO or EGF alone, demonstrating the requirement for EGFR-facilitated cancer cell invasion. Similarly, invasion of both MenaINV- and GFP-expressing cells was significantly reduced with needles containing α-CSF1R as compared with cancer cell invasion toward needles containing EGF + DMSO or EGF + IgG control antibodies, indicating the necessity of EGF production and signal propagation by macrophages. These results are indeed consistent with the requirement for co-migrating macrophages in cancer cell migration. Finally, these results demonstrate the requirement for paracrine signaling between MenaINV-expressing cancer cells and macrophages *in vivo*.

Both CSF1 secretion and EGF binding to the EGFR by cancer cells are essential components of the carcinoma cell–macrophage paracrine loop (Wyckoff *et al* 2004, Patsialou *et al* 2009). Real-time PCR was performed to determine the relative mRNA expression of CSF1 and EGFR in the Mena isoform-expressing cells lines in culture to investigate whether alterations in the expression of these signaling molecules may contribute to the differences detected in EGF-dependent *in vivo* invasion (Wyckoff *et al* 2004). As expected, Mena11a-expressingcells showed a four-fold decrease in CSF1 expression as compared with GFP-expressing cells, whereas CSF1 expression in Mena- and MenaINV-expressing cells was unaltered. Indeed, it has been reported that there is no difference in EGFR expression in cells expressing either Mena or MenaINV as compared with MTLn3 cells (Philippar *et al* 2008). However, cancer cells expressing Mena11a also showed no statistical difference in expression of EGFR compared with GFP- and MenaINV-expressing cells, indicating that altered receptor levels do not contribute to the altered EGF-dependent phenotypes observed in the different cell types. Thus, the inability of Mena11a-expressing cells to participate in macrophage-dependent invasion may be explained through the reduced CSF-1 expression, along with a reduction in direct response EGF.

In order to determine whether suppression of chemotaxis-dependent invasion of Mena11a-expressing cancer cells resulted from differences in the ability of these cancer cells to co-migrate with macrophages, a 3D invasion assay was performed that measured macrophage-dependent co-migration of cancer cells with macrophages in 3D collagen matrices (Goswami *et al* 2005). While addition of macrophages to GFP-expressing cancer cells significantly elevated 3D invasion, the addition of macrophages to Mena11a-expressing cells did not significantly increase cancer cell invasion. This result is consistent with the decreased response to EGF and the decreased CSF-1 expression levels in Mena11a-expressing cells that are below the threshold needed to stimulate pro-invasive macrophage behavior.

MenaINV-expressing carcinoma cell streaming requires
macrophages and paracrine signaling
The increased invasion of MenaINV-expressing carcinoma cells depends on paracrine loop signaling with macrophages and the paracrine loop has been found to be required for cancer cell migration in mammary tumors (Wyckoff *et al* 2004). In particular, it has been hypothesized that the paracrine loop also promotes carcinoma cell streaming *in vivo*. Using IVI, it has been observed that multiple carcinoma cells moved in streams among host cell shadows previously identified as immune cells and macrophages (Wyckoff *et al* 2000b, Wyckoff *et al* 2007).

Intravenous injection of Texas Red dextran during IVI indeed identified the presence of the host cell shadows in cancer cell streams as macrophages, because macrophages uniquely pinocytose dextran delivered intravenously into mammary tumors (Wyckoff *et al* 2007).

In order to investigate whether streaming required macrophages, mice were treated with clodronate liposomes 48 h prior to IVI to decrease the level of functional macrophages (Hernandez *et al* 2009). As expected, a 70% decrease in the number of Texas Red dextran-labeled macrophages was observed in animals treated with clodronate liposomes compared with those treated with PBS liposomes (control). IVI of primary tumors revealed that there were 90% fewer streaming cells in mice treated with clodronate liposomes as compared to controls, confirming the involvement of the macrophage–cancer cell paracrine loop in streaming. Moreover, to analyze the involvement of the paracrine loop in streaming, mice were treated with Erlotinib 2 h prior to IVI in order to block the EGFR on cancer cells (Zerbe *et al* 2008) or with a CSF1R antibody 4 h prior to IVI to block the CSF1R on macrophages (Wyckoff *et al* 2007). IVI of primary tumors demonstrated that there were 65% fewer streaming cells in mice treated with Erlotinib and 80% fewer streaming cells in mice treated with α-CSF1R as compared with tumors.

Expression of MenaINV in cancer cells increases transendothelial migration
Due to the data showing increased streaming and invasion in MenaINV-expressing cells, it has been hypothesized that expression of MenaINV might also increase intravasation and finally cancer metastasis. Following intravenous injection of Texas Red dextran, IVI of MenaINV-expressing tumors showed that cancer cell streaming was indeed directed toward blood vessels. Then the intravasation efficiency of tumors expressing MenaINV and Mena11a was quantified using IVI of photoconverted cancer cells adjacent to blood vessels to determine the percentage of cancer cells intravasating over a 24 h period (Kedrin *et al* 2008). The quantification of the alterations in the photoconverted tumor area 24 h after photoconversion showed that 95% of Mena11a-expressing cells remained in the converted area as compared with 75% of carcinoma cells expressing MenaINV. Additionally, the cancer cell blood burden to measure intravasation *in vivo* was evaluated (Wyckoff *et al* 2000b). Mice with MenaINV-expressing tumors had a four-fold increase in the number of carcinoma cells in circulation compared with mice with GFP- or Mena11a-expressing tumors. In line with this, Mena11a-expressing xenograft mice had similar numbers of circulating carcinoma cells as compared with GFP-expressing xenograft mice.

As previous studies showed that interaction between cancer cells and perivascular macrophages is required for intravasation (Wyckoff *et al* 2007) and that enhanced MenaINV cell streaming and invasion are paracrine-dependent, it has been hypothesized that increased intravasation in MenaINV-expressing cells might also be paracrine-dependent. In order to determine the minimum requirements for macrophage-assisted intravasation, a subluminal-to-luminal transendothelial migration (TEM) assay was used, in which the presence of macrophages could be varied to determine their need for carcinoma cell intravasation. Interestingly, less than 0.5% of cancer cells traversed the endothelium in the absence of macrophages, regardless of Mena isoform expression, indicating that Mena isoforms function only in the presence of macrophages. The addition of macrophages did not enhance TEM for cells expressing GFP or Mena11a. Importantly, in the presence of macrophages, 54% of MenaINV-expressing cells traversed the endothelium, which is a 200-fold increase in TEM compared to all other cell types.

In order to investigate the paracrine dependence of intravasation *in vivo*, the number of circulating cancer cells was determined following functional impairment of macrophages achieved by treatment of mice with clodronate liposomes or CSF1R-blocking antibody, or impairment of cancer cells by treatment with Erlotinib. Indeed, the tumors formed from injection of MenaINV-expressing cells revealed a significant reduction in circulating cancer cells following treatment with clodronate liposomes, CSF1R blocking antibody and Erlotinib as compared to controls (Roussos *et al* 2011).

MenaINV, but not Mena11a, increases intravasation,
dissemination and metastasis

In order to reveal the mechanistic consequence of enhanced transendothelial migration and intravasation by MenaINV expression or the suppression of invasion by Mena11a expression, the ability of these cells to extravasate, disseminate and metastasize to the lung was investigated. Thus, an experimental metastasis assay was used in order to measure of extravasation of cancer cells (Xue *et al* 2006). Then micro-metastases in the lungs were counted after intravenous injection of GFP-, Mena11a- or MenaINV-expressing cells. However, the metastatic burden was similar for all cell lines. Previous studies have shown that MTLn3 cancer cells forced to express MenaINV show a significant enhancement to metastases (Philippar *et al* 2008). Thus, it was investigated whether dissemination of single cancer cells to the lung, a step preceding growth of macrometastases, was also affected in xenograft mice derived from injection of cells expressing MenaINV or Mena11a. Mice with MenaINV xenografts showed significantly increased cancer cell dissemination to the lungs compared with animals bearing either Mena11a- or GFP-expressing tumors. Finally, mice with Mena11a xenografts had half as many cells in the lungs as mice bearing GFP-expressing tumors.

As MenaINV expression is known to increase cancer cell dissemination and it is known that that Mena11a expression reduces cancer cell dissemination, it has been hypothesized that Mena isoform expression may similarly modulate the final step in metastatic progression: the occurrence of spontaneous metastasis. Spontaneous metastases to the lungs have been detected in mice with mammary tumors at either three or four weeks after mammary gland injection of GFP-, Mena11a- and MenaINV-expressing cells. In particular, expression of MenaINV increased the incidence of metastasis compared with expression of GFP and Mena11a, whereas expression of Mena11a decreased metastases after three weeks of tumor growth. However, after four weeks of tumor growth, all primary tumors resulted in detectable metastases regardless of the Mena isoform expressed. In addition, MenaINV expression induced metastatic spread to the lungs with little effect at three weeks or no effect on primary tumor growth at four weeks or cell growth *in vitro* (Philippar *et al* 2008). Hence, the differences in tumor metastasis occurring in tumors with different Mena isoform expression are not an indirect consequence of tumor growth. These data clearly suggest that the increased incidence of spontaneous metastasis revealed in MenaINV-expressing tumors is due to metastatic events occurring prior to extravasation.

Enhanced Mena expression is correlated with metastasis in breast cancer patients (Di Modugno *et al* 2006). In particular, during invasion and migration of cancer cells the expression of MenaINV increases (positive regulator), whereas the expression of Mena11a decreases (negative regulator) (Goswami *et al* 2009). Invasion, migration and intravasation have been identified as crucial metastasis steps that are affected by the expression of MenaINV- and Mena11a-expressing cancer cells. A key characteristic of MenaINV-expressing cells is their contribution to cell streaming and increased intravasation as a result of the pronounced elevation in transendothelial migration. Another important result is the effect of MenaINV expression on cancer cell sensitivity to macrophage-supplied EGF and the subsequent enhancement of paracrine-facilitated invasion. Taken together, these findings ultimately suggest that the EGF-dependent enhancement of invasion and intravasation in MenaINV-expressing cancer cells ensures enhanced cancer cell dissemination and spontaneous metastasis to the lungs.

Conversely, it has been found that Mena11a-expressing cells do not show dramatically increased streaming and even fail to co-invade with macrophages, which indicates decreased paracrine signaling interaction. The decrease in CSF1 expression in Mena11a-expressing cells leads to impaired paracrine signaling and deficits are observed in activities that depend on this paracrine signaling loop *in vivo*, such as streaming, invasion, transendothelial migration, cancer cell dissemination and spontaneous metastasis to the lungs.

During invasion, cancer cells are known to reduce their expression of Mena11a and initiate the production of the MenaINV isoform (Goswami *et al* 2009). Moreover, it has been found that Mena11a expression is correlated with reduced EGF-induced *in vivo* invasion. In addition it has been shown that MenaINV-expressing migratory carcinoma cells are highly sensitive to EGF in their protrusion and chemotaxis activities, leading to significantly increased *in vivo* invasion. Finally, these activities can result in MenaINV-expressing cell migration toward and in association with perivascular macrophages, leading to enhanced transendothelial migration and intravasation.

In addition to decreased EGF-induced *in vivo* invasion of Mena11a-expressing cells, these cells have been found to express less CSF-1 mRNA. Data from patients revealed that CSF-1 and its receptor play crucial roles during the progression of breast cancer (Kacinski *et al* 1991, Scholl *et al* 1994) and that CSF-1 and the CSF-1R are coexpressed in more than 50% of breast tumors (Kacinski 1997). Elevated circulating CSF-1 has been proposed as an indicator of early metastatic relapse in patients with breast cancer, independent of the breast cancer subtype (Scholl *et al* 1994, Tamimi *et al* 2008, Beck *et al* 2009). Moreover, this suggests that lower levels of CSF-1 in Mena11a-expressing cells may lead to a reduced metastatic progression.

The decreased invasion, intravasation and dissemination of Mena11a-expressing cells is consistent with the reduction in expression of CSF-1 and the decreased sensitivity to EGF, which seems to force these cells to participate in a paracrine signaling loop with macrophages.

A major finding is that the expression of MenaINV increases a mode of coordinated cell migration not previously described, where cell migration is spatially and temporally coordinated between the cancer cells that are not connected by cell–cell junctions. This newly described mode of coordinated cell migration is called cell streaming. The streaming differs from previously described modes of coordinated cell migration, which require stable cell–cell junctions (Sahai 2005), while streaming cells require no cell–cell contacts and the migration velocities are 10–100 times more rapid. Previous studies have reported that *in vivo* MTLn3 cells express CSF-1 and EGFR, but express no CSF-1R on their cell surface and do not secrete EGF, while macrophages express CSF-1R and secrete EGF, but do not secrete CSF-1 or express EGFR on their cell surface (Goswami *et al* 2005). Thus coordinated parts of the paracrine signal transduction pathways are active in both cell types during cell invasion *in vivo* (Wyckoff *et al* 2004). Indeed, it has been demonstrated that streaming requires paracrine chemotaxis between carcinoma cells and macrophages. The ability of MenaINV-expressing cells to protrude and perform chemotaxis with 25- to 50-fold lower concentrations of EGF compared to parental cancer cells, and to suppress cell turning in streams, leads to a pronounced contribution to the extraordinary coordination and maintenance of high velocity migration of cell streams *in vivo*. Moreover, it has been proposed that the increased sensitivity of MenaINV-expressing cells to EGF in the EGF–CSF-1 paracrine loop is responsible for the enhancement in the streaming motility. This conclusion is further supported by the inhibition of streaming upon the inhibition of the EGFR by Erlotinibor of CSF-1R by an anti-CSF1R blocking antibody.

Invasive cancer cells from PyMT mice displayed increased MenaINV expression and decreased expression of Mena11a (Goswami *et al* 2009). Interestingly, studies using intravital imaging of mammary tumors in Mena-deficient PyMT mice revealed significantly reduced streaming motility of cancer cells, providing further evidence that Mena facilitates enhanced motility (Roussos *et al* 2010). Finally, mammary tumors derived from the human breast cancer cell line MDA-MB-231 contain a subpopulation of cancer cells that participate in macrophage–tumor cell paracrine-facilitated invasion (Patsialou *et al* 2009), and these invasive cancer cells have also been revealed to differentially up-regulate MenaINV and down-regulate Mena11a (Goswami *et al* 2009). Taken together, these findings suggest that paracrine-facilitated carcinoma cell streaming is a generalized cellular feature that occurs in rat, mouse and human models of breast cancer and that it is a clear consequence of the differential regulation of the Mena isoforms.

In summary, the suppression of invasion and streaming by the inhibition of paracrine signaling between macrophages and cancer cells *in vivo* and by decreasing macrophage function *in vivo*, has established the crucial role of macrophages during the coordinated migration of MenaINV-expressing cells (Wyckoff *et al* 2007, Hernandez *et al* 2009). It has also been demonstrated that macrophages are essential for the transendothelial migration of MenaINV-expressing cancer cells. Indeed, these results are consistent with previous work demonstrating that paracrine signaling between cancer cells and macrophages, and the presence of perivascular macrophages in the primary tumor, are required for invasion and intravasation, respectively (Wyckoff *et al* 2004, Wyckoff *et al* 2007). In particular, these results support other previous work that suggested that MenaINV- but not Mena11a-expressing cancer cells specifically contribute to the intravasation of breast cancer cells in humans by assisting cancer cells to assemble, leading to the formation of the macrophage-dependent intravasation compartment known as the TMEM (Robinson *et al* 2009, Roussos *et al* 2011).

In vivo, it has been shown that MenaINV-expressing cells invade toward very low concentrations of EGF in macrophage-dependent paracrine chemotaxis. However, *in vitro*, even low concentrations of EGF, such as that found in serum, lead to macrophage-independent 3D invasion of MenaINV-expressing cells (Philippar *et al* 2008), while completion of transendothelial migration requires EGF secreted by macrophages (Wyckoff *et al* 2004). The effects of MenaINV expression on EGF-dependent processes lead to enhanced cell invasion, intravasation, dissemination and metastasis to the lungs. These data suggest that drugs directed specifically to the inhibition of MenaINV-dependent increased EGF sensitivity will disrupt the paracrine interactions with macrophages required for metastasis and consequently result in the inhibition of metastasis in mammary tumors.

However, it is important to understand how the Mena isoforms differ functionally. The INV exon is inserted just after the EVH1 domain, which is primarily responsible for the subcellular localization of Ena/VASP proteins and interactions with several signaling proteins such as lamellipodin (Gertler *et al* 1996, Urbanelli *et al* 2006, Pula and Krause 2008). Thus, it is possible that the INV exon might influence the function of MenaINV by regulating its EVH1-facilitated interactions (Niebuhr *et al* 1997, Boeda *et al* 2007). By contrast, the 11a exon is inserted within the EVH2 domain between the F-actin binding motif and the coiled-coil tetramerization site. Indeed, the F-actin binding is crucial for nearly all known Ena/VASP functions, such as the localization to the tips of lamellipods and the ability to drive filopod and lamellipod formation and extension (Gertler *et al* 1996, Loureiro *et al* 2002, Applewhite *et al* 2007). *In vitro*, F-actin binding is necessary for the anti-capping activity of Ena/VASP and can be disrupted by phosphorylation at nearby sites (Barzik *et al* 2005), as it is the case in F-actin bundling. Because 11a is inserted into the analogous region of Mena, it will be interesting to investigate whether the barbed end capture activity is affected. Due to the phosphorylation of the 11a insertion (Di Modugno *et al* 2006), it is likely that inclusion of the 11a exon provides a regulatory mechanism for Mena11a.

In summary, research is required to investigate the molecular and biochemical mechanisms of action of the MenaINV and Mena11a isoforms and their potential as a prognostic marker for patient outcome and a therapeutic target for breast cancer metastasis.

Most cancer-related deaths result from cancer metastasis, thus it is important to increase knowledge of the molecular mechanisms of dissemination, including intra- and extravasation. Although the mechanisms of extravasation have been studied intensively *in vitro* and *in vivo*, the process of intravasation remains elusive. In particular, how cells such as macrophages in the tumor microenvironment facilitate cancer cell intravasation is still unknown. Using high-resolution imaging, it has been observed that macrophages facilitate increased cancer cell intravasation upon physical contact. In more detail, macrophage and cancer cell contact induce RhoA activity in cancer cells, triggering the assembly and exertion of actin-rich, degradative protrusions called invadopodia, enabling cancer cells to degrade and break through matrix confinements during cancer cell transendothelial migration (Roh-Johnson *et al* 2014). Interestingly, the macrophage-induced invadopodium formation and cancer cell intravasation also occurred in patient-derived cancer cells and *in vivo* models, suggesting a conserved mechanism of cancer cell intravasation. These results illustrate a novel heterotypic cell contact facilitated signaling role for RhoA and yield mechanistic insights into the capacity of cells such as macrophages within the tumor microenvironment to facilitate steps of the metastatic cascade (Roh-Johnson *et al* 2014).

In summary, understanding the mechanistic basis of specific steps within cancer metastasis is crucial for the identification of robust early prognostic markers. Knowing how cancer cells can leave the primary tumor and enter the vasculature (called intravasation) is a key step toward designing drug treatment strategies for the disease. Multiphoton-based intravital imaging of rodent mammary adenocarcinoma (Roussos *et al* 2011) and human tumors has uniformly revealed that cancer cells and macrophages cooperate during several key steps of the metastatic cascade. The cell polarization and subsequent motility of invasive cancer cells toward the blood vessels seems to be dependent on a paracrine loop of cancer cell signaling with macrophages (Wyckoff *et al* 2004, 2007). Cancer cells respond to macrophage-secreted EGF and, in turn, macrophages respond to cancer-cell-secreted CSF-1. Moreover, it has also been demonstrated that macrophages play a role in cancer cell entry into the vasculature *in vitro* and *in vivo*. In line with this, treatment with drugs that eliminate the macrophage function results in a reduced number of circulating cancer cells (Roussos *et al* 2011). As both the cancer cells' ability to migrate toward the blood vessel and the actual intravasation step are necessary for entry of cancer cells into circulation, how macrophages regulate cancer cell intravasation precisely remains elusive. Furthermore, several *in vitro* studies analyzing the mechanism of cancer cell intravasation have revealed that the presence of macrophages elevates cancer cell intravasation, while the mechanisms of this enhancement are still not well understood (Roussos *et al* 2011, Zervantonakis *et al* 2012).

Intravital imaging of the tumor microenvironment has revealed that cancer cells intravasate into the vasculature at sites in close neighborhood to macrophages (Wyckoff *et al* 2004). Due to these observations, a microanatomic landmark composed of a perivascular macrophage in contact with a cancer cell at blood vessels has been identified and termed TMEM. In a case-controlled study of metastastic and non-metastatic breast cancers, TMEM density in breast tumors at initial resection was found to be associated with the risk of metastasis (Robinson *et al* 2009), suggesting that macrophages, cancer cells and endothelial cells cooperate in providing cancer cell entry into the vasculature. However, the cell biological mechanisms that exist between these three cell types during intravasation are still elusive. It has been suggested that they are dependent on the particular cellular mechanical properties of these cell types (Mierke 2011, Mierke 2014).

In order to gain a closer look at the molecular mechanisms of cancer cell transendothelial migration, *in vitro* experiments focusing on investigating cancer cell and endothelial cell behavior have been widely performed. Models of cancer cell extravasation or the exit of cancer cells from the vasculature are used frequently, because the apical surface of the endothelial cell monolayer can be accessed easily by plating endothelial monolayers and seeding cancer cells on top of these monolayers (Reymond *et al* 2012a, 2012b, Jin *et al* 2012, Haidari *et al* 2011, 2012). From these studies, an extensive view of how cancer cells affect the endothelial cell architecture has been built up and the specific steps of cancer cell adhesion and intercalation during transmigration have been revealed (Reymond *et al* 2012a, 2012b, Voura *et al* 1998a, 1998b). A common downstream mechanism of cancer cell transendothelial migration is the opening of the endothelial monolayer. In more detail, endothelial cells lose their cell–cell adherence junctions and tight junctions open as gaps in the monolayer, allowing cancer cells to pass through. In addition, molecular pathways leading to adhesion dissolution have also been discovered (Voura *et al* 1998a, Stoletov *et al* 2007, Tremblay *et al* 2006). However, how cancer cells perform intravasation and whether intravasation and extravasation share mechanisms in common is less well understood. Cancer cells can withstand shear flow stresses in the blood vessels and, in addition, different cells are present at intravasation and extravasation sites; however, it can be predicted that different modes of cell–cell interactions between cancer cells and endothelial cells occur in the intravasation and extravasation steps of the metastatic cascade. Due to the fact that macrophages are recognized at sites of cancer cell intravasation *in vivo*, the link between the close association of macrophages and cancer cells should be characterized, as well as the subsequent cancer cell intravasation.

Using dissolution of endothelial cell–cell adhesion as a readout of cancer cell transendothelial migration, the relationship between macrophages and cancer cells can be assessed with an *in vitro* model of intravasation. In particular, the question of whether the association of macrophages and cancer cells reflects a cell biological mechanism promoting and increasing intravasation of cancer cells through the endothelial cell barrier should be addressed. High-resolution imaging has been used in combination with *in vitro* and *in vivo* models to investigate whether macrophages increase cancer cell intravasation, and to uncover the cell biological and signaling mechanisms that regulate this process (Roh-Johnson *et al* 2014). Taken together, these results reveal that macrophages facilitate cancer cell intravasation by the activation of the RhoA signaling pathway, which promotes the formation of invadopodia, enabling the cancer cells to initiate intravasation by penetrating through the basement membrane of the vasculature.

Further reading

Bailly M, Wyckoff J, Bouzahzah B, Hammerman R, Sylvestre V, Cammer M, Pestell and Segall J E 2000 Epidermal growth factor receptor distribution during chemotactic responses *Mol. Biol. Cell* **11** 3873–83

Bartocci A, Pollard J W and Stanley E R 1986 Regulation of colony-stimulating factor 1 during pregnancy *J. Exp. Med.* **164** 956–61

Bear J E, Loureiro J J, Libova I, Fassler R, Wehland J and Gertler F B 2000 Negative regulation of fibroblast motility by Ena/VASP proteins *Cell* **101** 717–28

Byyny R L, Orth D N, Cohen S and Doyne E S 1974 Epidermal growth factor: effects of androgens and adrenergic agents *Endocrinology* **95** 776–82

Kedrin D, Wyckoff J, Boimel P J, Coniglio S J, Hynes N E, Arteaga C L and Segall J E 2009 ERBB1 and ERBB2 have distinct functions in tumor cell invasion and intravasation *Clin. Cancer Res.* **15** 3733–9

Lichtner R B, Julian J A, North S M, Glasser S R and Nicolson G L 1991 Coexpression of cytokeratins characteristic for myoepithelial and luminal cell lineages in rat 13762NF mammary adenocarcinoma tumors and their spontaneous metastases *Cancer Res.* **51** 5943–50 PMID: 1718590

Lichtner R B, Kaufmann A M, Kittmann A, Rohde-Schulz B, Walter J, Williams L, Ullrich A, Schirrmacher V and Khazaie K 1995 Ligand mediated activation of ectopic EGF receptor promotes matrix protein adhesion and lung colonization of rat mammary adenocarcinoma cells *Oncogene* **10** 1823–32 PMID: 7753557

Lichtner R B, Wiedemuth M, Kittmann A, Ullrich A, Schirrmacher V and Khazaie K 1992 Ligand-induced activation of epidermal growth factor receptor in intact rat mammary adenocarcinoma cells without detectable receptor phosphorylation *J. Biol. Chem.* **267** 11872–80 PMID: 1318304

Raja W K, Gligorijevic B, Wyckoff J, Condeelis J S and Castracane J 2010 A new chemotaxis device for cell migration studies *Integr. Biol. (Camb.)* **2** 696–706

van Rooijen N and van Kesteren-Hendrikx E 2003 *In vivo* depletion of macrophages by liposome-mediated 'suicide' *Methods Enzymol.* **373** 3–16 PMID: 14714393

Wang W *et al* 2002 Single cell behavior in metastatic primary mammary tumors correlated with gene expression patterns revealed by molecular profiling *Cancer Res.* **62** 6278–88 PMID: 12414658

Welch D R, Neri A and Nicolson G L 1983 Comparison of 'spontaneous' and 'experimental' metastasis using rat 13762 mammary adenocarcinoma metastatic cell clones *Invasion Metastasis* **3** 65–80 PMID: 6677622

Wyckoff J, Gligoijevic B, Entenberg D, Segall J and Condeelis J 2010 High-resolution multiphoton imaging of tumors *in vivo* Live Cell Imaging: A Laboratory Manual 2nd ed R D Goldman, J R Swedlow and D L Spector (Cold Spring Harbor, NY: Cold Spring Harbor Laboratory Press) pp 441–62

References

Andresen V, Alexander S, Heupel W M, Hirschberg M, Hoffman R M and Friedl P 2009 Infrared multiphoton microscopy: subcellular-resolved deep tissue imaging *Curr. Opin. Biotechnol.* **20** 54–62

Applewhite D A, Barzik M, Kojima S, Svitkina T M, Gertler F B and Borisy G G 2007 Ena/VASP proteins have an anti-capping independent function in filopodia formation *Mol. Biol. Cell* **18** 2579–91

Asagiri M and Takayanagi H 2007 The molecular understanding of osteoclast differentiation *Bone* **40** 251–64

Barzik M, Kotova T I, Higgs H N, Hazelwood L, Hanein D, Gertler F B and Schafer D A 2005 Ena/VASP proteins enhance actin polymerization in the presence of barbed end capping proteins *J. Biol. Chem.* **280** 28653–62

Bear J E and Gertler F B 2009 Ena/VASP: towards resolving a pointed controversy at the barbed end *J. Cell Sci.* **122** 1947–53

Bear J E *et al* 2002 Antagonism between Ena/VASP proteins and actin filament capping regulates fibroblast motility *Cell* **109** 509–21

Beck A H, Espinosa I, Edris B, Li R, Montgomery K, Zhu S, Varma S, Marinelli R J, van de Rijn M and West R B 2009 The macrophage colony-stimulating factor 1 response signature in breast carcinoma *Clin. Cancer Res.* **15** 778–87

Boeda B, Briggs D C, Higgins T, Garvalov B K, Fadden A J, McDonald N Q and Way M 2007 Tes, a specific Mena interacting partner, breaks the rules for EVH1 binding *Mol. Cell* **28** 1071–82

Cecchini M G, Dominguez M G, Mocci S, Wetterwald A, Felix R, Fleisch H, Chisholm O, Hofstetter W, Pollard J W and Stanley E R 1994 Role of colony stimulating factor-1 in the establishment and regulation of tissue macrophages during postnatal development of the mouse *Development* **120** 1357–72

Chitu V and Stanley E R 2006 Colony-stimulating factor-1 in immunity and inflammation *Curr. Opin. Immunol.* **18** 39–48

Condeelis J and Pollard J W 2006 Macrophages: obligate partners for tumor cell migration, invasion, and metastasis *Cell* **124** 263–6

Condeelis J and Segall J E 2003 Intravital imaging of cell movement in tumours *Nat. Rev. Cancer* **3** 921–30

Dai X M, Ryan G R, Hapel A J, Dominguez M G, Russell R G, Kapp S, Sylvestre V and Stanley E R 2002 Targeted disruption of the mouse colony-stimulating factor 1 receptor gene results in osteopetrosis, mononuclear phagocyte deficiency, increased primitive progenitor cell frequencies, and reproductive defects *Blood* **99** 111–20

Di Modugno F *et al* 2006 The cytoskeleton regulatory protein hMena (ENAH) is overexpressed in human benign breast lesions with high risk of transformation and human epidermal growth factor receptor-2-positive/hormonal receptor-negative tumors *Clin. Cancer Res.* **12** 1470–8

Di Modugno F *et al* 2007 Molecular cloning of hMena (ENAH) and its splice variant hMena + 11a: epidermal growth factor increases their expression and stimulates hMena + 11a phosphorylation in breast cancer cell lines *Cancer Res.* **67** 2657–65

Egeblad M, Ewald A J, Askautrud H A, Truitt M L, Welm B E, Bainbridge E, Peeters G, Krummel M F and Werb Z 2008 Visualizing stromal cell dynamics in different tumor microenvironments by spinning disk confocal microscopy *Dis. Model. Mech* **1** 155–67

Ferron F, Rebowski G, Lee S H and Dominguez R 2007 Structural basis for the recruitment of profilin-actin complexes during filament elongation by Ena/VASP *EMBO J.* **26** 4597–606

Friedl P and Wolf K 2010 Plasticity of cell migration: a multiscale tuning model *J. Cell Biol.* **188** 11–9

Gaggioli C, Hooper S, Hidalgo-Carcedo C, Grosse R, Marshall J F, Harrington K and Sahai E 2007 Fibroblast-led collective invasion of carcinoma cells with differing roles for RhoGTPases in leading and following cells *Nat. Cell Biol.* **9** 1392–400

Gertler F B, Niebuhr K, Reinhard M, Wehland J and Soriano P 1996 Mena, a relative of VASP and Drosophila Enabled, is implicated in the control of microfilament dynamics *Cell* **87** 227–39

Giampieri S, Manning C, Hooper S, Jones L, Hill C S and Sahai E 2009 Localized and reversible TGFbeta signalling switches breast cancer cells from cohesive to single cell motility *Nat. Cell Biol.* **11** 1287–96

Gordon S and Taylor P R 2005 Monocyte and macrophage heterogeneity *Nat. Rev Immunol.* **5** 953–64

Goswami S, Philippar U, Sun D, Patsialou A, Avraham J, Wang W, Di Modugno F, Nistico P, Gertler F B and Condeelis J S 2009 Identification of invasion specific splice variants of the cytoskeletal protein Mena present in mammary tumor cells during invasion *in vivo Clin. Exp. Metastasis* **26** 153–9

Goswami S, Sahai E, Wyckoff J B, Cammer M, Cox D, Pixley F J, Stanley E R, Segall J E and Condeelis J S 2005 Macrophages promote the invasion of breast carcinoma cells via a colony-stimulating factor-1/epidermal growth factor paracrine loop *Cancer Res.* **65** 5278–83

Goswami S, Wang W, Wyckoff J B and Condeelis J S 2004 Breast cancer cells isolated by chemotaxis from primary tumors show increased survival and resistance to chemotherapy *Cancer Res.* **64** 7664–7

Gurzu S, Jung I, Prantner I, Chira L and Ember I 2009 The immunohistochemical aspects of protein Mena in cervical lesions *Rom. J. Morphol. Embryol.* **50** 213–6

Gurzu S, Jung I, Prantner I, Ember I, Pavai Z and Mezei T 2008 The expression of cytoskeleton regulatory protein Mena in colorectal lesions *Rom. J. Morphol. Embryol.* **49** 345–9 PMID: 18758639

Haidari M, Zhang W, Caivano A, Chen Z, Ganjehei L, Mortazavi A, Stroud C, Woodside D G, Willerson J T and Dixon R A 2012 Integrin alpha2beta1 mediates tyrosine phosphorylation of vascular endothelial cadherin induced by invasive breast cancer cells *J. Biol. Chem.* **287** 32981–92

Haidari M, Zhang W, Chen Z, Ganjehei L, Warier N, Vanderslice P and Dixon R 2011 Myosin light chain phosphorylation facilitates monocyte transendothelial migration by dissociating endothelial adherens junctions *Cardiovascular Res.* **92** 456–65

Hamilton J A 2008 Colony-stimulating factors in inflammation and autoimmunity *Nat. Rev. Immunol.* **8** 533–44

Hansen S D and Mullins R D 2010 VASP is a processive actin polymerase that requires monomeric actin for barbed end association *J. Cell Biol.* **191** 571–84

Hernandez L *et al* 2009 The EGF/CSF-1 paracrine invasion loop can be triggered by heregulin beta1 and CXCL12 *Cancer Res.* **69** 3221–7

Ilina O and Friedl P 2009 Mechanisms of collective cell migration at a glance *J. Cell Sci.* **122** 3203–8

Jin F, Brockmeier U, Otterbach F and Metzen E 2012 New insight into the SDF-1/CXCR4 axis in a breast carcinoma model: hypoxia-induced endothelial SDF-1 and tumor cell CXCR4 are required for tumor cell intravasation *Mol. Cancer Res.* **10** 1021–31

Kacinski B M 1997 CSF-1 and its receptor in breast carcinomas and neoplasms of the female reproductive tract *Mol. Reprod. Dev.* **46** 71–4

Kacinski B M, Scata K A, Carter D, Yee L D, Sapi E, King B L, Chambers S K, Jones M A, Pirro M H and Stanley E R *et al* 1991 FMS (CSF-1 receptor) and CSF-1 transcripts and protein are expressed by human breast carcinomas *in vivo* and *in vitro Oncogene* **6** 941–52 PMID: 1829808

Kedrin D, Gligorijevic B, Wyckoff J, Verkhusha V V, Condeelis J, Segall J E and van Rheenen J 2008 Intravital imaging of metastatic behavior through a mammary imaging window *Nat. Methods* **5** 1019–21

Kedrin D, van Rheenen J, Hernandez L, Condeelis J and Segall J E 2007 Cell motility and cytoskeletal regulation in invasion and metastasis *J. Mammary Gland Biol. Neoplasia* **12** 143–52

Le Devedec S E, Lalai R, Pont C, de Bont H and van de Water B 2010 Two-photon intravital multicolor imaging combined with inducible gene expression to distinguish metastatic behavior of breast cancer cells *in vivo Mol. Imaging Biol.* **13** 67–77

Le Devedec S E, van Roosmalen W, Maria N, Grimbergen M, Pont C, Lalai R and van de Water B 2009 An improved model to study tumor cell autonomous metastasis programs using MTLn3 cells and the Rag2(−/−) gammac (−/−) mouse *Clin. Exp. Metastasis* **26** 673–84

Levea C M, McGary C T, Symons J R and Mooney R A 2000 PTP LAR expression compared to prognostic indices in metastatic and non-metastatic breast cancer *Breast Cancer Res. Treat.* **64** 221–8

Lichtner R B, Julian J A, Glasser S R and Nicolson G L 1989 Characterization of cytokeratins expressed in metastatic rat mammary adenocarcinoma cells *Cancer Res.* **49** 104–11 PMID: 2461796

Lin H *et al* 2008 Discovery of a cytokine and its receptor by functional screening of the extracellular proteome *Science* **320** 807–11 PMID: 18467591

Loureiro J J, Rubinson D A, Bear J E, Baltus G A, Kwiatkowski A V and Gertler F B 2002 Critical roles of phosphorylation and actin binding motifs, but not the central proline-rich region, for Ena/vasodilator-stimulated phosphoprotein (VASP) function during cell migration *Mol. Biol. Cell* **13** 2533–46

Mierke C T 2011 Cancer cells regulate biomechanical properties of human microvascular endothelial cells *J. Biol. Chem.* **286** 40025–37

Mierke C T 2014 The fundamental role of mechanical properties in the progression of cancer disease and inflammation *Rep. Prog. Phys.* **77** 076602

Neri A, Welch D, Kawaguchi T and Nicolson G L 1982 Development and biologic properties of malignant cell sublines and clones of a spontaneously metastasizing rat mammary adenocarcinoma *J. Natl Cancer Inst.* **68** 507–17 PMID: 6950180

Niebuhr K, Ebel F, Frank R, Reinhard M, Domann E, Carl U D, Walter U, Gertler F B, Wehland J and Chakraborty T 1997 A novel proline-rich motif present in ActA of Listeria monocytogenes and cytoskeletal proteins is the ligand for the EVH1 domain, a protein module present in the Ena/VASP family *EMBO J.* **16** 5433–44

Pasic L, Kotova T and Schafer D A 2008 Ena/VASP proteins capture actin filament barbed ends *J Biol. Chem.* **283** 9814–9

Patsialou A, Wyckoff J, Wang Y, Goswami S, Stanley E R and Condeelis J S 2009 Invasion of human breast cancer cells *in vivo* requires both paracrine and autocrine loops involving the Colony-Stimulating Factor-1 receptor *Cancer Res.* **69** 9498–506

Perentes J Y, McKee T D, Ley C D, Mathiew H, Dawson M, Padera T P, Munn L L, Jain R K and Boucher Y 2009 *In vivo* imaging of extracellular matrix remodeling by tumor-associated fibroblasts *Nat. Methods* **6** 143–5

Philippar U *et al* 2008 A Mena invasion isoform potentiates EGF-induced carcinoma cell invasion and metastasis *Dev. Cell* **15** 813–28

Pino M S *et al* 2008 Human Mena + 11a isoform serves as a marker of epithelial phenotype and sensitivity to epidermal growth factor receptor inhibition in human pancreatic cancer cell lin *Clin. Cancer Res.* **14** 4943–50

Pixley F J and Stanley E R 2004 CSF-1 regulation of the wandering macrophage: complexity in action *Trends Cell Biol.* **14** 628–38

Pixley F J, P. Lee P S W, Condeelis J S and Stanley E R 2001 Protein tyrosine phosphatase φ regulates paxillin tyrosine phosphorylation and mediates colony-stimulating factor 1-induced morphological changes in macrophages *Mol. Cellular Biol.* **21** 1795–809

Pollard J W 2009 Trophic macrophages in development and disease *Nat. Rev. Immunol.* **9** 259–70

Pula G and Krause M 2008 Role of Ena/VASP proteins in homeostasis and disease *Handb. Exp. Pharmacol.* **186** 39–65 PMID: 18491048

Reymond N *et al* 2012a Cdc42 promotes transendothelial migration of cancer cells through beta1 integrin *J. Cell Biol.* **199**(4) 653–68 PMID: 23148235

Reymond N, Riou P and Ridley A J 2012b Rho GTPases and cancer cell transendothelial migration *Methods Mol. Biol.* **827** 123–42 PMID: 22144272

Robinson B D, Sica G L, Liu Y F, Rohan T E, Gertler F B, Condeelis J S and Jones J G 2009 Tumor microenvironment of metastasis in human breast carcinoma: a potential prognostic marker linked to hematogenous dissemination *Clin. Cancer Res.* **15** 2433–41

Roh-Johnson M, Bravo-Cordero J J, Patsialou A, Sharma V P, Guo P, Liu H, Hodgson L and Condeelis J 2014 Macrophage contact induces RhoA GTPase signaling to trigger tumor cell intravasation *Oncogene* **33** 4203–12

Roussos E T *et al* 2011 Mena invasive (MenaINV) promotes multicellular streaming motility and transendothelial migration in a mouse model of breast cancer *J. Cell Sci.* **124** 2120–31

Roussos E T *et al* 2011 Mena invasive (MenaINV) and Mena11a isoforms play distinct roles in breast cancer cell cohesion and association with TMEM *Clin. Exp. Metastasis* **28** 515–27

Roussos E T, Wang Y, Wyckoff J B, Sellers R S, Wang W, Li J, Pollard J W, Gertler F B and Condeelis J S 2010 Mena deficiency delays tumor progression and decreases metastasis in polyoma middle-T transgenic mouse mammary tumors *Breast Cancer Res.* **12** R101

Ryan G R, Dai M X, Dominguez M G, Tong W, Chuan F, Chisholm O, Russell R G, Pollard J W and Stanley E R 2001 Rescue of the colony-stimulating factor 1 (CSF-1)-nullizygous mouse (Csf1op/Csf1op) phenotype with a CSF-1 transgene and identification of sites of local CSF-1 synthesis *Blood* **98** 74–84

Sahai E 2005 Mechanisms of cancer cell invasion *Curr. Opin. Genet. Dev.* **15** 87–96

Sahai E, Wyckoff J, Philippar U, Segall J E, Gertler F and Condeelis J 2005 Simultaneous imaging of GFP, CFP and collagen in tumors *in vivo* using multiphoton microscopy *BMC Biotechnol.* **5** 14

Scholl S M, Pallud C, Beuvon F, Hacene K, Stanley E R, Rohrschneider L, Tang R, Pouillart P and Lidereau R 1994 Anti-colony-stimulating factor-1 antibody staining in primary breast adenocarcinomas correlates with marked inflammatory cell infiltrates and prognosis *J. Natl. Cancer Inst.* **86** 120–6

Segall J E, Tyerech S, Boselli L, Masseling S, Helft J, Chan A, Jones J and Condeelis J 1996 EGF stimulates lamellipod extension in metastatic mammary adenocarcinoma cells by an actin-dependent mechanism *Clin. Exp. Metastasis* **14** 61–72

Sherr C J, Rettenmier C W, Sacca R, Roussel M F, Look A T and Stanley E R 1985 The c-fms proto-oncogene product is related to the receptor for the mononuclear phagocyte growth factor, CSF-1 *Cell* **41** 665–76

Stanley E R and Heard P M 1977 Factors regulating macrophage production and growth. Purification and some properties of the colony stimulating factor from medium conditioned by mouse L cells *J. Biol. Chem.* **252** 4305–12 PMID: 301140

Stoletov K, Montel V, Lester R D, Gonias S L and Klemke R 2007 High-resolution imaging of the dynamic tumor cell vascular interface in transparent zebrafish *Proc. Natl Acad. Sciences USA* **104** 17406–11

Tamimi R M, Brugge J S, Freedman M L, Miron A, Iglehart J D, Colditz G A and Hankinson S E 2008 Circulating colony stimulating factor-1 and breast cancer risk *Cancer Res.* **68** 18–21

Tremblay P L, Auger F A and Huot J 2006 Regulation of transendothelial migration of colon cancer cells by E-selectin-mediated activation of p38 and ERK MAP kinases *Oncogene* **25** 6563–73

Urbanelli L, Massini C, Emiliani C, Orlacchio A and Bernardi G 2006 Characterization of human Enah gene *Biochim. Biophys. Acta* **1759** 99–107

van Furth R, Cohn Z A, Hirsch J G, Humphrey J H, Spector W G and Langevoort H L 1972 The mononuclear phagocyte system: a new classification of macrophages, monocytes, and their precursor cells *Bull. World Health Organ.* **46** 845–52 PMID: 4538544

Voura E B, Sandig M and Siu CH 1998b Cell-cell interactions during transendothelial migration of tumor cells *Microsc. res. tech.* **43** 265–75

Voura E B, Sandig M, Kalnins V I and Siu C 1998a Cell shape changes and cytoskeleton reorganization during transendothelial migration of human melanoma cells *Cell Tissue Res.* **293** 375–87

Wang J M, Griffin J D, Rambaldi A, Chen Z G and Mantovani A 1988 A Induction of monocyte migration by recombinant macrophage colony-stimulating factor *J. Immunol.* **141** 575–9 PMID: 3290341

Wang W, Goswami S, Lapidus K, Wells A L, Wyckoff J B, Sahai E, Singer R H, Segall J E and Condeelis J S 2004 Identification and testing of a gene expression signature of invasive carcinoma cells within primary mammary tumors *Cancer Res.* **64** 8585–94

Wang W, Mouneimne G, Sidani M, Wyckoff J, Chen X, Makris A, Goswami S, Bresnick A R and Condeelis J S 2006 The activity status of cofilin is directly related to invasion, intravasation, and metastasis of mammary tumors *J. Cell Biol.* **173** 395–404

Wang W, Wyckoff J B, Goswami S, Wang Y, Sidani M, Segall J E and Condeelis J S 2007 Coordinated regulation of pathways for enhanced cell motility and chemotaxis is conserved in rat and mouse mammary tumors *Cancer Res.* **67** 3505–11

Webb S E, Pollard J W and Jones G E 1996 Direct observation and quantification of macrophage chemoattraction to the growth factor CSF-1 *J. Cell Sci.* **110** 707–20

Wei S, Nandi S, Chitu V, Yeung Y G, Yu W, Huang M, Williams L T, Lin H and Stanley E R 2010 Functional overlap but differential expression of CSF-1 and IL-34 in their CSF-1 receptor-mediated regulation of myeloid cells *J. Leukocyte Biol.* **88** 495–505

Wiktor-Jedrzejczak W, Bartocci A, Ferrante A W Jr, Ahmed-Ansari A, Sell K W, Pollard J W and Stanley E R 1990 Total absence of colony-stimulating factor 1 in the macrophage-deficient osteopetrotic (op/op) mouse *Proc. Natl Acad. Sci. USA* **87** 4828–32

Wolf K, Mazo I, Leung H, Engelke K, von Andrian U H, Deryugina E I, Strongin A Y, Brocker E B and Friedl P 2003 Compensation mechanism in tumor cell migration: mesenchymal-amoeboid transition after blocking of pericellular proteolysis *J. Cell Biol.* **160** 267–77

Wyckoff J B, Jones J G, Condeelis J S and Segall J E 2000b A critical step in metastasis: *in vivo* analysis of intravasation at the primary tumor *Cancer Res.* **60** 2504–11 PMID: 10811132

Wyckoff J B, Segall J E and Condeelis J S 2000a The collection of the motile population of cells from a living tumor *Cancer Res.* **60** 5401–4 PMID: 11034079

Wyckoff J B, Wang Y, Lin E Y, Li J F, Goswami S, Stanley E R, Segall J E, Pollard J W and Condeelis J 2007 Direct visualization of macrophage-assisted tumor cell intravasation in mammary tumors *Cancer Res.* **67** 2649–56

Wyckoff J, Wang W, Lin E Y, Wang Y, Pixley F, Stanley E R, Graf T, Pollard J W, Segall J and Condeelis J 2004 A paracrine loop between tumor cells and macrophages is required for tumor cell migration in mammary tumors *Cancer Res.* **64** 7022–9

Xue C, Wyckoff J, Liang F, Sidani M, Violini S, Tsai K L, Zhang Z Y, Sahai E, Condeelis J and Segall J E 2006 Epidermal growth factor receptor overexpression results in increased tumor cell motility *in vivo* coordinately with enhanced intravasation and metastasis *Cancer Res.* **66** 192–7

Yamaguchi H, Pixley F and Condeelis J 2006 Invadopodia and podosomes in tumor invasion *Eur. J. Cell Biol.* **85** 213–8

Yoshida S I, Hayashi, Kunisada T, Ogawa M, Nishikawa S, Okamura H, Sudo T, Shultz L D and Nishikawa S 1990 The murine mutation osteopetrosis is in the coding region of the macrophage colony stimulating factor gene *Nature* **345** 442–4

Zerbe L K *et al* 2008 Inhibition by erlotinib of primary lung adenocarcinoma at an early stage in male mice *Cancer Chemother. Pharmacol.* **62** 605–20

Zervantonakis I K, Hughes-Alford S K, Charest J L, Condeelis J S, Gertler F B and Kamm R D 2012 Three-dimensional microfluidic model for tumor cell intravasation and endothelial barrier function *Proc. Natl Acad. Sci. USA* **109** 13515–20

Lightning Source UK Ltd.
Milton Keynes UK
UKHW050646060619

343842UK00004B/86/P